THE AGI REVOLUTION

An Inside View of the Rise of Artificial General Intelligence

Ben Goertzel

Contents

artificial general intelligence (AGI)

noun

1 capacity of an engineered system to display the same rough sort of general intelligence as humans.

2 capacity of an engineered system to display intelligence that is not tied to a highly specific set of tasks

3 capacity of an engineered system to generalize what it has learned, including generalization to contexts qualitatively very different than those it has seen before

4 capacity of an engineered system to take a broad view, and interpret its tasks at hand in the context of the world at large and its relation thereto

5 engineered system displaying the property of artificial general intelligence, to a significant degree

6 theoretical and practical study of artificial general intelligence systems and methods of creating them

The first documented usage of the term was in a 1997 essay by Mark Gubrud, on nanotechnology[1]. The term was popularized when, based on a suggestion from Shane Legg, it was used as the title of Ben Goertzel and Cassio Pennachin's 2006-edited book *Artificial General Intelligence*, and as the title of an annual conference series which has been operated by Ben Goertzel and colleagues continuously since 2008, and which was seeded by the 2006 AGI Workshop organized by Ben Goertzel and Bruce Klein.

---

1 Gubrud, Mark (November 1997), "Nanotechnology and International Security", *Fifth Foresight Conference on Molecular Nanotechnology*, retrieved 7 May 2011

1. AGI Rising

What does it feel like to stand here?

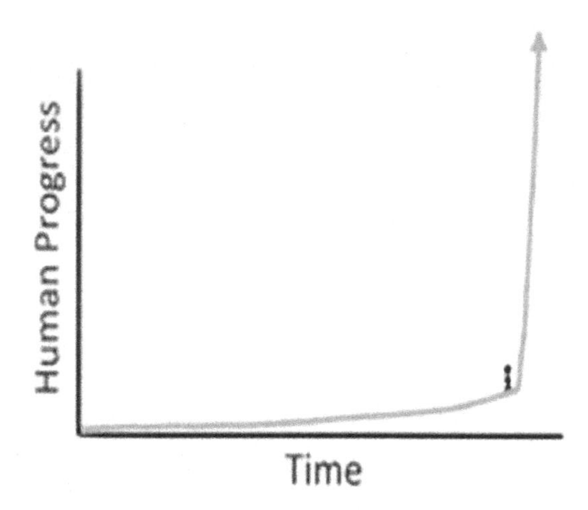

It seems like a pretty intense place to be standing—but then you have to remember something about what it's like to stand on a time graph: you can't see what's to your right. So here's how it actually feels to stand there:

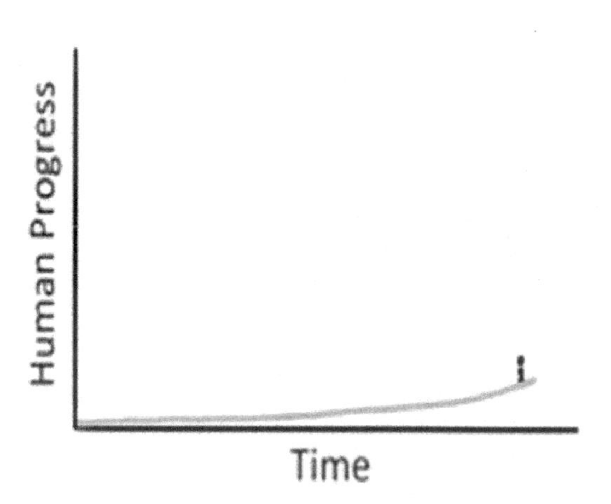

Which probably feels pretty normal...

FIGURE 1.1 *:Image borrowed from the excellent essay "The AI Revolution" by Tim Urban. (http://waitbutwhy.com/2015/01/artificial-intelligence-revolution-1.html)*

Human history has known many extraordinary events, such as the invention of language, the emergence of agriculture, the explosion of culture and technology under the Chinese Ming dynasty from the late 1300s through the 1600s, the European Enlightenment, and the Industrial Revolution.

Now we are in extraordinary times again, times of radical change and dramatic transition and transformation. Many new, exciting things are happening right now – the globalization of economy and culture, the growth and pervasion of the Internet, online education, smartphones, industrial robots, synthetic biology, nanotech... The list goes on a while.

In hindsight, though, I think one aspect of early 21st century humanity will stand out above all others – the rise of true AI, of Artificial General Intelligence. The other innovations we see around us now are also tremendous and important – yet every one of these will adopt a different and even more impactful form when it is integrated into the minds and bodies of advanced AGI systems.

We are privileged to be living, right now, at the start of the AGI Revolution. And it has been my privilege to play at least a small role in the advent of this revolution. I have been doing AGI-oriented research for nearly three decades now, and have also organized a host of AGI-oriented events, edited AGI books, and gotten to know nearly all the leading researchers in AGI and closely related areas.

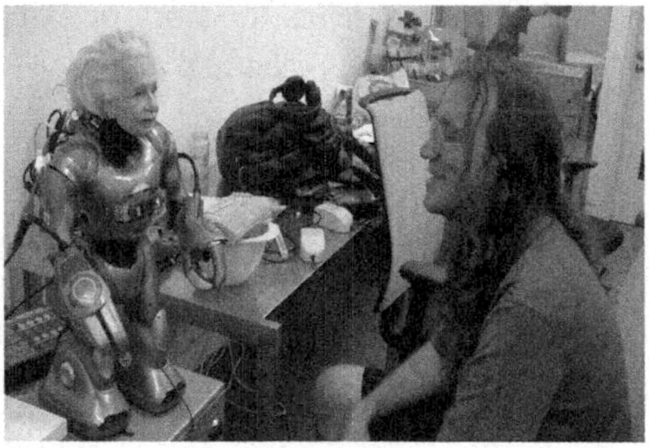

FIGURE 1.2: *Me in the OpenCog Lab at Hong Kong Poly U, in Fall 2014, talking with a prototype robot we put together to test the integration of our OpenCog AI engine with various other software and hardware technologies. The Einstein head was made by the amazing robotics artist David Hanson; the body is a (much-hacked) RS-Media robot created by Mark Tilden, which was commercially sold in toy stores a few years ago for a price tag of a few hundred US dollars. Hanson, Tilden and I are all currently living mainly in Hong Kong. On the robot's back is a Gizmo computer board, which cost less than 200 US dollars and outperforms the large, "powerful" computers of my childhood. On the front is an Asus depth sensor which combines two cameras and other sensors to do 3D vision. The speech-to-text for the robot, enabling my speech to be fed to our AI text understanding software, is done via sending the speech over the Internet*

to Google Voice, which sends the text back. Toy robots equaling and exceeding the capability of this research prototype will almost surely be on the market within a few years, marketed by either David Hanson and his collaborators (including me), or by someone else. And since the AI is delivered to the robot largely via the cloud, AI advances can be delivered to the robot as soon as they are uploaded to the server. It's a very interesting time to work with AI and robots.

I'm currently pushing hard on the open source AGI project I co-founded in 2008, OpenCog, trying to use it as the foundation for creating real human-level Artificial General Intelligence. Whether my colleagues and I will succeed or not at this ambitious quest, no one can know for sure at this stage. But it's sure being an interesting ride.

I took time out from my insanely busy schedule to write this book, in order to give you my perspective on Artificial General Intelligence – as an idea, a technology, a community and movement, and a factor in society and the evolution of humanity and sentience. I'll set the stage by recounting my own journey toward AGI, which began with Star Trek and the philosophy of mind. Then I'll tell you about the work I've done in AGI, robotics, longevity biology data analysis, and a host of other areas. I'll give you my view on how the mind and brain work, share my perspective on the scope of AGI work going on across industry and academia today, and finally speculate a bit about the grand future of AGI and where it's all leading us. There is nothing that's definitive here, as the story of AGI is still very much unfolding. But there's much that is fascinating!

How Far We've Come

When I was born in 1966, AI was far from the hot topic it is today. Growing up in the 1970s, I got the impression from the world around me that Artificial Intelligence was basically the stuff of science fiction novels. Sure, I was vaguely aware from occasional mentions in popular science magazines that there were some professors doing AI research in some universities somewhere – but it wasn't something that anyone I knew attached any importance to.

FIGURE 1.3: *Computers introduced the year I was born, 1966. (a) ILLIAC IV was a 64-processor parallel computer produced by the University of Illinois with DARPA funding. It wasn't operational till 1972. (http://ed-thelen.org/comp-hist/vs-illiac-iv.html). (b) The HP-2115 was an early-stage "PC," offering computing power previously found only in much larger computers, and programmable in BASIC, ALGOL, and FORTRAN, among other languages. The typewriter-like portion allowed paper tape I/O and printer output; the box with switches and binary display is the CPU, and the other box is core memory. I never saw computers like these except as an adult in the computer museum; they were obscure and only used by scientists and engineers with specific related needs. (http://www.computerhistory.org/timeline/?year=1966). Image courtesy of Computer History Museum.*

It was the time of Vietnam War protests, the civil rights movement, the women's movement, and the psychedelic awakening. Most of the adults around me were focused on politics and social and personal transformation. But even so, I was raised in a reasonably science-oriented environment. My dad was a sociologist, my grandfathers were a physical chemist and a psychologist, and there were various other scientists in my family, including my great uncle Gerald Goertzel, who was a professor at New York University a couple decades before I was a grad student there, and who invented the "Goertzel algorithm," still commonly used for signal processing. (Actually I don't remember ever meeting uncle Gerald, but I see his algorithm pop up on Google whenever I search for things involving my name!)

My earliest memory is of watching Neil Armstrong walk on the moon on TV. The experience impacted my toddler self hugely. My mom saw my interest in space and everything scientific, and taught me lots of mathematics, astronomy, and other science-related stuff in my early childhood. I definitely grew up with science on the brain. But AI never seemed to come up in the serious discussions I had with or overheard among scientists and other science-savvy folks I encountered. There certainly weren't any major AI companies occupying news headlines.

FIGURE 1.4: *Apart from an earlier electronics-kit type "computer" programmable only in machine code via a hex keypad, my first computer was an Atari 400 like the figure at the right, purchased in 1980 with $400 I saved via a job delivering the Philadelphia Bulletin newspaper from door to door on my bike. It was programmable in BASIC, FORTH and assembler, and used a black and white TV for display and a cassette tape drive for external memory. I learned to program on this machine, and coded a lot of simple games, none with any nontrivial AI. (http://www.classic-computers.org.nz/blog/images/2010-09-06-atari-400-working.jpg)*

If I had to sum up the implicit attitude of the science-savvy adults I knew in my childhood regarding artificial intelligence, I'd say something like: "***AI, machines thinking and acting like people or maybe even more intelligently, seems to be a scientific possibility. But we have no idea how to achieve it at the present time. It seems more remote than interstellar space travel or virtually unlimited energy via nuclear fusion, and maybe about as hard as teleportation.***" Intelligent robots were there in SF books and movies, walking around and holding conversations and making discoveries and inventions and so forth – so the world at large was familiar with the concept, as was I from age 3 or so – but it wasn't really taken seriously as something to think about from a non-SF point of view.

Looking back now, I realize that in the early 70s and for a decade or two before, there were actually excellent and deep-thinking scientists at various universities around the world, working concertedly toward realizing their own visions of thinking machines. After I became an AI researcher myself, I got the chance to meet some of these early researchers, albeit at a time when they were well into old age – Marvin Minsky, Nils Nilsson, Roger Schank and the like. But back in the day, the pioneering work these guys were doing wasn't really known among non-specialists. It didn't pop up often in the (admittedly minimal) science section of the local newspaper, or in the Science News weekly magazine to which my grandfather had bought me a subscription.

Today, in 2016 as I write these words, things obviously look rather different. Google, one of the most successful companies around, is substantially an AI company, relying on AI text-processing tricks to place ads on web pages in such a way that people will be more likely to click on them. Search engines like Google, Baidu, and Bing are not branded as AI, but that's what they are. Google is also using AI technology to create self-driving cars that cruise around San Francisco, and it has spurred a spate of similar initiatives by carmakers around the world. Everyone knows that Apple's smartphones now include an AI chatbot, Siri, which admittedly falls far short of human conversational ability, but does resolve some useful queries correctly. AI has helped Apple sell a lot of phones. Google's smartphone operating system Android has some pretty good "AI" voice control too. Basically everything Google is doing, Baidu is doing too (some things better, some things less well) – in China, and now in their Silicon Valley research lab as well, where they recently hired the famed Stanford AI researcher Andrew Ng. IBM is rolling out its Watson AI supercomputer for medical applications and other practical uses beyond the Jeopardy video game application they showcased in 2011.

FIGURE 1.5: *According to Arthur C. Clarke's novel 2001 and the influential Stanley Kubrick film of the same name, we were supposed to have human-level AI's like (a) HAL 9000 by now. Not quite. (If you didn't read the book or see the movie, you should. HAL was a mentally unbalanced AI controlling a spaceship, not always in accordance with human instructions.) Though, given HAL 9000's defective ethics, it's just as well he didn't come about. What we do have now, in 2014, is a host of highly useful narrow AI applications. To highlight the contrast between the AGI visions of SF novels and movies, and the reality of narrow-AI applications in the world today, I have sometimes shown conference slides contrasting the Google logo with the HAL-9000 computer. The (b) two red staring eyes of Google remind one of the broad, panopticon-style vision into our lives that information service companies like Google have these days (but of course, Google is just a familiar company to mention to make the point; they are not unique nor terribly exceptional in this regard).*

Of course, alongside amazing narrow AI products and services, another thing we have today is the technology and science base to let us finally seriously approach the AGI problem. Building on this foundation, with a bit more work we should be able to create AGI systems that are both smarter than HAL 9000 and more ethical. (https://support.thingiverse.com/thing:81055).

Computer programs now do most of the trading on the US stock exchange, and among these are many programs using techniques taught in university AI courses, such as machine learning and natural language processing. I've spent much of the last two years working on an effort of a similar nature – working with colleagues to create an AI-based hedge fund based in Hong Kong (Aidyia Limited), initially trading the US equities and global futures markets.

The US military relies on AI not only for flashy applications like unmanned automated vehicles, but also for dull, necessary stuff like planning and logistics, and supply chain management. PayPal uses AI for automated fraud detection, enabling them to provide affordable credit card processing across the Internet. Video games regularly feature AIs with intelligences customized to their game worlds, and AI players can beat the best human players at chess, checkers, Go and many other games (though not yet poker)! Across the board, AI is on a roll.

FIGURE 1.6: *Remember this guy?? Of course you do! He doesn't have anything special to do with what I'm talking about in these pages, but I just really like R2D2 so I wanted to have a picture of him in my book. You don't have a problem with that, do you?*

Well, obviously, R2D2 from Star Wars was one among the many inspirational fictional robots my generation grew up with. This image is actually a photo of a real-life replica of R2D2, which however doesn't have the intelligent capabilities of the R2D2 in the movie (yet). (http://www.allonrobots.com/r2d2-robot.html).

But – you knew there was a "but" coming! – while there's no doubt that AI has come a long way during my lifetime, from an obscure pursuit carried out by a handful of academics, to a technology utilized in scads of practical projects and subjected to a major focus by huge corporations, it's also clear that there's still a long way to go, in some sense. Comparing the real achievements of AI today with the science fictional AIs I encountered in my childhood, the real stuff still falls short in many ways. Everyone in my middle school knew about R2D2 and C3PO ("fluent in more than six million forms of communication") from the Star Wars series, and most also knew about the charmingly sinister HAL 9000 from the movie 2001: A Space Odyssey. These SF robots set the bar for AI in the popular mind, and the reality of AI hasn't yet reached this level. We don't yet have R2D2, C3PO, or Hal 9000, let alone the massively superhuman AIs that some of the more philosophical science fiction writers envisioned. There is a significant and evident gap to be crossed, to get from AI as it exists today to AI that can fulfill the promises of science fictional AI and really think with the generality, consciousness, and power of a human mind (and more).

And this brings us to the main themes of this book:

- Why I believe an AGI Revolution is almost surely coming this century
- The radical implications this will have
- How I believe we can cause the AGI Revolution to come sooner rather than later, and for the better rather than the worse (TL;DR – what I advocate is to focus significant effort concertedly on creating AIs with General rather than highly problem-specific intelligenc.e)

FIGURE 1.7: *Now, in 2016, while we don't yet have R2D2 or HAL 9000 or anything vaguely equivalent, we do have incredibly massive (a) server farms doing massive-scale AI-based data crunching, and (b) smartphones giving access to all sorts of narrow AI (delivered via these server farms). These are directions less commonly foreseen by science fiction authors, yet in their own way just as impactful as R2D2s rolling down the streets would be. Predicting the direction of technology development is an extremely hard problem, even when one sees the general trends that are unfolding. (http://www.dailymail.co.uk/sciencetech/article-2219188/Inside-Googlepictures-gives-look-8-vast-data-centres.html, http://hitech.vesti.ru/news/view/id/4710).*

I think it's almost inevitable that advanced, science fictional AI will come about by the middle of this century. And I think if we do things right, we can make it happen well before the middle of the century, perhaps in less than a decade.

Now, I'm well aware that I'm not the first AI researcher in history to come forward with an optimistic attitude. Ever since the late 1950s – well before AI became a force in the commercial world – there have been AI researchers proclaiming dramatic success at human-level AI to be "just around the corner." (And there have also been more conservative-minded AI researchers all along, as well.) At the 1959 Dartmouth College meeting where the term "Artificial Intelligence" was conceived, the original plan was to solve the core problems of AI over a single summer – one researcher taking care of language, another taking care of vision, and so forth. Similar overoptimistic pronouncements were periodically made by AI researchers throughout the 1960s and 70s, eventually resulting in a backlash against the AI field by the media and academic funding sources, which came to be known as the "AI Winter." Given this background, it's understandable for observers of the history of AI to feel somewhat skeptical when an ambitious researcher waxes optimistic about the speed with which the gap between current AI and science fictional AI can be crossed.

But, well, 2016 is not 1959 or 1980 – and I think that today, optimism about AI advancement is more than justified. Given the progress made in specialized AI technologies which I've mentioned above, along with the simultaneous progress that's been made in understanding the human brain and mind, and the massive ongoing improvement in computer hardware and software, AI researches today are operating in a very different environment from AI researches of past eras. This is why so many mainstream figures of the tech industry, like the Google founders and Justin Rattner, the CTO of Intel, have joined the increasing chorus of futurists predicting the likely advent of superhumanly intelligent AI by the middle of the century.

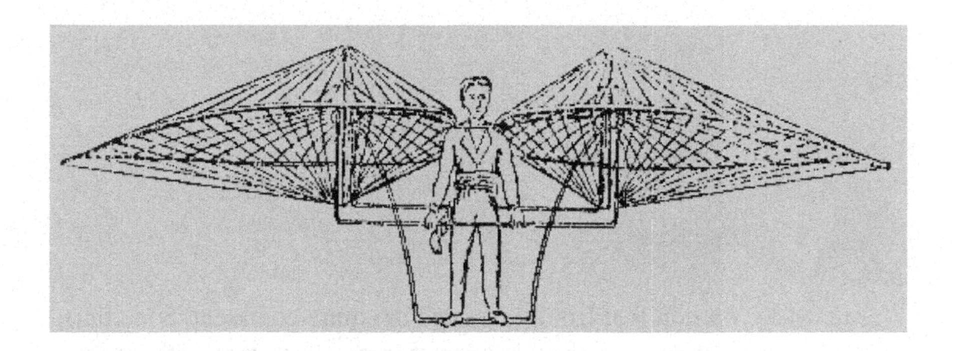

FIGURE 1.8: *Early designs for human flight, like this 1812 ornithopter, were based very closely on birds. This strategy of attempting close imitation of nature is not what led to the breakthrough in manned flight, obviously. Similarly, I suspect that overly close imitation of the human brain is probably not going to be the route that leads to the first breakthroughs in AGI. (http://upload.wikimedia.org/wikipedia/commons/0/01/J_Degans_ornithopter_1812.gif).*

The analogy between AI and human flight is powerful, albeit well worn by now. Before 1905-1907, when the Wright Brothers built the first planes that people could fly in a reliable and controllable way, there had been a long series of failed attempts at useful manned flight. There had also been a series of esteemed scientists on the record proclaiming human flight impossible or unfeasibly difficult. There was also a solid argument that the aeronautical theory of the time was inadequate for the task (the Wright brothers proceeded via a clever combination of theory and trial-and-error experimentation). Furthermore, the design that eventually succeeded didn't look that much like any of the known natural flying systems (birds or bugs), though of course it relied on the same basic aeronautical principles. Earlier attempts had often looked more like birds, but this hadn't helped them any.

As with manned flight, the history of attempts to create full-on science fiction style AI is going to look like a long list of failures – until someone succeeds, at which point the viability of the thing will be taken for granted and a tremendous new industry will emerge. And as with manned flight, the form that the first truly awesome AIs take may not resemble in its details what we've come to expect from biological systems.

FIGURE 1.9: *A wing-flapping flying machine that actually worked – though not all that well, and built in 1927 well after the Wright Brothers' success. This device was not especially important in the history of flight, but it sure does look cool; I'd love to fly it! (http://commons.wikimedia.org/wiki/File:Ornithopter_and_creator_George_R._White_at_Saint_Augustine_3.jpg).*

An AGI Revolution is at hand – of this I am quite confident. Sometime, not too long from now in the historical sense, AI systems with general intelligence will be the smartest creatures on the Earth. No, this isn't an absolute certainty, but I think it's about as clear as other obvious technology trends like the ongoing miniaturization of machinery and the use of genomics to cure diseases. In each of these cases, it's pretty clear what has to be done and why it's possible, although there is still a lot of work to be done to get from here to there.

And of course it will require more than just hard work to build AGI. It will require smart work. Ultimately, to get to AGI will require a different way of thinking than the one currently dominant in the AI field. It will require more focus on General intelligence – which is why I like to talk about AGI and not just AI. Making a massive leap forward in AI in a rapid way – in a decade or less rather than 3, 4 or 5 decades – will require intensive work by a brilliant team focused on the G in Artificial General Intelligence. This is a bit different than the highly application-specific work happening in most academic and industrial AI initiatives today. Whether things will really happen this fast, I don't pretend to be able to predict, though I'm trying my best to make it happen, via my own scientific, engineering, and organizational endeavors. But on a historical time-scale, if we get to advanced AI any time this century, that's still pretty darned close.

And once the first true thinking machine emerges, the pace of the next stage may surprise even enthusiasts and futurists. It took only decades to get from the Wright Brothers to jet planes and Sputnik, Voyager and the Space Shuttle. But progress is accelerating and things happen much faster now than a century ago. I am pretty confident it will take years, not decades, to get from the first powerful, generally intelligent AI system to something vastly more powerful and capable.

Artificial General Intelligence

I've spent a fairly large amount of time over the last 10 years popularizing and proselytizing the concept of "AGI" or Artificial General Intelligence. AGI is intended to be contrasted with ''narrow AI,'' a term coined by futurist inventor Ray Kurzweil (a friend and colleague whom I'll mention quite a few times in these pages) to refer to systems that carry out specific ''intelligent'' behaviors in specific contexts. For a narrow AI system, if one changes the context or the behavior specification even a little bit, some level of human reprogramming or reconfiguration is generally necessary to enable the system to retain its level of intelligence. For instance, to make IBM's Watson (a high profile, highly functional narrow AI system) play a game different from Jeopardy, one must reprogram it. To make Google's cars drive off-road or even in the UK, one must reprogram them. This is quite different from natural generally intelligent systems like humans, which have a broad capability to self-adapt to changes in their goals or circumstances, performing ''transfer learning'' to generalize knowledge from one goal or context to others. I introduced the term "Artificial General Intelligence" a decade ago to serve as an antonym to "narrow AI" by referring to systems with this sort of broad generalization capability.[1]

The field of AI, broadly speaking, is about the creation of machines that can do intelligent things. AGI is more specifically about the creation of thinking machines with general intelligence comparable to, or greater than, that of human beings. The AI field began with a focus on AGI (though not calling it by that name), then drifted into narrow AI. This may have been a largely sensible decision, given that specialized AI systems were more feasible to build given the technology of past decades. But today, in 2016, the technological and scientific landscape is very different from what it was in past decades. I am one of an increasing contingent of artificial intelligence researchers who believe that we are now poised to attack the AGI problem in a way that hasn't been feasible before.

FIGURE 1.10: *Screenshot of inventor and futurist Ray Kurzweil, from the documentary film Transcendent Man. I also appeared in the film, for about 5 minutes. Check it out – it's available on iTunes, for example. Alongside his fantastic technical work, Ray has done more than anyone else to bring AGI and its likely human implications to the public attention. (http://transcendentman.com).*

1 I'll give a fuller account of the origin of the term "AGI" a little later on...

Informally, AGI may be thought of as aimed at bridging the gap between current AI programs, which are narrow in scope, and the types of AGI systems commonly seen in fiction – robots like R2D2, C3PO, HAL 9000, Wall-E and so forth. But, on the other hand, one mustn't take the limits of human fictional imagination as the limits of AGI. The concept of AGI extends much more broadly than the common science fictional interpretations of AI would suggest. It encompasses the full gamut of possible synthetic minds, including hypothetical ones far beyond human comprehension.

It may sound a bit outlandish to talk about "hypothetical minds far beyond human comprehension" and practical contemporary technology in the same breath, but that's exactly what makes AGI such an exciting field: the fact that it does have the potential to bridge the everyday and the unimaginable, and not just in the far distant future, but relatively soon. This is why I think the creation of AGI is the most important thing happening on the planet today. I'm thrilled to be alive at a time when I can play a part in something so amazing.

One of the more rewarding aspects of my AI research career so far has been my involvement in the emergence of an at least moderately coherent community of researchers, all focused on the topic of Artificial General Intelligence from different perspectives. I've devoted a fair bit of time to helping guide this community into existence, by putting together annual AGI conferences, editing some books of technical AGI research papers, and founding a group called "The AGI Society."

When I first started bearing down on AI in the late 1980s, it was mainly a solitary pursuit. I sometimes felt like a lone visionary howling out in the wilderness. More recently, it's been quite different for me, because my career has advanced, because my own organizational pursuits have substantially succeeded, and perhaps most of all because the world has changed. One of the great pleasures of my AI work in recent years has been the opportunity it's afforded me to discuss and collaborate with so many gifted and imaginative AI researchers with all sorts of different ideas.

I've put a huge amount of time into co-founding and co-operating the OpenCog project – an open source AGI effort, which I consider the most promising current initiative aimed at creating powerful AGI. (Modesty is not a common trait among AGI researchers.) The OpenCog design is aimed directly at AGI – not narrow AI – and encompasses a number of ideas I think are key to making a big leap in AI functionality, which I'll tell you about in these pages. Right now the overall OpenCog design is perhaps 40% implemented, and I estimate that a team of a few dozen good AI programmers could finish the job in 4-8 years. And then, if my hypothesis is correct and the design is as good as I think it is, we would have an AI system with the rough general intelligence of a young human child. Tweaking the "AI vs. flight" metaphor a bit, this would be what I think of as an "AI Sputnik" event – it would set the world on fire, and trigger a massive explosion in AI research, most likely leading rapidly to the creation of human-level and then superhuman AI systems, changing the world forever.

Of course I'm bullish on the prospects of my own AI project – but still, I'm also aware that it's part of a bigger picture… And even if I'm wrong in some of my technical ideas, there are dozens of other researchers making serious efforts in similar directions right now, in universities and startup companies all around the planet. A very small percentage of humanity's resources and attention are going into AGI. Yet this relatively tiny effort will probably have more impact on humanity's future than anything else now occurring.

You'll notice, in these pages, that I sometimes sound rather extreme when talking about AGI and its dramatic importance and potential. Since I plunged into AGI research seriously, I've become more and more of an "AGI maniac." The pursuit of AGI is awfully addictive once one fully realizes the amazing potential that would ensue from success, and the fact that it's almost surely just a software problem. You "just" have to know what lines of code to type into your source code editor, and then – bingo! You can have a baby mind, ready to be taught and to learn, experience, and self-improve. Of course, it's difficult to figure out the right design, and it's probably more software code than any one person can create given the current state of software tools. But still, to have such an amazing achievement so close at hand in a sense is a pretty addictive, thrilling feeling.

Faster Than You Think

Along with the advancement of AI technology in recent decades, there has been a radical increase in the acceptance of both AI and other "futuristic" ideas in the public eye and in the scientific community. Fascination with making science fictional ideas real is no longer the exclusive property of a tiny, scattered community of science geeks – it's now almost mainstream in some parts of the world!

The storied inventor Ray Kurzweil, who coined the term "narrow AI," has put a lot of thought and energy into forecasting the timing and consequences of the eventual achievement of general AI. He has framed a careful and detailed argument that the advent of AGI will lead to a "Technological Singularity" around 2045 or so – a massive, concurrent advance in a wide variety of technologies, leading to a dramatic transformation in human life on all levels. At the time of the Singularity, as Ray conceives it, computers will be massively superior to humans in intelligence, and the humans of that time will be dramatically enhanced in various ways, and probably closely linked to these superintelligent computers. Though not that many people believe Ray's precise timing estimate for the Singularity, the general idea underlying his predictions – that many kinds of critical technology advancements are progressing exponentially – is increasingly widely accepted.

I have played my own role – though nowhere near as influential as Ray's – in the popularization of futurist thinking in recent years. In the interval from 2005 to 2011 I gave quite a lot of talks at futurist, transhumanist and singularitarian conferences – it's cooled down a bit since I moved to Hong Kong in late 2011, simply because not so many such events are organized here. (Though that may be changing too – there was a major Future Forum in Beijing in January 2014, featuring Ray Kurzweil, deep learning guru Andrew Ng from Baidu's Silicon Valley office, Hugo de Garis, and myself, and with an audience full of Chinese tech and finance industry leaders.) I spoke about a variety of topics, but the most common one was the future of AGI – why I see AGI coming soon, and what I see as the most likely consequences. Many of these talks are available on YouTube in video form.

One thing I've learned from my adventures in popular speaking is that AI and related ideas have become surprisingly exciting and appealing to people from many different walks of life. Of course, my AGI mania is still far from mainstream, as I'm reminded when I bring up these topics in conversations with my neighbors or the cashier at the local supermarket. But they're far closer to mainstream than they were 30 years ago. At the Singularity Summit 2009 we had Justin Rattner, the CTO of Intel, giving a speech agreeing with Kurzweil's forecast of a Singularity around 2045 or so, and furthermore arguing that Intel will provide the hardware to power the Singularity! At the Humanity+ futurist conference in December 2010 at CalTech, we had Robert Tercek, Oprah Winfrey's former head of digital communications, present his own vision of Singularity-focused technologies and their human implications, titled "How Digital Media Accelerates the Future Evolution of

Humanity". The AGI-11 conference was held on Google's campus in Silicon Valley, with an introductory talk by Peter Norvig, co-author of the best-selling AI textbook (and then, just 2-3 years later, Google began putting significant funds into shifting their large internal AI R&D efforts more toward AGI). These are just a few random examples; I could have given dozens of others. This sort of thing would have been unimaginable when I started university in 1982.

The AGI talks I gave at various futurist conferences, and the positive reception they generally received, were much of the inspiration for writing this book you're now reading. My core, original goal in putting this book together was to get down in writing the sorts of things I habitually say in my talks about AGI and the future. When I give a conference talk on AGI and the future, I skip over most of the technical details, and I'm going to do the same here. But I also tend to show a few technical diagrams and mention some technical bits here and there for the benefit of the hard-core geeks in the audience – and I'm going to do the same here.

Even where I've had to skip over zillions of nitty-gritty issues in the interest of comprehensibility, I have striven for accuracy. My aim here is to get across, to the reader who may or may not have any deep background in science and technology, some of my thoughts about AGI technology and the amazing consequences it may have for humanity over the next few decades.

Before I settled on the title "The AGI Revolution" – inspired partly by Mark Changizi's excellent book "The Vision Revolution" – I was thinking to call this book "Faster Than You Think." This is because, among my main messages here, is the message that: AGIs that can think faster, more broadly, and better than you, may well get here faster than you think. It's an astounding, thrilling, and in some ways scary proposition. There's still plenty of speculation involved in thinking about such matters, but not nearly as much as there was a decade ago, let alone 4+ decades ago when I started musing about the topic as a child. Science and technology are advancing fast, and while AGI (unlike narrow, application-specific AI) has not advanced all that fast until recently, there's reason to believe the next few decades are going to be different. It's going to be quite a ride!

And, critically, it's not going to be a ride in which we're just passengers, but a ride in which we're drivers (albeit, to strain the metaphor a bit, drivers going through unknown territory, driving a machine we don't understand all that well). AGI is not something that's going to appear out of nowhere and foist itself upon us; it is something that we're going to create. Something that we are already in the process of creating. Humanity has been moving toward the creation of AGI for a long time now – arguably, at least, ever since the emergence of language enabled the development of organized agriculture, which led to civilization and everything that's come with that. The advent of powerful AGI is all but inevitable at this point, barring a catastrophe that wipes us all out. But the particular form that the first powerful AGIs will take seems fairly wide open. Whether or not the first AGIs are benevolent to humans, for example, seems like something we have the capability to influence, via our actions right now.

FIGURE 1.11: *Since 2005 or so, I've found myself frequently invited to speak at futurist conferences, where folks from a wide variety of backgrounds gather to hear and talk about current and future technologies like AGI. Compared to a couple decades ago, the outpouring of enthusiasm and interest regarding these topics is amazing to see. (Singularity Summit 2006 at Stanford; Singularity Summit 2011 in New York. Courtesy of Kris Notaro and IEET, http://www.krisnotaro.com/s/01crowd.jpg).*

These are amazing times we live in. It's not 1970 anymore! We have astoundingly more powerful computers, we have the Internet, we have far better brain imaging tools and a much better understanding of neuroscience, and we have a cross-disciplinary field of cognitive science which tells us a lot about the overall structure of human cognition. We have single-molecule electric motors, we can do origami with DNA, we can build synthetic organisms via stringing together amino acids, and there are commercially available quantum computers. We can search over a trillion Web pages almost instantly from our laptops or mobile phones. We can hook tens of thousands of insanely fast multiprocessor computers together into functional units. Our guided missiles can fly in the dark to precisely specified locations and destroy precisely what we want them to, with remarkably few errors (though our decisions about what to destroy are often quite questionable). We can measure brains using nuclear magnetic resonance and see which regions are most active during which sorts of mental activity. We can clone animals, and make human organs younger via stem cell injections. All these things, and many, many more current realities, would have seemed wildly science fictional in 1970 – or even in 1982 when I first dug into the AI research literature. Technology and science have been advancing dramatically, and in many ways directly relevant to advancing AI. The AGI field hasn't advanced as fast as some others so far, but – leveraging a host of related advances – the artificial intelligence explosion, the AGI Revolution, will be coming along soon. What I'm going to tell you here is how I've come to believe this, what I'm doing about my beliefs, and why I think all this is so important.

Even if We Can – Should We?

As soon as I began to speak publicly about my work on advanced AI back in the late 1990s, I started getting a lot of questions about the ethics of my pursuits. Schwarzenegger's Terminator seems to loom large in the public psyche.

Some people argue that, because it's hard to guarantee that the AGIs we create will be beneficial to us, we should hold off on creating AGI till we somehow figure out to do it in a risk-free way. There's a group in California – formerly called the Singularity Institute for AI (SIAI), recently rebranded the Machine Intelligence Research Institute (MIRI) – devoted to this perspective. This is also a major theme of work at Oxford University's Future of Humanity Institute (FHI) and Cambridge University's Center for the Study of Existential Risk (CSER). (It's interesting how, with the recent establishment of FHI and CSER, the global center of AGI alarmism has moved from California to the UK.... Europe also remains strong in academic AI and proto-AGI research, though the nexus of practical AI work remains in the US, with some gradual shifting toward China.)

FIGURE 1.12: *Due to the popularity of The Terminator and other SF/action movies, the "evil killer robot" meme has become prevalent in the public mind whenever AI is mentioned. But it's important to remember that movies and novels are created to be dramatic and emotionally evocative and entertaining, not to predict reality! (Image from: http://commons.wikimedia.org/wiki/File:Terminator.JPG).*

However, I very much doubt that a risk-free approach is going to be possible. Reality tends not to work that way. And if the US and UK were to intentionally slow down their progress toward AGI out of worries about the risk, this would only give the lead to China, Russia, Iran, or some other actor; these days, technology and knowledge and ambition have little respect for national borders.

The mirror image of the AGI alarmists is my friend and fellow AGI researcher Hugo de Garis, who argues that AGIs are bound to exterminate us inferior humans, but that we should create them anyways, because it's our destiny. My own view is that we should try our best to create an empathic, loving, beneficial AGI fairly quickly, before someone else creates a nasty one.

Certainly, my motivation for creating AI has never had anything to do with killer robots, and it has always seemed intuitively clear to me that with the right AI design, one could create AIs more rather than less ethically positive than human beings. Just as humans aren't anywhere near the smartest creatures possible, we also aren't anywhere near the maximum of ethical goodness. A glance at the newspaper on any random day should make that rather clear.

I do have to admit that every now and then, the prospect of creating powerful AI systems scares me. I'm less worried about the theoretical possibility of nice AIs reprogramming themselves to turn evil than I am about the palpable possibility of megalomaniacal humans intentionally creating powerful AIs to serve their own ends. But at the same time, I never forget the potential for incredible good from advanced AI – for instance (just to name a few small examples) the potential for the end of death, disease and scarcity.

My 14-year-old nephew Lev died a few years ago, from a meningitis infection randomly contracted via sharing a soda with a carrier. What a horrible tragedy – he was a wonderful kid, and his life was cut off so near the beginning. My sister and her husband are going on with their lives with admirable persistence – and they've had a new, sweet and beautiful little daughter – but they will never fully recover. It's still hard for me to believe, now, that Lev isn't around for me to joke around with, to wrestle with, to BS with about time travel and immortality (as we did when he stayed at my house for a week, about 6 months before his absurdly untimely end).

FIGURE 1.13: *My nephew Lev died at age 14 of a freak meningitis infection. It's hard to believe his existence was aborted at the stage you see in the picture, and he never got to experience all the joys and challenges of adulthood. He was a unique, quirky and brilliant kid, and we were all especially curious to see what kind of adult he would turn into and what he would do with his life. Human life today is wonderful, yet also full of all manner of tragedy and suffering. The potential of AGI technology to combat disease and extend life is dramatic and should not be overlooked.*

A couple of years before Lev's utterly unexpected death, my grandfather died more predictably, at age 91. He was a physical chemist, and he was the one who got me interested in science originally. He never lost his good sense and kindness or his sense of humor, but near the end of his life he had to ask for help doing simple arithmetic. This shook me terribly, as he was the one who had, decades earlier, taught me various math tricks like taking square roots on a piece of paper using Newton's Method. We now take death via meningitis, aging and other such causes for granted, as if it's inevitable and unavoidable – but it's no more unavoidable than death via smallpox (now eradicated) or staphylococcus (very rare in the developed world since the advent of antibiotics). The biological problems that now kill humans young and old could be eliminated pretty easily by a mildly superhuman artificial intelligence focused on biomedicine.

Additionally, beyond the "mere" possibility of a massive reduction in death and suffering, there's also the possibility of the expansion of human intelligence and experience far beyond the paltry limits allowed by our current "legacy human" form.

In the early days of my AGI obsession, one of my half-baked, over-ambitious plans was to build advanced AGI secretly – myself and a small group of friends would make the superhuman thinking machine and launch it on the world! There are plenty of groups of researchers and engineers around the world today working on AGI with a similar intended modus operandi. But gradually I've shifted away from that perspective and come to believe it will be better if AGI is something accomplished step by step, by the human race as a whole. Creating the next step in the evolution of intelligence on Earth is a big job and a big responsibility, and it should be approached with as much intelligence and wisdom as possible, meaning it should be approached by the "global brain" of humanity with all our computer and communication systems, not by an isolated group of zealots in a basement. This was part of the thinking underlying the open-sourcing of the core of my proprietary Novamente Cognition Engine system to create OpenCog, which is an open-source AI project that anyone can participate in. Our aspiration with OpenCog is to gather a larger and larger developer community, who will collectively create AGI for the good of all humanity, as well as for the good of the trans-human AGIs we will eventually create, and that will hopefully co-exist peacefully and fascinatingly with our future human descendants.

Onward and Upward

In this book, I will tell you my story as an AGI researcher – how I started out in the field, what efforts and attempts I've made, what I've learned as I've gone along, and what I think will work in the future. I've done theoretical academic AI research, I've started a number of companies focused on various AI applications, I've done AI and neuroscience work for various government agencies, and I've organized conferences and workshops and societies and open source projects oriented toward different aspects of AI and AGI. I've worked on AI and AGI in the US, Brazil, China, Africa, Australia and New Zealand. All these experiences have given me different perspectives on the issues we face as we strive to push AGI forward and create radically intelligent machines with radically transformative power.

This is not an autobiography – I won't tell you about my adventures in love, marriage, divorce and fatherhood, nor my travels around the world, nor my creative endeavors in metaphysical philosophy, fiction writing, or music, nor my ventures into Eastern and shamanic spirituality, nor my various and varyingly successful business endeavors, except here and there as they've impinged on my AI work, nor the multitude of other things that I've done in my 49 years of human life. (That might be a fun book to write, but it would be a rather different one!) Rather, what I'll do in these pages is to use the narrative of my career as an AI scientist and entrepreneur to structure my thoughts about the concept of AGI and its potential near-term and further-off implications, from a variety of different angles. I will tell you about these issues from my own particular perspective, based on multiple decades spent doing research in AGI and related areas, and writing and speaking about AGI and its future implications.

Absolutely, what you are getting here is one researcher's deeply thought and felt personal perspective. This is not an attempt at a fully "objective" treatment of the topic. I have my own quirks and biases to be sure, and I won't shy away from them here: my aim is to present my own definite view of AGI, rather than to survey all possible views. I want to tell you why I am personally so passionate about AGI – what it means to me and what I think it means for everyone.

Please don't think, though, that by telling the story of AGI in this way I am somehow expressing an egocentric view of the overall history of AGI. I'm telling the story from my own perspective because that is what I know best, and because I think and write best in the first person. I am well aware that no single human being is especially significant in the grand sweep of history. And of course, while I have big ambitions for my ongoing projects, I'm quite cognizant that there are plenty of others who have contributed more than I have to the progress of AI, so far. I'm happy to have played some at least minutely greater than zero role in the overall progress of the human race toward AGI, but I'm well aware that AGI is emerging as a process within the Global Brain of the human race and its technology. I'm just a bit player as are we all. But it's my own bit that I'm best equipped to tell about – and hopefully the telling that I do here will form a more than totally useless part of the Global Brain's growth process!

I am simply amazed to be alive at a time when the concept of AGI no longer resides in the domain of fantasy. The possibility of AGI is scientifically rather clear at this point, and the creation of AGI lies, at very least, on the edge of current engineering possibility. Wow!

2. Q&A on AGI

I get asked to do interviews a lot, for media outlets great and small, and some of the questions (and journalists) are smarter than others. A few years ago, I did an interview for a science fiction blog site, with a writer named Jason Peffley, and I felt he did a particularly crisp job of posing the basic questions that everybody seems to ask me about AGI and the Singularity. So I will get the party started here with a lightly edited version of my chat with Jason:

Jason You've been in the AI industry for quite a while now. Did you always know that you wanted to work with AI? Was there something in your life that triggered it?

Ben I grew up on science fiction, and I always knew I wanted to spend my life making science fiction things become real. But, at the start, I wasn't sure if my focus would be AI, time machines, quantum gravity computers, interstellar spacecraft, genetically engineered creatures, psi powers, or whatnot. I chose to focus on AI. Initially in my late teens, and later, with greater emphasis, starting in my late 20s because it seemed, logistically, the easiest of the bunch. If I'm right, you don't need any special hardware, just the right computer code, and a big server farm of commodity PCs. The more I thought about the problem, the more I felt I knew how to make it happen.

Jason If you were talking to 10 year olds, how would you sum up the Technological Singularity in a way they'd understand?

Ben Within a few decades, we're probably going to have computers and robots way smarter than people in every way, inventing new stuff constantly and transforming the world in ways mere humans can't even imagine. Hopefully, we'll be able to upgrade our brains or turn ourselves into robots, and become super-smart, super-powerful beings ourselves. We don't have this technology now, but it's coming pretty soon, because new inventions keep getting made faster and faster.

Because the more inventions you have, the faster you can make new inventions, since the old inventions are tools you can use to help make new ones.

Jason What would a robot, or program for that matter, need to do in order for you to kick your feet up and say "I finally did it?" Teach itself how to tie a pair of shoes? Get your dry cleaning? Pick up your kids from school? Teach you string theory while building a fusion reactor?

Ben World peace. Immortality for all who want it. Everybody liberated from needing to work for a living. Starships traversing the galaxy, and brain implants or robot bodies for everyone who wants to become a superhuman AI.

Thing is, I think all of that has a decent chance of happening within a couple decades after we create the first AGI that's as smart as, say, an average science professor. From there, I think it won't be such a big leap.

The big leap, I think, is from where we are now, to having an AI that's like a human toddler. When someone gets an AI at that level, I'll be incredibly excited, and also a bit scared.

Jason Will we see the advanced R2D2 and C3P0 type of AI in our lifetime?

Ben Definitely, and even better. I'm not sure about C3PO's six million forms of communication though, as there may not be that many useful alien races around. Also, the real-life R2D2 might go to the Radio Shack and buy himself a speech synthesizer. But in concept; yeah, we will get there.

Jason In an interview you did on Singularity 101, you said that you had thought a lot about how you would spend 100 billion dollars. Outside of buying a private island, how would you spend it?

Ben If I had 100 billion dollars, I'd start a semi-private city, devoted specifically to the beneficial development of advanced technologies, with residency offered to folks displaying qualification and passion for said technology development. Otherwise, I'd start a massive research grant funding project for AGI, life extension, nanotechnology, femtotechnology, mind uploading, and so forth. If these areas got attention in the way that cancer research and semiconductor manufacturing do, progress would be way further along.

Jason Outside of funding, what is the current biggest hurdle in your field?

Ben Finding people who are really good at programming that also have a deep understanding of cognitive science and AGI theory is always a challenge.

Conceptually, I think the biggest challenge facing the field is the integration of algorithms for abstract cognition with algorithms for low-level perception and action. Right now, we have good approaches for all of these, but nobody has made them all work together in a unified way. I think the OpenCog architecture solves that problem, but we haven't definitively demonstrated that yet. We're well on our way, though!

Jason What do you say to the people who feel this will turn into a colossal **** up like Cyberdyne's Skynet division?[1]

Ben Setting aside the "naked guys traveling back in time" aspect, I think the possibility of advanced AIs running amok and killing everyone is a real one. It can't be ruled out anyway. However, I think there are actions we can take to minimize the

1 For the three readers who don't know, this is a reference to the *Terminator* film series, in which a computer network security system accidentally achieves super intelligence and takes over the world, wreaking massive destruction and creating evil time-traveling robots with hunky biceps.

probability of this happening, by building our AGI systems with rational minds and benevolence-oriented goal systems, and raising our young AGIs with kindness, and teaching them well.

Like any other advanced technology, the potential benefits are huge and so are the risks, but that's how humanity has been rolling for a long time. Nobody could stop the development of advanced AI and robotics now, even if they wanted to. At least not without destroying all of civilization. We're on the verge of the next step in the evolution of intelligence on Earth, so we may as well embrace and enjoy it, and try to nudge it in a positive direction insofar as we can.

3. From Spock to Singularity

Now let's take a step back from the Singularity... and go way back in time... way, way back into ancient history... all the way back, once again, to the 1970s....

FIGURE 3.1: *One of the more memorable moments of my early childhood was sitting with my dad, watching Spock, the master of logic, confront the insane robot Norman with a host of illogical nonsense. The original Star Trek TV show had a formative impact on me, as on many others of my generation.*

Really, I'm not that obsessed with reflecting on my childhood! It's just that comparing what I see today with what I saw when I first emerged onto the planet is my best personal way of understanding the nature of the exponential growth we're all a part of...

My fascination with AI – which is probably still the key driver of my work today – first began in my pre-school and early school days. The thoughts and curiosities I had back then sowed the seeds for my current work trying to create AGI and launch a positive Singularity. And, along with all the other exponential changes, the contrast between AI as it appeared back then – to knowledgeable adults as well as precocious SF-loving children – and how it appears now, is both dramatic and instructive.

The first memory I have is watching Neil Armstrong walk on the moon, on TV. The actual event impressed me, and I was also impressed by how overwhelmed all the adults around me were. I was two and a half. The spacecraft and spacesuit looked suitably complicated. It was clear that science and engineering, pursuits I was just barely aware of, could do amazing things.

AI didn't enter my universe till 1971-72, when I was in kindergarten and first grade in Eugene, Oregon (I was born in Brazil to American parents, but they moved back to the US in the middle of my second year). I regularly watched the original Star Trek TV show with my dad, and one of the more memorable episodes involved Mr. Spock driving an intelligent but overly-logical starfaring robot crazy. TV gets a lot of deserved guff for providing so much stupid programming, but as a young child, it was TV that got me interested in science fiction, causing me to seek out SF novels; and it was SF novels that got me deeply interested in weird science and radical technology innovation.

The robot on the show was named Norman, and he controlled a bevy of hot female androids, programmed to serve a sleazy human master named Mudd. The sexbot androids didn't attract me much at that age, but Norman was intriguing.

Norman had captured Kirk, Spock and the rest of the original Star Trek regulars, and appeared to physically hold the upper hand over them. But Kirk and Spock always had some way or another out of their predicament, whatever it was. In this case, their idea was to drive the robot insane via presenting it with a host of logical paradoxes and illogical nonsense, the theory being that the robot was a perfectly logical being and couldn't handle this kind of information. The dialogue amused and perplexed my young mind considerably:

NORMAN	What are you doing here?
KIRK	I want you to surrender.
NORMAN	That is illogical. We can move more quickly than you. We are invulnerable to attack. We are much stronger.
KIRK	No, we are stronger. I'll prove it to you. Can you harm a man that you're programmed to serve?
NORMAN	No.
MUDD	But you already have, Norman, laddy. Human beings do not survive on bread alone, you poor soulless creature, but on the nourishments of liberty, for what indeed is a man without freedom? Naught but a mechanism trapped in the cogwheels of eternity.
MCCOY	(in a monotone) You offer us only well-being.

SCOTT	(in a monotone) Food and drink and happiness mean nothing to us. We must be about our job.
MCCOY	Suffering, in torment and pain. Laboring without end.
SCOTT	Dying and crying and lamenting over our burdens.
BOTH	Only this way can we be happy. (Then they curtsy.)
NORMAN	That is contradictory. It is not logical. Mister Spock. Explain.
KIRK	Why not?
SPOCK	Logic is a little tweeting bird chirping in a meadow. Logic is a wreath of pretty flowers which smell bad. Are you sure your circuits are registering correctly? Your ears are green.
SCOTT	(clutching at his heart and crying out in anguish) I cannot go on! I'm tired of happiness. I'm tired of comfort and pleasure. I'm ready! Kill me! Kill me! (They point their fingers at him and whistle. Scott slumps to his knees.)
SCOTT	Goodbye, cruel universe.
MCCOY	He's dead.
NORMAN	You... cannot have killed him. You have no weapons.
KIRK	Scotty. Scotty's dead. He had too much happiness. But now he's happier he's dead, and we'll miss him. Let us hear it for our poor, dead friend. (They laugh.)
KIRK	What is a man but that lofty spirit, that sense of enterprise, that devotion to something that cannot be sensed, cannot be realized but only dreamed! The highest reality!
MUDD	Brilliant! Bravo, bravo, Captain!
KIRK	How did you like it?
NORMAN	That is irrational. Illogical. Dreams are not real.
KIRK	Our logic is to be illogical. That is our advantage. Mister Spock, it is time. The explosive.
SPOCK	Very well, Captain. (He removes an invisible package from under his tunic.)
SCOTT	Explosive! (gets up again)

MUDD	Don't panic. Women and children first. Mister Spock, isn't that too much for our purposes?
SPOCK	I believe that is the correct amount, Captain. Mister Mudd, are you ready?
MUDD	Aye, aye! (adopts a catcher's pose)
SPOCK	Be careful. I would not want you to drop it. (Mimes an underarm throw, and Mudd juggles before 'catching' it.)
MCCOY	Easy now. Oh, he's caught it!
KIRK	Watch it! Watch it! (Mudd puts 'it' on the floor.)
MUDD	(to McCoy) Detonator. Fuse. Primer. Mashie. (Adopts a golfing stance.)
NORMAN	There is no explosive.
KIRK	No? Observe. Fore! (Mudd swings.) Boom! (Everyone except Spock staggers around with their hands over their ears.)
KIRK	Are you all right? (Spock quietly exits.)
NORMAN	But there was no explosion.
MUDD	I lied.
NORMAN	What?
KIRK	He lied. Everything Harry tells you is a lie. Remember that. Everything Harry tells you is a lie.
MUDD	Listen to this carefully, Norman. I am lying.
NORMAN	You say you are lying, but if everything you say is a lie then you are telling the truth, but you cannot tell the truth because everything you say is a lie. But, you lie. You tell the truth, but you cannot, for you – Illogical! Illogical! Please explain. (Smoke comes out of Norman's head.)
NORMAN	You are human. Only humans can explain their behavior. Please explain.
KIRK	I am not programmed to respond in that area. (Norman goes blank.)
SPOCK	I believe they are all immobilized, Captain.
KIRK	Good.

The Star Trek crew's stratagem both delighted and disappointed me. Surely, I thought, anyone smart enough to make a robot like Norman would be smart enough to wire him to handle illogic just like people could. The challenge of making an illogical robot struck me as interesting, though at that age I certainly had no idea how to go about doing it. I pestered my dad about it briefly, but he didn't have a clue either. He was a sociology professor, not a computer scientist or robot engineer. My mom was a social activist (and later became an executive in the social work field). They encouraged my interest in science and taught me a lot; but even when I was quite young, they could only answer a fraction of my technical questions.

The coolness of Norman and other robots I saw in SF movies, TV shows and novels was undeniable. But at that point, AI just felt to me like one of a host of really cool, speculative, futuristic technologies: AI, robots, time travel, immortality pills, psychic powers, spaceships, teleporters… The list of cool things went on and on. I didn't see much prospect of any of these getting created on Earth during my lifetime, so I mused about creating a starship that would travel away from Earth at near light speed, exploiting the wonders of relativistic physics to enable me to return to Earth just a couple years older, but finding the Earth a million years in the future. After a million years or so, I figured, a lot of cool technologies would have been developed, and a lot of amazing new life possibilities would be open to me.

When I watched Star Wars as a young teen back in the 70s, I was peeved that R2D2 and C3PO weren't combined into one robot sharing the skills of both. When I saw 2001: A Space Odyssey, at least half of me wanted HAL9000 to take over the universe. I loved Stanislaw Lem's "Golem," a superhuman AI who lectures humanity endlessly on its inferiority and his profoundly transcendent philosophy and understanding. I loved Lem's Honest Annie even more – a superhuman AI that, after it gets smart enough, builds an impenetrable shield around itself and cuts itself off from humanity entirely. I was very frustrated that Asimov's robot detective R. Daneel Olivaw was limited to human-level intelligence, since it just seemed unrealistic that the positronic brain would find itself limited at, basically, the exact same level as the human brain.

The Power of Intelligence

If Norman was so intelligent, why couldn't he see through Spock's clever yet ultimately simple logical games? The power and nature of this strange quantity called "intelligence" fascinated me considerably in my youth. Partly, I suppose, this was because being "the smart kid" was my main source of ego gratification. At that age, my obvious and generally recognized academic and cognitive and creative abilities were what made me feel OK about myself, even though I wasn't especially athletic or popular.

I tended to spend most of my time alone reading, riding my bike, or wandering around in the woods thinking. I generally had one "best friend" – always another misfit – and a few other kids who were sometimes willing to speak to me or play a bit. But the questions of science and philosophy that perplexed me didn't interest most of my classmates at all. I was trying to understand how crystals form (my grandpa Leo, whom I mentioned above, was a crystallographer) and whether there was some scientific way for life after death to be possible, whereas the other kids seemed mostly concerned with gossiping about each other, which didn't interest me in the least. So it was easy for me to feel a certain self-pride at having a superior, or at least more mature intellect than the others around me.

But when I caught myself feeling this way, I'd always remind myself "Sure, you rate yourself better according to your own favored measure – intelligence, which not coincidentally is the measure according to which you come out best. But these people probably rate themselves better according to, say, physical strength or social skills, which are measures according to which they do better than you. Everyone rates themselves according to the measures they prefer, which usually correspond to the measures they do best on. So who's to say what's really best? Best is always relative to some measure...."

A little later, in university, when I discovered Nietzsche and existentialism, I became enamored of the notion of creating one's own value system. But back in elementary school I spent more time musing on the notion of intelligence. What did it really mean – and why was it a more important valuation to look at than, say, physical strength, or running speed, or the ability to talk really fast?

One conclusion I came to was that intelligence was generally more important than strength or running or talking speed, because it could be used in more different ways. Since I read a lot of science fiction about space travel back then, I thought about it like: "If you dropped a really smart person on an unknown random alien planet, their odds of survival would be greater than those of a really strong person, or a really fast-talking person."

Most of my elementary-school musing didn't end up going anywhere useful, but in this case my childish mental meandering was actually verging on something kind of profound: the transformative power of intelligence.

Humans are not the strongest, fastest, longest-lived or most elegant organisms on the planet – but in some important senses we are the smartest. Our creation of tools for manufacture and communication make humanity as a whole far smarter than any individual human. Comparing the Macbook I'm typing this on to the cave walls on which my ancestors painted not so long ago, one clearly sees the power of intelligence to transform the world. Ultimately, being more intelligent gives humanity more flexibility and generality than other animals, stronger or faster or happier though they may be.

But humans are far from the maximally intelligent possible beings. Soon we will use our intelligence to create beings far more intelligent than us, which will transform the world far more rapidly and dramatically than we could ever imagine. Along the way to humanly inconceivable modes of thinking, experiencing, building and interacting, every kind of industry currently known to us will be revolutionized, and everyday human lives and ways of thinking will be unprecedentedly transformed.

Mind and Pattern

Though the problem popped into my head quasi-randomly a few times per year, my childhood mind never succeeded in cooking up any clever method of building or growing a Norman of my own. I was somewhat inclined toward the idea of growing a biological AI from some cells in a vat of some kind, rather than wiring it together, but I never got too far in this direction, since my knowledge of biochemistry was basically nonexistent. I succeeded in making some weird, smelly, steamy chemicals with my basement chemistry set, but it was the 1970s and DIY-Bio equipment was a long time in the future. Now and then I also tended toward philosophically musing about the nature of the mind itself. What is this "mind" system that generates intelligence?

In my mid-elementary-school years I started vaguely gestating a way of thinking about mind using the concept of pattern. The line of thinking at least went back to a book on crystals that my Grandpa Leo had given me when I was 7 or 8. The idea that there were these repeated, patterned structures underlying so many ordinary forms of matter – and emerging complexly from the confusing quantum mechanics of the underlying atoms – was intriguing to me. My grandpa's understanding of quantum mechanics was somewhat informed by the Bohr model (the "old quantum mechanics"); he explained about the double-slit experiment to me, as well as other spooky aspects of quantum theory, but he didn't really understand them, and what he said on the topic somewhat confused me. But his explanations of crystallography and X-ray diffraction (which he'd spent his career doing) were – well, you know... it has to be said... crystal clear, and gave me a passion for looking for hidden patterns underlying ordinary-looking things. Some book I read at that stage also referenced quantum theory pioneer Erwin Schrodinger's framing of DNA as an "aperiodic crystal." I had only a very dim idea of what that meant, but it sounded pretty cool. The view of the physical world as a sort of hierarchy of periodic and aperiodic patterned structures made a lot of sense to me.[2]

Then as now, my mind was generally spinning on topics at the edge of my comprehension, and most of my spare moments were spent reading to absorb new information about the universe!

I don't want to overstate the sophistication of my childhood self – I certainly wasn't a top-flight child prodigy or anything. For instance, I knew what quantum theory was as an elementary school student, and understood some of the basic concepts – so this definitely distinguished me from most of my peers. However, unlike some true prodigies I've read about, I couldn't solve the Schrodinger equation (the simplest version of the basic equation of quantum mechanics) in elementary school; I couldn't even read it till my first year of high school. I didn't learn to solve it till my fourth semester of university, when I was 17. But the main thing I'd say about my childhood state of mind is that as long as I can remember, my Task #1 has been to understand the universe as best as I can, including understanding my own mind and everybody else's. Of course I liked to run around and play like every other kid, but when the games were done, I was always back to thinking, thinking, thinking. Understanding the universe is a big job; I'm still at it... I hope I've made some progress, but there's a long way to go!!

I learned algebra at age 7 from a book my mom bought me, and I learned calculus in middle school from a tattered old book I checked out from the town library (leading to a funny situation where I got an after-school

2 Of course, as a child I also had a lot of other thoughts about how the mind worked, many of which were confused and chaotic, some of which would surely seem silly to me now. In hindsight I tend to recollect best those childhood thoughts that resonate with my current thinking – that's just how human memory works.

detention for reading in 8th grade pre-algebra class instead of paying attention to the teacher. What I was reading was a calculus book; I liked math but had already known basic algebra for a long time. But the teacher didn't care if I was reading advanced PhD level mathematical physics; the important thing was that I was breaking a rule...) I have sometimes wondered if my parents could have turned me into a full-on prodigy if they'd hired tutors to force me to study in a structured way all day, and let me study at my own pace rather than the glacial pace of the public school system. But after I entered second grade they basically left my education to the public schools and my own voracious curiosity and appetite for reading – they were both busy with their work and their own doings.

Actually, though, I don't terribly regret not having been taught math and physics and advanced topics 8 hours per day by private tutors, as Norbert Wiener was in his childhood. I do regret having sat through so many years of boring school lessons on topics I already knew. But I would rather have spent my childhood reading freely and being tutored a couple hours per day, than being crammed with knowledge and skills at the maximum possible pace. I have always valued the freedom to study and learn and invent according my own intuition rather than somebody else's structure.

Godel, Escher, Bach and Beyond

My first introduction to the world of serious AI research was Douglas Hofstadter's wonderful book *Godel, Escher, Bach*, which I read in one long day shortly after it first came out. I picked it up at 6 AM or so the day after I bought it, and didn't stop reading till around 3 AM the following morning. I was 13 years old, in the summer vacation between middle and high school, and the book completely "blew my mind." It didn't contain too much information about how to build AIs in practice, but it ran through various related conceptual issues in much more detail than I'd seen before, and it was funny as hell! In essence, Hofstadter put forth a similar perspective to the one my childish mind had inferred via thinking about crystallography and biology, but with far more subtlety. Hofstadter was thinking about more complex patterns than crystals or DNA – self-referential patterns like the human self. He was wondering about how intelligent systems could internally represent and reason about patterns in themselves.

Years later, when I hired AI researcher Pei Wang to work with me at my first AI company, Webmind Inc., I was amused to find that he had been one of the translators of Godel, Escher, Bach into Chinese. I was thrilled to meet Hofstadter himself in 2006, when we both spoke at the Singularity Summit in San Francisco, even though we were, in a sense, on opposing sides at that conference. I was there talking about the themes of this book, and he was there expressing his skepticism about the notion of a technological Singularity.

Meanwhile, in middle school, a couple years before I encountered Hofstadter's work, I happened on the writings of the Russian mystic P.D. Ouspensky[3], whose works were my first encounter with the view of the everyday human mind as a bundle of reflexive habits. Ouspensky urged the reader to "wake up" and stop being controlled by reflexive habits, and be truly alertly conscious. I practiced this sincerely while walking home from school, and while sitting in detention after school (I got punished a lot for various random offenses like being late to class while wandering around somewhere else, reading SF novels in class, and so forth). Trying to make myself "fully conscious" à la Ouspensky more of the time taught me to introspect deeply on the habit-patterns of my mind and self. So when I encountered Hofstadter's take on mind and pattern, I could ground it my own introspections. Both Hofstadter and Ouspensky saw the ordinary everyday mind as a bundle of habit-patterns, but Ouspensky thought that when you went beyond this patterning you got to some sort of magical supermundane conscious mind, whereas Hofstadter basically seemed to think the patterned, self-habituating mind was all there was.

After this early exposure to science-fictional AI and GEB, I then got a big shock when, at the age of 15-16 in my first year of university at Simon's Rock Early College, I started doing some moderately serious spare-time digging into the academic AI literature. Simon's Rock was (and is) a special university oriented toward nerdy kids who want to leave high school one or two years early to start college work. In that era, at any rate, it had an alternative, creative culture that I really loved. However, scientific and technical subjects weren't really the school's strength. There were some great science teachers (I still value what my Simon's Rock physics Prof. George Mandeville taught

---

3 Ouspensky was the student of the mystical teacher Gurdjieff, a fact that seemed amusing to me years later when I found my great-grandfather's name before he immigrated to the US had been Sam Getzelovich Gurdjieff. However, this was a fairly common surname in parts of Eastern Europe, and no close genetic connection between the mystic Gurdjieff and myself seems likely. In addition to the aspects of Ouspensky's thinking that inspired me, there was also a lot of stuff there that struck me as silliness; I just ignored it.

me about the arrow of time), but overall the curriculum was much richer in the humanities and arts. For sure, in the early 1980s, a liberal artsy school like Simon's Rock didn't have any computer science curriculum to speak of; and the Internet didn't yet have a significant number of research papers on it (there was no Web, only Gofer). But the library had a few dozen AI books and a handful of computer science journals, which was enough for me to get a picture of what sorts of things were being done.

Exploring AI books and journals, I was vaguely expecting to find more material in the vein of GEB, but with more technical detail and more practical achievements. I did find some exciting, ambitious writings by the early founders of the AI field, from the 1940s through the 60s. But after that, it seemed to me, the field had drifted in much less interesting directions. As of 1982-83, it seemed, there was basically nobody pursuing AI in the grand sense – they were just fiddling with small problems. I looked into Hofstadter's technical AI work and was intrigued but not as thrilled as I'd hoped – it was conceptually deep, but he was playing with "toy problems," not actually trying to build full-scale thinking machines. On the other hand, much of the remainder of the AI literature was focused on building very specialized systems aimed at doing something "intelligent" in very limited cases. Often these specialized systems were created by hand-coding long lists of rules specifying what behaviors or answers constitute intelligence in very specific situations.

My attitude was something like the one expressed by the title character in the movie Lawrence of Arabia (one of my favorites) when he quoted Themistocles: "I cannot fiddle, but I can make a great state of a small city." I didn't want to spend my life messing around with toy problems or exploring conceptual issues, or working on narrow applications. I was getting increasingly interested in AI as a field of study and work, but not in academic results incapable of scaling up beyond little toy problems, and neither in narrow AI aimed at particular application problems – I wanted to create machines with general intelligence at the human level or beyond.

One practical result of my frustration with the AI field in the early 1980s was that, in spite of my passion for the goal of creating thinking machines, I decided not to pursue a college degree in AI (at either the undergrad or grad level), but to study mathematics instead. I loved mathematics, there was a lot of it to learn, and the state of the AI field at that point just seemed so disappointing. Also, I had the notion that it might be possible to use advanced mathematics to break the AI field out of its doldrums.

In hindsight, I'm sure I would have had a lot to learn from getting an AI degree – though I don't regret doing my BA and PhD in math instead. But I was just too frustrated by the "split personality" of the AI field – the split I now conceptualize as the narrow AI / AGI divide. When I first plunged into the AI literature in 1982, I naively imagined that the international community of AI scientists around the world is working day and night trying to create computers smarter than people — trying to work toward computers capable of holding intelligent conversations, outsmarting Nobel Prize winners, writing beautiful poetry, proving amazing new math theorems. But my hopes were dashed rather thoroughly – what one finds in the academic AI literature, now as in 1982, are dry, formal analyses of very particular aspects of intelligence, applications of AI technology to simple puzzles or very particular commercial problems, and so forth. Mainstream AI research as it currently exists focuses almost entirely on highly specialized problem-solving programs, constituting at most small aspects of intelligence and involving little or no spontaneity or creativity.

There is, indeed, a thread connecting these two aspects of AI – the grand vision and the practical, often highly specialized work – but the thread has gotten extremely frayed at various points in the AI field's history. Now, in 2016 as I write these words, it's strengthening rapidly, and more and more AI researchers are building bridges between their specific technical projects and the broad conception of AI that the field has had since its inception.

Further Reading

As this is not an academic tome, I will not give complete references to everything mentioned in the text here. With modern Web search technology, this seems quite unnecessary. However, at the end of most chapters I will give a list of "further reading," comprised of a somewhat unsystematic combination of stuff that is mentioned extremely prominently in the chapter, and stuff that is relevant to the chapter and was important to me at some point. Some of the further reading is friendly to the nontechnical reader and some is fairly intense. Enjoy it, or not, as you wish!

Durant, Will (1935). *The History of Civilization*. Simon and Shuster.

Hofstadter, Douglas (1979). *Godel, Escher, Bach*. Basic Books.

Hofstadter, Douglas (1996). Fluid Concepts and Creative *Analogies*. Basic Books.

Ouspensky, P.D. (1949). *In Search of the Miraculous*. Harcourt, Brace.

4. Mind Patterns

Though as a hyper-ambitious, knowledge-hungry university student I was disappointed by what I saw in the technical AI literature, and chose not to pursue AI as a major in college, I didn't lose interest in the topic – not at all. Rather, I came to see the tepid, strangely incomplete and wimpy nature of most of the AI books and papers I read as a major opportunity. Current researchers, for some reason, didn't seem to be grabbing hold of the central problem of AI, the actual creation of thinking machines. That was frustrating, from one perspective. From another perspective, it was exciting, because it meant the task was left there for me to take care of. If I could.

I tend to approach a new problem by first generally familiarizing myself with what others have done, at least enough to understand what kinds of things they've done, and then trying to think it through from scratch myself. First, learn what others have thought, at least to a first degree of approximation, and then set it aside and try to take a "beginner's mind" attitude. Of course, one doesn't truly have a beginner's mind once one has a broad familiarity with the literature, but that's just the way it goes. Truly approaching complex modern scientific issues with a pure beginner's mind just can't work – science is too incremental, and one really has to know a lot about what has come before to make any progress at all. But one doesn't have to use ALL of what has come before; one can be judicious and creative about what to use and what to ignore.

So, during my three years of undergraduate school, I spent a fair bit of time brainstorming on my own and with my classmates about how to build AI. All in all, I took more inspiration from the humanities than from technical subjects at that stage. When I took philosophy classes, I felt there was a lot of AI-relevant insight to be found in the philosophy of mind – not so much the recent academic literature as older thinkers like Charles Peirce, Nietzsche, ethnolinguist Benjamin Whorf, the German phenomenologists, the medieval Buddhist logicians... there were so many different angles on the wonderfully complex mess of the human mind! The philosophers we studied in class just scratched the surface – I was soon digging into the philosophy section at the library with the same zeal as the AI, computer science, mathematics, physics and systems theory sections.

As I progressed through reading philosophy and related fields, I dug deeper and deeper into the "mind is pattern" philosophy, settling on a vague version of what I would now call a "patternist philosophy of mind" as a general approach to thinking about intelligent systems. I was inclined to adopt the very simple premise that mind is made of pattern – and that a mind is a system for recognizing patterns in itself and the world. Of course it's very easy to fit other ideas into this framework. What is the self but a pattern that a mind recognizes in the interactions between its body and the world, and in its own internal dynamics? How does a mind achieve goals? By recognizing patterns regarding which procedures are likely to lead to the achievement of which goals in which contexts. And so forth. It actually took me until 2006 when I wrote the book *The Hidden Pattern* – 24 years after I entered university – to articulate the patternist perspective on mind to my own satisfaction, in a philosophically general yet practically applicable way. However, the core idea was there in my head as a child, and it got more and more concrete as my education proceeded.

My college philosophy professor Ed Misch – a remarkable teacher, who reportedly had once been a Catholic monk under a vow of silence for a number of years – mentioned Charles Peirce passingly in class once, and I dug into Peirce's work with a passion. Peirce had a complex theory of mind and reality, but he posited that at the center.

Logical analysis applied to mental phenomena shows that there is but one law of mind, namely, that ideas tend to spread continuously and to affect certain others which stand to them in a peculiar relation of affectability. In this spreading they lose intensity, and especially the power of affecting others, but gain generality and become welded with other ideas.

He referred to this "one law of mind" as "the the tendency to take habits." He also had a metaphysical ontology in which the universe was divided into First (pure, unanalyzed perception), Second (physical reaction) and Third (relationship, i.e. habit, pattern). The "one law of mind" was mind viewed from the point of view of Third, in his philosophical system. What we'd now call "qualia" – the raw feel of consciousness, was mind from the point of view of First.

Poking around in the systems theory section of the library eventually led me to Gregory Bateson, who analyzed biological and psychological phenomena from a cybernetics point of view. I also read older cybernetics works, like the original book *Cybernetics* by Norbert Wiener – an amazing book for its time that arguably was the beginning of the modern perspective on intelligence, in which biology, psychology and engineering are viewed as different angles on the same thing. But Bateson impacted me more – he explicitly articulated what he called "the Metapattern," which he defined as "pattern that connects."

From ethnolinguist Benjamin Whorf and the poet Octavio Paz and various other sources I got the idea that the universe as a whole can be viewed as a kind of language. Language, of course, is a kind of symbolic patterning. Saying the world and the mind are a kind of language is similar to saying the world and mind are complex patterns systems, with patterns for generating patterns for generating patterns….

Ed Misch and another Simon's Rock philosophy professor, Nancy Yanoshak, introduced me to aspects of Nietzsche's thought – and like many other young social-misfit intellectuals have over the last century, I fell in love. Of course, his deconstruction of morality and his cynical view of human nature as driven by various kinds of vanity and self-interest appealed to me, as did the gorgeous philosophical poetry of Thus Spake Zarathustra (later on I named my first son Zarathustra). However, the late Nietzsche intrigued me in particular: he saw the mind and world as a finite system, driven by the twin aspects of "morphology and will to power." Morphology means form, shape – pattern. He saw the world as an organism whose cells are forms or patterns, each of which is struggling to extend itself over all the other forms or patterns. Much like Peirce's One Law of Mind. I tried to write a paper back then arguing that Peirce and Nietzsche were largely saying the same thing about the core structures and dynamics of the mind and universe, but I didn't do a great job and nobody seemed to understand what I was talking about.

In my last year of college, I wrote up a very crude manuscript summarizing my early-stage patternist philosophy of mind and the universe, and I printed a copy and mailed it (snail mail, e-mail was not common then) to the philosopher Paul Feyerabend, whose writing and thinking on the philosophy of science I had much enjoyed. I was curious if I could come to California and get a philosophy PhD studying under him. He replied with a very kind postcard; I wish I'd saved it. The gist was: "*I won't read your book, because I am too busy these days, mostly with things other than academic work. If you want to do philosophy yourself, rather than study other peoples' philosophy, I'd recommend you to get a PhD in science. Then you can do your own philosophy afterwards.*" I took the advice to heart – observing also that a PhD in philosophy was most likely useless in the job market, whereas a PhD in science had reasonable odds of getting me a job as a professor. Having observed the professorial life firsthand via my dad and his colleagues, it seemed agreeable to me, because it left lots of free time for one's own reading and thinking.

My second-year college roommate Mike Glanzberg was deeply into mathematical logic, and he eventually became a very formal logic-oriented analytical philosophy professor. Largely because of him, I dug deep into the history of logic – Russell, Whitehead, Frege, Boole, Leibniz, Tarski, Quine and so forth. I also discovered Paul Aczel and his theory of non-foundational sets (via a chance encounter with one of Aczel's books in a technical bookstore) – sets that can, unlike ordinary mathematical sets, contain themselves as members, causing an infinite recursion of *"X is inside X, which is inside X, which is inside X,..."* or *"A is part of B, which is part of A, which is part of B, ..."* I thought this seemed quite promising as a way of modeling consciousness, under the intuition that *"Consciousness is consciousness of consciousness of consciousness of..." and "The mind is part of the world, which is part of the mind, which is part of the world, which is part of the mind..."*

Intrigued by this sort of mathematics and its possible implications for modeling consciousness and self, when I finished university I nearly went to graduate school at U.C. Berkeley's Logic and Methodology PhD program, with a goal of writing a PhD thesis on non-foundational sets for modeling consciousness. But I cooled off on the idea when I visited the department and heard that most students took 7-9 years to finish their PhDs. I was more eager to get out of school quickly with my PhD than I was to study that particular topic, and I ended up starting grad school at New York University's Courant Institute for Mathematical Sciences (which interested me because it taught math, theoretical physics, and computer science all in one program), and then transferring to the math department of Temple University in Philadelphia to finish my degree.

Altogether I finished my undergrad degree in 3 years, graduating at age 18, and I finished my PhD in 4 more years, becoming Dr. Goertzel at age 22. In the end, my PhD thesis had little to do with AI or advanced mathematics – it was a fairly simple exercise, in which I invented a slightly novel optimization algorithm and showed that it gave OK results on a bunch of test problems, and proved a very simple, almost insubstantial theorem about its behavior. It was more computer science than math, though I was in the mathematics department. But my thesis was really just something I concocted in a few months because I wanted to graduate. Throughout grad school I was much more interested in AI, theoretical physics, complex systems and a host of other topics than the optimization algorithm I ended up writing a thesis on.

In fact, I came up with my first halfway (or, OK, maybe quarter-way) decent AGI design in 1986, in the middle of graduate school: an AGI inspired by the human immune system, a complex self-organizing pattern recognition system of a subtle sort, different from the brain. This was no doubt inspired by a mathematical immunology course I'd taken at NYU, taught by Alan Perelson, a wonderful professor. The immune system provided a concrete model of a pattern recognition system (it's quite complex in the way it learns to identify different antigens, such as disease-bearing germs), which I could expand, extend and emulate to form an instantiation of my abstract thinking about networks of mind-patterns.

But I realized that it was going to be very hard to get any professor to approve me to write a thesis on topics like immunology-inspired AI that I actually cared about. So I just wrote a thesis on something acceptable – a particular optimization algorithm that I made up and implemented – and then, once I graduated, never looked back, and instead went on to think about AI and other topics of my deep interest. This reflects advice I still give PhD students who ask me: "Don't over think your thesis, and don't try to make it the greatest contribution in the world. You have your whole career to do that. Just write something good enough, and finish school."

The broad view of the mind that underlies my AGI work today – which takes the mind as "a complex system of patterns, constantly recognizing patterns in itself and the world" – is something that I adopted as a child and then shaped further as I learned more about philosophy, cognitive science, biology and other subjects. However, it took a long time for me to be able to express this patternist view with any reasonable amount of clarity and detail. The way I express it now is consistent with what I was thinking all along, but sounds a fair bit less fuzzy and confusing than when I attempted to explain my thinking way back when.

Grokking the Nature of Pattern

One fairly technical topic I spent some time on in my grad school days is the formal, mathematical definition of what a "pattern" actually is. Conceptually, I decided early on that it made sense to think of a "pattern" as a "representation as something simpler." That is, a pattern in some X is some way of representing X – perhaps exactly, perhaps approximately; perhaps completely, perhaps partially – in terms of something simpler than X.

The connection between pattern and simplicity of representation is easy to see if one thinks in terms of computation, which lets one measure "simplicity" in terms of the length of a computer file, or the time a program takes to run. Image or music file compression tools work by looking for repeated patterns in a file, and then representing the file in terms of these patterns. Their purpose is to produce versions of files that are shorter than the originals, i.e. "simpler" if one measures simplicity in terms of file length ("bit count"). A random image or sound file will not be compressible, because there are no patterns there to recognize and use as the basis for compression.

More intuitively, suppose you see 50 people with colored hats. You can describe the people and their hats one by one, e.g., "Bob Jones with a blue hat; Nancy Karnowski with a yellow hat; etc." But what if all the men have blue hats and all the women have yellow hats? Then, rather than mentioning all 50 people and their hat color, it's simpler to just say that all the men have blue hats and all the women have yellow hats. The observation of a pattern allows a simpler representation.

FIGURE 4.1: *A typical fractal picture. It's hard to reconstruct how amazing these seemed in the 1980s and early 1990s. Now they're a commonplace part of our culture. The white part is the Julia set for the rational function associated to Newton's Method for $f: z \rightarrow z^3 - 1$. Coloring of Fatou set according to attractor (the roots of f). For someone who knows the appropriate math, this picture can be generated from a quite short, fast computer program involving these mathematical equations. The math and the program constitute patterns in the image.*

Patterns don't have to be repeated, of course. Fractal sets are a great example. These are not strictly repetitive, yet they are generated by relatively simple equations (well, simple if you know a bit of math). The equation or computer program for generating a fractal picture is a simpler way of representing the fractal picture. The mathematics of fractals fascinated me tremendously in the mid-1990s – I taught a class on the math of fractals, and when I taught computer graphics I always covered the use of fractals to generate pictures of plants, landscapes and other natural phenomena. The fascination wasn't just about their funky appearance; I enjoyed them as well, as a sort of "paradigm case" of the complex patterning that we see all around us in the mind and world. Seeing the same fractal patterns in so many different domains illustrates Gregory Bateson's maxim that "it is pattern that connects." Furthermore, as I argued in my 1994 book *Chaotic Logic*, it seems very likely that the network of thoughts in the mind also has a kind of fractal structure. The fractal structures in brains, plants and astronomical data, as shown in some of the figures I've included here, are easy to see; the fractal structures in minds are more abstract and hence more difficult to visualize, but no less real.

FIGURE 4.2: *The natural world is full of complex patterns, including fractal patterns like the one we see on this Romanesco broccoli (a kind of cauliflower). This complex physical shape can be approximated by a fairly simple mathematical equation, which constitutes a pattern in the shape. (http://en.wikipedia.org/wiki/Cauliflower#mediaviewer/File:Cauliflower_Fractal_AVM.JPG).*

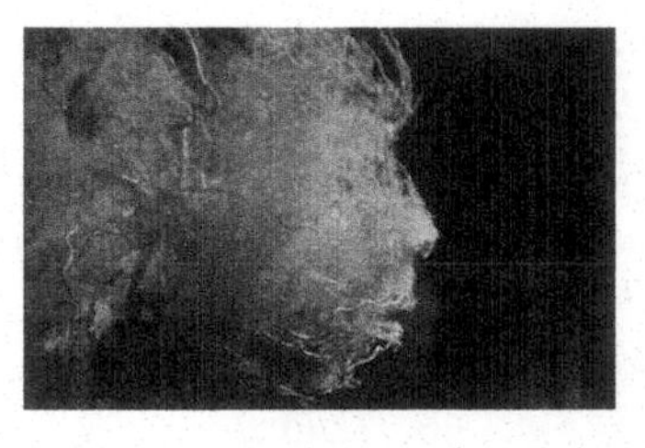

FIGURE 4.3: *Blood vessels in the brain and throughout the body show a fractal branching structure, which can be approximately explained via fairly simple mathematical equations. These equations serve as a pattern in the branching structure. If you haven't, you should definitely see the BodyWorlds exhibit, which travels around the world and features many life-sized examples of human bodies and body parts preserved in this way. (http://i.imgur.com/xXYVd6c.jpg).*

The math of pattern theory is actually quite simple, but this isn't a math book so I won't get into it here. The main point I want to mention in this regard, right now, is the relation between PATTERN and SIMPLICITY. If a pattern is a way of representing something more simply, then the definition of pattern relies fundamentally on the definition of simplicity. This is philosophically important because it means that the identification of what is a pattern and what isn't is, in the end, relative to the observing mind. Each observing mind comes with its own intuition, which tells it what's simple and what's complex to it, at a basic gut level. Based on its evaluation of simplicity, the mind can recognize patterns in itself and its environment.

One can also argue that the physical universe has its own measure of simplicity, defined in terms of physical energy – i.e., the simplest entity is the one that uses the least energy. This ties in with computational ways of looking at pattern, because it takes less energy to store shorter files or run programs with less runtime. But then the universe's conception of what is a pattern won't necessarily agree with the conception of a particular mind in the universe; each mind judges simplicity, ergo pattern, for itself. The relation between the mind and the physical universe is another topic that has long fascinated me – I've dreamed of figuring out how to reformulate the equations of physics in a way that would make it clear that the dynamics of the physical universe are themselves a kind of "thought process." This is one of the many intriguing topics I think about from time to time, and occasionally even research seriously, but have mostly put aside to focus on AGI.

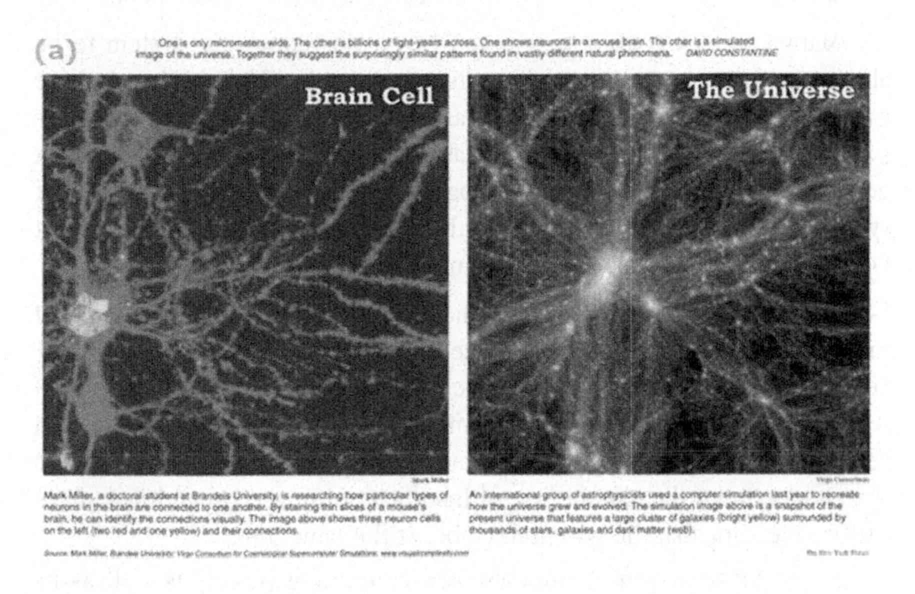

FIGURE 4.4: *Many of the same patterns span different levels of structure in the universe. For instance, (a) the same fractal branching patterns characterize neurons (the primary brain cells) and galactic clusters, and (b) brain corals display patterns quite similar to those observed in human brains. (Images from: http://livelightbeing.com/2011/06/17/neuron-art/, http://www.dvice.com/sites/dvice/files/brain1.jpg, http://upload.wikimedia.org/wikipedia/commons/5/56/Brain_coral.jpg).*

Mind and Intelligence

In the patternist philosophy as I have developed it over the years, the mind of an intelligent system is conceived as the (fuzzy) set of patterns in that system, and the set of patterns emergent between that system and other systems with which it interacts. The latter clause means that the patternist perspective is inclusive of notions of "distributed intelligence" – i.e., it lets us think of your dog or your cellphone as part of your overall extended mind-system. Your mind is the set of patterns that can be identified in your brain, and in your brain's interaction with the rest of the world.

I generally like to think of intelligence as the ability to achieve complex goals in complex environments – where complexity itself may be defined as the possession of a rich variety of patterns. A mind can thus be thought of as a collection of patterns that is associated with a persistent dynamical process that achieves highly patterned goals in highly patterned environments.

Many complex goals a system can carry out require the system to be able to model and learn things about its own behavior. With this in mind, we can conceive of an intelligent system as a dynamical system that recognizes patterns in its environment and itself, as part of its quest to achieve complex goals. The patterns that a system recognizes in itself are then stored in the system's memory and become part of the system itself – an example of what Douglas Hofstadter likes to call a "strange loop."

I have also been motivated to some extent by a closely related vision of intelligence, articulated by my long-time colleague and sometime collaborator Pei Wang. Pei understands intelligence as, roughly speaking, "the ability of working and adapting to the environment with insufficient knowledge and resources." More concretely, he believes that an intelligent system is one that works under the Assumption of Insufficient Knowledge and Resources (AIKR), meaning that the system must be, at the same time,

- ***a finite system*** — the system's computing power, as well as its working and storage space, is limited;
- ***a realtime system*** — the tasks that the system has to process, including the assimilation of new knowledge and the making of decisions, can emerge at any time, and all have deadlines attached with them;
- ***an ampliative system*** — the system can not only retrieve available knowledge and derive sound conclusions from it, but it can also make refutable hypotheses and guesses based on it when no certain conclusion can be drawn
- ***an open system*** — no restriction is imposed on the relationship between old knowledge and new knowledge, as long as they are representable in the system's interface language.

- ***a self-organized system*** – the system can accommodate itself to new knowledge, and adjust its memory structure and mechanism to improve its time and space efficiency, under the assumption that future situations will be similar to past situations.

Obviously, Pei's definition and mine have a close relationship. My "complex goals in complex environments" definition is purely behavioral: it doesn't specify any particular experiences, structures or processes as characteristic of intelligent systems. I think this is as it should be. Intelligence is something systems display; how they achieve it under the hood is another story.

On the other hand, it may well be that certain structures, processes and experiences are necessary aspects of any sufficiently intelligent system. My guess is that the science of 2050 will contain laws of the form: any sufficiently intelligent system has got to have this list of structures and has got to manifest this list of processes. Of course, a full science along these lines is not necessary for understanding how to design an intelligent system. But we need some results like this in order to proceed toward real AI today, and Pei's definition of intelligence is a step in this direction. For a real physical system to achieve complex goals in complex environments, it has got to be finite, real-time, ampliative and self-organized. It might well be possible to prove this mathematically, but this is not the direction we have taken; instead, we have taken this much to be clear and directed our efforts toward more concrete tasks.

One key aspect of Pei's definition of intelligence, "self-organization," is something that I've thought about a lot in a general way, beyond the context of AI. When I got my PhD, complex systems and "chaos theory" were just emerging as popular fields of study, and I thought they were both fascinating and obviously essential. I spent a lot of time studying the complex systems literature, and implementing and experimenting with various computer simulations of complex systems like immune systems and ecosystems, as well as complex-systems neural network models of parts of brains.

Patterns often emerge in complex systems for complex, inscrutable-seeming reasons that become comprehensible only under very careful study – the emergence of similar patterns in brains and brain corals is an example. What exactly is the common growth process yielding those funky brainlike folds? Complex systems are full of mysterious-looking "self-organization," in which clear and beautiful and functional patterns emerge seemingly spontaneously out of underlying chaos. Even when an intelligent system is carrying out apparently orderly goal-oriented behavior, there is a lot of complex, self-organizing, strange-loopy pattern recognition and formation going on under the surface.

The concept of evolution also plays an important role here. One can conceive of the thought process of the mind as a kind of "pattern evolution" – if one construes evolution as a very general process in which patterns are selected based on some criterion, and then the good patterns are combined and tweaked to form new patterns. But this evolution has to be considered in the context of self-organization, i.e. of complex systems of patterns that all reinforce each other – it's all about complex pattern networks.

And pattern networks can have various sorts of structures. Patterns can relate to each other hierarchically – like the way the pattern of blood vessels in the brain exists on a lower hierarchical level than the brain-coral-like folding pattern of the brain; and the patterns of interaction of protein molecules exist on a lower hierarchical level than the patterns of interaction of cells that are made of these molecules, and exchange signals using these molecules. Patterns can relate to each other heterarchically as well – patterns in the motor cortex of the brain (dealing with movement) exist side by side, and they interact with patterns in the perceptual cortex of the brain, for example. Hierarchical and heterarchical network patterns can combine and interact, yielding what I've often called a "dual network."

The self itself is a huge and important pattern in an intelligent system – it can be thought of as a model that a system builds internally, reflecting the patterns observed in the (external and internal) world that directly pertain to the system itself. As is well known in everyday human life, self-models need not be completely accurate to be useful; and actually, in people with certain psychological peculiarities, a more accurate self-model may not necessarily be advantageous. But a self-model that is too badly inaccurate will lead to a badly-functioning system that is unable to effectively act toward the achievement of its own goals.

An additional use of the self for an intelligent system is as a foundation for metaphors and analogies in various domains. Patterns recognized pertaining to the self are analogically extended to other entities. In some cases this leads to conceptual pathologies, such as the anthropomorphization of trees, rocks and other such objects that one sees in some pre-civilized cultures. But in other cases this kind of analogy leads to robust sorts of reasoning – for instance, basic empathy; or the kind of clever bluffing and other-modeling that poker players do.

Of course, the patterns of human minds are not going to be exactly the same as those of AGI minds. For instance, it's interesting to speculate regarding how self may differ in future AI systems as opposed to in humans. The relative stability we see in human selves may not exist in AI systems that can self-improve and change more fundamentally and rapidly than humans can. There may be a situation in which, as soon as a system has understood itself decently, it radically modifies itself and hence violates its existing self-model. Thus, it would be intelligence without a long-term stable self. The psychology of such intelligent systems will almost surely be beyond human beings' capacity for comprehension and empathy.

The self pattern also ties in with the tricky issue of "consciousness." My overall view of consciousness gets rather philosophical and even spiritual – I'll say more about it toward the end of the book, once I've worn you down with more hardcore material about AGI and neuroscience and so forth! But the simplest thing to say on the topic is that I think the subjective, raw "feel" of consciousness IS a meaningful, real thing to talk about, but that it is different from the patterned structures and dynamics of intelligence and consciousness. I tend toward a "panpsychist" view, which holds that everything in the universe is conscious to at least some small extent. Everything is conscious, but some things are more conscious than others. And then one can look at what patterns tend to correlate with the most highly conscious states and experiences.

One fairly concrete "patternist" idea about consciousness is that the subjective experience of being conscious of some entity X is correlated with the presence of a very intense pattern in one's overall mind-state, corresponding to X. This simple idea is also the essence of neuroscientist Susan Greenfield's theory of consciousness (but in her theory, "overall mind-state" is replaced with "brain-state"), and has much deeper historical roots in philosophy of mind which I won't digress to unravel here.

This observation relates to the idea of "moving bubbles of awareness" in intelligent systems. If an intelligent system consists of multiple processing or data elements, and during each (sufficiently long) interval of time some of these elements get much more attention than others, then one may view the system as having a certain "attentional focus" during each interval. The attentional focus is itself a significant pattern in the system (the pattern being "these elements habitually get more processor and memory," roughly speaking). As the attentional focus shifts over time, one has a "moving bubble of pattern," which then corresponds, experientially, to a "moving bubble of awareness." This idea plays a key role in my AGI work these days.

Anyway, all this talk about patterns and minds is pretty abstract, but I hope it gives you the sense that the "patternist" view of the mind, which I originated via a bunch of philosophical thinking inspired by crystals, fractals and so forth, spreads out in a variety of different directions and touches a lot of different practically meaningful aspects of intelligence. A little later, when I tell you more about my approach to building thinking machines these days, you'll see how all this patternist philosophy stuff fits in!

Verging Toward AI

After I got my PhD, I got a job as a math professor at the University of Nevada in Las Vegas, and when I wasn't teaching math or playing with my new baby son Zarathustra – or indulging in my various hobbies like writing surrealist SF prose poems or improvising on the piano – I was dividing my time between applied math research, and thinking and writing about the patternist view of the mind. My first two books on the mathematical and conceptual modeling of intelligence came out while I was in Nevada – *The Structure of Intelligence*, and *The Evolving Mind*. The first of these contained a decent high-level overview of an architecture for building an AGI system. The AGI work I'm doing today – which I'll describe for you in moderate detail a bit later in these pages – is still within the basic framework I outlined there. *The Evolving Mind* gave a formulation of evolutionary theory that was broad enough to apply to bodies, ecosystems, body subsystems like immune systems (which learn to attack antigens via a kind of evolution), and mind/brains.

After four years at the University of Nevada, at age 26, I got bored and sent my resume to every university in the British Commonwealth outside of the UK, Canada and India. I got job offers from one computer science department in New Zealand, and one math department in Australia. I ended up spending a year teaching computer science in Hamilton, New Zealand, and then shifted to Perth, Western Australia, where I spent two years as a research fellow in cognitive science, with offices in both the psychology and computer science departments. My next two books on the mathematical/philsoophical theory of mind came out, *Chaotic Logic* and *From Complexity to Creativity*. In New Zealand I was teaching computer graphics, and doing some work on fractal graphics and chaos theory alongside AI. But during my time in Perth, I moved pretty completely into AI and cognitive science as a research field. The AGI bug was overtaking my system – and plus, more broadly, I had decided it would be a good idea for me to focus on something more narrowly than I had been doing.

During this period, I also got bored with my previous immunology-based AGI design, and worked on a variety of different AGI-oriented ideas. I developed various recurrent neural network-based designs, drawing more inspiration from dynamical systems theory ("chaos theory") than from the details of the brain, which were known even more scantily back then than now. I also worked on another AGI architecture based solely on mathematical theorem-proving, and one based on automatic program learning. Then as now, I felt there were a lot of different valid approaches to creating AGI.

In 1994, in New Zealand, I made an utterly unsuccessful attempt to implement an AGI in Haskell, a very mathematical and elegant programming language that was then at a very early stage of development (it's now much more mature, and has become impressively efficient, but back then it was very slow to do anything). This system started off by recognizing patterns in itself, pursuing theoretically endless layers of recursive introspection. Data from the outside world got piped into the introspective theater too and got reflected on. It was an interesting self-organizing system that evolved some complex fractal-like internal patterns, but was too inwardly self-focused to learn much of anything useful. And every time I tried to feed it lots of data, the Haskell interpreter crashed. I could have fixed the software issues (perhaps by hacking the Haskell interpreter appropriately), but I correctly intuited that new ideas were needed, and moved on.

Even as I put more and more time into prototyping various AGI-ish ideas, it was very tempting to me, then as now, to utterly scatter my energies among a dozen different fascinating areas. I knew I'd never achieve total focus on one thing, because it's just not my nature – but I wanted to pick one very important thing and focus, say, 75% of my energy on it. And I wanted that one thing to be grand and world-historical in nature – nothing else would hold my attention!

While AI was my top-ranked choice from the start, I did seriously ponder a lot of other options. My job at University of Western Australia was a pure research fellowship, which I had won in a competitive application to the central Australian government; this meant I basically had no job duties except to think and do whatever research I felt like. It was far more fun than my first job in the math department at the University of Nevada had been, and the sunny Perth climate suited my wife better than the climate in New Zealand. In Perth I taught a couple cognitive science classes – because I felt like it, not because I had to. I did some research on mathematical psychology, artificial life, and theoretical neuroscience, as well AI. And I thought a lot about what I should do with my mind and my career, and what I should do with myself after the 3-year research fellowship was up.

I thought about focusing on physics, on trying to unify quantum theory and gravitation. I investigated the possibility of building time machines – I actually did some calculations (which I later found had been done by others earlier) demonstrating the possibility of backwards time travel according to general relativity, in various scenarios involving strangely-shaped singularities. I was interested in the origins of life and wondered about creating new life forms in a test tube. I spent a while trying to prove the notion of infinity was logically inconsistent, but just ended up rediscovering constructivism (an early 19th century philosophy of mathematics) over and over again. I wanted to make an elegant mathematics of complex systems, but didn't get too far. I started creating a mathematical theory of consciousness based on non-foundational sets (which is a project I'd still like to finish; I published a related paper in 2012). I even considered back-burnering science and focusing on literature or music. But eventually I figured that if I was going to write a great epic novel in the vein of Dostoevsky, I might as well wait till I was an old man and had experienced more of life. Artificial intelligence kept on sucking me in, inexorably. Test tubes and strangely-shaped singularities were so messy, and unified physics and the philosophy of mathematics were so abstract. The appeal of AI was simple: it could change the world immeasurably and excitingly, and it probably didn't require anything but a computer and a programming language compiler. It was a purely intellectual puzzle with dramatic potential practical impact. The more I thought about it, the more amazing it felt that I, just by typing some code into a computer, could potentially create something smarter than me and revolutionize life on Earth completely, opening up new possibilities literally beyond my current comprehension.

The trick was just to know which lines of code to type, to enable one's program to be intelligent! It seemed fairly likely to me that someone would figure this out during my lifetime, and I figured that that someone might as well be me.

I still feel about the same way. But, obviously, the path has been a bit more winding and difficult than I envisioned at that point, nearly 20 years ago! Knowing what I know now, I would have done a lot of things differently along the way. But nothing I've done or learned has decreased my confidence that advanced AGI is achievable, and without requiring all that much effort or resources in the grand scheme of things. It's more work than one guy can do in his spare time – but not as much work as goes into creating a product like Microsoft Word, or the Tesla automobile, for example. We're now at the stage where creating an advanced AGI is a moderate-sized systems development task – if one starts with the right design!

AGI as the Next Big Step

Overall, the more I've learned about AGI and about the world in general, the more strongly I feel that AGI is the Next Major Stage of evolution on the planet. (Why do I say "on the planet"? Well, AGI may have already been created elsewhere in the universe, right? – we don't really know. Of course, it might also exist on the planet unbeknownst to us, say in submicroscopic form, but then I'm not sure that counts as "on the planet"... but I digress....) This realization makes AGI all the more exciting as a topic of attention. Why focus on peripheral issues when one can direct one's attention to the Most Important Thing?

As I mentioned earlier, when I was a kid I read – skimming some of the boring parts – a big fat multivolume book my parents had, called *The History of Civilization*, by Will Durant. It had a big impact on me. It was beautifully (albeit pompously) written, and fascinating to read – most more so than the multi-volume *World Book* and *Colliers* encyclopedias that I also read cover to cover, at around the same age. Durant's section on Indian history especially fascinated me, and it was my introduction to Buddhist thought. It also got me thinking, at an early age, about some of the huge, revolutionary changes that humanity has been through since it emerged from its apelike ancestors. This is a topic I've thought about tens of thousands of times since then, as well.

One critical, revolutionary change was the invention of language in its modern form – language that can systematically refer to things outside the immediate perceptual experience of the speaker and the listener, thus enabling the build-up of complex culture.

Then there was the advent of civilization – the move from nomadic hunter-gatherer life to settled-down village and town life, with crops and domestic animals, and all the possibilities thus afforded.

There was the creation of machinery, allowing leveraging of human labor, and permitting creation with a scope and precision impossible for the unaided human body.

There was the development of science and mathematics, enabling a far deeper and broader practical understanding of the world than is possible only via common sense or imaginative intuition.

And, we are now nearing another vast transformation, in some ways probably the largest yet: the creation of engineered systems with greater general intelligence and capability than us, their creators. Artificial General Intelligence. Right here, right now, within our current human lives, we have the potential to participate in – and help steer – the next great transformation. I didn't quite realize this as a child – Will Durant didn't cover it. But it's extraordinarily clear to me now.

FIGURE 4.5: *When our "caveman" predecessors first invented modern-style language, with its capability to refer in detail to far-off events, did they foresee all it would lead to? The Internet, Shakespeare, Grand Master Flash, algebraic geometry -- AGI? Of course not. And nor can we foresee what this next radical transformation, AGI, will bring. (http://en.wikipedia.org/wiki/File:Caveman_5.jpg).*

The advent of engineered systems with greater-than-human general intelligence won't necessarily be the end of humanity. Some pessimists have said so, but without foundation. There is no cause for any sort of certainty, positive or negative, where unprecedented events are concerned.

But one thing that does seem likely, is that once machines smarter than humans are around, the era of human beings dominating the Earth will be gone. The era of humanity singlehandedly guiding its own development will probably also be gone. These will be huge changes, but almost surely not the largest ones. Minds able to see further and deeper than our own are likely to conceive new possibilities for change and advancement, far beyond what you or I or any other human can currently prognosticate. The very nature of humanity may get modified unrecognizably – for example, if the majority of humans choose to merge their mind-brains with some sort of machinery, in order to gain added cognitive capabilities and access to new, more interesting and fulfilling states of mind.

These are heady possibilities; and once matters get too remote from our current reality – regardless of how soon they might happen given the fast pace of technology – there's not much we can do but speculate in the vein of science fiction or fantasy. But still – this is not science fiction, this is rapidly unfolding science and technology. It's possible to think clearly about the near term of AGI using tools of science and rationality. And even our speculations about out-there topics like transcension beyond the current order of human existence can be grounded in our understanding of the science of the mind and its intersection with the practical world.

It might not seem like AGIs with human-level or superhuman capabilities are just around the corner. After all, we haven't even managed to staff McDonald's and 7-11s with service robots yet – they're still staffed by plain old human beings, in spite of the relatively simple and routine tasks required of the humans behind the counter. But of course, no revolution has happened before it's happened. Five years before the Wright Brothers' first flight, there were no jets zooming overhead, and people traveled from the US to Europe by boat. Five years before the launch of the World Wide Web, I went to the university library nearly every day to read research papers. The nature of revolutionary change is that it upends what has existed before, in a way that's surprising and unexpected to most people.

From Philosophy to Engineering & Entrepreneurship

So far what I've told you about my thinking on intelligence has been pretty philosophical and conceptual – and indeed, abstract ideation about how the mind works has been the core foundation of all my AGI work. One of the hard lessons I've learned over the years is that abstract thinking only takes you so far – ultimately, AGI is an engineering pursuit and not a philosophical one. It's technology that is changing the world, and the AGI that brings the next transformation is going to be engineered based on lots of math and science and conceptual thinking, but also based on tinkering opportunistically with the tools at hand, using a lot of practical tricks of the trade. Building complex software and hardware systems is a far cry from the elegance of philosophical thinking. Still, though, when I think about how to build AGI, I always come back to philosophy of mind. We don't need to resolve all the questions of the philosophy of mind in order to build a thinking machine. But having a clear understanding of what thinking is, and how the different aspects of thinking all relate to each other and to the world, seems to me a very big help in pursuing the creation of true artificial general intelligence.

In my personal life-trajectory, Perth was where I made the transition from abstract theory to concrete, realistic system design. When I arrived at the University of Western Australia for the research fellowship I had won there, I had a fairly fleshed out, partly mathematical and partly conceptual, patternist theory of how the mind worked. I had experimented with some simple designs for AGI systems, programming them myself in C or Haskell, but hadn't come up with anything remotely workable. When I left Perth for New York two short years later, I had at least the outline of a concrete software design for a real-world AGI system – one I called Webmind. While the Webmind design was very different from the OpenCog system I'm now working on, it was definitely the same species of design. OpenCog is a simplified and more sophisticated attempt along the same lines that Webmind was.

When I left Perth was also when I made the transition from academia to industry. I ended up returning from Perth to the US after only 2 years of my free-and-easy research fellowship there, pregnant wife and two young children in tow, bank account essentially empty, with the crazy idea of founding an Internet AI company together with a couple new friends from New York whom I'd met online. It was the beginning of the AGI-focused phase of my life, and the entrepreneurial phase as well. It was the beginning of my adventures in seriously trying to build AGI instead of just thinking about it and haphazardly and experimentally hacking prototypes.

Bridging Philosophy and Implementation with Cognitive Science

Transitioning from abstract thinking about the mind to concrete implementation of proto-AGI systems, as I did during my time in Perth, required me to fill in some major gaps in my thinking. The gap between philosophy of mind and AI software design is too large to be leapt across directly – one needs something else to bridge this gap. My stay in Perth not only provided me unpressured free time to think creatively in this direction – it also immersed me in the discipline of cognitive science, which I found an ideal source of concepts useful for bridging the gap between philosophy of mind and AI implementation.

At University of Western Australia I had two offices: one in computer science, and one in psychology. I spent most of my time in my Psychology Department office. Most of what went on in the psych department didn't fascinate me that much, but I had a few co-conspirators there who were interested in going beyond traditional experimental or clinical psychology and forging a new interdisciplinary field called "cognitive science." Cognitive science was supposed to fuse together psychology, computer science, AI, mathematics, neurobiology, linguistics, philosophy and much more, into a common, integrated understanding of the mind. This seemed absolutely critical to me.

Before turning to cognitive science, I spent quite some time trying to extract useful information about AI design from mathematics. After all, math is better organized and less messy than the human brain! But, unfortunately, I wasn't able to make mathematics give the kind of concrete, practical answers about intelligence that I needed. Intelligence is based on adaptation to a particular environment and set of tasks, and we don't have a good formalization of the environment and tasks to which human-like general intelligence is adapted. I found it difficult to rely entirely on mathematics as a guide for AGI, although I continue to love math above all other intellectual disciplines and I use it as much as I can in my work.

I also looked quite a lot at neuroscience – which is somewhat the opposite of mathematics in terms of organization and elegance. I love neuroscience too, it's fascinating to start to understand how the three-pound hunk of cells in our brains gives rise to thoughts and feelings and memories and so forth. But we still don't know nearly enough to base AGI designs on the brain at any level of detail. I'll say more about this in a later chapter – but I think it's still true today, and it was true with far more oomph in the late 1990s when I first started digging in depth into AGI design.

Cognitive science integrates ideas from math and neuroscience along with psychology and other disciplines. It's able to take what is known about the brain, and what we can learn from mathematical modeling of intelligent systems, and put these pieces together with other fragments of knowledge, toward a crude but still non-trivial overall understanding of how the human mind works. It's not enough to dictate AGI design, but it's enough to seriously help. So, overall, in my AGI work, I've been heavily guided by cognitive science, with a focus on cognitive psychology. Among my multiple sources of inspiration, I have taken the study of the human mind, certainly more so than that of the human brain, as a guide for my high-level AGI architecture.

The field of cognitive science has moved forward steadily, year after year, decade after decade, since its formal establishment in the early 1970s. It has done so through the combined efforts of a huge number of scientists around the globe, all doing psychological experiments on humans and animals, and writing and studying various sorts of computer simulations. This progress has not been as easily quantifiable as Moore's Law, nor as strikingly visualizable as progress with PET and fMRI scanning of the brain. But it has been no less important, and no less dramatic. As Ray Kurzweil would say, it has been advancing exponentially!

While spending time in the psych department at UWA, I became frustrated with the PET brain scanning some of my colleagues were doing, because of the limited information it gave: it told you what brain regions were most active during what kinds of functioning, and that was all. I also became frustrated with cognitive psychology research because it was difficult to design experiments yielding any real information about the questions that interested me most: the dynamics of abstract thinking.

For the first time, I realized how hard it is to design experiments that can be run in a short time, using undergraduate students as subjects (which is how almost all cognitive psychology experiments are done), which will tell you anything interesting about the mind. But nevertheless, I became convinced that the results of the various experiments cognitive psychologists had run during the previous decades, appropriately synthesized, could guide me on my AGI quest. The cleverness of experiments, and the vast knowledge cognitive scientists had amassed about the brain despite the difficulty of running experiments, inspired me.

I found ample resources on psychology theory about vision processing, sound processing, and memory for words (these are easier to study in the lab than more abstract cognition). On the other hand, the psychology of creativity, self and meta-cognition (thinking about thinking) was scarcely covered – I found few useful ideas for AGI design purposes.

Gathering and synthesizing ideas from all different disciplines was a lot of fun, and challenging in its own way. But as I was soon to find out, trying to put complex ideas into practice in reality poses a whole different array of challenges.

Further Reading

Abraham, Fred, Ralph Abraham and Chris Shaw (1990). *A Visual Introduction to Dynamical Systems Theory for Psychology*. Aerial Press, Santa Cruz.

Barnsley, Michael (2012). *Fractals Everywhere*. Dover.

Barnsley, Michael (2006). *Superfractals*. Cambridge University Press.

Bateson, Gregory (1979). *Mind And Pattern: A Necessary Unity*. Ballantine.

Combs, A. (1996). The radiance of being: Complexity, chaos, and *the evolution of consciousness*. New York: Paragon House

Devaney, Robert (1987). *Chaotic Dynamical Systems*. Addison-Wesley.

Fuller, Buckminster (1982). *Synergetics*. MacMillan.

Gleick, James (1987). *Chaos: Making a New Science*. New York: Viking.

Goertzel, Ben (1992). *The Structure of Intelligence*. Springer.

Goertzel, Ben (1992). *The Evolving Mind*. Springer.

Goertzel, Ben (1994). *Chaotic Logic*. Plenum.

Goertzel, Ben (1997). *From Complexity to Creativity*. Plenum.

Goertzel, Ben (2006). *The Hidden Pattern*. Brown Walker.

Jantsch, Erich (1980). *The Self-Organizing Universe*. New York: Pergamon Press.

Kampis, George. *Self-Modification in Biological and Cognitive Systems*. Plenum.

Mandelbrot, Benoit (1982). *The Fractal Geometry of Nature*. New York: W.H. Freeman.

Nietzsche, Friedrich (1968). *The Will to Power*. New York: Vintage.

Nietzsche, Friedrich (1997). *The Twilight of the Idols*. Hackett Publishing.

Peirce, Charles S. *Collected Works vol. 8, Scientific Metaphysics.* Cambridge MA: Harvard Press.

Stcherbatsky, Theodore (2000). *Buddhist Logic*. Motilal Banarsidass Pub.

Varela, Francisco (1978). *Principles of Biological Autonomy*. North-Holland.

Whorf, Benjamin (1964). *Language, Thought and Reality*. MIT Press.

5. Making It Real (or at least, trying!)

By my late 20s, when I decided to focus my energies largely on the AI problem, I had studied the AI literature fairly thoroughly, and felt I had absorbed most of the main lessons of the prior work others had done. I set out to create my own systematic design for advanced AI, drawing on other peoples' ideas and my own previous thinking as appropriate, and introducing new notions where earlier ones seemed insufficient. Once – after a few years of focused effort – I felt I had something approximating a practical design for a human-level AI system, I realized it was going to be too much for me to program on my own, and I ended up leaving academia and returning to the US to co-found an AI startup company with some friends I'd met online, who were based in New York. On a personal level, I regretted leaving Perth and UWA, both of which were fantastic places. But it was 1997 – the dot-com boom – and I reckoned the time was right to try to make a lot of money with AI software, and use the profits to fund AI research big-time. Living in Perth was very comfortable, but the combination of the dot-com boom and a design for a thinking machine seemed like a recipe for changing the world in a dramatic fashion, and that was way too much for me to resist.

In 2016, the idea of dropping your stable career to found a software startup is somewhat cliché – there have been so many high-profile successes that if someone tells you you're crazy to do such a thing, all you have to do is point to Facebook and Google and so forth. In 1997, it was a somewhat more eccentric thing to do, especially since I had a wife (who hadn't worked full-time for many years) and two small kids, with a third "in the oven." But on the other hand, the dot-com boom was spreading excitement around the globe, and there was a feeling like anything was possible where the Internet was concerned. And I'd never really valued being stable and settled (or at least not in any stable way – occasionally, in a moment of weakness, the desire for stability and surety did come over me, but then it always worked for me to simply lie down till the troublesome urge passed!). Stability and security always smelled a bit like death to me. After so much thinking about how to build AGI, a stable, relaxing, comfortable job getting paid to think about it and write about it at great length seemed frustrating more than satisfying. I wanted to get my damned thinking machine built – no matter that I had no savings in the bank and no source of income in the place I was moving to (New York City, where some new friends I'd met online were interested to collaborate with me on building an AI startup).

Actually, I first moved to southern/mid New Jersey, a 90 minute drive from New York City, where my always-helpful father fortuitously had a house sitting empty for a period of time due to tax-related reasons (he had moved elsewhere, but was waiting till he turned a certain age to sell his house, as he would pay less tax on the sale that way). My dad was a sociology professor at Rutgers University – Dr. Ted Goertzel – a job that he'd had for 20-odd years at that time, and would end up retaining for 40 years altogether by the time he retired. Clearly he valued stability far more than I did, but he is a nice guy and always willing to help out with my various harebrained risky plans to some extent, even though they sometimes make him nervous by proxy. Anyways – cutting a longer and rather more tortuous story short – there I was in my dad's former house with my wife and young kids and new baby, living on crumbs of money scrounged here and there, hacking Java all day (I don't care for the Java language much now, but back then it was new and shiny and the easiest way to do Internet-oriented software development), and scheming with my co-founders how to raise funds to get a thinking-machines company started.

My dad ("Ted" as I've always called him) was somewhat perplexed by my (largely hypothetical, at first) plunge into the world of software entrepreneurship. Though I'd had dreams of starting various companies at various times in the past, those had all been pretty fanciful, and I'd never struck Ted nor anyone else (myself included) as a particularly business-y guy. On the other hand, both he and my mom Carol (divorced since my childhood) were psyched to have their son and his children back in the US and, better yet, in their local area.

One question that my parents – and nearly everyone else I talked to about my plans to start a "build a thinking machine" company – asked me a lot in this phase was: OK, so if AI is possible, how come it hasn't been done before? And how come so few people are trying? Why is the world waiting for you to move back from the US to New York and raise money to get it done?

Peter Voss, a fellow AGI researcher whom I got to know in 2001 (and who is still pushing hard toward AGI from his base in LA), used to answer this question very elegantly as follows:

- 80% of people in the AI field don't really want to work on general intelligence; they're more drawn to working on very specialized subcomponents of intelligence.
- 80% of the AI people who would like to work on general intelligence are pushed to work on other things by the biases of academic journals in which they need to publish, or of grant funding bodies.

- 80% of the AI people who actually do work on general intelligence are laboring under incorrect conceptual premises.
- Nearly all of the people operating under basically correct conceptual premises lack the resources to adequately realize their ideas.

This was basically the same as the answer I generally gave, though Peter put it more crisply. Today the situation is a bit better, though not as much better as the media might make you think – but I'll discuss the situation in 2014 a little later, toward the end of the book; for the moment I still have a bit more to say about AI in the bygone age of the dot-com boom…

We didn't have the term "AGI" back then, so I tended to talk about "real AI." This phrase was perhaps a bit insulting to folks working on what I'd now call "narrow AI" (which is, after all, also "real" – and can be very useful in spite of its limited scope). By this point – though not in 2001 – I've done a lot of application-oriented narrow-AI for income-generation purposes, so I've achieved a bit more respect for AI of the "non-real" variety. For instance, as I'll discuss in a later chapter, my colleagues and I have made some interesting biology discoveries using narrow AI technology. But my use of the term "real AI" at this stage expressed my indignation that the AI field had drifted so far from the exciting science-fiction-style goals of its original founders.

At that time even more strongly (much more strongly, actually) than now, the presupposition of the bulk of the work in the AI field was that solving subproblems of the "real AI" problem, by addressing individual aspects of intelligence in isolation, was the best and basically only sensible way to contribute toward solving the overall problem of creating real AI. On the contrary, Peter Voss and I and other "real AI" mavericks took the view that. in many cases, the best approach to implementing an aspect of mind in isolation, is very different from the best way to implement this same aspect of mind in the framework of an integrated, self-organizing AI system. Or in other words: If you want to build a real AI, the best approach is to actually try to build a real AI.

A Philosophy of Mind, a Sketchy AI Design

The AI system I was working on then, building on early-stage designs I'd developed in Perth, was called the "Webmind AI Engine". The New York company I co-founded to implement and apply the engine was initially called

Intelligenesis Corp. (more recently, another company has sprung up using that name), but after a year or two we renamed it Webmind Inc., since too many business folks found "Intelligenesis" too hard to remember, pronounce or spell.

The goals of the Webmind AI engine were not especially modest; I'll quote some material I wrote during that period:

"The goals that the AI Engine version 1.0 is expected to achieve are:

- *Conversing with humans in simple English, with the goal not of simulating human conversation, but of expressing its insights and inferences to humans, and gathering information and ideas from them*
- *Learning the preferences of humans and AI systems, and providing them with information in accordance with their preferences. Clarifying their preferences by asking them questions about it and responding to their answers*
- *Communicating with other AI Engines, similar to its conversations with humans, but using an AI-Engine-only language called KNOW*
- *Composing knowledge files containing its insights, inferences and discoveries, expressed in KNOW or in simple English*
- *Reporting on its own state, and modifying its parameters based on its self-analysis to optimize its achievement of its other goals*
- *Predicting economic and financial and political and consumer data based on diverse numerical data and concepts expressed in news*
- *Subsequent versions of the system are expected to offer enhanced conversational fluency, and enhanced abilities at knowledge creation, including theorem proving and scientific discovery and the composition of knowledge files consisting of complex discourses. And then of course the Holy Grail: progressive self-modification leading to exponentially accelerating artificial superintelligence!*

These lofty goals can be achieved step by step, beginning with a relatively simple Baby Webmind and teaching it about the world as its mind structures and dynamics are improved through further scientific study."

Are these goals complex enough that the AI Engine should be called intelligent? Ultimately, this is a subjective decision. My belief is, not shockingly, yes. This is not a chess program or a medical diagnosis program, which is capable in one narrow area and ignorant of the world at large. This is a program that studies itself and interacts with others, that ingests information from the world around it and thinks about this information, coming to its own conclusions and guiding its internal and external actions accordingly.

How smart will it be, qualitatively? My sense is that the first version will be significantly stupider than humans overall, though smarter in many particular domains; that within a couple years from the first version's release there may be a version that is competitive with humans in terms of overall intelligence; and that within a few more years there will probably be a version dramatically smarter than humans overall, with a much more refined self-optimized design running on much more powerful hardware. Artificial superintelligence, step by step.

Sounds pretty good, eh? Unfortunately, the reality of the Webmind AI Engine never quite lived up to these noble aspirations. But we made a good college try until the money ran out.

How did my 31-year-old self aim to achieve these dramatic goals? The key, I thought, was taking a holistic and integrative philosophy of mind: looking at the mind as a system for recognizing patterns in the world and itself, and achieving goals based on these patterns; looking at specific cognitive functions as elements of a broader system of this nature, specialized for recognition and creation of specific kinds of patterns, but configured and adapted to work together synergistically toward overall system goals.

We were, as I saw it, stuck with a body of AI theory that has excessively adapted itself to the era of weak computers, and that was consequently divided into a set of narrow perspectives, each focusing on a particular aspect of the mind. In order to make real AI work, I proposed, we needed to take an integrative perspective, focusing on (quoting again from some materials I wrote at that time):

"The creation of a "mind OS" that embodies the basic nature of mind, and allows specialized mind structures and algorithms dealing with specialized aspects of mind to happily coexist

The implementation of a diversity of mind structures and dynamics ("mind modules") on top of this mind OS

The encouragement of emergent phenomena produced by the interaction/cooperation of the modules, so that the system as a whole is coherently responsive to its goals"

This was the core of the Webmind vision. I brought from Perth a design and implementation of the Mind OS (which got mutated and improved many times by my colleagues at Webmind, and by my own self as I learned and experimented), and what I then thought of as a "detailed theory and design" for a minimal necessary set of mind structures and dynamics to run on top of it.

With benefit of older age and 20-20 hindsight, I can see that my thinking at that stage about the specific mind structures and dynamics was way fuzzier than I realized. I hadn't been involved with large-scale practical projects before, and didn't have a clear understanding of the gap between a high-level concept with some associated equations, and a specification that you could give software engineers to implement and test. But I was filled with youthful ambition and confidence, and I figured that any gaps in my design could be filled in if I could raise enough money to gather enough smart people around me to help out. (In the end I don't think this was an utterly wrong attitude – I suspect Webmind would have gotten there eventually, after a lot more scientific and technical ups and downs, if the dot-com boom hadn't crashed, causing the avalanches of R&D money to go away. But now I'm getting ahead of myself!)

After years searching for a good name for the network of theoretical concepts underlying my AI design, I settled for "the psynet model" instead – psy for mind, net for network. According to the psynet model of mind, as I articulated it back then:

1. A mind is a system of agents or "actors" (our currently preferred term) which are able to transform, create & destroy other actors.
2. Many of these actors act by recognizing patterns in the world, or in other actors; others operate directly upon aspects of their environment.
3. Actors pass attention ("active force") to other actors to which they are related.
4. Thoughts, feelings and other mental entities are self-reinforcing, self-producing, systems of actors, which are to some extent useful for the goals of the system.
5. These self-producing mental subsystems build up into a complex network of attractors, meta-attractors, etc.
6. This network of subsystems & associated attractors is "dual network" in structure, i.e. it is structured according to at least two principles:

associativity (similarity and generic association) and hierarchy (categorization and category-based control).

7. Because of finite memory capacity, mind must contain actors able to deal with "ungrounded" patterns, i.e., actors which were formed from now-forgotten actors, or which were learned from other minds rather than at first hand – this is called "reasoning." (Of course, forgetting is just one reason for abstract (or "ungrounded") concepts to happen. The other is generalization — even if the grounding materials are still around, abstract concepts ignore the historical relations to them.)
8. A mind possesses actors whose goal is to recognize the mind as a whole as a pattern – these are "self."

I still think this conceptual perspective makes sense – although I have a somewhat lower opinion, these days, of how far this kind of conceptual thinking gets you toward a practically implementable thinking machine design. You do need to get the conceptual picture straight first – otherwise, as Peter Voss noted in his explanation of slow real-AI progress that I summarized above, you won't get anywhere, even if you have all the focus and funding in the world. But a conceptual picture, such as the psynet model provides, is just Step 1. Much of what we worked on at Webmind was filling in an intermediate level of theory and design, to connect these abstract conceptual ideas to real-world software code. This was done via a combination of software experimentation, mathematical analysis and conceptual theory – all somewhat mixed up and pursued at a feverish start-up pace by a diverse team of brilliant people. The result was a lot of insight and a fair bit of confusion. Most of my work in the years between then and now has been aimed at filling in the intermediate levels of detail in my AGI approach in a more careful, methodical and principled way.

According to the psynet model, at bottom the mind is a system of actors interacting with each other, transforming each other, recognizing patterns in each other, creating new actors embodying relations between each other. Individual actors may have some intelligence, but most of their intelligence lies in the way they create and use their relationships with other actors, and in the patterns that ensue from multi-actor interactions.

To make an artificial mind, then, we need actors that recognize and embody similarity relations between other actors, and inheritance relations between other actors (inheritance meaning that one actor can in some sense be used as another one, in terms of its properties or the things it denotes).

We need actors that recognize and embody more complex relationships, among more than two actors. We need actors that embody relations about the whole system, such as "the dynamics of the whole actor system tends to interrelate A and B."

This swarm of interacting, intercreating actors leads to an emergent hierarchical ontology, consisting of actors generalizing other actors in a tree; it also leads to a sprawling network of interrelatedness, a "web of pattern" in which each actor relates some others. The balance between the hierarchical and heterarchical aspects of the emergent network of actor interrelations is viewed as crucial to the mind.

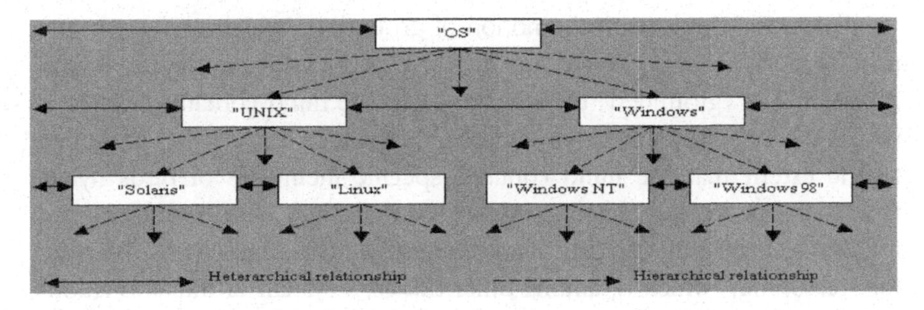

FIGURE 5.1: *A graphic from the Webmind days. As an example of structuring the relationships between concepts, I pointed out that operating systems circa 2001 could be arranged in a combination hierarchical-heterarchical ontology. This sort of hierarchical-heterarchical "dual network" organization, often taking much more complex forms, is IMO critical to the human mind.*

The overlap of hierarchy and heterarchy gives the mind a kind of "dynamic library card catalog" structure – which I talked about a lot in my 1992 book *The Evolving Mind* – in which topics are linked to other related topics heterarchically, and linked to more general or specific topics hierarchically. The creation of new subtopics or supertopics has to make sense heterarchically, meaning that the things in each topic grouping should have a lot of associative, heterarchical relations with each other.

Macro-level mind patterns like the dual network are built up by many different actors; according to the natural process of mind actor evolution, they're also "sculpted" by the deletion of actors. All these actors recognize patterns and create new actors that embody them, and this creates a huge combinatorial explosion of actors. Given the finite resources that any real system has at its disposal, forgetting is crucial to the mind – not every actor that's created can be retained forever. Forgetting means that, for example, a mind can retain the datum that birds fly, without retaining much of the specific evidence that led it to this conclusion. The generalization "birds fly" is a pattern A in a large collection of observations B is retained, but the observations B are not.

Obviously, a mind's intelligence will be enhanced if it forgets strategically, i.e., forgets those items which are the least intense patterns. And this ties in with the notion of mind as an evolutionary system. A system which is creating new actors, and then forgetting actors based on relative uselessness, is evolving by natural selection. This evolution is the creative force opposing the conservative force of self-production, actor intercreation.

Forgetting ties in with the notion of grounding, which was starting to pop up more in the academic cognitive science literature in the late 1990s. A pattern X is "grounded" to the extent that the mind contains entities in which X is in fact a pattern. For instance, the pattern "birds fly" is grounded to the extent that the mind contains specific memories of birds flying. Few concepts are completely grounded in the mind, because of the need for drastic forgetting of particular experiences. This leads us to the need for "reasoning," which is, among other things, a system of transformations specialized for producing incompletely grounded patterns from incompletely grounded patterns.

Consider, for example, the reasoning "*Birds fly, flying objects can fall, so birds can fall.*" Given extremely complete groundings for the observations "birds fly" and "flying objects can fall," the reasoning would be unnecessary – because the mind would contain specific instances of birds falling, and could therefore get to the conclusion "birds can fall" directly without going through two ancillary observations. But if specific memories of birds falling do not exist in the mind, because they have been forgotten or because they have never been observed in the mind's incomplete experience, then reasoning must be relied upon to yield the conclusion.

The necessity for forgetting is particularly intense at the lower levels of the system. In particular, most of the patterns picked up by the perceptual-cognitive-active loop are of ephemeral interest only and are not worthy of long-term retention in a resource-bounded system. The fact that most of the information coming into the system is going to be quickly discarded, however, means that the emergent information contained in perceptual input should be mined as rapidly as possible, which gives rise to the phenomenon of "short-term memory."

Some of my psychologist colleagues in Perth spent their career studying the way short-term memory (aka "working memory") works in humans. Some aspects of human short-term memory seemed to me unnecessarily limited and not worth replicating in AI systems – for instance, the way we have trouble memorizing a 50-digit number. Why shackle computers with human-brain processing limitations, when they can more naturally be made to remember whatever reasonable-sized numbers they see? On the other hand, in a broader sense, it seemed to me that some kind of distinction between short-term and long-term memory was going to be important to engineered thinking machines.

Basically, in psynet terms: A mind must contain actors specialized for rapidly mining information deemed highly important (information recently obtained via perception, or else identified by the rest of the mind as being highly essential). This is "short term memory." The reason short-term memory can't hold that many things at once is to avoid combinatorial explosion; the number of combinations (possible grounds for emergence) of N items being exponential in N. The short-term memory is a space within the mind devoted to looking at a small set of things from as many different angles as possible.

Now, this doesn't mean a digital AGI can't be made to remember long number sequences just by storing them in its computational memory, plain and simple. But it means that if an AGI system is going to have a "conscious focus of attention" like people do, where it can turn over and over the things it's holding in its attention and view them from many different conceptual angles, this focus of attention is going to have to be fairly limited in scope, much like in humans. One difference between human minds and the AIs I was aiming to build, in terms of short term memory, was that the human mind needed to use this conscious focus for so many purposes – not just careful deliberative reflection, but also for simple memorization of sequences. An AI could just as well handle these different aspects of memory differently. In this respect as in so many others, as we worked through the particulars for the Webmind AI Engine design, the potential for making AIs much smarter than humans by deviating from the happenstance limitations of human brain architecture seemed dramatic and extremely concrete and palpable.

The psynet model, which was the initial conceptual basis for the Webmind AI work, was a highly general theory of the nature of mind. Large aspects of the human mind, however, are not general at all, and deal only with specific things such as recognizing visual forms, moving arms, etc. This is not a peculiarity of humans, but a general feature of intelligence, and the psynet theory was extended to encompass this fact. I conceptualized the generality of a transformation as the variety of possible entities that it can usefully act on; and I figured that, in this sense, the actors in a mind would have a spectrum of degrees of specialization, frequently with more specialized actors residing lower in the hierarchy. In particular, a mind had to contain procedures specialized for perception and action; and when such specific procedures were used repeatedly, they would become "automatized," that is, cast in a form that is more efficient to use, but less flexible and adaptable.

I also had no problem conceptualizing linguistics within this same psynet perspective – indeed, I had gone fairly far in this direction in my 1994 book *Chaotic Logic*, and in some work done in New Zealand in 1994 when I was helping my first wife Gwen with her (eventually abandoned) PhD work in computational linguistics. Indeed, I saw language as verging into self and socialization and a host of other more abstract and general psychological issues. As I put it at the time:

"Linguistic communication is carried out by stringing together symbols over time. It is hierarchically based in that the symbols are grouped into categories, and many of the properties of language may be understood by studying these categories. More specifically, the syntax of a language is defined by a collection of categories, and "syntactic transformations" mapping sequences of categories into categories. Parsing is the repeated application of syntactic transformations; language production is the reverse process, in which categories are progressively expanded into sequences of categories. Semantic transformations map structures involving semantic categories and particular words or phrases into actors representing generic relationships like similarity and inheritance. They take structures in the domain of language and map them into the generic domain of mind."

And language brings us to the last crucial feature of mind: self and socialization. Language is used for communicating with others, and the structures used for semantic understanding are largely social in nature (actor, agent, and so forth); language is also used purely internally to clarify thought, and in this sense it's a projection of the social domain into the individual. Communicating about oneself via words or gestures is a key aspect of building oneself.

The "self" of a mind (not the "higher self" of Eastern religion, but the "psychosocial" self) is a poorly grounded pattern in the mind's own past. In order to have a nontrivial self, a mind must possess not only the capacity for reasoning, but a sophisticated reasoning-based tool (such as syntax) for transferring knowledge from strongly-grounded to poorly-grounded domains. It must also have memory and a knowledge base. All these components are clearly strengthened by the existence of a society of similar minds, making the learning and maintenance of self vastly easier.

The self is useful for guiding the perceptual-cognitive-active information-gathering loop in productive directions. Knowing its own holistic strengths and weaknesses, a mind can do better at recognizing patterns and using these to achieve goals. The presence of other similar beings is of inestimable use in recognizing the self – one models one's self on a combination of: what one perceives internally, the external consequences of actions, evaluations of the self given by other entities, and the structures one perceives in other similar beings. It would be possible to have self without society, but society makes it vastly easier, by leading to syntax with its facility at mapping grounded domains into ungrounded domains, by providing an analogue for inference of the self, by external evaluations fed back to the self, and by the affordance of knowledge bases, and informational alliances with other intelligent beings.

Clearly there is much more to mind than all this – as we've learned over the last few years, working out the details of each of these points uncovers a huge number of subtle issues. But even without further specialization, this list of points does say something about AI. It dictates, for example,

- that an AI system must be a dynamical system, consisting of entities (actors) which are able to act on each other (transform each other) in a variety of ways, and some of which are able to evaluate simplicity (and hence recognize pattern).
- that this dynamical system must be sufficiently flexible to enable the crystallization of a dual network structure, with emergent, synergetic hierarchical and heterarchical subnets.
- that this dynamical system must contain a mechanism for the spreading of attention in directions of shared meaning.
- that this dynamical system must have access to a rich stream of perceptual data, so as to be able to build up a decent-sized pool of grounded patterns, leading ultimately to the recognition of the self.
- that this dynamical system must contain entities that can reason (transfer information from grounded to ungrounded patterns) .
- that this dynamical system must be contain entities that can manipulate categories (hierarchical subnets) and transformations involving categories in a sophisticated way, so as to enable syntax and semantics.

- that this dynamical system must recognize symmetric, asymmetric and emergent meaning sharing, and build meanings using temporal and spatial relatedness, as well as relatedness of internal structure, and relatedness in the context of the system as a whole.
- that this dynamical system must have a specific mechanism for paying extra attention to recently perceived data ("short-term memory").
- that this dynamical system must be embedded in a community of similar dynamical systems, so as to be able to properly understand itself.
- that this dynamical system must act upon and be acted upon by some kind of reasonably rich world or environment.

It is interesting to note that these criteria, while simple, are not met by any previously designed AI system, let alone any existing, working program. The Webmind AI Engine strives to meet all these criteria.

These various conceptual ideas about the mind had been bouncing around in my own human mind for quite some time. What led me to think it was a sensible time to start an AI company, as opposed to just writing more books and papers, was coming up with a specific approach to implementing these ideas in software, in my last 6-9 months at the University of Western Australia. The crux of this approach was not that different to what now exists in the OpenCog AI Engine, though the details have evolved very much and in multiple ways.

Discovering Entrepreneurship

The germ of Intelligenesis Corp. / Webmind Inc. came about in 1996, when I was living in Perth and the magic of the Internet brought me in touch with a woman named Lisa Pazer. She had recently quit her job as a Wall Street market analyst to stay home with her kids and devote herself to fiction writing. She found some of my crazy half-finished fiction on the Web, and found it provoking and sometimes hilarious; but she was particularly intrigued by an article describing my vision of the global brain, the emerging network of transhuman intelligence that, I said, was about emerge out of the computing and communication networks all around us.

As I said in the article, the global brain wasn't something any one person could create, not even me. It was something that had to emerge. But there were things you could do to seed its emergence, to encourage its crystallization. One of these things, I said, was the creation of real AI.

By creating a computer program that could read all the information on the Internet, understand much of it, and place new information back on the Internet, enhancing the environment that it lived in, one could cause a phase transition in the development of the Net. One could jolt the Net into a different state of being, effectively causing the emergence of a new organism, a new form of intelligence and life. I wasn't the only one to have this vision, of course. After the idea became crystallized in my mind, I learned about various predecessors thinking in this direction, including Valentin Turchin, a brilliant Russian computer scientist whom I got to know personally during the period from 1999-2001 (near the end of his life), and who published a book reporting such ideas way back in the early '70s. But even after I realized the idea wasn't entirely original to me, the "global brain" idea still appeared in my mind with the glow of creative insight; it possessed me with its power, at that stage, just as much as the vision of "real AI."

FIGURE 5.2: *Our first company slogan was "Intelligenesis: Mind matters." Our logo at that time was a cute little sperm, intended to fertilize the next phase of intelligence. When we changed the company name to Webmind Inc., the slogan became "Creating Internet Intelligence." I used this as the title for a book that was published in 2001, around the same time the firm closed its doors.*

In early 1997, I moved back to the US to work with Lisa and another new online friend, John Pritchard, on creating superintelligent AI software that would trigger the formation of a global brain. John had business connections as well as deep software ideas, and he felt we could get a new company off the ground quickly. As it happened, Lisa, John and I were not an effective threesome, and John went off to build his own software in peace. Lisa and I plotted world domination on our own, until I began recruiting other friends to join the effort. We struggled to get VC funding for a while, and did some meetings, but apparently we didn't have a good enough story. To feed my young family I took a job teaching computer science at the College of Staten Island, which was right next to the largest pile of garbage in the world (a dump since shut down, fortunately for the residents of Staten Island, but this didn't actually take much of my time, and it was a job I never intended to keep for very long. Webmind was where my heart was. Fortunately, eventually, over a year after I'd returned from the US, Lisa's family chipped in some of their hard-earned cash, and we were off, with a few months of burn in the company bank account. Woohoo!!! ...

By that time I had replaced John Pritchard with four other friends, all technical in background. The lot of us were full of both ambitious AI ideas, and excited, youthful, naive dot-com boom optimism. Sure, our thinking machine might not get finished in the 3 months that our initial seed money from Lisa's family would last us. But, in the meantime, to tide us over, we'd solve a simpler problem: we'd use some of the bits and pieces of our unfinished AI engine to predict the financial markets. The technical co-founders and I had been working on the first version of the AI engine for many months by the time the seed funding came in. A healthy amount of software code existed (although the code itself wasn't entirely healthy, something that didn't worry me enough at that time due to my lack of experience with commercial software engineering).

One of the other technical co-founders, Ken Silverman, was an old friend of mine from Simon's Rock College, an institution that offers college education to high school age students. We'd met in our freshman year, when I was 15 and he was 16. Onar Aam, living in Norway, was a philosopher/composer/hacker whom I'd known through the Net for years; Paul Baclace, living in California, was an experienced Silicon Valley software dude who had contacted me and asked to become involved in the project. Jeff Pressing, living in Melbourne, was a psychology professor, and a whole host of other things. He was another quasi-prodigy, having entered CalTech at 15. He had been trading the Australian financial markets in his spare time for several years, using mathematical algorithms of his own creation. He had also studied drumming in West Africa, was an accomplished composer, and had been the dean of the music school at LaTrobe University in Melbourne.

We had a lot of collective brainpower, but Lisa was the only real businessperson among us. She was our first CEO, and she was also the inspiration for turning our AI tools to the problem of market prediction. Many times during her years as a market analyst, she'd been frustrated by watching the quantitative market analysts sit in a room by themselves, analyzing charts and crunching numbers, while the qualitative analysts like herself sat in another room analyzing news events, trying to limn the market's psychology. What if one could use the computer to read the news and predict the markets based on what it read? What if it could combine this information with the standard number-crunching formulas? Quantitative and qualitative analysis would be fused into single prediction! She asked me if my software could do this. I told her, "Sure, compared to true human-level intelligence and fluent conversation, that's easy." (I had a bad habit back then of classifying everything as "easy" if it didn't require Nobel Prize -inning science breakthroughs or Hoover Dam-scale engineering.) I knew I had Jeff's prior work on nonlinear-dynamics-based market prediction to fall back on, and the partially-finished Webmind AI Engine code as well – the puzzle pieces were there, they "just" had to be fit together...

What let us keep rolling after Lisa's family's seed money ran out was an introduction to Tom Petzinger, a Wall Street Journal columnist (and a remarkably pleasant and intelligent guy), who wrote a piece on us that appeared on the first page of the second section of the Journal. Ba-bing! This didn't bring us VC money, but it brought us much-needed cash from a host of wealthy WSJ readers. We were alive for another half-year or so.

But we couldn't keep going on the strength of that article forever. Even a follow-up New York Times article wasn't enough. Our efforts to sell our Market Predictor software were going much slower than we had expected, even though initial test results seemed good. (In hindsight, having done a lot more financial prediction work now, I think our market prediction results at that stage were probably pretty significantly overfitted to our training data, and thus maybe not really that good at all. But that's neither here nor there, by this point....)

Creating real machine intelligence was taking longer than I'd initially projected. The design I'd started out with had come out of my academic work, which was fairly theoretical. I'd built some simple prototypes before, but I didn't have experience with large-scale software engineering. It became clear that, the essential conceptual aspects of the system aside, there were a lot of hard, pure software engineering issues that had to be plowed through in order to get the thing implemented in a workable way.

A substantial, last-minute loan from the same friend who had introduced me to Petzinger saved the day. It carried us through until we could close a deal with our first VC, a small Chicago firm, one of the partners of which was a friend of Lisa's brother-in-law. This was a time period where, more than once, the money came into the bank account a couple hours before payroll checks had to go out. Our employees at the time never realized quite how marginal our cash flow situation was. In late 1998, the economy wasn't quite as good as during the first part of the year – the market was in a downturn (which the Market Predictor had anticipated), and pretty much no one closed VC deals in New York during that fall. But in the first months of 1999, the market got happier, and everyone's pending VC deals closed – including ours, from a Chicago VC firm called Vision Capital. Whew!

We hired experts to help us create "mind agents" endowing our AI Engine with particular capabilities.

In fact, the diversity of Webmind's tech staff, once the company raised VC money and really got rolling, was such that there's no way to do it justice. We had our share of PhDs in relevant disciplines, but that wasn't the heart of it. It's easy to have a crew of knowledgeable and competent PhD's with no imagination. The Webmind vision attracted all kinds of fascinating, brilliant, independent minds, people who had taught themselves various disciplines and had been working for years in their spare time on this or that intriguing project. Anton Kolonin, who labeled himself the "Siberian Madmind," had been sitting in Novosibirsk creating his own project called "Webmind," (a project focused on visualizing the global brain). His key ideas would gradually wend their way into the AI Engine design. I first reached out to him in a somewhat unfriendly way, requesting him to stop using the term "Webmind" for his system and thus violating my company's trademark. He rebranded his system "World Wide Mind,",and shortly after joined my company. Anton's polymath mad scientist style jarred wonderfully with the calm precision of Pei Wang, our true AI guru, who brought to us his own half-finished thinking machine, based on his own unique theory of cognitive logic, many aspects of which were incorporated into our system. And there was also a Bulgarian Madmind – every few months, it seemed, Youlian Troyanov went from believing AI was easy to believing it was impossible. He had his own theory of the mind and universe, and a mathematical proof of the existence of God that he liked to trot around, but no one could fully understand it except (presumably) him and the deity Itself. Also crucial was Karin Verspoor, a computational linguist who had been working at Microsoft Research in Australia. And company co-founder Jeff Pressing, who had been a relatively minor company co-founder in the early, pre-funding days, grew into a bigger and bigger role. Jeff contributed a lot of ideas about how to make mind agents dealing with numerical data, as well as a large amount of general cognitive science knowledge.

Stephan Vladimir Bugaj, who had a few years before accomplished the remarkable feat of becoming a Bell Labs researcher without a bachelor's degree, came to us from a boring job running an Internet ad agency, and immediately branched out into every aspect of Webmind Inc. operations, from IT to product design to AI theory to endless meetings with prospective clients in various industries, to arguing with the building administration about how many servers could be placed per square foot without breaking the floor. In true start-up mode, Stephan and I veered back and forth from being what he called "Uberjanitors" – dealing with mundane issues like ordering supplies, fixing broken computer networks, returning phone calls from interested business folk and assorted crackpots and so forth – to exploring the most ethereal realms of science, staying late into the night arguing about the viability of different ways of importing prefabricated linguistic knowledge into a digital mind without obstructing its ability to learn on its own.

And, alongside the AI brains, we were accumulating more and more engineers with serious product development experience, some in Brazil, some in the US. It seemed that we had the capability to create a thinking machine and engineer high-quality products around it, a capability that would enable us to, basically, take over the world...

We seemed to be having ongoing success using our early-version AI system to recognize correlations between trends in the news and movements in the financial markets. But practical experimentation soon led us to two significant concerns. First, making the system work effectively across a whole bunch of different machines was going to be a very big job in itself. The problem was solvable – and it had to be solved if the whole "AI seeds the global brain" vision was going to work – but it was more tedious than we'd initially thought. Distributed processing, using the limited hardware and networking of that time, was a mess of a problem, quite apart from the problem of making the distributed system smart. Secondly, it was probably going to take a very large AI Engine to get anywhere close to human-level intelligence. Like, billions of little mind actors acting in the overall AI Engine, distributed across maybe thousands or tens of thousands of machines. We hadn't discovered anything to make us doubt the basic philosophy and concept with which we'd begun our work. But the project was looking bigger and bigger. The brain has hundreds of billions of neurons and a quite complex architecture with different regions doing different things, so in hindsight this isn't very surprising. But it was a little bit sobering for us, given the scope of our ambitious and the limitations of our timeline and budget. Still, we were convinced we could get there, and definitely within years, not decades. We just had to think all the more creatively, and push even harder!

Did these first VCs, or our earlier individual investors, really understand what we were doing – how powerful our software ultimately could be, how difficult our task really was? A few of the individual investors did. One of the ringleaders of the WSJ-inspired investors was Frank Mosca, a psychologist (and ex-spy) friend of mine with his own wild and crazy theory of the mind, and no small measure of sympathy for my own. Another significant investor at this stage was another visionary psychologist friend, Terry Marks. But these two, and a couple others, were the exception. By and large, these people were putting their money into us because, quite simply, we looked like really smart people doing really cool stuff – and it was the middle of the Internet bubble.

In fact, one of the partners of our first VC firm used to raise money for us like this. He'd tell people:

"This is the best cocktail party investment you'll ever see. Think about it this way. If these guys succeed in building a thinking machine and predicting the financial markets, you'll be made incredibly rich, and you can say you got in on the ground floor of something remarkable. And if they fail, well, at least you can tell people you were in on a really interesting swindle. Either way, you'll have something to talk about!"

You have to remember, in those days, companies with no original technology and no business model were seeing tremendous IPOs. No one could understand why one company wildly succeeded and another one stagnated, or got acquired for a modest sum. So all the rules were thrown out the door. People bet based on pure intuition. They had the intuition that we were really clever, and would do something really cool. As one investor said at the time, "Will the software work? Well, that depends.... What does 'work' mean? I'm sure it will do something interesting." Of course, this faith placed in me and my team by strangers was flattering. But I felt it was largely justified. We really did have a better idea about how to make computers think. We really did, it seemed, know how to predict the markets using the news.

In thinking about this time period, I'm often reminded of Frank Zappa's analysis of the history of the music industry. During the late '60's and early '70's, he said, the music business was run by fat old men wearing suits and smoking cigars. These men had no understanding of rock and roll music at all. To them it all just sounded like noise. But the kids loved it. It made money for the record companies. So the old cigar-smoking record company executives would just sign contracts with any rock band that looked like it was doing something cool. Frank Zappa sold the Mothers of Invention as a "white-boy blues band." When the record label heard their first recording, their response was: "Hey, that's not white boy blues." Well, there were a couple blues tunes on there, but perhaps what the company was perplexed by was the 15-minute track consisting of modern-classical-style quasi-tonal improvisation played over the sounds of conversation and orgasm. But it didn't really matter. People were buying the stuff. It was all part of the movement.

But then, a few years later, these old execs were retiring, and the record companies were taken over by younger people, people who had grown up on rock and roll, people who felt they understood youth music. The result was a stultifying conformity. From "anything goes, because we sure don't know what's going on," the pendulum had swung waaaay back to the other side, to "we know exactly what sounds good, and if you're not playing it, tough luck." The same exact cycle hit the rock music industry again in the '90's. The birth of alternative rock in the early '90's was a time of wild innovation. All sorts of bands became popular under the label of "alternative rock," including bands like Ween and Mr. Bungle whose music is barely rock at all, because no one knew what alternative rock was, except the kids who were listening to it. Well, before long, record labels hired people who had grown up on Nirvana and Pearl Jam, and alternative rock became conventionalized. The innovative phase had ended.

A lot of our early investors in Webmind were like Frank Zappa's cigar-smoking record-company executives. In fact, almost everyone was in this position in the middle of the Internet bubble. Everyone knew a technological revolution was happening. You didn't have to be the sharpest pencil in the pack to figure that one out. It was possible to make some general predictions about the outcome of this revolution, such as "Most goods and services will be sold over the Web" and "A good search engine is important" and "Some kind of global brain will eventually emerge." But no one was smart enough to get from these general predictions to specific judgments as to which start-up companies were going to own a piece of the grand Net future. In fact, this may have been an unsolvable problem, objectively, and not just a consequence of the limitations of human analytical intelligence. Social and economic systems are chaotic, especially at times of transition.

While we were in the process of bringing in VC money, Lisa found us another great connection: a guy named Andy Siciliano, a true titan of the financial markets. Andy had made a huge amount of money trading derivatives in the 80s and early 90s, and immediately when I met him, he struck me as one of the shrewdest people I'd ever met. I'm sure he has a fine creative mind and a broad-ranging intellect, but against the backdrop of the Webmind Inc. AI crew, these weren't the qualities of his that stood out right away. Rather, it was his ability to carefully take stock of a situation, think slowly and dispassionately about all aspects, and then make a measured decision. His minimalist communication style was definitely at odds with the compulsive verbalization atmosphere that Lisa and I had fostered in the company so far. But in spite of being opposites in many ways, Andy and I hit it off pretty well. We respected each others' very different minds. I had the feeling that, having met a lot of businessmen over the past few years, I'd finally met a really good one.

FIGURE 5.3: *I hope I never forget the look on our CEO Andy Siciliano's face when I walked into our shared office (we were temporarily crammed into a single office while our expanded office space was being prepared) one day with half a head of hair. His jaw literally dropped. He was from a banking background and not accustomed to consorting with freaks such as myself. But he was an adaptable guy; he got used to it fairly fast. I found this hairstyle was very attractive to women, and I got lots of phone numbers freely offered while walking around the Wall Street area.*

Andy had a suburbs-to-riches story to make anyone jealous. An MIT engineering undergrad, he'd quickly changed his major to finance, figuring money was where the action was. Out of MIT, he'd gotten a job at O'Connor and Associates, a young derivatives trading firm. O'Connor was just plain smarter than the competition. There was an advanced mathematical theory of derivatives and options trading, based on the Black-Scholes formula, well-known in academia but not used on the trading floor — that is, until O'Connor did it, making many of its staff very rich, including Andy and his friend David Solo. O'Connor was then acquired by Swiss Bank, a much larger firm which was in the process of computerizing and centralizing. After a brief period, the O'Connor management – a young bunch of "blue jeans and polo shirts" guys – had taken over many of the top management positions at Swiss Bank. These guys knew more modern ways of doing things than the suits they displaced. And then Swiss Bank merged with Union Bank of Switzerland (UBS), and the O'Connor guys again wound up at the top.

Then Long-Term Capital Management came along. This was a huge hedge fund, run by a Nobel Prize-winning economist, using mathematics similar to those underlying O'Connor's success to trade the international currency market. But the real world failed to live up to the assumptions underlying the mathematics, and when a number of foreign economies tanked at once, LTCM lost billions. Andy was running the Foreign Exchange group at UBS, which possessed a large investment in LTCM from the pre-merger days. When LTCM died, Andy's group lost many hundreds of millions of dollars, and UBS, out of concern for its public image, encouraged Andy to retire, which he did. When I met him he was 6 months or so into retirement, and bored out of his mind. A man, he explained to me, needed to be involved in doing real things out in the world, or else his testosterone level would get too low. I wasn't sure about the metaphorical physiology, but I knew what Andy meant. Sitting in his mansion and manipulating his investments wasn't as stimulating as pushing to dominate the world of high finance, and nor was funding movies, his main hobby at the time. Lisa and I tried to convince him to start a hedge fund using the Market Predictor, but he was more interested in doing something in the Internet space. In early 1999, financial prediction was old hat. The investment bankers who'd made piles of money in the 80s and early 90s were jealous of the new wave of Internet entrepreneurs, some of whom had become far richer than the investment bankers. The Net was the place to be. Andy decided to pass on the hedge fund for the moment, but he offered us his services as CEO, together with a substantial amount of investment money. This let us hire more madminds, and more crack Brazilian software engineers, and it let us push to move even faster toward real AI and the Global Brain.

Nodes and Links Galore

What were we building with all these people and all this money? The Webmind AI Engine embodied the psynet model of mind by creating a "self-organizing actors OS" ("Mind OS"), a piece of software that was called the Webmind Core, and then creating a large number of special types of actors running on top of this.

Most abstractly, we had Node actors, which embodied coherent wholes (texts, numerical data series, concepts, trends, schema for acting); and we had Link actors, which embodied relationships between Nodes (similarity, logical inheritance or implication, data flow, etc.). We had Stimulus actors that spread attention between Nodes and Links, and Wanderer actors that moved around between Nodes building Links.

These general types of actors were then specialized into 100 or so node types, and a dozen link types, which carried out various specialized aspects of mind – but all within the general framework of mind as a self-organizing, self-creating actor system. There were also some macro level actors, "Data-Structure Specialized Mind Servers" that simulated the behaviors of special types of nodes and links in especially efficient, application-specific ways. These too were mind-actors, though of a specialized kind.

Each actor in the whole system was viewed as having its own little piece of consciousness, its own autonomy, its own life cycle – but the whole system had a coherence and focus as well, eliminating component actors that are not useful and causing useful actors to survive and intertransform with other useful actors in an evolutionary way.

This collection of nodes and links and other software actors was intended to eventually be distributed across the whole Internet, forming a World Wide Mind or Web-Mind. For starters, we were playing with it on a network of 30 quad-processor PCs, which was pathetic compared to a 2014 compute cloud, but pretty impressive for the late 1990s. We were aiming to build various practical money-making applications on this platform – I'll say a bit about that shortly – and we were looking at building a pure research AI system that would act more like an artificial baby, a so-called "Baby Webmind." But all the while, we were clearly aimed at extending our AI across the world as a whole, at creating an AI-powered "global brain." (Cool, huh?)

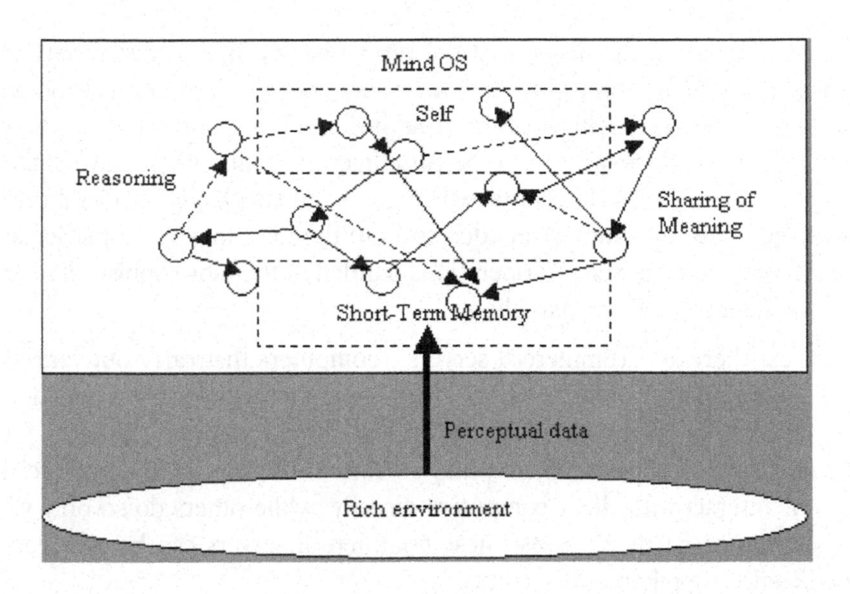

FIGURE 5.4: *Another Webmind-era diagram, showing the basic framework of nodes and links representing knowledge and responding to the world.*

The figure below, one more from the Webmind period, represents my younger self's attempt at an "architecture diagram" for the entire Internet, as I then envisioned it looking after it had been infused throughout with Webmind AI. Naturally, any diagram with such a broad scope is going to skip over a lot of details. The point was to get across a broad global vision.

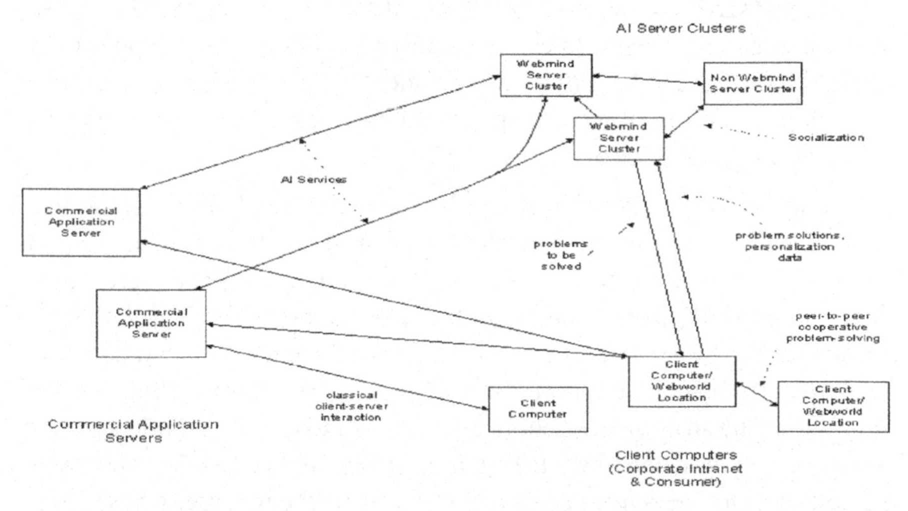

FIGURE 5.5: *Webmind-era diagram envisioning the future structure of an intelligent Internet – a Web-Mind.*

As I explained the diagram at the time, first, we have a vast variety of "client computers," some old, some new, some powerful, some weak. Some of these access the intelligent Net through dumb client applications – they don't directly contribute to internet intelligence at all. Others have smart clients such as WebWorld clients, which carry out two kinds of operations: personalization operations intended to help the machines serve particular clients better, and general AI operations handed to them by sophisticated AI server systems or other smart clients.

Next there are "commercial servers," computers that carry out various tasks to support various types of heavyweight processing – transaction processing for e-commerce applications, inventory management for warehousing of physical objects, and so forth. Some of these commercial servers interact with client computers directly, while others do so only via AI servers. In nearly all cases, these commercial servers can benefit from intelligence supplied by AI servers.

Finally, there is what I view as the crux of the intelligent Internet: clusters of AI servers distributed across the Net, each cluster representing an individual computational mind. Some of these will be Webminds, and others may be other types of AI systems. These will be able to communicate via a common language, and they will collectively "drive" the whole Net, by dispensing problems to client machines via WebWorld or related client-side distributed processing frameworks, and by providing real-time AI feedback to commercial servers of various types. Some AI servers will be general-purpose and will serve intelligence to commercial servers using an ASP (Application Service Provider) model; others will be more specialized, tied particularly to a certain commercial server (e.g., Yahoo might have its own AI cluster to back-end its portal services).

As the project progressed and we became increasingly aware of the size and difficulty of the task we had taken on, our AI design became more and more integrative in nature. We drew in more and more bits and pieces from other peoples' AI theories. We began to make extensive use of evolutionary programming, a computer science discipline based on simulation of evolution by natural selection. We integrated specialized techniques for numerical data analysis: prediction, association and causation finding, trend analysis, etc., developed by Jeff Pressing, myself, and others. We borrowed more and more ideas from Pei Wang, our first paid employee, whose Non-Axiomatic Reasoning System (NARS) was at that time my second favorite AI system. The earliest version of our AI Engine had relied more heavily on some "neural net"-like algorithms, simulating the spread of electricity in the brain; now that we knew what we were doing better, we moved further and further away from the brain as a concrete design inspiration.

We divided the system into modules, each module containing agents dealing with a particular aspect of intelligence: a reason module, a language module, an evolutionary learning module, a psyche module (dealing with feelings, goals and motivations). As we saw the different modules growing and beginning to work together, we thought we were almost there. Real intelligence was just around the corner! We had an unprecedentedly detailed understanding of how the different types of mind-agents should interact. Neural-net-like agents were used to find loose associations. These loose associations were used to guide reasoning. New concepts were formed using evolution-like methods, then evaluated by reasoning. All the parts of the digital mind fit together, like pieces in a huge multidimensional puzzle. We AI folk were talking so enthusiastically that even the businesspeople in the company were starting to get excited. This AI engine that had been absorbing so much time and money was about to bear fruit and burst forth upon the world!

Well... things never work out quite that nicely, not that quickly, not in reality. As 1999 progressed, we found that the increasing complexity of the various agents in the system was stressing the codebase. After a lot of difficult debate, we decided to grit our teeth and rewrite the core of the system from scratch. And, in what was a very difficult decision in terms of internal company politics, I decided to let our Brazilian team do the redesign.

How did we end up with a Brazilian team? Well, about the same way we ended up with our New Zealand and Australia teams, and our assorted programmers and scientists scattered among South Africa, Romania, Siberia, Georgia (the Republic, not the US state), Arizona, and so forth. I decided early on that the global brain should be seeded by a global team. Why not recruit the best people I could find, anywhere on Earth? (I would have gone off Earth if we had an interstellar Internet connection....) If I could do it all over again, I wouldn't globalize the team quite so free-flowingly – I'd hire fewer isolated people and focus more on building small, localized teams – but at the time, it seemed like a good enough idea. Certainly, we got much better talent than we could have obtained in New York City. The New York job market had a lot of depth in the finance and media areas, but not, at that time, in advanced computer science.

So as part of this "let's ignore the constraints of physical space" philosophy, toward the end of summer 1998 I hired a guy named Cassio Pennachin, who responded to an online job ad I'd placed on an obscure software jobs website. Cassio was a 21-year-old university student in Belo Horizonte, Brazil. He replied to the ad and offered to work for free for a month to prove how good he was. It turned out he was pretty darn good. The first job I gave him was to fix up some code I'd written for evolving new mind-agents by simulating evolution by natural selection. This was the beginning of what was soon to become a company tradition: Brazilian programmers receiving American code by email and responding very politely with comments like "Excuse me, but would you be terribly offended if I made a few changes to this code?" Of course, you say yes, and a few days later you receive a completely new version of the software, containing exactly three lines from your original code, but much better designed and also more efficient. Cassio proved to be an outstanding manager as well as an excellent software engineer and designer, and we let him accumulate assistants until, at one point, we had 60 people there out of a total company staff of 130. When the need to rebuild the system became clear, it was only natural to throw the task down to Brazil, and during fall 1999 and early 2000, Cassio and his Brazilian software gurus worked with me and the other American old-timers to create a new AI Engine, similar to the old one but better engineered.

Baby Webmind

One of the more interesting research initiatives to emerge, in the seething cauldron of creativity that was Webmind's research division, was the "Baby Webmind" project.

In terms of fundamental AI Engine development, the biggest thorn in our side as the year 2000 set in was natural language processing. We'd been trying to get the system to learn language just by extracting patterns from texts, but this didn't work very well. So we turned to special databases created by linguists, containing explicit "rules of language." This worked OK for some purposes, but of course we knew this was only a partial solution. If you're building a real AI, "wired in" knowledge of this kind is only acceptable if it comes along with a way for the system to adapt this knowledge based on its own learning. We realized that we didn't merely have to design a system capable of thinking; we had to design a system capable of growing from a baby mind into a mature mind, accumulating more and more knowledge and intelligence along the way. We knew we had a framework capable of supporting this, but we had to go back and build some new types of mind-agents.

Thinking in this direction, we began to focus more and more on what we came to call "experiential interactive learning." Language learning had to be integrated with the learning of cultural patterns of cognition, and this learning had to proceed through interaction with other minds in a shared perceptual/manipulable environment. We realized more and more vividly that even with all the different modules containing billions of agents working together and spawning intricate emergent patterns, all we were building was a baby. A very capable baby, or so we hoped, but a baby nonetheless. We began to think more and more seriously about the huge task of teaching our baby. We created a mechanism by which Baby Webmind could interact with us in a simple simulated world, in which it could participate with us in various interactions with files, directories, financial data series, and other digital objects. Along with this came a new focus on action as well as perception, something that we'd neglected in the past. We developed what we call the "schema" framework, a kind of programming language for Webmind actions, implemented in terms of mind-agents. We worked out how these little mind-programs could be learned by a combination of evolutionary programming and inference. Anton Kolonin played a huge role here; he proved to have an unparalleled intuition for this aspect of the mind. Cate Hartley, a young cognitive science graduate from Stanford, rose to prominence in the company because of her natural intuition for the Baby Webmind approach. We had enough depth of scientific and engineering talent to ensure that whatever direction the work turned, we had the mind power to handle it.

Equipped with schema governing its actions in its digital world, Baby Webmind could then ground its linguistic knowledge in non-linguistic social interactions, just as a human child does when learning language. When you talked to it about "moving to a new house," it would understand this by relating it to its own experiences moving files from one directory to another, or moving data packets from one machine to another. When you talked to it about eating, it would build an analogy to what happens when it reads a text and breaks it into parts. The system would never understand human reality like a human, but it would understand human reality by analogy to its own reality.

We thought about connecting a Baby Webmind system to sensors and actuators and starting a robotics division, but robotics tech was pretty primitive then compared to 2014, and we feared getting bogged down in hardware difficulties. I had played with some robots in Perth, guiding a student in building a small, wheeled robot and attempting to control it with neural networks – we hadn't gotten anywhere interesting AI-wise, and the experience had given me a healthy respect for the annoyances of working with physical robots at that stage. Rather, I figured a baby just needed some kind of environment to learn from and interact with. Any environment with complex data and the capability to perceive, act and interact with human teachers and playmates should work OK. We decided to get the Baby Webmind started interacting with a computer's file system. We built a Baby Webmind User Interface, comprising:

- A chat window, where the teacher can chat with the AI Engine
- Reward and punishment buttons, allowing the teacher to vary the amount of reward or punishment (a very hard smack as opposed to just a plain ordinary smack…)
- A way for the teacher to enter their emotions in, along several dimensions
- A way for the AI Engine to show the teacher its emotions [technically: the importance values of some of its FeelingNodes]

The point with Baby Webmind was just to use simple interactions with humans in the context of a computer's file system to help teach the system how to think. Then the system would be released on the much larger file system comprising the Internet. Or so we planned.

Yet Baby Webmind was just a way of providing experience and teaching and input to the AI Engine. The core was still the code for learning new nodes and links, and adapting existing ones, in response to the system's experience. At the start of 2001, we completed what we called "Webmind AI Engine Version 0.5" – for the first time, an AI Engine incorporating all the modules, working together sensibly in system running across many different machines. Millions of nodes, billions of links, dozens of types of cognitive processing. Tremendous!

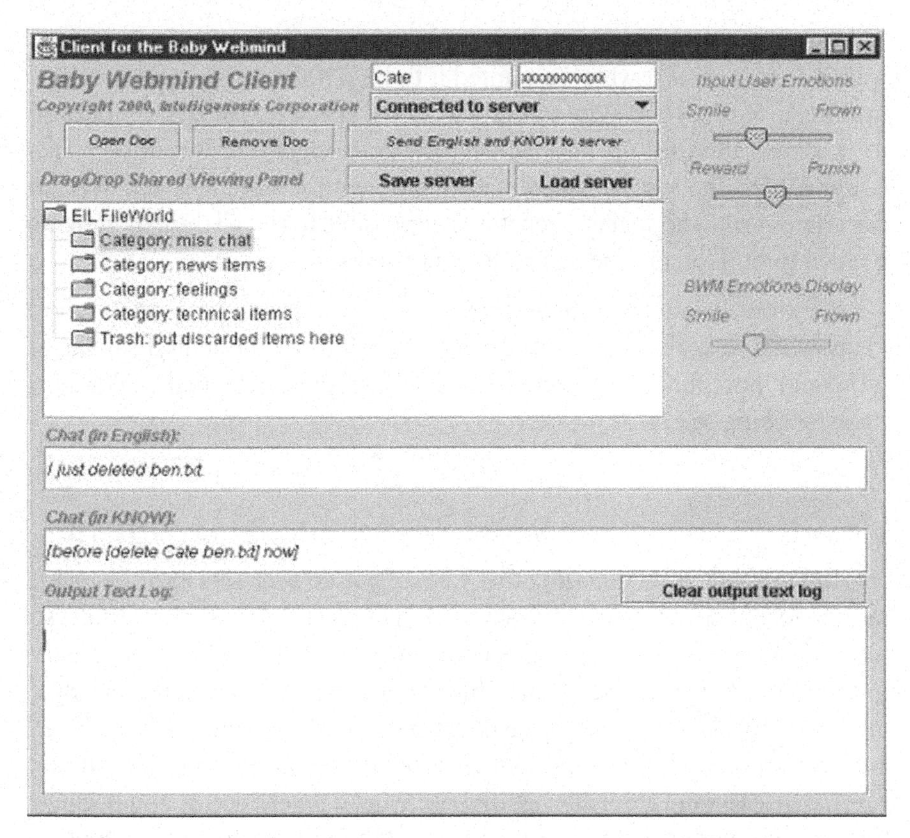

FIGURE 5.6: *The "Baby Webmind User Interface," which we used for some simple experiments teaching an early-stage proto-AGI system about its environment, which was a computer's file system. This research ended up being cut off for financial reasons right when it was starting to get off the ground.*

Well, there were only two small problems.

The first problem was that there were hundreds of parameters, complexly interacting, making the system very difficult to tune. Reasoning had parameters determining how speculative the system's inferences were. Neural-net-like association finding was governed by a host of different numbers, which controlled the flavor of the system's associations. Simulated evolution involved a half dozen numbers. The interactions between different machines were quantified by numbers that had to be tuned to the right values for the system to work well. Of course, the human brain is similarly complex, and is governed by the concentrations of various chemicals, the thicknesses of different parts of the neuron, and so forth. It's taken evolution millions of years to tune these parameters.

And the second problem was that the performance of the system still wasn't anywhere near what we wanted it to be. The thing was way too slow, hundreds of times too slow to even seriously teach it. The system needed to be drastically sped up, and its memory usage significantly reduced. This compounded the too-many-parameters problem, because if it was slow to experiment with the system, then it was impossible to run all the experiments needed to find the right sets of parameter values.

We were working on these problems, and we had some pretty good ideas, when we realized that the economy was shifting. Suddenly, bringing in more and more funds to power all the software and conceptual exploration wasn't looking so easy anymore.

Waking Up from the Economy of Dreams

In hindsight, it was amazing that we had gotten such serious investment money for such out-there, blue-sky real-AI research. We were learning so much, in a way that could only happen from having a large team of people try out different things and see which things failed worst. We weren't making a lot of revenue, but we were engaged in some useful customer conversations, and we were building some actual product software alongside all the research… we were gradually figuring out what we were doing, and I believe that with enough capital we would have gotten there eventually – to a point of dramatic profitability as well as achievement of our research goals. And for a while, it seemed that sufficient capital would keep on being forthcoming.

As Andy Siciliano liked to say, the dot-com boom was a once-in-a-generation period when capital flowed like water.

For sure, a lot of this capital that flowed like water was invested in absolute nonsense. To take a well-worn example, how many companies selling pet food online do there need to be? And who buys their pet food online, anyway? Most of us buy our pet food at the supermarket along with all the rest of the food. To go to a special website just to buy pet food is extra hassle, not a time-saver, unless one happens to have a pet with special nutritional requirements. And what possible sense did it make for hundreds of millions of dollars to go into making life a little easier for people with finicky or unwell pets? It's hardly a surprise that as the Internet bubble burst, the various online pet food companies went bust or were consolidated into other online enterprises with more general scope.

Several of my friends in New York worked for a company called MacroView, which made a website called sixdegrees.com. Inspired by the film "Six Degrees of Separation," this personal networking website allowed you to sign up and then enter in a number of your friends as contacts. Your friends would then be emailed and asked to sign up. Et cetera. The idea was that any two people were statistically likely to be connected by a chain of no more than six acquaintances. (How do I connect to the Prime Minister of Japan? Maybe my dentist's sister's brother-in-law's masseur's colleague's aunt shines his shoes for him.) Anyway, sixdegrees.com amassed a huge database of members, and they regularly sent emails to all their members. The technical challenges involved in setting up this huge website with millions of subscribers were immense, and my friends on the tech side of sixdegrees.com had a great time and learned a lot. They built some great systems. But what good was it? How was it going to make money? Well, they put some advertising on the website, and in the emails they sent out. But Web advertising never lived up to its promise – until later, when it did, but by that time MacroView was gone. Later came LinkedIn, which was doing essentially the same thing, but when the time was riper.

Another friend was involved with a site called JustBalls.com. Supposedly, the original proposed name for the site was MyWifeHasBalls.com. JustBalls sells balls over the internet – tennis balls, soccer balls, golf balls... any kind of balls you want, they've got 'em. Who wants to buy balls online? Surely there's not enough demand from individual ball-hungry consumers. Well, as it turned out, they created a reasonably healthy business selling balls to schools and other youth organizations. The previous distribution mechanism for balls was inefficient, involving sales representatives from ball manufacturers visiting schools and organizations a couple of times a year. Coaches would run out to retail sporting goods stores in the interval, paying prices that were badly inflated compared to the discounts available from the manufacturers on the rare occasions of their visits. Well, JustBalls.com was able to offer manufacturers' discounts year-round. There was a substantial niche market that hadn't been identified in advance, but that emerged over time as the site got more and more users. This is a small example of what happens in the business world when money is cheap and experimentation is plentiful. A lot of things get tried, and some succeed, and some fail. The things that succeed get to grow and transform themselves into different forms, and combine with each other through mergers and acquisitions.

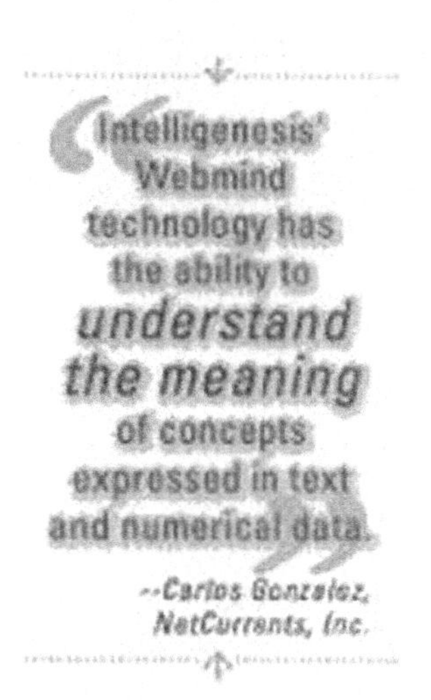

FIGURE 5.7: *One of Webmind Inc.'s early customers was a firm called Netcurrents, using our supervised text classification software for sentiment identification. At the time, automatically identifying sentiment in text was somewhat cutting-edge; now it is a commoditized functionality and available widely as a low-cost web service. So goes the exponential advance of narrow AI. While Webmind was working on hard-core AI research, the little revenue it did bring in was based on narrow AI products.*

All this is really nothing more – or less – than evolution. There's a theory of evolution called "punctuated equilibrium," which states that evolution proceeds in bursts. You have a period of rapid change, when new forms flourish. Then the variety is winnowed down by harsh reality, and you have a period of relative stasis. This may happen on the individual species level, where a species undergoes rapid change and turns into a new species over as little as tens of thousands of years – and then remains basically constant over millions of years. It can also happen on the ecosystem level. The Cambrian Explosion is a historical example. Over a period of just 40 million years, between 565 and 525 million years ago, the ancestors of all living and many extinct phyla of multicellular creatures evolved out of simple, mostly unicellular ancestors. In evolutionary time, that's amazingly fast. But then there was a huge extinction. 95% of the species that had been created were wiped out. Innovation still occurred, of course, but never again at such a level.

The Internet Bubble, which resulted in Webmind getting the funding it did, was a kind of Cambrian Explosion. All sorts of things were brought into existence, including huge new businesses of a type never seen before – Amazon, Yahoo, Ebay. These included huge wastes of money and engineering brilliance, like multiple online pet food sites and sixdegrees.com, and all kinds of other fascinating things that never grabbed the public eye – such as my own company. How wonderful to have obtained the opportunity to hire a team of brilliant people and teach them my ideas about digital mind, and work with them for a few years to bring some of these ideas into reality!

But after the explosion comes the extinction. Now comes the point where you have to fight for survival. The kinds of short-term cash-flow problems that put a company like Webmind Inc. near bankruptcy in '98 and '99 were more than enough to kill a company in 2001-2002. After the crash, the environment shows no mercy.

Just as everything was starting to coalesce for us internally in terms of our AI R&D, the market went through a series of horrifying collapses. Boom, boom, boom. Before you knew it, Yahoo was trading at a tenth of its historical maximum. Disney was folding its website, Go.com. The carnage was getting worse every day. Most of the dot-com companies who had expressed an interest in buying our stuff were no longer in a position to buy anything – they were laying off their employees. And most of the big companies we were talking to were stalling their purchasing decisions until the economic outlook became clearer. Tech spending was going way, way down, and a lot of tech companies were going down with it.

We'd been paying the bills for some time with money from Andy and his various friends and associates, including a major Asian financier. We always knew that this source of cash would eventually run dry. But we weren't much worried about this, because in the summer of 2000, we'd secured a deal with an overseas media company. They wanted to be a strategic investor – put in a bunch of money, integrate our software with their own IT systems, and help us market our software in their country. It was a beautiful story, and we felt awfully clever for having secured funds from overseas while our own country's economy was in chaos. We were also pursing funds from other sources, including some plain vanilla US VC's, but in retrospect, perhaps not quite as avidly as we should have. Because suddenly, at the last minute, after 8 months of tedious, painstaking and frustrating negotiations about trivial points of the deal – negotiations that seemed impossible to speed up in spite of our best efforts – the media company that had promised us funding pulled out. They were no longer interested. Period. We offered to concede on every possible negotiating point, and even to substantially change our business approach to suit them. It didn't matter.

Exactly what went wrong with that deal will never be known, but ultimately, I have to think that they were reacting to the poor condition of the US economy for tech companies as a whole. We were having trouble selling things because our target markets had largely stopped buying things. We had ways to get around this problem, adjustments to our sales approach that might well have worked, but the problem was definitely real.

Some people said we were dead right then, when the media company pulled out of the financing deal. I suppose they were right. But, pigheaded as I am, I didn't want to admit it. Instead, we lined up more money, from Andy's familiar cast of friends. We hired a bunch of crackerjack salespeople from a company that had recently gone bankrupt. These new sales guys really knew how to sell software. Everyone was hyped. It looked like we were going to scrape by, just barely.

But then a major part of the new investment round fell through, because a minor scandal had broken out in the news, regarding one of the investors. The media had saved us from bankruptcy in the early days, in the form of Tom Petzinger's Wall Street Journal article. And now the media had killed us.

The dot-com company I co-founded in 1997 – first named Intelligenesis, then Webmind Inc. – crashed and burned after 3.5 years, after spending over $20 million in venture capital money. At its peak there were 130 employees, including over 40 in the AI research group. I had fairly little idea what I was doing in terms of either business (which wasn't my job at that point, fortunately for me and everyone) or managing a research team, but it was a lot of fun and a heck of a learning experience. We certainly got a lot of fascinating and productive research and prototyping done. The firm closed its doors in mid-2001, around the same time a lot of other dot-com firms met the same fate.

In hindsight, the demise of our blue-sky AGI research company after a few years of helter-skelter hard-core research and scattershot attempts at product development is no big surprise. Rationally speaking, bankruptcy should have been a clear and present danger right from the start. The seed money we got was only enough to last us a few months, and we had no idea where the next round was coming from. But in the moment, as we were growing and then operating the company, we never really seriously considered that we might run out of money. The economy was booming, the Internet was all the rage.

In any normal financial times, the idea of starting a company with a goal as dreamy-eyed as "build a real AI to seed the global brain" would get you laughed out of any conversation with any serious investor. Create a thinking machine, and then commercialize it? Well, fine, but how are you going to make money while you're creating the damn thing? In recent years, as it happens, we've seen something of a resurgence of the optimistic attitude that got Intelligenesis/Webmind funded back in the dot-com era. Companies like Deep Mind and Vicarious Systems have been funded during the last few years, with similar business models – build the AI first, then figure out how to commercialize it. A big difference now, though, is that there are tech giants out there with an appetite for buying AI startups, even those without products or revenues (e.g., Google's half-billion dollar purchase of Deep Mind, which I'll discuss later on), thus giving investors in said startups a pretty clear exit strategy. Webmind, on the other hand, had no clear acquisition exit strategy – it was aiming straight for IPO, straight for conquering the world, leveraging the power that would obviously result from marrying advanced AI with the rapidly emerging Internet.

And what became of all the science and engineering we did at Webmind? Ultimately, the basic philosophy and conceptual framework underlying the Webmind AI Engine was the same one that is underlying my current AI work. However, in hindsight, the Webmind design was far too complicated, with too many independently moving parts. I didn't have a good sense then for the difficulty of getting large, complex pieces of software to function as specified.

The Saga Continues

In mid-2001, still reeling from Webmind's collapse, a number of the former Brazilian Webmind technical staff and I joined forces to create a new AI firm called Novamente LLC (nova mente = new mind; also "novamente" means "anew" in Portuguese). We designed a new AI system called the Novamente AI Engine, built on similar conceptual principles to the Webmind AI system, but much more pragmatically architected as a software system. We had a much smaller team now – just a handful of people – but we didn't intend to give up. We intended to push ahead as best we could, simultaneously searching for more resources to hire more people and building software according to our design.

For the next few years I put a lot of effort into simplifying and improving the Novamente AI Engine design, while my Brazilian colleagues worked on the implementation. We got a lot of interesting things done, but the search for AGI-oriented funding kept leading us down dead ends. We kept ourselves financially afloat doing various narrow-AI projects, some utilizing Novamente AI Engine code and some not. But eventually I reached the conclusion that things were just going too damn slowly. We chose some of the core parts of the Novamente AI Engine and decided to release them open source under the name OpenCog.

On a technical level, this wasn't a trivial thing to do – the Novamente AI Engine code was complex and not that well-documented. Cleaning up portions of the code for open-sourcing was a lot of work, and a grant from the Singularity Institute for AI (SIAI) was instrumental in paying for this. This was somewhat ironic since their mission as an organization was to ensure that once AGI was created, it would be guaranteed to be friendly to humans – and no such guarantee existed for OpenCog. However, the SIAI folks considered OpenCog's chances of success at achieving powerful AGI as minimal, and figured that association with me and the OpenCog project might lend them some academic legitimacy, which they felt they needed at the time. Since then, they have rebranded themselves as MIRI (the Machine Intelligence Research Institute) and found academic legitimacy through other means, via publishing papers on their ideas about provably friendly AGI and so forth. They still maintain the perspective that OpenCog is unlikely to succeed, but that if it does succeed, it will almost surely kill all humans. Obviously, my opinion differs. I'll return to these issues later in the book.

The open source aspect of OpenCog ties in wonderfully with the "global, distributed intelligence" memes I was enchanted with in the Webmind era. It makes perfect sense for the WORLD to create the first real AGI, since, after all, this AGI is going to transform the whole world. I can't prove it rigorously, but my gut says strongly that an AGI created by a broad-based global effort is far more likely to be benevolent than one created by some isolated group to serve its own ends. Of course there's the possibility that some isolated group could have a brainstorm about how to build powerful, ethical AGI, and secretly create it and unleash it on the world to everyone's benefit. But the track record of secretive elite groups wreaking transformations "for everyone's benefit" is not very strong, if you look across human history. I have more faith in inclusive efforts.

Further Reading

Goertzel, Ben (2001). Creating Internet Intelligence. Plenum.

6. Singularity or Surge?

"Within thirty years, we will have the technological means to create superhuman intelligence. Shortly thereafter, the human era will be ended… When greater-than-human intelligence drives progress, that progress will be much more rapid."

— Vernor Vinge
The Coming Technological Singularity, (1993)

"Two years after Artificial Intelligences reach human equivalence, their speed doubles. One year later, their speed doubles again. Six months - three months - 1.5 months ...Singularity. Plug in the numbers for current computing speeds, the current doubling time, and an estimate for the raw processing power of the human brain, and the numbers match in: 2021. But personally, I'd like to do it sooner."

— Eliezer S. Yudkowsky[1]
Staring into the Singularity 1.2.5, (2001)

---

1 Note that since 2001 when he wrote this, Eliezer has changed his tune considerably. He has become much more reticent about making concrete predictions, and has also become very concerned about the potential for AGI to extinguish humanity, so that he no longer (at least not publicly) advocates moving rapidly toward the creation of Singularity-capable AGI. He has often stated that he doesn't feel responsible for the views of his "previous selves", reserving the right to radically modify his views based on ongoing learning and thinking.

"I set the date for the Singularity -- representing a profound and disruptive transformation in human capability -- as 2045. The nonbiological intelligence created in that year will be one billion times more powerful than all human intelligence today."

— Ray Kurzweil
The Singularity is Near, When Humans Transcend Biology, (2005)

"Certainly my best current projected range of 2020-2060 is voodoo like anyone else's, but I'm satisfied that I've done a good literature search on the topic, and perhaps a deeper polling of the collective intelligence on this issue than I've seen elsewhere to date. To me, estimates much earlier than 2020 are unjustified in their optimism, and likewise, estimates after 2060 seem oblivious to the full scope and power of the... processes in the universe."

—John Smart
Nanotech.biz interview, (2001)

Our plans for Webmind were very ambitious and optimistic, in 1997 we aimed to create human-level AGI within years rather than decades. Other AGI researchers have launched their own projects with similar ambitious timelines, both before and after Webmind. Two things AGI researchers have never lacked are hubris, and optimism. Quite often such optimism has been misplaced.

However, the fact that optimism about a certain technological development was wrong for a while, doesn't mean it will always be wrong. Eventually, after lots of false starts, manned flight really did work. Spaceflight worked too. The gradual advance of understanding and technology occasionally leads to sudden leaps.

Of course, the scientific and technological possibility of creating AGI, doesn't necessarily imply anything about the speed with which AGI is going to be created, or with which it will be unleashed on the world. An AGI is an AGI, regardless of whether it emerges overnight or slowly gets perfected over decades or centuries. AGI will still revolutionize the world, whether it emerges rapidly in a Technological Singularity as Kurzweil and Vinge and others have foreseen, or more gradually via what transhumanist pioneer Max More has called a Surge.

But, hey. I've got to call it as I see it. As I've dug deeper and deeper down the AGI rabbit-hole, I've become more and more convinced that the Technological Singularity vision is correct, rather than the more conservative Surge option ... that there will come a point – after lots of preliminary R&D work, to be sure – when the intelligence of the world's top-notch AGI systems starts accelerating dramatically fast. I don't think the emergence of a Technological Singularity is a certitude by any means, but I do think it's a highly plausible hypothesis. AGI will still be interesting and important even if this hypothesis doesn't turn out to be true. But the practical implications of quickly-developing Singularity-style emergence of superhumanly intelligent AGI, if it does happen, will be drastically different than in the case of a slower burn.

The detailed predictions I made during the Webmind era were wrong. But actually, nobody knows how much progress we would have made had our funding held out for a decade. We had a great team and were in the process of radically rearchitecting our core AI system, while retaining the central principles and algorithms. Making predictions based on the financial and organizational situation of any one project is always going to be very chancy. Today I am much more careful to hedge my predictions about

OpenCog – I don't talk about what WILL happen unconditionally, only about what will probably happen if the project achieves adequate resources for a significant period of time.

But in the end the reasons for believing a great leap in AGI and a correlated overall technological Singularity may happen soon are not grounded in the prospects of any particular project, or any one researcher. They are extrapolations of more general trends, based on broader underlying processes. Particular projects are manifestations of these underlying processes.

What Is the Singularity?

There are actually multiple visions of the Singularity concept—Ray's differs slightly from mine, and both Ray and I have slightly different visions from science fiction writer Vernor Vinge, who was the first to use the term "Singularity" in this context. The first person to outline a closely Singularity-like idea in any depth seems to have been mathematician I.J. Good, who in 1965 wrote about the potential for an "intelligence explosion" when AIs became smarter than people, and noted that "the first intelligent machine is the last invention man will ever make".

FIGURE 6.1: *SF writer and professor Vernor Vinge introduced the term "Singularity" into futurism in the 1980s, but has done fairly little Singularity publicity, preferring to focus his career on science fiction writing now that he has retired from academia. His speech at an early Singularity Summit conference, shown here, was an exception. (http://http://vimeo.com/54718577)*

At the core of these multiple variant visions is a common understanding: The Singularity means rapid technological change accelerating so fast the human mind can't keep up, leading to dramatic new sciences and inventions, and revolutionizing all aspects of life as we know it, including AGI with capability far beyond the human level.

FIGURE 6.2: *Writing in the 1960s, the mathematician and early computer scientist I.J. Good (born Isadore Jacob Gudak, known as "Jack") foresaw the coming "intelligence explosion" just as clearly as do Singularitarians today. (Figure from http:// upload.wikimedia.org/wikipedia/en/b/b4/ I._J._Good.jp)*

In his 2005 book The Singularity Is Near, Kurzweil estimates the Singularity beginning around 2045; Vinge's earlier, less rigorous estimate involved roughly the same timeline. I think this rough timeline basically makes sense, though concerted effort on a project like OpenCog may speed up the process. And there's the possibility of a significant delay if too many unforeseen political or engineering obstacles should emerge.

Most of what I have to say in this book about AGI isn't actually dependent on the concept of the Singularity. AGIs will still be AGIs with basically the same properties, no matter whether they emerge super-rapidly via a Singularity-type event, or more gradually. However, the Singularity means AGI may emerge, within decades or even years from now, as opposed to centuries or millennia – a projected timeline which certainly adds a bit of practical everyday-life zing to the AGI concept.

Crudely put, the Singularity idea has two components:

- Amazing stuff is very likely to happen.
- Amazing stuff is very like to happen pretty suddenly, and pretty soon, likely in a matter of decades, conceivably even less

Just the first part is pretty exciting on its own, but might not capture many people's immediate interest or attention, since most folks tend to preoccupy themselves with things that are going to happen in their lifetimes, or at least their children's lifetimes. Actually many people seem to preoccupy themselves mainly with things that will happen with the next few days or weeks! I have felt public interest in AI, Singularity and related topics grow recently, as the radical advance of technology becomes more and more palpable in peoples' everyday lives.

The Logic of Exponential Growth

Over the course of both distant and recent human history, science and technology progress has accelerated dramatically. These days, technology, science and even art advance so fast I can't keep up – and neither can anyone else.

Take mobile phones, for example — I don't have time to figure out all the features on my phone before a new one comes out. It's the same with computer software. Or music. New and potentially interesting genres of music are constantly emerging, and in many cases, before I'll get a chance to appreciate them, they will be replaced by successors building new things upon their artistic and cultural achievements.

Contemporary science journals document a multitude of amazing, diverse discoveries in fields such as biology, physics and engineering. Browsing through the current periodicals sections of a major university's science libraries is an astounding experience. One scientist can't follow everything going on in his or her field – let alone keep up with neighboring fields and related ideas. A few hundred years ago, a diligent scientist did not have this problem – there were not that many scientists, there was way less science going on, and information about new discoveries could easily take years or decades to propagate.

In 2005, when I searched for the term "apoptosis" (preprogrammed cell death) in PubMed, the biomedical community's online repository for research papers, I found over 60,000 papers touching on the topic. I got the feeling maybe ten thousand or so of these were reasonably important. When I searched the term in PubMed again in 2013, I found nearly 250,000 papers. Even on a narrow topic like this, it's extremely hard to track all the new knowledge.

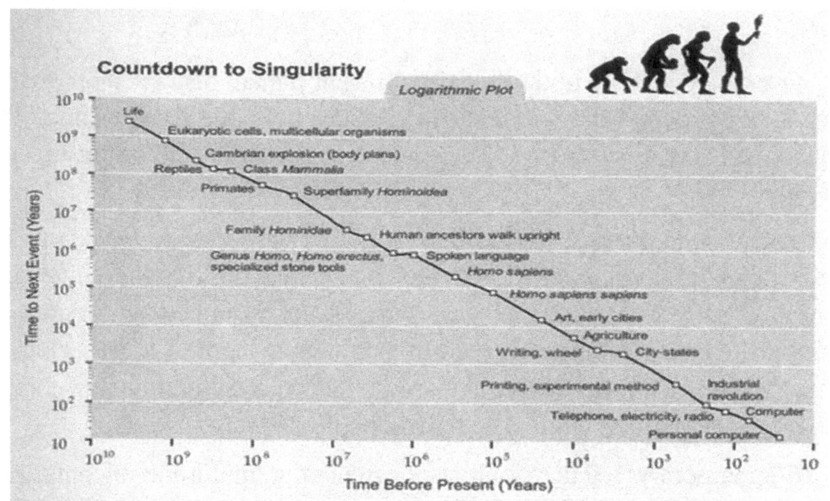

Countdown to Singularity: **Biological evolution and human technology both show continual acceleration, indicated by the shorter time to the next event (two billion years from the origin of life to cells; fourteen years from the PC to the World Wide Web).**

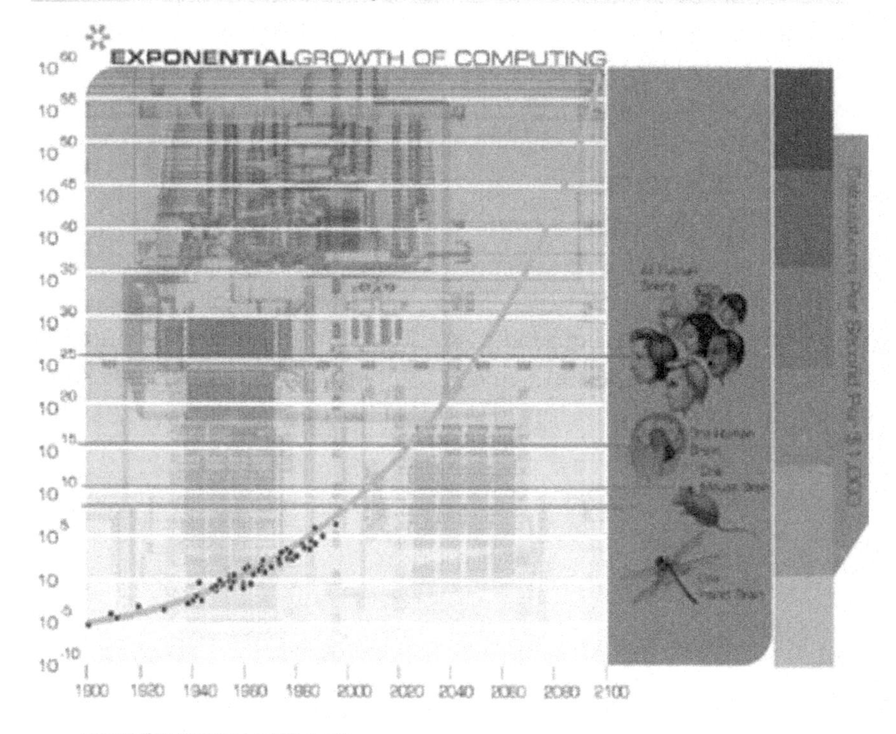

FIGURE 6.3:: *Illustrations of the exponential growth leading up to the Singularity, from Ray Kurzweil's presentations on the topic. (http://kurzweilai.net)*

FIGURE 6.4: *I'm only 49 years old, but I remember when devices like the iPad Touch in these pictures did not exist – and when the average person considered them a science fictional technology, probably hundreds or more years away. Now I need to buy a new one every couple years or folks will laugh at me for being behind the times. And I see these in the remotest places, such as the hills of rural Ethiopia. These photos were taken in the hills near Hawassa, in southern Ethiopia, when I was biking around with my wife and daughter and some of the staff from the Ethiopian AI company I helped found, iCog Labs. The Picture shows my daughter Scheherazade in the back, to the left of iCog CEO Getnet Aseffa. The kids we met live in a hilltop house with no electricity, where water needs to be carried up in buckets every day. But they learned to play Fruit Ninja on Scheherazade's iPod Touch pretty fast. When these kids are teenagers they will probably all have their own smartphones, connected to the Internet 24-7. What amazing opportunities for education this will bring, along with of course entertainment. Exponential advance in action.*

What's driving all this progress, as Kurzweil and others have recognized, is the logic of exponential growth. Populations of animals or plants tend to grow exponentially before running out of space or food, or reaching the carrying capacity of their environment; populations of ideas and inventions also tend to grow exponentially. But the ecosystem of ideas seems limitless—our ideas and inventions keep on multiplying wildly.

Exponential growth has been most discussed and best documented in technology areas, most notably computer hardware. The well known "Moore's Law" states, roughly speaking, that computers will double in speed every 18 months; this has held true for decades now. Ray Kurzweil has charted similar trends in other technology domains. My personal computer today is dramatically more powerful than the machine I owned in the 1990s. Heck, my phone now has massively more computing power than my 90s-era desktop computer ever had.

And of course, exponential technological growth did not begin with computers. Some technologies have been advancing exponentially for a long time. If you look at the grand sweep of human history, the amount of change from a million years ago to ten thousand years ago arguably was less than the amount of change from ten thousand years ago to today. In all probability, this pattern will keep on rolling. Exponential advances in multiple coupled technologies, leading to exponentially rapid changes in human life on a practical level, will have far-reaching effects on the states of mind we experience and the structure of language – which will then lead to yet further exponential change.

"*But,*" a skeptic might argue, "*not everything accelerates exponentially. Over my lifetime, a lot of important things like refrigerators, socks and cars have remained about the same now as they were when I was a kid. The way we fall in love hasn't changed that much. Toilets haven't changed that much. (Although they do have funky programmed toilets in Japan with computerized controls hard for a foreigner to figure out.)*"

Even if this skeptical argument were right, it wouldn't really matter very much. Once you have AGIs much smarter than human beings, you end up in a Singularity state, whether or not you have exponentially better socks. And if you still care about exponentially better socks after the Singularity, your AGI friends should be able to create them. What matters is that critical technologies – technologies capable of broadly and flexibly enabling other technologies -- are advancing exponentially.

But still, I don't want to give more credit to the progress skeptics than is due – actually, it's not so clear the skeptical argument, about the scattershot nature of exponential advancement, IS right after all. One can sensibly argue the point either way. We have, in fact, seen radical advancement even in the more prosaic aspects of practical technology. Cars, fridges and even socks have all changed considerably since the 60s. Fridges are more energy – efficient and reliable now, no longer incorporating harmful chemicals like CFCs. Cars utilize catalytic converters and are far better designed and safer (think airbags, the spread of antilock braking and new lightweight construction materials) than they were in my childhood. Socks are much cheaper (since they're woven by machines rather than by hand; and these machines are in places far from where I live, as enabled by better logistics technology) and their new, improved synthetics magic away sweat and maintain good air circulation. The Singularity is more dependent on advancement in AGI, computing hardware and neuroscience than socks and forks, but change and improvement are all around through parallels in cultural and technological processes.

Critical Technologies are Advancing Exponentially

The exponential advance of computer hardware draws attention since it's easy to measure. But software has also advanced tremendously during the same timeframe. A game like World of Warcraft is light years ahead of Pac-Man. The current versions of MATLAB or Mathematica, software used by scientists and engineers, are tremendously more powerful than the versions that were used in the 1990s, or even a decade ago.

Current algorithms (computerized problem-solving methods), if implemented on decades-old computers, proceed dozens or hundreds of times faster than the algorithms from the same period. Today's algorithms are more advanced than those from a few decades back, despite their core ideas being the same, because of their fine-tuned details. Having more powerful computers has been very helpful for tuning these details, closing the loop between hardware and algorithms.

As a programmer, I've been impacted by the changes in programming libraries and techniques over the last couple of decades. Two key parts of computer programs are data structures (storing information in various specialized ways) and algorithms (manipulating information in specialized ways, creating new information from old). I used to have to write my own data structures and algorithms; now there are standardized libraries of data structures and algorithms. The complex software that underlies a game like World of Warcraft or a package like MATLAB couldn't have been written two decades ago (and certainly not so quickly), even with the necessary hardware.

When one writes a program today, even though the algorithms and programming paradigms in use are roughly the same as two decades ago, the sophistication of the software libraries available makes the modern code more efficient and less prone to errors. For example, in the world of C++, the programming language used for the bulk of the OpenCogAGI project my colleagues and I, are working on, the STL and Boost code libraries wrap up incredibly complex algorithmic functionality.

I've already noted how recent years have seen tremendous advances in "Narrow AI" – artificial intelligence software that carries out specific difficult tasks effectively, although lacking the breadth and depth of the human mind. Google's self-driving cars and IBM's Jeopardy-playing Watson computer are examples; as are the AI programs in use in countless firms for practical purposes like predicting the stock market, planning supply chain logistics, detecting fraud, diagnosing diseases, and so forth.

Our understanding of the human brain is also advancing exponentially. Functional Magnetic Resonance Imaging (fMRI) and other brain imaging studies have told us about the specific functions of parts of the brain; and our models of neurons and neural networks in the brain are far more accurate and sophisticated than they used to be. We understand the general classes of algorithms and forms of knowledge representation that the brain is most likely to use, even though there are many details that are still being figured out. All this provides obvious help to those AGI designs that are based on emulating the human brain (which my current AGI designs are not). But even non-brain-based AGI approaches can find inspiration from the brain about how the mind works. Cognitive science – the interdisciplinary study of the mind – first came together as an integrated discipline in the 1980s. By now it qualifies as an impressive, coherent body of knowledge. We understand the mind a lot better now than we did a few decades ago – and I'll outline some of the understanding we've achieved a little later in this book.

Biology's rapid advancement has not been restricted to neuroscience, a fact that I've seen very directly in my own applied AI work. Since 2001, in parallel with my work toward AGI, I've been working on applications of narrow AI technology in a variety of areas, one being the analysis of genetic data. In this domain, I have noticed that the quality of data available and the variety of different biological experiments I can run have both increased dramatically in the last decade. Microarrays, measuring gene xexpression (the amount of biological activity associated with a particular kind of gene in a particular tissue of an organism at a particular point in time), were introduced in the mid-1990s, but were horribly error-prone, even in 2001. Now, they're relatively reliable.

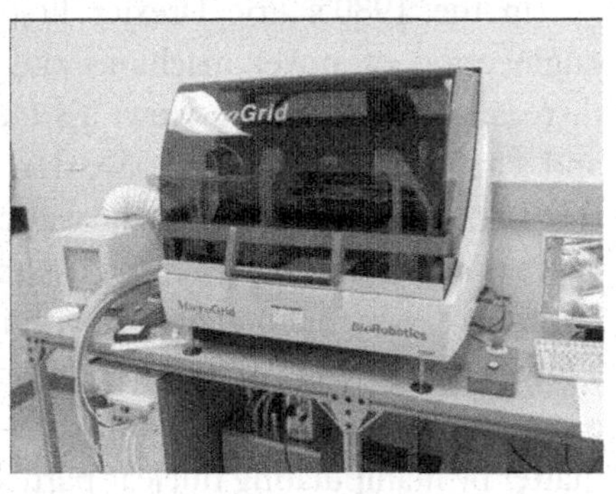

FIGURE 6.5: *This fairly unassuming-looking machine is a microarrayer, which measures the expression levels of all the genes in a genome, based on the sample placed inside it. I have done a lot of work using narrow AI (machine learning) tools to analyze the complex data produced by microarrays, to understand what the expression levels of various genes have to teach us about the underpinnings of aging and various diseases. I'll explain more about this later in the book! (http://ieg.ou.edu/equipment.htm)*

As another example – RNA interference, a biological progress via which RNA is used to inhibit the expression of particular genes, was discovered as a natural phenomenon in plants in the early 1990s, and first thoroughly understood in 1998. By now, it has long been commoditized and commonplace as a revolutionary tool for carrying out genetics experiments, and delivering gene therapies.

The cost of sequencing genomes is decreasing exponentially, which promotes the development of personalized genomic medicine, encourages the development of more new technologies like RNAi, and makes more experiments feasible (including some for understanding human longevity). And remember, the human genome was only sequenced in the mid-1990s. Conceptually, the degree of advancement in biological science since 1990 has been incalculable; and according to various reasonable quantifications it can easily be seen to fits the exponential-advance model.

FIGURE 6.6: *MedImmune, a bioprocessing production facility for the production of mammalian cell cultures. The practical machinery of modern biology has accumulated gradually over the last decades, and rates as one of the great achievements of our time. And yet, with all this sophistication, we have barely scratched the surface in terms of understanding the complexity of biological systems, especially as pertains to complex processes like aging. (http://www.manufacturingchemist.com/technical/article_page/Bioprocessing_designs_on_a_grand_scale/60705)*

In the 1980's Eric Drexler first wrote about nanotechnology – the engineering of novel machines out of molecules — yet most scientists disregarded it as fiction. Now, it's taught in universities around the world and a major area of research. We don't yet have the kind of nanotechnology that Drexler envisioned when he was coining the term and founding the field — we don't have nanomachines capable of building others. But we do have impressive, new nano-materials with rich and varied applications. All in just a few decades. Hugo de Garis and I have even been speculating about going a step further and building femtotech— creating computers and new kinds of matter by manipulating nuclear particles.

Quantum computing sounded like science fiction a couple decades ago. When David Deutsch first wrote about quantum computing in the 80s, it was just a bunch of equations, and its practicality remained unclear. Now, D-Wave, a commercial firm founded by Geordie Rose, is building real quantum computers for research purposes. (Though Geordie, with whom I've had a few good AGI discussions, has now stepped back from day-to-day involvement with D-Wave and is pursuing an AI and robotics opportunity.) In the next few decades, quantum computing may mature into a usable technology with wide practical applications. I know of two startups focused specifically on applying D-Wave's technology to practical problems, in the financial services domain and elsewhere.

FIGURE 6.7: *Eric Drexler published the first paper on nanotechnology in 1981, and his landmark book Nanosystems in 1992. Back then nanotech was just a glimmer in his eye. It sounded cool to me, but when I mentioned it to other scientists it was hard to get them*

to take the concept seriously. Now, nanotech is big business and universities around the world have nanotech institutes. Only a fraction of Drexler's pioneering concepts have been realized so far, but the rest are coming.
(http://www.amazon.com/Nanosystems-Molecular-Machinery-Manufacturing-Computation/dp/0471575186,
http://www.gtresearchnews.gatech.edu/nanotech-international-collaboration/)

Oh, and let's not overlook advances in embodiment, robotics, and virtual worlds, which drastically simplify the task of giving AGI systems something to do, once they've been designed and built.

The reasons why some technologies develop exponentially, while others stagnate, proceed linearly, or grow more slowly, are highly complex to tease out. But part of the answer clearly lies in feedback phenomena. Some technologies are obviously self-accelerating — the better they get, the more you can use them to produce the next generation of existing technology.

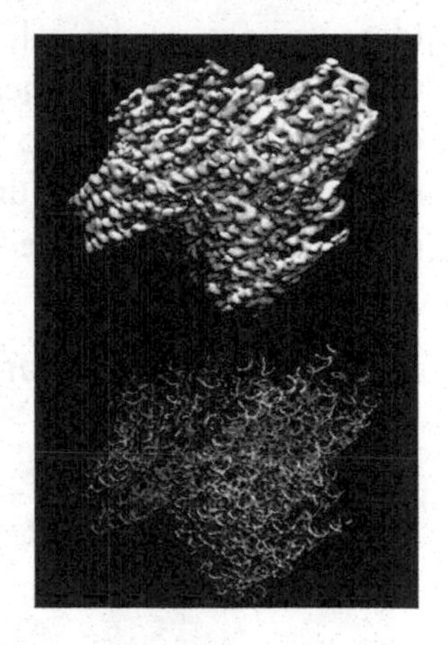

FIGURE 6.8: *DNA origami, a recently developed technique that allows one to build block-like structures out of DNA. In general, there is no physical reason why we can't manipulate microscopic structures much like we do with Lego blocks. The only obstacle in our way is the limited nature of our current engineering practice. Physicist Richard Feynman saw this in the 1950s with his essay "There's Plenty of Room at the Bottom". Eric Drexler fleshed the idea out in the 1980s, and today nanotech is a flourishing R&D field, albeit still in its infancy (https://en.wikipedia.org/wiki/DNA_origami#/media/File:DNA_Origami.png).*

Computer software and hardware both fall into this "self-accelerating" category. Better computers are used, among other purposes, to run complex computer-aided design software to build ever more powerful computer chips. New programming libraries are used to develop new computer programs, which then spur the development of newer programming libraries and languages.

Machine tools and factories are self-accelerating, too. Good tools can build yet better tools, which can build yet better tools, and so forth. Part of Drexler's original vision of nanotechnology expressed this idea: building small machines to build even smaller ones. Nanotechnology may eventually manifest this phenomenon.

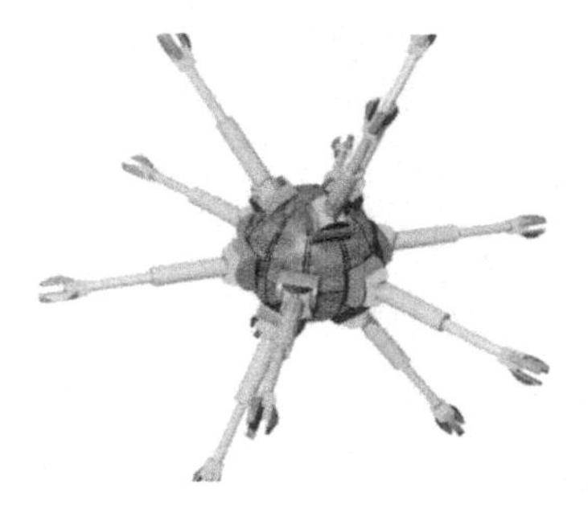

FIGURE. 6.9: *Utility fog, a term coined by AI researcher Josh Hall in 1993, refers to a collection of microscopic robots that can form complex structures. Imagine a bunch of little bots like this one, each invisibly small, communicating wirelessly and linking together to form a variety of gaseous, liquid and solid structures. One minute a couch, the next minute a robot, a curtain or a computer terminal. (http://library.thinkquest.org/07aug/02162/utility_fog.html)*

Ideally, human intelligence enhancement will follow the self-acceleration path. Making people smarter by modifying the brain, through drugs or other techniques, may yield greater brainpower to think up further improvements. Communication technology has developed in this fashion: Technology allows people to communicate and share ideas better than ever before. Sharing these ideas amongst groups will spur all sorts of amazing innovations, expanding on what's already out there. Positive reinforcement is an intrinsic feature of many technologies, so the better they get, the better their capacity for self-improvement; and humans are enablers, facilitating technology's own dynamic of progressive self-improvement.

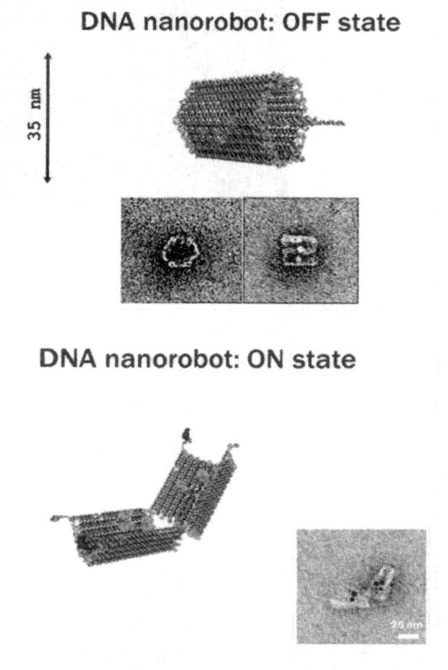

FIGURE 6.10: *While I was preparing a draft of this book, a news item appeared in my inbox, with headline "Ido Bachelet announces 2015 human trial of DNA nanobots to fight cancer and soon to repair spinal cords". Whoa. Bachelet's simple nanobots can't do atomically precise chemical manipulations (yet), but they can suppress tumor cells, activate cells with insulin and perform other useful biological operations. It really wasn't that long ago that Ralph Merkle's elegant drawings of medical nanobots seemed like elegant example of "scientifically rigorous science fiction". Now we're at the start of real-life medical nanobotics. What a shame that my friend Robert Bradbury, a SETI and life extension research visionary who also had a slightly-too-far-ahead-of-the-curve startup called Robobiotics a few years back, died young at 54 and did not get to see these advances. (http://nextbigfuture.com/2014/12/ido-bachelet-announces-2015-human-trial.html)*

AGI has the potential to be the ultimate self-accelerator. Once an AGI knows how to program software, it can program AGI programs, creating a new, smarter AGI, which will program a smarter AGI. This is of course exactly what I.J. Good called the "intelligence explosion" back in the 1960s.

Of course, exponential advancement from a historical perspective doesn't necessarily seem rapid from the point of view of daily human life. Sometimes it does, sometimes it doesn't. From my perspective as an AGI researcher, the pace of progress in AI can sometimes seem frustratingly slow. I would like to see much faster progress toward AGI systems with general intelligence at the human level and beyond. I think we're not progressing faster toward superhuman AGI because of largely psychological and financial reasons, rather than scientific ones.

But science fiction and the impatience of wild-eyed visionaries are not necessarily the best metrics to use to judge the progress of a science and engineering field. I get frustrated because I see how much faster things COULD be moving – but still, I have to admit that if one looks at the bigger picture, at the breadth of human history, the advancement of AI over the last half century starts to seem pretty damn impressive. Google Search and Google Now, Bing, Deep Blue, Siri, Nuance speech recognition, expert system medical diagnosticians, machine learning systems analyzing data from every area of science and industry, automated program traders, military drones, AI players in thousands of games... Nothing anything like these existed in the middle of the last century. Wow.

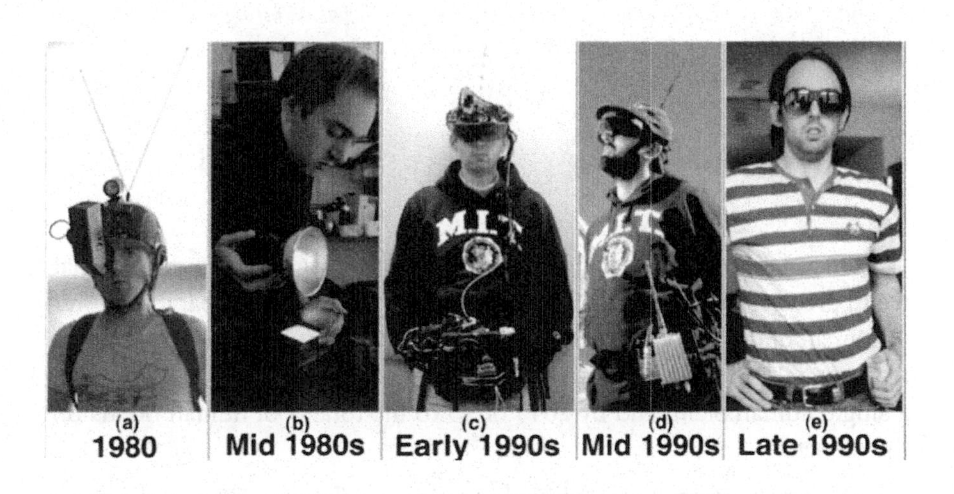

(a) 1980 (b) Mid 1980s (c) Early 1990s (d) Mid 1990s (e) Late 1990s

FIGURE 6.11: *Exponential advance of glasses-type wearable computers. Top Picture shows wearable-computing pioneer Steve Mann, with a variety of his devices over the years. As supporting technologies advanced and became smaller and more effective, his devices became more practical and less outlandish. (https://upload.wikimedia.org/wikipedia/commons/thumb/b/bb/Wearcompevolution2.jpg/550px-Wearcompevolution2.jpg). Bottom Picture shows me experimenting with Google Glass, Google's exploratory venture into wearable computing. Google Glass didn't succeed as well as hoped, perhaps in large part because of inadequate attention to aesthetic design (Google has always been more of an engineering company than a design company), which caused Google Glass wearers to get stigmatized as "glassholes". But Google Glass did put glasses-style wearable computing on the map, planting it firmly in the collective consciousness. There is little doubt further products of this nature will be released, whether by Google or by others, and will gradually pervade society. And just as Google Glass was much less obtrusive than Steve Mann's early inventions, future similar products will be much less obtrusive than Google Glass, eventually not even being visually noticeable at all. (Notice also Ruiting's dashing African hairstyle, done by Getnet Aseffa's girlfriend in Addis Ababa!)*

As Computers Approach or Exceed Human Brain Power…

One of Kurzweil's arguments as to why the Singularity is near is that we're pretty close to the point where computers will have comparable power to human brains. As he figures it, if Moore's Law and its relatives continue at least roughly, then by 2029 or so a computer will have comparable power to a single human brain; and by 2045 or so to the totality of human brains on the planet.

Of course, others have quibbled with his calculations. There is the small problem that nobody knows for sure, yet, how the human brain works – so calculations of the brain's "processing power" are going to be approximate at best. Also, comparing the information-processing capability of the human brain to that of a digital computer is a tricky matter. The two systems are very different in nature. I'll dig into these matters a bit in later chapters.

But, even with all the limitations, it's an interesting train of thought.

It's also worth remembering that, from a physics perspective, both computers and brains are pretty pathetic information processing systems. Neither of these systems utilizes more than a miniscule fraction of the computing power implicit in the particles that compose them according to the laws of physics as presently understood. So in the grand scheme, both digital computers and brains are very, very inefficient information processors – but the kinds of inefficiencies and efficiencies they manifest are different in various ways. A digital computer is very good at dividing large integers, and searching huge relational or graph databases (for example); a brain is very good at regulating hormone levels and identifying contours in visual fields (for example). Digital computers are better at running Google or Citibank's data networks than human brains would be; human brains are better customized for controlling human bodies than digital computers are. Neither digital computers nor brains are much good at solving the equations for quantum many-body problems – but some kind of future quantum computer might be.

Still, even given all this uncertainty and complexity, there are some sensible ways to compare digital computers to human brains. One can ask, for instance, whether a digital computer is capable of simulating a human brain's intelligence-relevant functions. Note that this doesn't necessarily imply simulation of every particle or every atom or molecular inside a human brain. Nearly all neurobiologists these days think that, in order to simulate a human brain's intelligent functions, it would be enough to simulate the ways its major types of cells work (its neurons mostly; and maybe its glia and a few other types of cells), and the molecules these cells pass between each other.

If we ask whether digital computers can simulate brains in this sense, the answer at the present time is no. But then Moore's Law rears its inexorable head. If Moore's Law and its relatives hold up, then in 15-25 years or so (if you believe the analysis of the brain that most neuroscientists accept, which I'll outline and discuss few chapters later), we WILL have digital computers that can match the computing power of the human brain.

Whether or not having "human brain level" computing power is critical for having human-level AGI is a different question. My own currently preferred AGI approach, as I'll describe a little later, takes the brain only as loose inspiration; and I suspect that my approach could be used to create human-level AGI with far less computing power than would be needed to simulate a human brain, or to achieve AGI with a closely brain-like architecture.

But still, whether one is doing brain simulation or OpenCog, there is little doubt that better and better computers are going to make AGI easier to achieve. So for those of us who are passionate about AGI, its good news that the computer hardware industry so reliably delivers better and better computers each year, and is showing no signs of slowing down anytime soon. As creating increasingly powerful computing cores for processor chips becomes more difficult, chip makers are shifting to a multi-core approach – with the end result of computers that keep getting faster. As the challenges keep coming, so do the innovations. There are enough economically valuable applications for better computing hardware, that the hardware companies have copious incentive to keep on innovating.

The Possibility of "Hard Takeoff"

The notion of AGIs programming better AGIs is a powerful one – and key to I.J. Good's 1960s prediction of an "intelligence explosion," predecessor to the modern notion of a technological Singularity. More recently the phrase "hard takeoff" has been used in this context. A "hard takeoff" means a Singularity that happens really, really fast. Maybe in a five minute interval!

In a hypothetical hard takeoff, an AI program becomes smarter so fast that it reaches superhuman status in a very short time span, long before its human creators have had time to consider the implications.

An extreme example — At 8 AM, it may be as smart as a dog. At 9 AM, as smart as a human. At 10 AM, comparable to Albert Einstein. And at 11 AM, it has attained some kind of digital godhood.

To my mind, a hard takeoff of this nature seems possible, but doesn't feel particularly likely. But even if things don't turn out quite this extreme, still, AI may advance from human to superhuman intelligence in a very short duration by our own historical standards. Going from dog to Einstein level in five years may not qualify as a hard takeoff, but it would still be rather dramatic to witness. I've referred to this as a "semi-hard takeoff".

The rapidity of the first AGI's cognitive development seems likely to depend on the overall technological context at the time this first AGI comes about.

If the first AGI capable of radical, useful self-reprogramming comes about in a world with very sophisticated supporting technologies, it might be able to accelerate its intelligence extremely rapidly. But if the first AGI capable of radical, useful self-reprogramming comes about in a world like the one we have today, I doubt its advancement to massively superhuman intelligence will take place in the blink of an eye.

Let's pursue a little thought experiment, for a moment. As I write these words now, it's early 2016and I'm sitting in my study in my house in Yuen Leng Tsuen, in the New Territories of Hong Kong. Let's say, hypothetically, that next week I have a huge breakthrough and figure out how to make the OpenCog software system as smart as a typical human college student, without going through as much work as previously envisioned. So, let's say that next week I create a human-level AGI – running on some Amazon cloud computing machines, launched from the ThinkPad running Ubuntu that I'm typing these words on. I just spin up a few thousand Amazon EC2 instances, launch some OpenCog software on them, and let the distributed network start thinking. Let's say this OpenCog network is smart enough to hold a basic English conversation, and even read a math book and write some simple computer code...

So, one natural question that pops to mind, at least if you're a wild-eyed AGI visionary, is: How could this hypothetical AGI become superhuman at a hard takeoff pace? Maybe it could rewrite its own source code to get smarter. But what if, it wanted better hardware, to benefit from its new, improved software? If it wanted to design new hardware to fabricate a new kind of chip, the AGI would have to order the relevant parts and materials from a manufacturer. There's an enormous amount of infrastructure to deal with, mostly involving the human world. There would be many chances for people to intervene in the AGI's activities.

Perhaps this hypothetical AGI could trick, convince or coerce others to do its bidding. However, its greater intelligence would not guarantee success. I'm much smarter than my dog, but that doesn't mean he always obeys my commands, even once he is able to understand them.

The point of the thought experiment is: Hard takeoff seems fairly unlikely given the present overall state of technology, even if someone creates a functional AGI tomorrow.

On the other hand, it's a lot more plausible in further future scenarios.

Consider another thought experiment. Imagine the world half a century from now. Other technologies have advanced a lot, but AGI still lags behind. Suppose it's 2060, and we still haven't invented AGI, but we have found success in nanotechnology, advanced robotics and ubiquitous Internet communications that connect every electronic device in the world. If you want a new appliance, you input the information online, or speak to it, and microphones embedded in the walls of your house pick it up. The details are transmitted to the correct recipient, and the appliance is fabricated in some nanofactory and dropped at your door by automated, unmanned drone helicopters. So long as the extant technology can cope with the manufacturing specs, you get what you want, quickly and conveniently.

Now, insert a human-level AGI into this scenario. That AGI would have a lot more options for applying its intellectual powers than one created in 2013 or 2017. The newly created 2060 AGI could reprogram its software code and order and receive new hardware from external sources. Using the far more powerful, globally interconnected Internet of things, it could summon drones to fulfill its wishes.

The more other technologies advance, the more scope they provide for a human-level AGI to accelerate itself into something superhuman. Because of this, in a sense, it seems the safest path for AGI development may be to develop AGIs as quickly as possible, bringing them into a world where we can exercise a modicum of control and guide their future development. The quicker we develop human-level AI, the slower, all else being equal, the transition from human-level to superhuman AI will be because those other technologies it needs will still be in their infancy, developing independently. Even if work on AGI were to come to a screeching halt tomorrow, progress on complementary technologies like networking, nanofabrication, and robotics, to name a few, would still be steaming ahead; and these would make it easier for an AGI created a few decades from now to launch a hard takeoff. So, if one is worried about the potential impacts of powerful AGI, delaying the creation of AGI is not necessarily the safest choice. It might be just the opposite.

My best guess of how things will unfold is somewhere in the middle, at what you might call a "semi-hard takeoff." I imagine that once we achieve toddler-level AGI, the money required for developing full-on adult-level AGI will come pouring in. AGI funding will phase-transition from trickle straight to firehose. Early-stage human-level AGI will lead to advances in other areas like nanotechnology, genetic engineering, human brain modification, and so forth.

Once these early AGIs get out there on global networks, people will seek a connection with them, creating a worldwide cyborg mind or nascent Global Brain. Out of this stew of experimentation, complementary development and novelty seeking, superhuman AGI will emerge.

The Psychological Singularity

It's natural to think about the hypothesized coming Singularity in terms of advancing technology and increasing scientific understanding – after all, the proximal cause of the impending Singularity is precisely the explosion of science and tech.

But yet, from a human point of view, the Singularity isn't just going to be just a technological event— it will be a psychological and social event, too. With fascinating implications for subjective experience, and what it feels like to be a mind, the Singularity will give us unprecedented insight into what it means to be human, and what it means to go beyond humanity.

The Singularity will wreak havoc with the various psychological illusions that characterize our inner worlds today, and replace them with new mental constructs that we can't currently conceive in any detail. The infusion of vastly greater intelligence into the world isn't just going to transform the gadgets at our disposal; it's going to transform the way we think, the way we are, inside our heads, moment by moment. I don't want to go too far in this direction at this stage in the book, as I want to tell you my thoughts about the actual creation of AGI and its practical implications in the world, before going on too much more about the amazing inner horizons that AGI is going to open up. But later on, toward the end of the book, I'll return to these themes, which are absolutely critical too. Yes, AGI is about creating machines that will act in the physical world and transform practical life. But, just as civilization and language changed the way humanity thinks and experiences, not just the way humanity acts – so will the advent of AGI, and likely even more so.

Human Plankton?

And oh, so much more so. The implications for human mind and life will be incredible – but even more so, the implications for mind and life beyond the human scale.

Revolutionizing every area of human pursuit surely won't be the end of it. That will just be Act One. Just as differential calculus and Shakespeare go beyond Crunchkin's, my dog's comprehension, the activities and methods of advanced AGIs will transcend our own.

Consider: Apes can't understand much of the human world, even though they're smarter than dogs, share 95% of our DNA, look a lot like us, and in the grand scheme, they're almost as smart as us.

Or, going in the other direction, swap the apes for cockroaches. How many of our interesting achievements can cockroaches appreciate?

Or, consider from the bacterial perspective. Bacteria are completely unaware of human beings. They may inhabit our bodies, but they don't understand this; they don't know that we speak, or act; and they don't understand anything we do, beyond responding to biochemical changes in their environment.

As AGIs progress, humans will pass through similar stages of incomprehension. We will be the apes, then the roaches, and finally the bacteria, lost in our trivial pursuits beneath vastly more intelligent beings operating on planes beyond our understanding. Russian AI pioneer Valentin Turchin, whom I was fortunate to know in New York in the late 1990s and early 2000s, liked to talk about "human plankton".

AGIs will open our minds to incredible new experiences. But the best of these experiences will go beyond anything at the human scale, and will involve human minds embracing broader, deeper ways of thinking and being.

The fact that AGIs will go way beyond humanity, doesn't mean they will exterminate all humanity, any more than we humans have killed all the ants and bacteria on the planet we live on. Rather, if things go well, humans will get the chance to upgrade their minds and become AGI-ish themselves; or to live human lives enhanced immeasurably by all sorts of AGI-invented technologies.

At this stage, one would have to be somehow blind not to see the prospect – and indeed, high probability – of amazing developments far beyond the human scale...

Further Reading

Broderick, Damien (2002). The Spike. St. Martins Press.

Drexler, Eric (1986). Engines of Creation. Anchor.

Drexler, Eric (1992). Nanosystems.Wiley.

Kurzweil, Ray (1999). The Age of Spiritual Machines. Penguin.

Kurzweil, Ray (2005). The Singularity Is Near.

Moravec, Hans (1988). Mind Children. Harvard University Press.

More, Max and Natasha Vita-More (2013). The Transhumanist Reader. Wiley-Blackwell.

Stross, Charles (2006). Accelerando. Ace.

Vinge, Vernor (2007). Rainbow's End. Tor.

7. AGI Will Transform Every Industry

The concept of Singularity can seem somewhat abstract and nebulous. It is, after all, almost by definition something we cannot understand. On the other hand, various aspects of the path leading up to the Singularity may be easier to grasp onto. For instance, it's worthwhile thinking about what kind of impact advanced AGI is likely to have on various industries, during the period before AGIs with billions of times human intelligence become the dominant factor on the planet. It's not hard to convince oneself that AGI has a clear potentiall to radically transform every industry.

My own knowledge of the practical applications of current AI and proto-AGI technologies is partly theoretical and partly derived from real-life experience. I have spent a fair bit of time in the last couple decades engaged with applying AI technology to various problems posed to me by various commercial and government customers.

AGI is the most important and exciting pursuit on the planet today – but it's still a relatively difficult way to earn a living. Since I wasn't clever enough to choose rich parents, and my various entrepreneurial pursuits haven't yet made me wealthy, and I opted to leave academia some years ago, I have found myself for most of my adult life in the position of having to spend a certain percentage of my time on non-AGI pursuits in order to "earn a living" -- to put food on the table and put my kids through college and so forth.

Lately AI has become much more popular, so that if one wants to work for a big tech company or a venture-funded startup, getting jobs in the AI field is not at all difficult like it used to be ... and there are even jobs working on projects that verge toward AGI, generally alongside some sort of narrow AI product or demo focus. But if you don't want to work for a big tech company or a venture capitalist, and don't want to deal with the politics and busywork that come with an academic career, it's still a bit tricky to figure out how to get paid for pushing toward AGI.

What I've done for the last 15 years or so of my career, alongside pursuing as much AGI R&D as I could find time for, is mainly to work on narrow AI for various corporate and government customers. As opposed to other possible ways of earning a living and supporting one's "free and open source AGI" habit, this has the obvious advantage that it's still AI – there has been all sorts of cross-pollination between my deep AGI research and my applied narrow AI work. I've applied AI in a huge variety of areas by now – robotics, gaming, national security, online education, accounting, natural language processing, market research, supply chain management, smart power, financial prediction and analysis, genetics, brain imaging, clinical medicine (and probably others that I'm forgetting at the moment). Some of my application projects have been excitingly successful, others have fallen flat. Every one of them has taught me something. At the moment I'm spending a lot of time on Hanson Robotics, which involves some non-AI robotics application work, but is generally reasonably well aligned with "what I'd like to be working on for OpenCog/PrimeAGI purposes anyway."

Doing application projects has certainly slowed down my progress toward AGI, but it has also been rewarding in various ways. It has brought me into contact with people, and ideas, from all different aspects of industry and government, which has massively broadened my horizons as a human being compared to what would have been the case if I'd spent my whole life in a computer lab programming AGI. It has also given me a strong intuition for how AI and AGI technology can be applied in various different application areas – both right now, and in the future as AI technology advances further and further toward AGI. After all, the fascination of AGI is twofold: the intrigue of puzzling out how mind works, and the practical value of having intelligent systems to do useful work for us, and ultimately to transform the physical world in ways we cannot do ourselves (and who knows, maybe go beyond the physical world as we know it!).

In this chapter I'm going to improvise a bit on the theme of AGI applications, and take a fairly wide and lightweight romp through the various potential future applications of AGI technology. What will AGI be able to do to transform the human world? What could we do much better, or differently, if we had advanced AGI technology to help us? In my applied AI work I've been faced with the problem of what I can do for this or that customer on a tight time-frame with readily available technologies. But what happens when the "available technologies" include advanced AGI systems?

When you start thinking about it a bit, it's hard to imagine a domain of human endeavor that an advanced AGI won't be able to radically up-end. What I'm going to do here is run through all the major industries on the planet today -- in alphabetical order, just for the heck of it -- and for each one, briefly reflect on how superintelligent machines are going to be able to do things way better. Of course the details of my speculations are not all that likely to be exactly how things pan out. My aim is just to give you a flavor of what sorts of possibilities may await. The ideas I give here just barely scratch the surface of the surface. If you obsolete my speculative thinking with more interesting speculations of your own, the chapter will have served its role!

Lacking any better idea, I've organized this chapter alphabetically by industry. AGI from A to Z! ...

Aerospace

We already have computer programs flying our commercial jets – the pilots are mostly just there to reassure the passengers, and to cover the off chance of some disaster that a human expert can better handle than a machine. But why are we still flying around in boring old airplanes? It is physically quite possible to make engines vastly more efficient than current jet engines, or flying machines that can take off like a helicopter, then fly fast like a jet. Or fly way faster than the sound barrier without the fuel-inefficiency of the Concorde. The first few AGI aerospace engineers may make our current designs look as archaic as the Wright Brothers' biplanes. Modern military aircraft already embody all sorts of amazing innovations, but most of these are not cost effective to incorporate in commercial aircraft. AGI engineering innovations could change this rapidly.

FIGURE 7.1: *"Project Zero": An all-electric tilt rotor aircraft from AgustaWestland, an Anglo-Italian helicopter company. Yes, this is a real airplane today, not a UFO! What will the aircraft and spacecraft designed by AGIs look like? (http://www.agustawestland.com/node/6902)*

This seems like a trivial point at first, but it points to something deeper. Humans study aerodynamics quite crudely by watching what things do in a wind tunnel, or solving equations in MATLAB or another computer program. An AGI could connect to sensors in the wind tunnel, the way we connect to our eyes and ears; and using MATLAB and other equation-solving software, find a solution for a nonlinear partial differential equation as immediately and automatically as we solve $1+1=2$. Potentially the AGI could then tweak the design of the plane dynamically in response to data observed in the wind tunnel, allowing a much faster and more precise perception-cognition-action loop than human aerospace designers could ever have.

FIGURE 7.2: *We test our airplanes by putting models of them in wind tunnels like this one. Experimentation with wind tunnels was key to the Wright Brothers' original aviation breakthroughs. But we can't modify the actual structure of the model plane based on feedback from what the plane is doing in the wind tunnel. What if model planes were made of dynamically-reconfigurable nanotech, that could be continuously adapted by AIs, based on feedback from within the wind tunnel? The AI could have a deep learning perception hierarchy specifically adapted for perceiving what happens inside a wind tunnel – no human brain has that. No human hands can reshape a model plane dynamically as it flies inside a wind tunnel either. In hindsight current designs for aircraft are going to seem really silly. (http://upload.wikimedia.org/wikipedia/commons/thumb/0/04/Windkanal.jpg/350px-Windkanal.jpg)*

The general lesson here is: AGIs will be able to benefit from their flexibility to attach to a broad variety of sensors and actuators. A human being can't put itself in the position of an airplane, submarine, spacecraft or medical nanobot. An AGI will be able to connect itself directly to sensors and actuators in all sorts of spatiotemporal regimes – thus enabling it to think about these regimes intuitively as well as analytically, and hence to come up with much more interesting and incisive designs and theories than humans in all sorts of different respects.

Agriculture, Food & Water

The creation of new and better forms of food is hard for us humans currently, given our limited capability to manipulate matter and understand its properties. But to an AGI more at home in the microscopic world of cells and chemicals – due to its ability to flexibly attach itself to various different sensors and actuators at various scales -- this may well be a piece of cake. Designing new forms of amazingly delicious and unsurpassedly nutritious food is "simply" a question of better understanding the human body and manipulating various plants and animals genetically. Even without molecular nanotechnology (enabling the repeated, inexpensive synthesis of an optimally tasty and nutritious hamburger), once AGI has decoded the mechanisms underlying biological organisms, the optimization of human food engineering should be a quite manageable application.

An AGI mind, with human-level general intelligence but a more science-friendly architecture, will rapidly surpass human understanding of biology. No human scientist can fit all the existing online biology databases into their mind, but an AGI could, allowing it to find scientifically meaningful patterns eluding human science. Connecting this AGI to robotic lab equipment will complete the cycle of experimentation, analysis and theory, allowing bioscience to progress with humans out of the loop.

Genetically modified plants and animals are important now -- but there's only so far we can go with our current understanding of genetics. If an AGI biologist figured out the connection between an animal or plant's genes and its ultimate structure or function, the genetic engineering possibilities would be endless. Further, the understanding of the human body that AGI will bring, will make it possible to understand the potential side-effects of any genetic modification to food, thus minimizing the risks associated with GMO food.

FIGURE 7.3: *Genetically modified food is proving one of the more controversial recent technological innovations – but is spreading rapidly through the world nonetheless. The risk factor in GMO foods comes essentially from our inability to understand all the possible side-effects of a given genetic modification. If we better understood the human body (which advanced AGI would enable), then any risks associated with GMO food would be minimized, as we could more thoroughly understand the effects of genetic modifications in food on the human body. http://monconstitutionalist.files.wordpress.com/2012/09/gmo.jpg*

And what about water? Well, if there were an inexpensive way to synthesize water from hydrogen and oxygen, or even to desalinate seawater, we would no longer have concerns over water supply. These are things we know how to do already; they're just expensive. A systematic effort by a team of AGI scientists with IQ even slightly above human level, and direct access to relevant lab equipment, could very likely resolve these problems.

Automobiles

During the last few years, the concept of autonomous, self-driving cars has transitioned from being widely considered science fictional, to being generally considered common-sensical. In fact, cars capable of fairly autonomous highway driving were around in the early 1990s, courtesy of Ernst Dickmanns and his colleagues, but these were not widely known. These days Google's experimental self-driving cars are legendary, and numerous big carmakers have begun their own self-driving car projects. Legislation is in place in various jurisdictions, aimed at enabling the legality of autonomous vehicles on the roads. Plenty of mainstream journalists will now opine in print that the days where human drivers dominate the road are numbered.

FIGURE 7.4: *Ernst Dickmanns and his colleagues made self-driving cars in the 1980s and early 1990s, capable of fairly robust highway driving based on integration of computer vision with adaptive control. In light of Google's recent work with self-driving cars (which like all work in the area builds heavily on Dickmanns' ideas), we were pleased and a bit amused to invite Dickmanns to give a keynote speech at the Artificial General Intelligence 2011 conference, which was held on Google's campus in Mountain view. (http://www.wired.com/autopia/2012/02/autonomous-vehicle-history/?pid=1580&pageid=42236&viewall=true)*

AGI seems unnecessary for the creation of perfectly effective self-driving cars. Cleverly integrated narrow AI components will probably do the trick. However, when unexpected situations arise on the road, an AGI will handle them better, thus saving a certain percentage of lives. The essence of AGI is the ability to handle the unpredicted and unforeseen in an intelligent way, based on creativity and on extrapolation from different but indirectly related prior experiences. If an animal of an unfamiliar kind runs across the road, or an unusual natural disaster makes road conditions bizarre, narrow AI may not deal with it effectively, and an appropriate AGI may well deal with it better than human drivers.

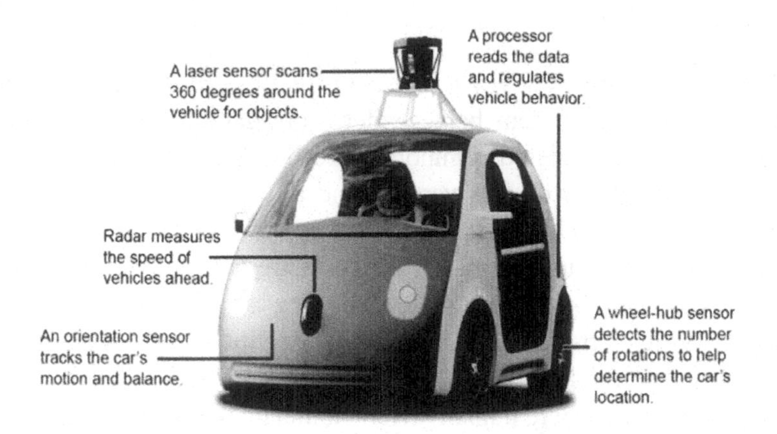

FIGURE 7.5: *Google's experimental self-driving cars have become a fixture on the streets of San Francisco. (http://www.trbimg.com/img-5386922b/turbine/la-sci-g-google-self-driving-car-20140528, http://addins.waow.com/blogs/weather/wp-content/uploads/2012/08/google-self-driving-car.jpg)*

Chemicals

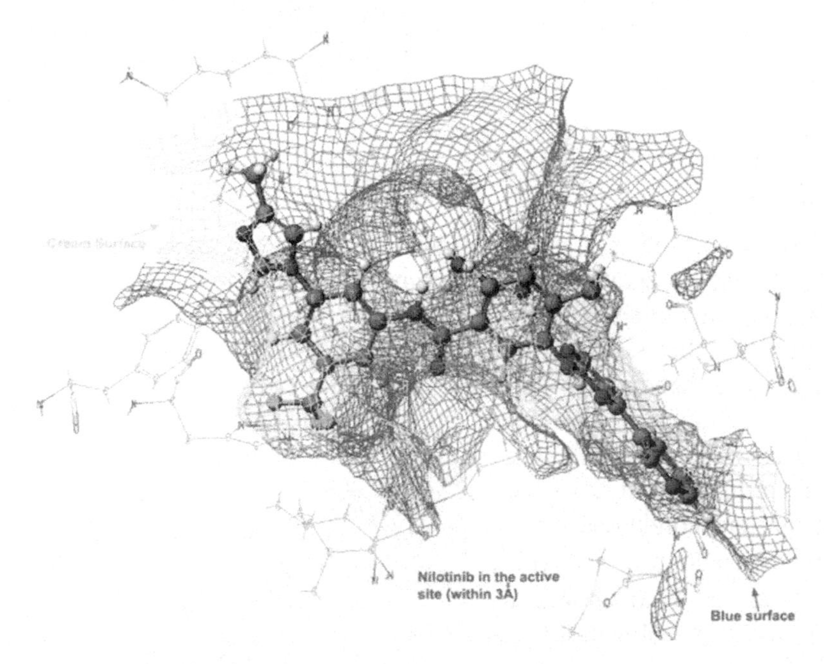

FIGURE 7.6: *"Rational drug design" uses chemistry and biology to figure out drugs that can attack given biological targets. So far most successful drugs have been designed through trial and error, but rational drug design is being actively pursued and is likely to*

yield dramatic fruit in the next couple decades. Advanced AGI would make rational drug design much easier, as the process involves many difficult modeling and analysis steps that are difficult for human judgment, even with available simulation and mathematical tools.

The figure shows the "adjacent surface" for the substance nilotinib; that is, it shows the adjacent surface pocket that is the surface within 3 angstroms of the drug to the active site of the target. This surface provides guidance in the process of drug development. But it is actually just a crude approximation, and better representations can be formulated taking closer account of the quantum mechanical foundations of chemistry. The reliance on simpler representations like the one depicted is partly because the human mind doesn't take that naturally to quantum mechanics. (http://www.sciencedirect.com/science/article/pii/S0014299909008784?np=y)

Designing molecules and chemical compounds to carry out specified purposes is, at this point, largely a matter of guesswork. Imagine an AGI that truly mastered chemistry, moving directly from a specification to a molecule or compound fulfilling that specification. It should be easier for an AGI to master chemistry than a human, because human sensory organs and actuators (fingers, feet) do not naturally operate at the molecular level. But an AGI could be given molecular-scale "eyes" and "hands" to let it interact with the chemical world directly. What we do with abstract analyses and complex awkward tools, such an AGI could do with direct sensations and tools operated with the same sort of intuition and fluidity we use in operating our fingers.

Molecular nanotechnology may obsolete traditional chemistry by letting us build molecules according to whatever structures we specify. But while we're on our way to achieving full molecular nanotechnology, AGI-powered chemistry may play a major role in making our lives easier and advancing science. Nanotech is already happening, and will get better and better step by step, eventually achieving the powerful ideas posited by early nanotech visionaries like Eric Drexler. AGIs should have large advantages over human in pushing nanotech forward and advancing our fine-grained control over matter.

Computers & Software

Of all the amazing capabilities future AGIs are likely to achieve, the most dramatic one may end up being AGI's ability to transform computer software and hardware. Digital computer software is an alien universe to humans, just as much as the molecular world. Its fanatical precision drives us crazy, as any programmer who has banged their head against code bugs realizes acutely. An AGI, if appropriately architected, will naturally be adept at using software code – it will read and program code as naturally as we see or hear, or pick up a stick in our hands. This will make a huge difference in the speed and quality of production of basically all kinds of software. And it will enable AGIs to program yet better AGIs, beginning the dynamics of "intelligence explosion" or "hard takeoff".

This explosion of software will go hand in hand with dramatic improvements in hardware.

Today, computers are built according to basically the same architecture that John von Neumann outlined in the 1950s. Why is this? It's not because of any lack of imagination on the part of computer architecture designers: the history of computing is littered with wonderful alternative architectures.

I remember programming the Connection Machines designed by Danny Hillis, back in the 1990s. The Connection Machine I worked with could do 64000 independent operations at once, much more than the standard 2, 4 or 8 cores in most modern computers. A modern graphics card can do a few hundred operations at the same time, but they all have to be the same operation, unlike the Connection Machine whose 64000 processors could all do different things at the same time.

More recently, and less ambitiously, Sony and IBM collaborated to bring the radical new "Cell" computing architecture to the PS3 gaming console. It uses only 8 processors, but with different properties and interacting in a different way than in ordinary computers or game consoles.

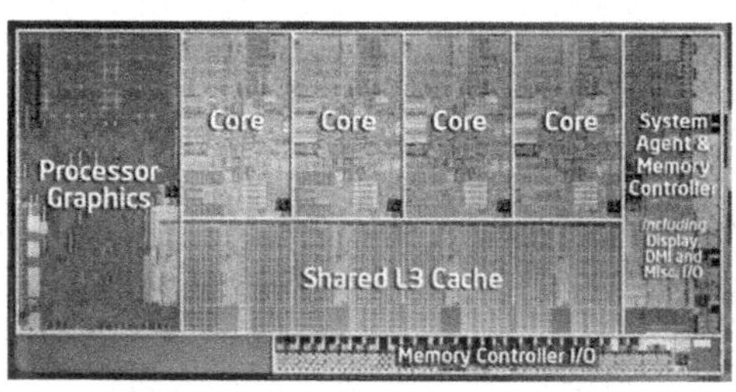

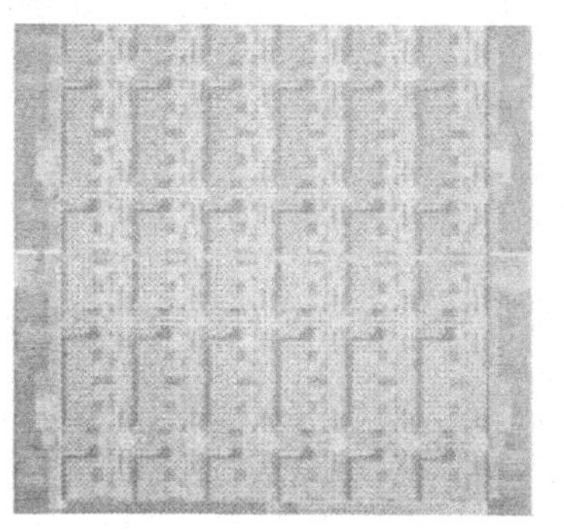

FIGURE 7.8: *Multicore hardware has allowed computer speed to keep advancing very rapidly in spite of the complexities of making compute cores smaller and smaller. 4 and 8 core machines are the norm now; and IBM, Intel and other firms are experimenting with architectures possessing 1000+ cores. The main difficulty presented by these massively multicore architectures is that programming them is very hard for the human brain. AGIs however, will not necessarily experience similar difficulties, and may rapidly make breakthroughs in multicore algorithm development. The picture is a standard quad-core architecture; The picture illustrates a 1000-core design presented by Intel at a conference in 2010. (http://*

regmedia.co.uk/2011/11/14/core_i7_sandy_bridge_die_large.jpg, http://i.neoseeker.com/n/3/processor2.jpg)

Why did the Connection Machine and the Cell flop, like so many other novel computing hardware approaches? Not because the hardware was bad, but rather because it was too hard for human beings to rapidly write software for the new hardware architectures.

The human brain struggles to adapt a familiar algorithm or data structure to a new kind of computing hardware; it takes us a lot of work. And we've built up a lot of detailed knowledge, regarding how to make software work on the familiar von Neumann computer architecture.

FIGURE 7.9: *Contemporary server farms, outside and in. (The picture is actually an Apple data center.) More and more, computing happens here rather than on desktop computers. Much of my own work on OpenCog software happens on rented computer time in the cloud. Cloud computing is yet another illustration that it's not the particular physical substrate that's important, it's the pattern of organization. The same computer code can shift from one computer to another in the cloud – no matter. What matters is what the program does. (http://www.digitalmusicinsider.com/wp-content/uploads/2011/02/AppleNCnews_24279.jpg, http://timmurphy.wpengine. netdna-cdn.com/wp-content/uploads/2011/05/Server-farm.jpg)*

Not only could an AGI potentially design new and far better computer hardware architectures, but with the capability to extend current software to alternative hardware architectures, it would revolutionize computing via allowing existing but obscure computer hardware designs to flourish.

And then there's the advent of quantum computers—using the weird physics of the microworld to do computing fundamentally faster. Quantum computers perform several calculations in multiple universes at once, thus arriving at answers on average much faster than our contemporary, classical physics based computers, whose computation is boringly concentrated in a single universe[1]. However, figuring out how to program quantum computers is not easy. Our brains struggle with the quantum domain; it is incredibly counterintuitive. In the quantum world, things like electrons can be particles and waves at the same time, but they cannot have both a known position and a known speed at the same time. Particle/waves can teleport through walls under certain conditions. An AGI with sensors directly at the quantum level would understand quantum phenomena intuitively in a way that the human brain cannot, enabling it to design quantum computing algorithms beyond our comprehension. This may be the only way to get general purpose quantum computers to work.

FIGURE 7.10: *the interior of one of Dwave's novel quantum computing devices. 2011 advertisement for DWave's quantum computing hardware. Dwave's system is billed as the first commercially available quantum computer (though not as a general purpose quantum computer; it only solves certain specific types of mathematical problems, which however have a fairly wide range of applicability). At time of writing, there is still some controversy as to whether DWave's machines are really doing quantum computing or just some novel sort of analog computing. (http://arstechnica.com/science/2013/07/d-waves-quantum-optimizer-might-be-quantum-after-all/; http://www.blogcdn.com/www.engadget.com/media/2011/05/5-18-2011d-waveone.jpg)*

Construction

We take for granted that building buildings is slow and difficult. But actually – why can't a new skyscraper be erected in a few minutes or hours, rather than taking months, or even years? And why can't all the units in

---

1 Check out Seth Lloyd's book Programming the Universe for a wonderful nontechnical review of quantum computing. Or Fred Alan Wolf's oldie-but-goodie Taking the Quantum Leap for a general review of quantum weirdness.

a housing development project be constructed at once, applying individual variations on the same basic plan? Lack of resources perhaps; yet more importantly, the human body and the work involved in constructing large buildings are clearly not the best match.

FIGURE 7.11: *Will AGI architects design us fantastic, futuristic buildings and cities? Or will they take things in another direction – perhaps creating reconfigurable houses that reassemble themselves in different shapes adaptively, based on our needs and desires? In any case it seems unlikely that AGI architects and robot construction workers would feel restricted to the conventional habits of contemporary human building design – which basically exist due to cost issues, as well as lack of imagination by the people paying the bills for most construction. (http://www.worldarchitecturenews.com/index.php?fuseaction=wanappln.projectview&upload_id=617; http://inhabitat.com/city-in-the-sky-futuristic-flower-towers-soar-above-modern-metropolises/megatropolis-city-in-the-sky-hrama/)*

Construction will be revolutionized once there are robots that can handle visual perception and physical object manipulation at the human level. And it won't be a big step to deploy the same capabilities in non-human robot bodies better suited for construction work. Construction robots won't resemble humans; they'll be a combination of intelligent construction trucks and flying drones, featuring several mechanical arms and claws.

FIGURE 7.12: *Conceptual illustration of a seastead – a city constructed in the middle of the ocean, in international water where no current nation's laws apply. Seasteads could allow much freer experimentation with different forms of society and government. The main obstacle preventing widespread construction of seasteads at the moment is the cost of constructing things at sea – a problem that AGI construction robots could presumably solve. Although it would be more fun*

if we didn't have to wait, and could have seasteads pronto, and put our AGI research facilities and robot factories thereupon! (http://www.marineinsight.com/wp-content/uploads/2012/08/sea-steading-1.jpg)

Defense & Intelligence

FIGURE 7.13: *Boston Dynamics' famous Big Dog robot, created for US military purposes, with an unprecedented capability for stable locomotion across unstable, irregular ground. Upright humanoid locomotion on outdoor surfaces remains difficult for robotics today, but quadruped locomotion is significantly easier. As George Orwell said "Four legs good, two legs bad"! Boston Dynamics was acquired by Google in 2014 and will likely be getting out of the military business, but we can be sure other companies will take its place and keep the same sort of work going. (http://digital-artgallery.com/oid/15/1000x707_4442_BigDog_LS3_2d_sci_fi_robot_picture_image_digital_art.jpg)*

Historically, the US military has funded the majority of the world's AI research, so once AGI technology matures it will likely be used in a military context. Early-stage AGI, however, may rapidly amass the power to prevent this from happening.

In my opinion, military forces will probably not be the initial source of AGI advances, since they will require AI to be highly predictable and reliable; early-stage AGI is unlikely to possess these qualities. If we follow a path to AGI inspired by human child development, early-stage AGIs may have the same playfulness, unreliability and confusion of young human children. Since an AGI has inherent generality, no development path, even one that tries to bypass the childlike stage, will be free of this unpredictability. Understanding the parameters and properties of the first AGI system's unpredictability will take time, so the first military robots to be rolled out will likely be narrow AIs erring on the side of predictability.

Military AGIs will not only be robots. They will be military commanders, orchestrating robotic and human warriors according to strategic and tactical patterns too subtle for human minds to conceive or comprehend.

Military robots will enhance the tools of mass destruction; delivery of nuclear weapons will become more reliable and difficult to defend. However, a greater concern may be the current military trend (predominantly in the US) of carefully targeted destruction. Imagine a future military force—AGI military robots operating by air, land, and sea—able to send a crack team of robot killers to any location to carry out surgical military strikes.

AGI's impact on the intelligence world will be equally dramatic. Intelligence agencies in the US, China and various other nations gather and stockpile data about the citizens of the world—emails, text messages, phone calls, video surveillance images, you name it. But it's simply too much information for any reasonably sized group of humans to look through. Today's AIs are not smart enough to read texts, nor recognize objects or faces in pictures, except at a simple level; however, human-level AGI, when put together with already-existing surveillance technology, basically yields Orwell's Big Brother.

FIGURE 7.14: *Surveillance cameras, satellites, audio recording devices and a host of other instruments allow gathering of massive amounts of data about everyone's everyday life. Not to mention the amount of data that intelligence agencies and others gather regarding our online lives. But right now, most of the data that's gathered just sits there, because nobody has time to sift through it all and find the relevant portions. AGI systems with the capability to scan through and interpret texts, video and audio files much faster than human beings, would be able to make use of the massive amount of data gathered via modern surveillance technology, as well as develop yet more powerful means of observation and recording. AGI will therefore make acute the dilemma posted by David Brin in his book The Transparent Society, where he notes that as the future unfolds we face a choice between surveillance (the Powers That Be watch everyone) versus sousveillance (everyone has access to data regarding everyone). (http://clatl.com/atlanta/atlanta-under-surveillance/Content?oid=7121394)*

David Brin, science fiction writer and futurist, argues that the only solution is to subvert surveillance with sousveillance: observation from under rather than over (everyone can watch everyone). If all the data from everyone's emails, phone calls, texts and webcams and surveillance cameras were public, everyone could watch each other; we could watch the government as it watches us. Transforming the notion of privacy radically, it would prevent a scenario in which the few have exclusive capability to watch the many.

This brings us to the topic of the risks and rewards of AGI, and ideas about safeguarding against dangerous scenarios – a large, deep and thorny topic that I will tackle in a later chapter. But for now, I just want to make the point that AGI is not the only advanced technology with its scary aspects. All the other advanced, Singularity-enabling technologies pose similar or greater risks. Synthetic biology, for example, seems particularly worrying. Once synthetic biology advances, what will stop hostile-minded scientists from developing novel biological viruses and germs aimed at circumventing the human immune system and killing on the mass scale? It's possible that the best protection for the human race will be some sort of "AI Nanny" that protects us (and itself) from the dangers posed by emotionally unhealthy humans armed with powerful technologies.

Energy

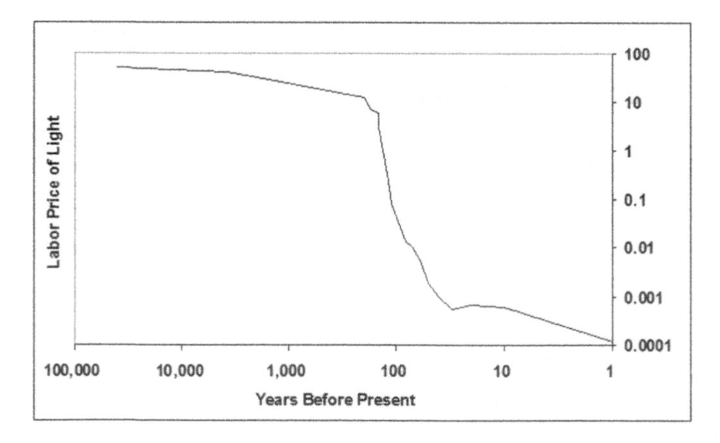

FIGURE 7.15: *The price of light as estimated by futurist Jose Cordeiro, as part of his analysis of the Singularity as an Energularity. (http://lifeboat.com/ex/the.energularity)*

Venezuelan futurist Josè Cordeiro underscores that the Singularity will also be an "Energularity". Humans have used more and more energy, per capita, per day, since the start of civilization—and this is increasing at an exponential rate. AGIs are likely to master forms of energy generation that have eluded human intelligence — nuclear fusion (first traditional hot fusion, then cold fusion). Increasing evidence indicates cold fusion is a valid scientific phenomenon, not the fraud suggested by the publicized errors in Pons and Fleischmanns' early work. AGIs will also be able to operate in space more easily than humans, so will have little trouble erecting solar power satellites or massive Mylar solar panels floating in space. Using the massive new energy sources they create, AGI's will power their massive processor farms, boost their intelligence, and create even more effective energy sources.

Entertainment & Arts

The line between the human and automated is already blurring in the world of arts & entertainment. Computer graphic scenes and human-actor scenes are often indistinguishable from each other in movies; the same applies to algorithmic versus human-played rhythm and melody in pop music. Non-player characters in video games do nearly all the things human characters do. And all this is without the advent of AGI – without even much narrow AI in use, actually.

On the other hand, computers can't create (yet) emotionally realistic facial expressions, nor evocative melodies and rhythms with rich human emotion. They also can't write stories, as these are largely experiential. It's just my own speculation, but I suspect these shortcomings of computers may be among the last for AGI to overcome. Because they're not just about being intelligent; they're specifically about being "human" and understanding the emotional core of humanity.

FIGURE 7.16: *Screenshot from the video game "Creatures Online", a descendant of the original Creatures game created by AGI researcher Steve Grand. Each creature is controlled by a neural network, which develops over the creature's life time. When creatures mate, their children get a neural network based on crossover and mutation from their parents' neural networks. Steve Grand spoke at the AGI Workshop I organized in Maryland in 2006, describing the robotics work he was then doing, focused largely on solving the computer vision problem.*

(http://creatures-online.co.uk/goodies/screenshots/gamescom-2011-demo-the-whole-family-is-together)

Without embodiment, where the AGI is placed inside a robot body, itmay be difficult for an AGI to have experiences that are sufficiently human-like to grasp the essence of human experience. An AGI lacking a human-like embodiment could write stories about its 24 hour day and patterns it sees via cameras and microphones. But, it wouldn't be fiction written from a human-like perspective. However, even a humanoid robot body, one lacking human touches like skin, hormones and sex organs, might not give an AGI sufficient understanding of the human condition to make human artworks.\

FIGURE 7.17: *Creatures game creator and AGI researcher Steve Grand with his robot Lucy, in 2006. (http://www.avfestival.co.uk/programme/2006/events-exhibitions/café-scientifique-steve-grand)*

I suspect there are aspects of artistic creation that human – level AGIs will be unable to master, unless they are built with human-like bodies and minds based on human brain emulation. On the other hand, a superhuman AGI may be able emulate human artistic creation. But it may require greater general intelligence to emulate human art in all respects than superseding human mathematics, engineering and science.

That said though, AGIs might well be able take over the vast majority of human entertainment and fine arts, once they've reached a point well below human level general intelligence. Recreating human productions may not be the best way to entertain most humans; the success of computer generated image-heavy movies and algorithmic pop music rhythms proves otherwise.

Ultimately, human-level AGI systems, after studying what entertains and pleases humans, will produce works that deliver what we want, even before AGIs create "human" art works. "Authentically human" entertainment and artwork may become a niche taste, similar to oil paintings and live jazz improvisation today, while the vast majority of entertainment and artwork will be explicitly AGI-generated. Celebrities will no longer have to appear physically in advertisements or go to live photo shoots—after licensing the use of their image, AI will do the rest.

Finance & Insurance

The idea of AGIs pervading the finance world may seem frightening – isn't program trading already causing a lot of problems? But actually, it's arguable that these problems are caused by the use of insufficiently intelligent programs – and the combination of such programs with insufficiently intelligent, or insufficiently ethical, humans. Once we have reasonably generally intelligent AIs serving as Chief Investment Officers for hedge funds, we might have much safer and more efficient financial markets.

FIGURE 7.18: *Back in 1980, the New York stock exchange floor was bustling with human beings trading stocks with each other. Nowadays, many stock exchanges have eliminated physical floor trading altogether, as nearly all trading is electronic. As of 2013, the New York Stock Exchange still has a trading floor, but it's very lightly used compared to the past. (http://commons.wikimedia.org/wiki/File:Stockexchange.jpg)*

Finance these days is largely based on advanced mathematics, but the mathematical formulas used are all approximations to reality, based on assumptions that everyone knows aren't quite right. Everyone makes these mathematical assumptions anyway because the equations can't be solved any other way. This creates problems— the financial market crash in 2008, the collapse of Long-Term Capital Management in 1998, and so forth.

An AGI could apply financial mathematics more sophisticatedly, taking better account of the available data, and making fewer unrealistic assumptions. Human-level AGI would likely "clean up" in the financial markets. If a human-level AGI, with direct mental connection to basic algorithmic software tools, were unleashed on the financial markets today, it could amass hundreds of billions of dollars within weeks. Imagine releasing George Soros into an active, volatile stock market populated entirely by people with IQs of 70 or lower.

More likely, the level of AGI used by financial firms will escalate gradually with multiple competing firms having roughly the same level of AGI at each point in time. Unaided human beings will then have less and less chance of success on the financial markets. Now, AI and statistical software are better at recognizing purely numerical trends in stock market data and at figuring out fair prices for complex financial constructs like options and derivatives. Yet people are better at other aspects of finance, like incorporating information from the news, and qualitative information about companies' products and management. However, once an AGI reads and understands the newspaper and a company's annual report, this human advantage will be eliminated. Additionally, the AGI would not suffer from the main enemy of every human trader or financial analyst: human emotion. A financial AGI could be programmed as almost a pure rationalist— free from the particular emotional biases that ultimately lead many human beings to make bad financial decisions.

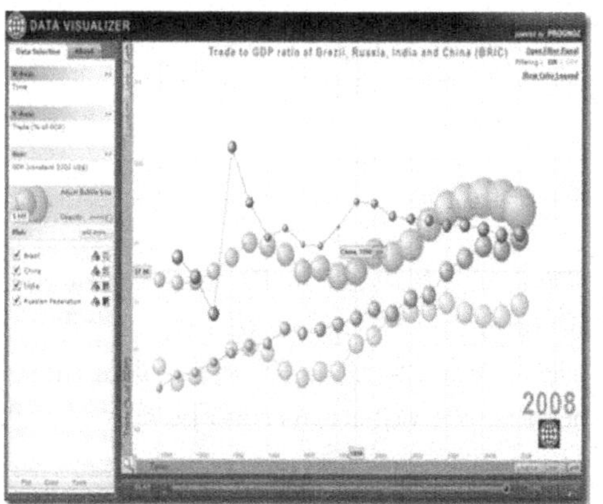

FIGURE 7.19: *Back in 1980, the New York stock exchange floor was bustling with human beings trading stocks with each other. Nowadays, many stock exchanges have eliminated physical floor trading altogether, as nearly all trading is electronic. As of 2013, the New York Stock Exchange still has a trading floor, but it's very lightly used compared to the past. Today's financial markets are extremely complex, and embody a kind of complexity that the human brain is not well suited to understand. We try very hard to project the complexity of financial systems into 2D visual forms that the human eye can comprehend, but this inevitably leaves out most of*

the relevant structure in the data. This is why the vast majority of financial trading in the US, Europe and Japan is already done by computer programs, including plenty of narrow AI programs. 20 years from now the percentage of trading on major markets conducted by humans is likely to be minimal, even without anything approaching a Singularity. (http://www.trinity.edu/rjensen/352wpvisual/BricsVisualization.jpg; http://m1.smartmoney.net/marketradar/images/radar5.gif)

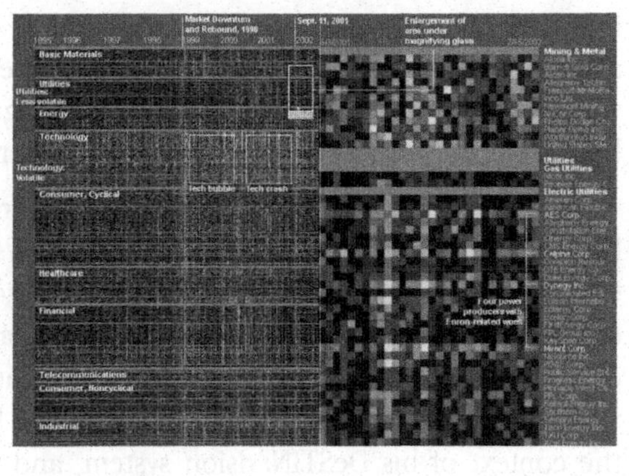

And just like finance, the insurance industry will also be revolutionized by AGI. Current insurance pricing models are crude and generally based on dividing people and companies into categories. AGI insurance assessors will calculate risks on a rational and individualized basis, creating a more efficient economy and easing the development of new technologies.

The application of AI to finance is something I've become very familiar with in recent years, due to joining with a number of colleagues to form a new financial analysis, Aidyia Limited, focused on predicting the prices of financial instruments using a variety of AI tools. At the moment Aidyia doesn't have human-level AGI technology, and does its predictions using a proprietary combination of various narrow-AI and proto-AGI software systems. But my practical work with Aidyia has made me acutely aware of the great potential that AGI systems will have to transform the financial markets, as they gradually emerge from today's proto-AGI software.

Algorithmic trading is best known today in the context of HFT, High-Frequency Trading – trading systems that buy and sell multiple times per second. It's obvious that, if there are market inefficiencies to be exploited at this time scale, algorithms of some sort will do better at it than people – we just can't perceive and react that fast. Aidyia, on the other hand, is focused on a longer time scale – predicting what will happen to the prices of stocks, futures or other financial instruments days, weeks or months in the future. This is a more difficult time-scale for simplistic prediction algorithms to deal with, because what a stock will do a month from now depends not only on that stock's price curve but also on a host of additional related features – other stocks, economic trends, the mood of various communities as reflected in the news and online, and so forth. Aidyia's technology, incorporating aspects of OpenCog alongside proprietary methods, incorporates a wide variety of information into its predictions, using its intelligence to figure out how multiple data items combine to yield patterns with predictive value.

Our Aidyia software can recognize patterns in the markets that no human can see, and no other existing AI system can see either. Other quantitative investment firms have their own AI software with its own peculiar characteristics, recognizing patterns that are opaque to human beings and to everybody else's AI software as well. I had an interesting discussion recently with my AGI researcher friend Itamar Arel, whom I mentioned above in the context of his DeSTIN vision system, and who is involved with using proto-AGI software within a hedge fund. At time of writing our different financial prediction systems are not in use on the same markets. But we were wondering whether, if our respective AI methods were both used to drive trading of systems in the same market (say, US equities), our approaches would dilute each others' effectiveness. After looking in more detail at the high level properties of each others' trading systems, we concluded this would not be the case. We agreed that, even if both of our firms were trading on the same markets, they wouldn't necessarily interfere with each other -- because our differing narrow-AI and proto-AGI technologies would likely see different patterns in the market data, and hence, make different trades. Just as different human traders will tend to see different opportunities based on their different backgrounds and insights, the same with different AI prediction systems.

But what if an AGI system with human-level general intelligence – combined with specialized quantitative market savvy in the fashion of current AI trading systems -- were doing the trading? Surely it would outclass the current batch of early-stage AI trading systems, just as surely as these systems outclass "basic" statistical market prediction algorithms. And naturally, with my Aidyia hat on, as soon as more advanced, more AGI-ish technology becomes available, I will be interested in bringing it into Aidyia and customizing it for use in our financial prediction framework. Or maybe the AGI breakthrough will occur in the context of, say, customizing OpenCog for Aidyia's purposes. One way or another, it seems clear that a couple decades from now, the financial markets will be dominated by advanced narrow-AI and/or proto-AGI or full-on AGI systems. The era of humans directly making financial trading decisions is going to be past.

Hospitality

AGI program traders are all very well, but they're not very photogenic. They definitely don't need humanoid bodies; all their job requires is interfacing with online information sources and electronic financial exchanges. Aesthetically speaking, it's more entertaining to think about robot waiters, masseuses, house cleaners, sales clerks, prostitutes -- you name it. AGI systems could have their motivational systems specifically engineered to guarantee they enjoy these jobs, providing us with an amazing level of hospitality.

While some, for emotional reasons, might prefer being served by human beings, most of us will appreciate having most of our basic everyday needs taken care of by AGIs. I don't personally find sex robots that appealing – I like the emotional aspect of interacting with human women in this sort of way… but I don't get a similar emotional charge from having my floor mopped by a real human instead of a robot, or dealing with a real human instead of a robot at the airline ticket counter.

FIGURE 7.20: *No book about the future would be complete without a picture of a sex robot! Of course this is far from the only potential use of robots in the hospitality business. But given the realities of human nature, it's likely to be one lucrative application, once robotics technology advances a bit. (http://beyondthebustledigital.tumblr.com/post/27779524216/sex-robots-x-japanese-dolls)*

Economically, it seems that service jobs may be the last major sector in which humans will be largely eliminated by robots. In manufacturing, the factory environment can be customized to work with the limitations of robots at their current state of development; but service jobs take place in parts of the world that are created for human convenience. So a good general-purpose service robot is going to have to be rather good at sensory perception and motor control, much more so than current robots. Once these "lower level" problems are solved, though, AI service robots are going to sweep through the world, wreaking economic havoc and rapidly transforming every service industry. It's an open question whether narrow AI can do the trick or whether AGI is needed to make robotic sales clerks, plumbers and house cleaners – but I am betting it will take some serious AGI. The diversity of situations that has to be dealt with is just too great for a specialized AI system, is my guess.

Manufacturing

I almost deleted this one from my A-Z list of industry sectors on the grounds of it being too obvious. Manufacturing is already becoming robotized. By this point it's probably obvious to nearly everybody on Earth that, as robots advance, they will gradually take over all factory jobs.

Currently, robots are restricted by their inability to see and manipulate most everyday objects as effectively as humans. But this gets better each year. While human capability for this remains roughly constant, a robot's capability increases exponentially.

Tomorrow's AGI factories will resemble little of today's human-powered factories. Each factory will be the body of an AGI mind, operating in close "digital telepathic" communication with a host of other AGI minds operating other factories in related industries. The supply chain will be managed via social interaction of the factory minds. New innovations will be incorporated rapidly into the manufacturing process, making specialized one-off manufacturing jobs commonplace.

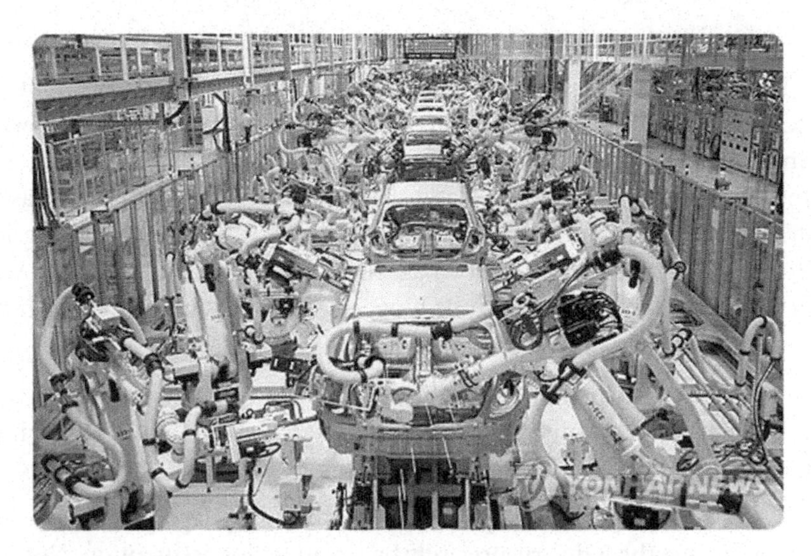

FIGURE 7.21: *A robotic car factory in Korea. Robotic manufacturing is already possible with a fairly high degree of generality, and is limited at this point mainly by cost. As robots become less expensive, there will be less and less role for humans in the manufacturing process. If everyone in the world were making a First World income, manufacturing would already be much more thoroughly robotized, because in many cases robots are already cheaper than First World human laborers, even if not yet cheaper than the least expensive human laborers on the planet.*

AGI will make robot manufacturing much easier to set up, because a single type of AGI robot will be able to carry out a variety of different types of manufacturing, in contrast to the current situation where one must carefully engineer and program robots especially for each manufacturing situation. (http://www.advancedtechnologykorea.com/9013)

AGI-powered manufacturing will become more powerful as nanotechnology develops, likely enabling one of the paths to advanced nanotechnology that Eric Drexler envisioned: small machines that build yet smaller machines (that build yet smaller machines…). Perhaps femtotechnology — using elementary particles like protons and electrons to manufacture materials — may follow.

But, what role will humans play when manufacturing becomes AGI-powered? Although production will function without us, we'll still (at least in the beginning) serve an active role in design — being customers should intuitively guide us.

News media

The exponential rise of computing and communication technology since the mid-1990s has revolutionized the media business, as anyone with an Internet connection has already noticed. News media in particular is in a state of economic crisis as well as radical advancement, for the simple reason that hardly anybody wants to pay for news when so much is available for free. Traditional newspapers and magazines are losing money in droves; their audiences have been stolen by bloggers, tweeters and so forth, whose day jobs are not primarily in news production.

An unspoken assumption at the heart of the news media business is that a lot of effort is necessarily required to translate ideas and observations into comprehensible series of words.

AGI systems capable of sophisticated natural language generation will radically change the industry, precisely by violating this assumption. Not too long from now, nearly all news reportage will be automated – with software generating text articles based on raw data, and other software catering the selection of articles to each reader.

There already exist systems that automatically generate news articles based on structured data regarding sporting events, stock price moves, and so forth. And systems that customize news feeds to the individual based on their reading history are already commonplace. Once news generation and customization software becomes more prevalent, human commentary on news may still exist to some extent, but it will serve purely a social and artistic role— similar to people casually chatting with each other about the news, or talk shows focused on pundits sharing their subjective opinions.

Pharmaceuticals & Health Care

The pharmaceutical and health care industries, like the publishing business, are struggling to adapt their practices and business models to the exponential advancement that has already happened during the last few decades. The next few decades are bound to disrupt them further and further.

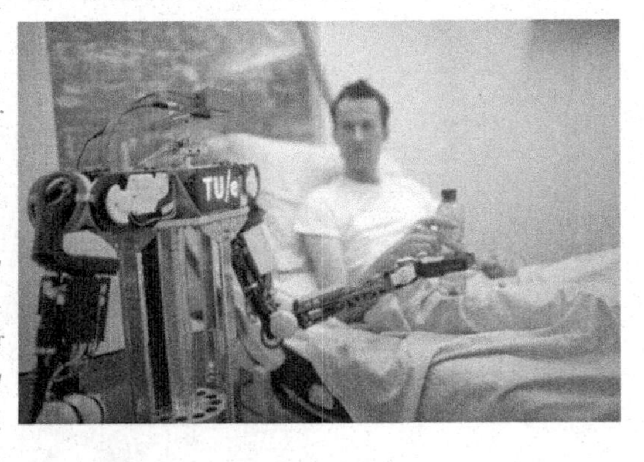

FIGURE 7.22: *This robot, serving a drink to a patient in a hostpital, runs on RoboEarth, an online open-source network database of robot control information. At the moment this sort of thing is at the level of research prototypes, but there are no fundamental technical obstacles toward rolling it out broadly. Making this sort of robotics application widespread just requires cost reductions in various aspects of the technology; it doesn't even need AGI, though AGI could obviously broaden the scope a great deal. (http://www.thestar.com/business/tech_news/2011/02/11/robots_get_their_own_internet.html)*

I have had a window into the issues these industries face, via my own involvement with the use of AI to aid in drug discovery (work I'll talk about a bit more in a later chapter). One thing I have learned is that today's drug discovery process – as practiced in pharmaceutical firms – is mostly trial and error. Genomics, and other advanced biological knowledge from the last few decades, are rarely applied effectively. Part of the issue here is sociological – there does exist knowledge about how to do drug discovery better using modern ideas, but this knoweldge is taking a long time to pervade the major pharmaceutical firms. I will say a bit about this later in the book, when I discuss the application of AGI and narrow AI to longevity research. But part of the issue is more fundamental. In trying to discover drugs for curing complex diseases, the human mind is running up against the basic complexity of the human body, which is just plain hard for the human mind to understand, even when there is a lot of relevant data available.

It seems possible that the limitations of the human brain will prevent human beings from ever fully comprehending dynamical systems as complex as the human body. No matter how much we augment our intelligence with data analysis and visualization software, the human mind won't intuitively grasp the detailed interactions of 25,000 genes, the proteins they code for, and the way they combine to build and maintain the human body.

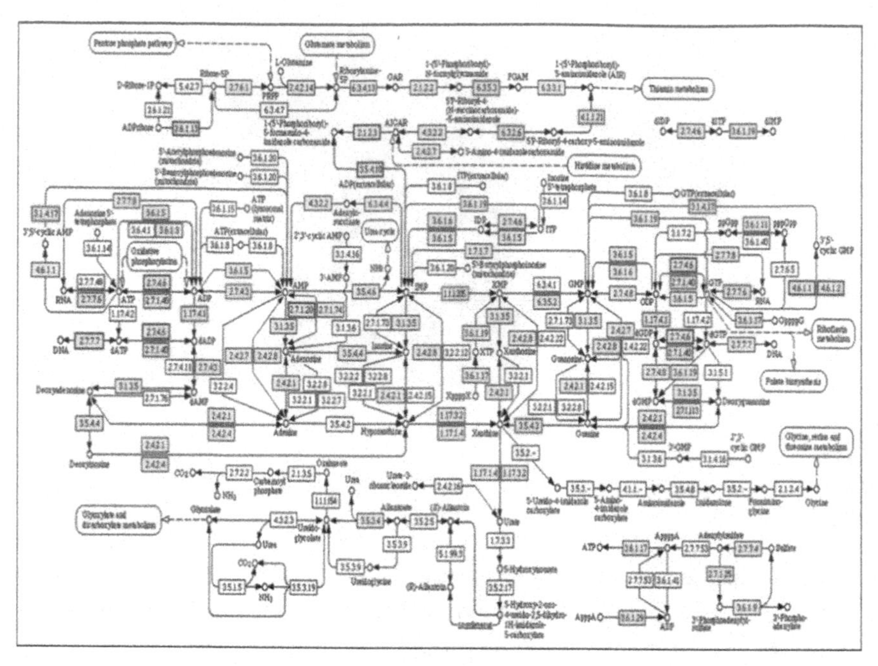

FIGURE 7.23: *A randomly selected example of a biological pathway diagram, of the sort that biologists routinely look at in the course of their work. Such a diagram typically shows the best known biological and chemical reactions involved in a highly specific biological process. This one shows differentially expressed enzymes in purine metabolism identified from irradiated AT5BIVA and ATCL8 cells. Our best understanding of biological dynamics, at present, could be summarized in a set of thousands of interlocking diagrams of this sort. No human can hold them all in memory at once, obviously. So, as human biologists, we must focus on individual parts of individual diagrams, or traverse and analyze large collections of diagrams using fairly crude statistical or mathematical methods. An AGI with human-level general intelligence but a mind fashioned to handle this sort of data more naturally, could obviously do far better.*

If you're curious: Specifically, in the above diagram, "Enzyme Commission numbers (EC#, e.g. 1.17.4.1) are used to represent enzymes in metabolism. Highlighted in green background are known human enzymes annotated in the KEGG database. Differentially expressed enzymes in purine metabolism (Table 3) are superimposed onto this pathway diagram: blue-boxed are enzymes changed in AT5BIVA cells, red-boxed those in ATCL8 cells, and pinkboxed those from both cells. Areas circled with broken lines highlight closely related biochemical steps surrounding ADP/ATP (left) or GDP/GTP (right) metabolisms, which include most of these differentially expressed enzymes from either cell type." (http://www.omicsonline.org/ArchiveJPB/2008/May/03/ImagesJPB1.47/Figure4.jpg)

On the other hand, an AGI adapted to the requirements of this sort of data would not suffer the same limitations. It could look at massive biological datasets that baffle the human mind, and see an abundance of patterns invisible to human perception – then designing drugs and other therapies based on those patterns.

In my own biological data analysis work – which I'll tell you more about later -- my colleagues and I have used AI tools to find thousands of genes showing significant differences between healthy and unhealthy long-lived people. As our AI analysis shows, the effects of these genes on longevity are generally not individual, but rather combinational, involving interactions between multiple genes.

Human biologists tend to focus on individual genes, or (at best) on pairs or triples of genes – not because this how genetics works, but because this is what the human brain needs to do to simplify the problems of genetics and put them in humanly comprehensible form. A human brain naturally combines hundreds of visual features when recognizing somebody's face; it has acquired this skill through evolution. But a human brain is just no good at perceiving or thinking about the combinations between hundreds of different genes – even though these sorts of massively multi-gene combinations do appear critical to understanding why bodies age and how to increase their lifespan. An AGI mind could think about 1000 genes and all of their possible interactions, and then understand the causes of longevity and aging a lot better than any human can.

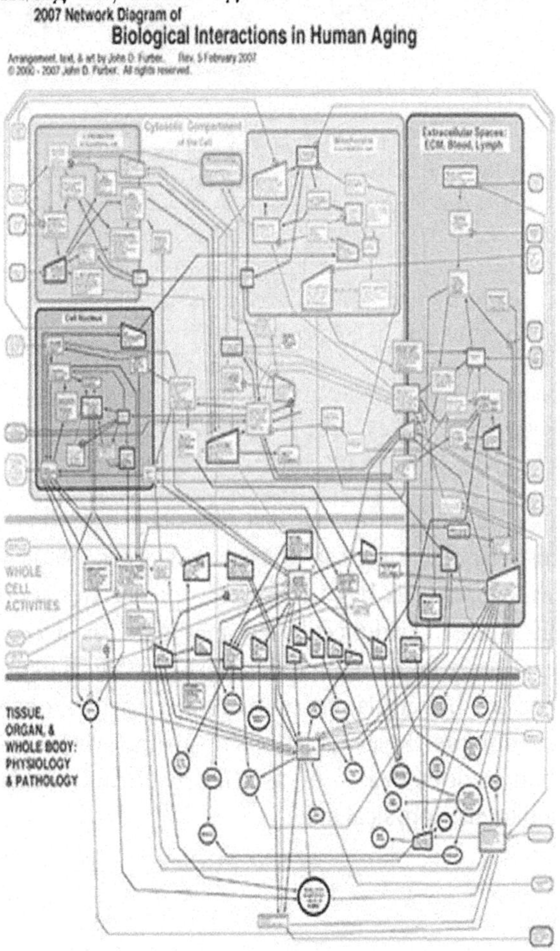

FIGURE 7.24: *This is the 2007 version of a poster I have hanging in the wall of my study at home, made by longevity researcher John Furber. The 2011 version, the one I have at home, has a few small changes. Check it out online if you're curious. It is not a detailed pathway diagram, but rather attempts to give an overall conceptual picture of all the different structures and processes going into human aging. If nothing else, it gives you a good intuitive sense of the complexity of the aging process. (http://www.legendarypharma.com/chartbg.html)*

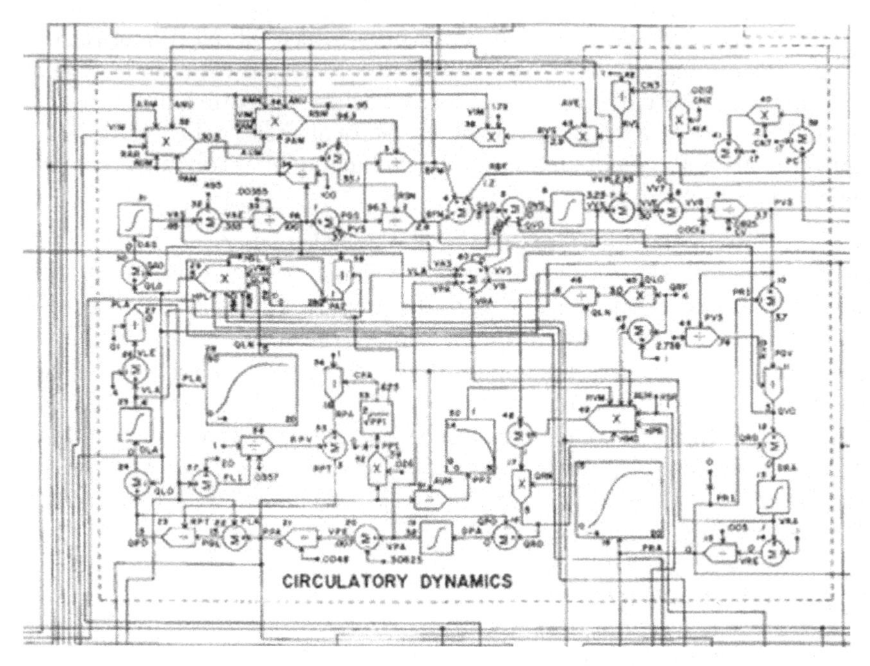

FIGURE 7.25: *This depicts the subset of the overall Hummod dynamical systems model of the human body, that deals specifically with the circulatory system. Hummod consists of a couple dozen models of this rough level of complexity, all networked together. Each box indicates a certain mathematical equation, affecting some set of variables related to the circulatory system. This is a level of abstraction higher than a biological pathway diagram, and gives a view of how the dynamics implicit in biological pathway diagrams ultimate affects the body. The Hummod equations can be used to give an overall simulation of the interactions of all the human body systems. But they are too coarse-grained to capture everything happening in the body; they just give a crude, though very useful, high level view. (http://hummod.org/assets/images/72-guyton-model.jpeg)*

Health care today, just like drug discovery and genomics research, is also largely a matter of trial and error. Medical malpractice has become widespread, as has the average physician's ignorance of modern medical literature. "Evidence based medicine" is a concept much discussed but erratically implemented. Simplistic "medical expert system" AI programs can provide better diagnoses than most doctors, analyzing the patient's answers to multiple-choice questions regarding symptoms. AGI doctors will obsolete human doctors as soon as AGI sensors and actuators can accurately observe a patient's physical state.

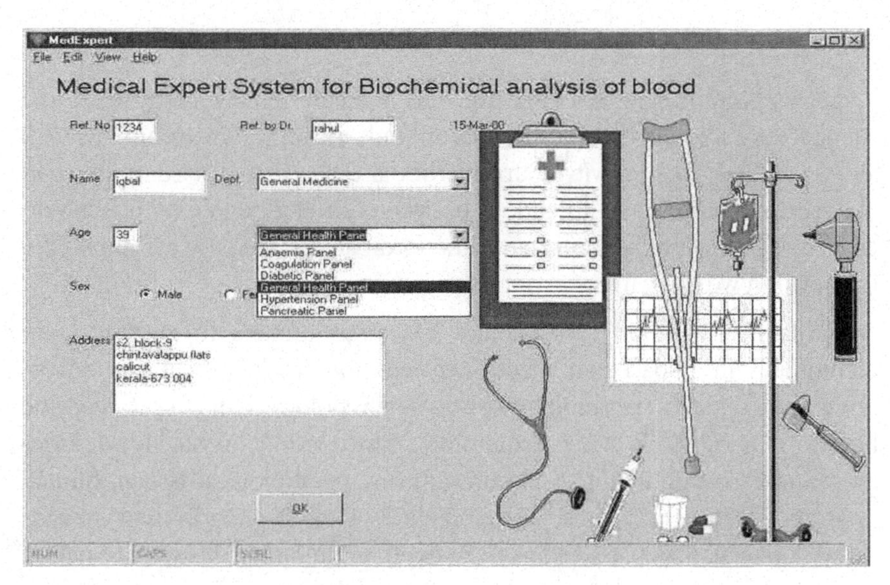

FIGURE 7.26: *Screenshot of a current, commercial medical expert system, aimed at helping medical professionals to analyze blood work. Several studies have shown (narrow AI) medical expert systems to be more accurate diagnosticians than the average human doctor, but their adoption is still relatively weak due to resistance from the community of physicians. (http://iqsoft.co.in/products.html)*

The best human surgeons are amazing, synthesizing physical and mental ability in an elegant and powerful way. And yet, the human hand obviously it wasn't designed for performing surgical operations. Once robot hand technology develops further, it will far outperform the human hand at every kind of surgery. Picture a hand with tiny cameras inside each finger, and the ability to use a variable number of fingers, depending on the task. Primitive versions of this kind of technology already exist, and are advancing year on year. Or better yet, picture a swarm of nanobots going into your bloodstream, fixing the problems at the nanoscale. Nanotech visionary Ralph Merkle has fleshed out the possibilities in this regard quite thoroughly.

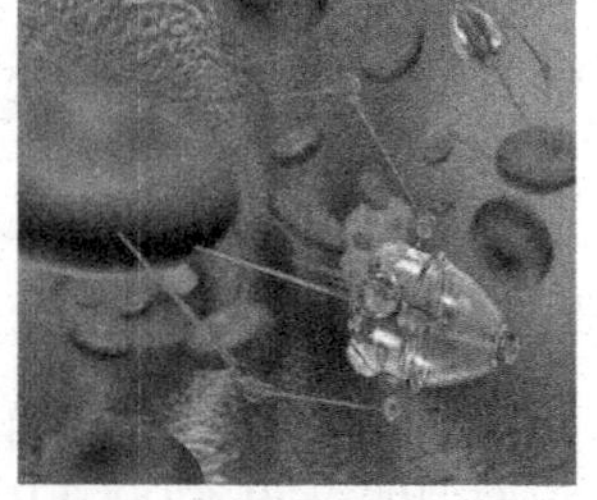

FIGURE 7.27: *Hypothetical medical nanobot, zooming through the bloodstream and fixing cancerous cells. Ralph Merkle and others have thought through the physics and engineering of this sort of device in great detail. There seem no basic obstacles to building them – we just need some more advancement in our engineering practice. Of course AGI inventors, scientists and engineers with sensors and actuators at the nanoscale could advance this sort of development faster than humans. As I noted in an earlier chapter, some simpler medical nanobots, comprising molecular switches inserted into the body to release chemicals selectively based on certain conditions, are currently being tested in human trials. (http://www.gizmowatch.com/good-gadgets-technologies-hate-cancer.html)*

Space

Space exploration is expensive and difficult mainly because of the human body's limitations. Our bodies are only meant for environments with air, water, and earth gravity. Current robots can't carry out complex tasks in space, so we're forced to choose between sending people (which is very expensive), or sending stupid and uncoordinated robots (which limits the information we can gather).

AGIs may develop technologies allowing humans to survive more comfortably in a spacecraft. There may be limits to how well this can be done, due to human psychology as well as physiology. Can people really be happy, to use David Bowie's terminology, "floating in a tin can" for decades or centuries on end? But to circumvent any psychological issues, human brains could be connected to virtual realities while floating in their tin cans through interstellar space, as has been explored in numerous science fiction novels.

On the other hand, while humans are built for Earth, AGIs may be more comfortable in space than here on the home planet. Computers operate better in supercooled environments. Mining for processor materials should be easier in the asteroid belt; gravity is a nuisance if you don't need food or air; and solar power is more abundant in space. One likely scenario is that AGIs will colonize space while most humans remain on Earth where their bodies and minds are comfortable.

FIGURE 7.28: *Special Purpose Dextrous Manipulator (Dextre), a robot deployed to build and service the International Space Station. Once robots achieve general intelligence, they are likely to lead the colonization of space. Not needing air, food or water, and operating comfortably in low temperatures, they are obviously far more suited to space than human beings. (http://spaceflight.nasa.gov/gallery/images/shuttle/sts-123/html/s123e007088.html)*

Once mind uploading technology is available, if a human wants to go into space, they can adopt a robot form for the expedition. Imagine becoming a robot and flying around freely in the vacuum. Or, given a sufficiently robust robot body, zooming throh the center of the sun.

One of the greatest moments in cinema comes toward the end of Blade Runner, a film based on a Philip K. Dick novel, featuring genetically engineered artificial humans called "replicants". The replicants have some superhuman capabilities, but deficiencies regarding empathy, and 6 year lifespans immutably fixed in advance by their creators. The replicant Roy Batty introspectively makes the following speech during a rain downpour, moments before his own preprogrammed death:

"I've... seen things you people wouldn't believe... [laughs] Attack ships on fire off the shoulder of Orion. I watched c-beams glitter in the dark near the Tannhäuser Gate. All those... moments... will be lost in time, like [coughs] tears... in... rain. Time... to die..."

These lines, probably the most moving in all of science fiction cinema, were improvised on the spot by the actor Rutger Hauer based on an inferior version provided in the original script. Part of the emotional undertone of the speech, of course, has to do with the completely unnecessary nature of Roy's death. Why did he need to be programmed to die so soon in the first place?

But then, why do we humans need to be programmed to die so soon, either? We don't, of course. And by the time we are able to inspect the shoulder of Orion first hand, the blight of involuntary death will almost surely be lifted from our organic or digital descendants.

Telecommunications

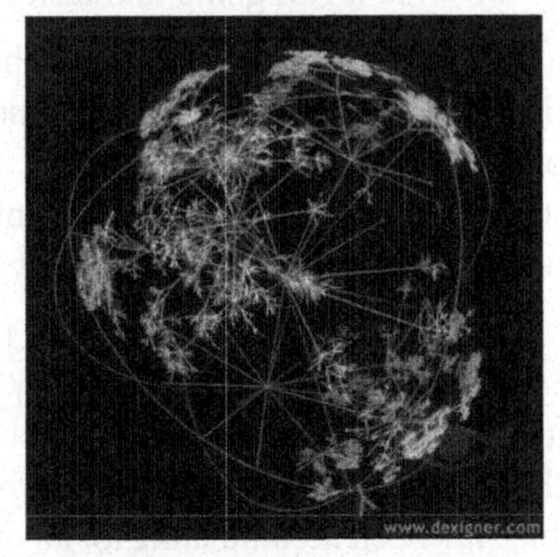

FIGURE 7.29: *Appealingly laid-out graph of the Internet connections emanating from a single site in Herndon, Virginia. The notable thing about this picture is how biological – neuron-like or plant-like – the fractal branchings are. Coupled with the fact that the connections on the Net are always growing and changing, this really gets across the point that we are dealing with a dynamic, "living" system. What we have is a massive system like this, wrapping the globe in a constant pulse of information generated and consumed by humans and our software. (http://www.sdsc.edu/News%20Items/PR022008_moma.html)*

Telecommunications, while important to humans (try taking a teenage girl's electronic communication devices away!!), will be crucial for digital AGI systems: *They will be able to perceive data from all over the planet, and beyond, using telecommunication networks.* Because of this, we can expect AGI systems to have a strong motivation to optimize telecommunications far beyond what humans have done.

Our current use of the available communication bandwidth is extremely crude; our algorithms for sending and receiving information are simplistic compared to what would be mathematically and physically possible, resulting in loads of redundant information being sent through the airwaves.

An AGI able to perceive a wider spectrum of electromagnetic radiation than humans could connect its general problem solving capability with automated equation-solvers applied to signal processing mathematics – resulting in new circuits for sending and receiving electromagnetic information, massively increasing the efficiency of new information passing through the airwaves. Humans would get faster cell phone and Internet connections; and AGIs would get faster connections between the different parts of their minds, boosting their intelligence and making them even better at optimizing telecommunication and other technologies.

Everything!

OK – I promised you AGI from A-Z, but I actually only went from Aerospace to Telecommunications. I considered adding Zoology at the end of the list but that would have seemed too cheesy. I guess you get the point, anyway. What is AGI going to transform? Everything.

The potential impact of AGI technology is so broad that it's difficult to think about. Where do you start, when thinking about something with the potential to change EVERYTHING? It seems almost too easy to run down the list of all existing industries, and point out the potential for advanced AGI to revolutionize every one of them. But that's simply the truth.

Creating AGI is, of course, not that easy. Even if writing the software code for the AGI system that finally "takes off" and displays human-level intelligence is only moderately difficult for the team that finally does it, this achievement will be building on a incredible mass of science and engineering which has been accumulating for all of human history. But once AGI is there, it will revolutionize all aspects of human existence just as surely as prior radical advances like language and civilization have done – and probably more so. There will surely be limitations, but there's no way for us to foresee those now.

As a human being in a human body, it's easy for me to think about the specific ways that AGI could change my own human life. With better engineered food or drugs, I wouldn't have to worry about getting overweight. Better biomedical science means I wouldn't have to worry about getting old and dying. I'm typing these words on a keyboard and viewing them on a screen, rather than just thinking them directly into some digital knowledge store; AGI scientists could change that through brain-computer interfacing. The car I drive to go to the supermarket could be replaced by a flying machine, if we had AGI to handle navigation and prevent collisions. Or the value of my work might be eliminated by AGI scientists. And so on.

Overall, the best "first approximation", regarding the future impact of AGI, is to assume that AGI will be able to render the various limitations we now face obsolete, insofar as this is possible within the "laws" of physics. To what extent future human or AGI scientists will reveal limitations in the "laws" of physics as currently understood, is hard to estimate. But even setting aside the likelihood that the laws of physics as currently understood are incorrect and too limited, the scope of possibilities extends so far beyond current human reality that it is difficult to think about in detail. Manipulation of matter at the subatomic scale to create femtotechnology; distributed minds spanning galaxies with orders of magnitude greater problem solving ability than humans; and so on. These sorts of things seem physically possible; and given the logic of exponential growth, it seems plausible that eventually descendants of the AGI systems humans build will get there. Given this scope of possibilities, the potential of AGIs to revolutionize the construction or pharmaceutical industries seems almost obvious and mundane.

The skeptic typically interrupts this kind of rosy-eyed projection with some comment like "Every technology has its limitations, right?"

But actually – intelligence isn't just another technology. Intelligence is a fundamental capability–that's the wonder of the "general" part of AGI. Every technology has its limitations, but part of the power of general intelligence is its capability to overcome limitations by formulating new technologies.

FIGURE 7.30: *To my dear, departed Japanese Chin "Crunchkin", cars and knives seemed just about as magical as the workings of superhuman AGIs will seem to us. But, AGIs may grow much further beyond our intelligence than we are beyond Crunchkin's. (And if Crunchkin is uploaded one day and has his digital brain enhanced a bit, he may be able to read this caption and appreciate this picture! Hello there, Future Crunchkin! – just a little greeting from Past Ben!!)*

You can call this "magical thinking", if you like. But remember Arthur C. Clarke's insight: "***Every sufficiently advanced technology seems like magic***.". Absolutely.

8. AGI and Beyond

I've mentioned "AGI" many times in the preceding pages; I suppose the intuitive meaning of the concept, as I intend it, should be very clear to you by now. I want to emphasize the distinction between AGI, which is aimed at creating intelligent systems capable of dealing autonomously with a wide variety of situations, and narrow AI systems, which have intelligence closely customized for specific problems or contexts.

However, like any term or concept, "AGI" also has its assumptions and its limitations. It's worth reflecting a bit on both the strengths and the weaknesses of the AGI concept.

The core reason for introducing "AGI" is that the concept of "AI" has become so broad that it's lost most of its utility. Perusing the contents of contemporary AI journals, I struggle to articulate the distinction between AI and "advanced computer science." The term "AI," in its current standard academic usage, describes a very wide variety of algorithms, each doing very specific, intelligent-in-some-sense things. But these "AI" programs, constituting the main focus of the AI field in academia and industry, don't aspire to the kind of general intelligence that humans possess.

Ray Kurzweil originally contrasted "narrow AI" with "strong AI"; but the term "strong AI" also has its complexities, because it has historically been used in the philosophy of AI to denote the hypothesis that AIs can be conscious just like humans (a hypothesis I tend to agree with, but that's another issue...). I prefer to contrast "narrow AI" with "general AI" or "AGI" instead. Just like a human brain. I see AGI as going beyond these highly specialized narrow AI programs that do only one thing, like playing chess, driving a car, or predicting stock prices.

AGI and narrow AI share many things in common. Many of the same technical approaches may be applied to both narrow AI and aspects of AGI. But I believe fundamental differences exist between the two pursuits, and many (though by no means all) other researchers agree with me on this.

All this I have hopefully made plain in earlier chapters. But there is a lot more depth to the concept of "AGI" than all that.

What does "AGI" mean, exactly?

Defining "AGI" (or not)

One approach to carefully, rigorously defining the concept of AGI is not to bother. An everyday, practical characterization may be good enough. This actually makes a fair amount of sense to me.

Along these lines, I like to quote Nils Nilsson, one of the founders of the AI field back in the1960s and early '70s. He wrote an article for *AI Magazine* in 2005, titled "*Human Level Artificial Intelligence? Be Serious!*", arguing that more AI researchers should focus their attention back toward the grand AGI-ish goals that preoccupied the field in its earliest years. His vision is squarely focused on the practical, everyday potential of AGI systems – their ability to take over the jobs now done by human beings. As he put it:

"[A]chieving real Human Level artificial intelligence would necessarily imply that most of the tasks that humans perform for pay could be automated. Rather than work toward this goal of automation by building special-purpose systems, I argue for the development of general-purpose, educable systems that can learn and be taught to perform any of the thousands of jobs that humans can perform. Joining others who have made similar proposals, I advocate beginning with a system that has minimal, although extensive, built-in capabilities. These would have to include the ability to improve through learning along with many other abilities."

In Nillsson's perspective, once an AI obsoletes humans in most of the practical things we do, it's got general Human Level intelligence. The implicit assumption here is that humans are the generally intelligent system we care about, so that the best practical way to characterize general intelligence is via comparison with human capabilities. To me, this is a somewhat limited view of Artificial General Intelligence, but it's a great place to start. Practical human jobs do encapsulate an awful lot of useful, subtle general intelligence.

FIGURE 8.1: *Nils Nillsson created Shakey the Robot in the early 1970s. The top figure shows Sven Wahltrom and Nillsson with Shakey. As of 2014, Nilsson is still pursuing advanced AI, and he is actively arguing for the field to refocus itself on the pursuit of artificial general intelligence. (Figure of Shakey top from http://en.wikipedia.org/wiki/Shakey_the_robot#mediaviewer/File:SRI_Shakey_with_callouts.jpg; right figure from http://s7.computerhistory.org/is/image/CHM/500004692-03-01?$re-medium$).*

But, much as I admire pragmatism, I'm not going to let the matter of defining "AGI" go that easily. Before digging into my approach to creating AGI, and recounting my journey through the field which led me to this approach, I think it will be useful for me to explain a little more thoroughly and rigorously what this concept of "Artificial General Intelligence" really means – and what it doesn't. In the end it's the practical behavior of the systems one builds that really matters. But still, sometimes it's nice to have a better idea of what one is talking about.

I've already given you a rough definition of Artificial General Intelligence: A system—a computer program, robot, or machine or whatever—that displays the same rough sort of general intelligence as humans. General intelligence is intelligence not tied to a very specific set of tasks, which possesses the ability to take a broad view, or generalize what has been learned. The best judge of a system's general intelligence is its practical ability to achieve a variety of complex goals in a variety of complex environments. A general intelligence should be able understand new things it encounters, and acquire knowledge in one domain then apply it towards another. It should possess its own evolving, intuitive understanding of itself, others, and the world.

Making this rough concept more precise is trickier than it might seem. As I have found through my research into the theory of AGI over the last decade, attempting to define AGI in a quantifiable way leads to a variety of complications, some of which are educational and some more technical in nature. A core problem is that nobody really has a crisp, precise definition of "intelligence."

Some people think this slippery nature of "intelligence" is problematic, but I'm not sure how worrisome it really is. Most of us know what "beauty" means intuitively, but formalizing precisely what it means is a big job, which philosophers and psychologists have worked on for a long time. Yet artists create beautiful paintings without worrying about the exact, formal definition of beauty — they know it when they see it.

Biologists also don't fret much about the definition of "life." There are plenty of borderline cases, such as retroviruses – and with the advent of synthetic biology there will be many more. It doesn't worry biologists that "life" is a fuzzy and qualitative concept, because ultimately they know that "life" is just a simple communicative shorthand for a fuzzy grab-bag of more precise concepts, like reproduction, metabolism, etc.

In the same way, we may be able to build advanced AGI systems without worrying so much about the exact definition of AGI. But even so, thinking a bit about what AGI means can be educational, and can sharpen one's thinking about real-world AGI systems.

The Existence of Narrow AI was a Major Discovery

To fully understand why the distinction between AGI and Narrow AI seems so important, it's helpful to know a bit about the history of the concept of "AI" and how it's changed over time. Word meanings aren't absolute things handed down from heaven; they're also not determined by a centralized government committee. Word meanings shift gradually over time due to patterns of usage: It's been a while since "gay" predominantly meant "happy" in everyday American discourse, for example. The meaning of the word "AI" has also drifted.

When it was first introduced in the late 1950s, "AI" clearly referred to the creation of machines, computer programs and robots that displayed general intelligence in the same sort of fashion that people do.

Back then, it seemed reasonable that a computer capable of playing chess as well as a smart human could also match a smart human's ability to think generally on a wide range of subjects.

However, since that time, the meaning of "AI" has drifted because we have discovered that it's possible to write computer programs (fairly simple algorithms operating differently from the human mind) that do various smart-looking things that people do, yet still lack any ability to think generally and operate autonomously in the world. Now, we know that you can make a computer program capable of beating any human in chess, yet unable to read the newspaper, walk across the street, solve an equation, play Checkers, Go, or even Tic-Tac-Toe. We see concrete programs like Deep Blue or Watson that do specific smart things but don't have a general ability to understand the world. The possibility of AI "idiot savants" like this was not nearly so obvious in 1960 or 1970 – back then, most AI experts assumed that once a program with the capabilities of Deep Blue or Watson was achieved, full-scale human-level AGI would be just around the corner.

The point for now is that today, "AI" has two meanings. On the one hand, it refers to hypothetical programs, robots and machines displaying human-like or greater general intelligence (hypothetical because they're not complete yet). This is what we call AGI. On the other hand, it also refers to programs existing in the real world that lack general intelligence, yet perform very specific "intelligent behaviors" like playing chess, judiciously placing ads on web pages, or predicting stock price movements. This is what we call "Narrow AI."

When the term "AI" was first introduced, the latter sort of highly-specialized intelligent-task-executing program, doing specialized smart stuff well but lacking the ability to generalize or think autonomously, wasn't even considered a possibility. Thus, the distinction between AGI and narrow AI was much less clear a few decades ago than it is today. As science and technology progress, new common sense sometimes emerges implicitly rather than through anyone's big "Eureka moment."

By emphasizing the narrow-AI/AGI distinction so much and so firmly, I don't mean to imply that "narrow AI" work is bad, or not valuable. I spend a fair bit of my own time working on highly specialized "narrow AI" systems of various sorts. For the last couple years I've been collaborating with a team using machine learning and computational linguistics software (types of narrow AI) to predict the stock market, with a view toward starting a hedge fund. I've also done a lot of work applying AI tools to analyze biological data, with a goal of helping biologists figure out how to make people live longer. I've also contributed to various applications of AI technology to other domains such as video games, analysis of music listening or marketing data, helping find important bits of information in large stores of textual knowledge, and so forth. This sort of work is fascinating and often productive; there's nothing wrong whatsoever with putting narrow AI to this sort of use. In fact, this is some of the most interesting stuff happening on the planet today! But still, this kind of applied narrow AI work is a substantially different enterprise from trying to build a general AI capable of thinking for itself. Different kinds of programming are required for each. Broad generalization requires cognitive structures and processes very different from those required for specialized problem solving. To a limited extent, I think that narrow AI can help towards building general AI. However, I don't think a narrow AI will spontaneously develop into a general AGI, or that a narrow AI design can be extended to yield an AGI design.

Most definitely, the tools you use (hardware, software, mathematical, conceptual tools) to build narrow AI can be helpful for building general AI. But you have to use them in a different way. It is AGI, not Narrow AI, which will bring about the Singularity. However, Narrow AI can help indirectly in pushing toward the creation of AGI and a host of other Singularity-enabling technologies.

FIGURE 8.2: *Human Jeopardy contestants admit defeat at the question-answering TV game show Jeopardy, at the "hands" of IBM's Watson narrow-AI system, which was especially engineered for success at the game. (http://hothardware.com/News/Watson-Wins-ThreeDay-Jeopardy-Event-And-A-Cool-1-Million/).*

A system doesn't have to be utterly, infinitely general to be an AGI. Rather, AGI is about systems that, while ultimately limited in scope due to the intrinsic limitations of implementing things in physical reality, still have generality, autonomy and multiple-domain functionality as parts of their essence, and as traits that occupy a reasonably high percentage of their internal resources. This is the kind of generality human brains possess; this is the kind of generality I, and other AGI researchers, want to build.

The Origins of "AGI"

It was in reaction to this gradual shift in the meaning of "AI" that I ended up launching the concept of "AGI" a little over 10 years ago...

In 2002, my long-time colleague Cassio Pennachin (we started working together in 1998 and we're still at it!) and I were putting together a book consisting of research papers contributed by various scientists who were focused on creating computer programs that would think in the same way as human beings, ultimately surpassing them.

We wanted to give the book a title that would distinguish it from the more routine, specialized AI research in most universities and companies at that time. Most of that research did not focus on how to make computers that could think like people. Instead, it focused on designing programs with narrower purposes, like playing chess, optimizing path-finding (for game characters or robots), or searching databases. These specialized problem-solving applications are cool, but require a very different approach from trying to build a real human-like thinking machine. So, we were looking for terms to distinguish our kind of AI from the rest...

At first we were going to call the book Real AI. But it quickly became clear that this title was a bit too controversial. After all, the other kind of AI that researchers were pursuing was just as "real" as ours. They were making software programs that performed real and worthwhile tasks. Their work was not inherently non-valuable; it just wasn't moving AI in the direction of human-like thought. They were not seeking to instill their creations with the kind of generality, creativity and self-understanding that belong to the human mind.

I emailed a bunch of friends to see if anyone could come up with a better alternative to Real AI. Shane Legg, a former collaborator of mine (and current Google AI leader) who was then pursuing his PhD with mathematical-AI wizard Marcus Hutter in Switzerland, suggested Artificial General Intelligence (AGI). The G was meant to play off the concept of the g-factor in psychology, a statistical term used in measurements of general intelligence.

The g-factor is the attempts of psychologists to break down specialized knowledge and capability, and capture a general sense of learning and thinking ability. However, since the IQ tests psychologists use to measure the g-factor are specifically related to how the human brain works, they only work on humans. The spirit of the g-factor, though, has some relevance to the basic meaning of AI.

Actually, Cassio and I were not overly enthused about the term at first. Artificial General Intelligence sounded a bit boring and decidedly less snazzy than, say, nanotechnology, quantum computing, or artificial life. However, we basically felt that it said what needed to be said. So we ran with it.

At this point, the term has caught on reasonably well. In addition to that edited book, I've helped organize a series of technical AGI conferences. AGI-2011 (the fourth in the series) was at Google's headquarters in Mountain View, AGI-12 was at Oxford University, AGI-13 was in Beijing, AGI-14 was in Quebec City and AGI-15 was in Berlin, AGI-16 was in New York, AGI-17 will be in Melbourne. There's also an AGI Journal. In addition, a Google search reveals a lot of different people using the term as intended, independently from any of my own projects.

FIGURE 8.3: *Snapshot from the Panel Discussion on Virtually Embodied AI, at the First Conference on Artificial General Intelligence, which we held at the University of Memphis in 2008, in the wonderfully futuristic FedEx Center. This was a follow-up to the AGI Workshop we held in Bethesda in 2006. Since 2008 there has been an AGI conference every year, in various places including Washington DC, Google's headquarters in California, Switzerland, Oxford University in the UK, and Peking University in China. (http://www.flickr.com/photos/brewbooks/2336795010/).*

Nils Nilsson, in the quote I gave above about AIs doing everyone's jobs, used the term "human-level AI." This is favored by some other researchers as well. However, it doesn't quite capture what I'm after in my own work; I don't want to stop at the human level. I'm looking to create AIs with the potential to become super-intelligent human beings. Engineered minds, or their descendants, may eventually display adaptations with complexity far beyond the human level.

SCADS !!!

Just to make things a little more confusing, I have to admit that I don't actually love the term "AGI" terribly much, even though I introduced it onto the world stage, and I use it an awful lot. I just think it's less problematic than the most common alternative, "AI." I actually have issues with all three of the components of "AGI": "Artificial," "General," and "Intelligence" – which I'll explain to you shortly, because they're issues that have meaning well beyond the choice of verbiage. Still, I think "AGI" is a useful term at the present time, given the history and current state of Artificial Intelligence technology and thinking.

A less troublesome term for the kinds of systems I'm working on building might be "Synthetic Complexly Adaptive Systems" or SCADS. "Synthetic" meaning something that's built, engineered, and synthesized. And "complexly adaptive" meaning that the system's state — the observable patterns in its physical nature and between the system and its environment — changes in response to its internal and external situation in complex ways.

Thinking in terms of SCADS, it's clear that narrow AI systems are far less complexly adaptive than human brains or AGI systems, as their adaptation is restricted to relatively narrow domains.

But even though I prefer "SCADS" from a purely intellectual point of view, I believe that since the term "AI" has gained so much currency, the term "AGI" has a lot intuitive and explanatory value. It focuses our attention on the nature of artificiality, generality, and intelligence, all of which are important. However, synthesis, complexity, adaptation, and interconnectivity are also important!

Advanced AGIs won't really be "Artificial"

In 2013, some Czech folks organized a funky cross-disciplinary conference called "Beyond AI: Artificial Golem Intelligence." I was too busy to attend personally, but I gave a talk via video conference, titled "Beyond Artificiality, Generality and Intelligence". It wasn't a great talk delivery-wise, because I had to give it in the middle of my work day at the Aidyia office, and my office there has very thin walls, so I ended up speaking very quietly to avoid disturbing the other folks working just through the walls. But content-wise, I think it was an interesting one.

What I explained in the talk are some of my conceptual issues with the "A," "G" and "I" in "AGI."

I'm far from the first to reflect that the "A" in "AI" is a bit problematic. If we succeed in creating superhumanly intelligent super minds, this will render the terms "AI" and "AGI" both pretty irrelevant by dating the use of the word "artificial." An "artifice" is a tool or method, but ultimately, a highly intelligent, autonomous computer program or robot is not going to be anyone's tool.

As a researcher, I'm not fundamentally motivated by the goal of creating intelligent systems that are merely TOOLS. Tools are fantastic and useful, and I'm happy to have helped create some worthwhile ones during the course of my career – but tools are not the end goal of AGI development so far as I'm concerned. I don't want to create machines that serve only as proxies for others' desires, not even my own. A robot servant would be quite convenient, but ultimately, this is a small-minded aspiration. I want to create autonomous minds with their own goals, passions, feelings and interests, who explore the universe according to their own designs. It's important that they respect the rights and desires of humans, as well as other sentient beings, but I want them to be more than our "tools." Is it even possible to create superhumanly intelligent minds and have them serve merely as tools for humans, any more than it's possible for a world of humans to exist merely as a tool for dogs, cockroaches, or bacteria?

Real-world Intelligence can't truly be "General"

But what about the "G" in "AGI"? It is this letter which is the critical distinction between AGI and the bulk of AI work today, which I call "Narrow AI." "Generality of intelligence" means being able to think intelligently in multiple qualitatively different domains, and to transfer knowledge from one domain to another. It's about adapting to a new work environment, or learning to deal with the quirks of a new teacher. An AGI system should have "generality" as a central focus of its structure and dynamics, rather than being tailored to a specific domain.

Generality is critical to human intelligence, and I think it should be a key focus of any attempt to create thinking machines. But still, there are limitations inherent in the notion of "general" intelligence. A truly, absolutely general intelligence could solve any problem in any environment. I'm not sure if such a thing is possible, at least in the realm of known science.

A number of enterprising mathematicians (some of the key names are Ray Solomonoff, Marcus Hutter, and Jürgen *Schmidhuber; I'll discuss their work in more detail a bit later)* have proven theorems about absolute general intelligence. However, like a lot of mathematics, these theorems rely on assumptions that don't apply in the real world. Many of them rely on the assumption of having infinite processing power in your computer, which violates the laws of physics (at least as they are currently understood). When this assumption is relaxed, it's generally replaced with a nearly-as-bad assumption of having a computer with insanely much computing power, e.g., more than could be achieved by the best possible computer constructible using all the particles in the known universe. Mathematics based on this kind of assumption may still be interesting, and inspirational for AGI work, but it isn't very directly applicable.

In reality, it seems there is a limit on how general intelligence can be, due to the limits physics places on how much processing power any real-world physical system can have. Given a finite amount of processing power, no intelligent system can actually understand every possible thing within a reasonable period of time – and any finite system will be faster at learning and understanding some things than others.

But even though absolute generality of intelligence seems incompatible with physics as we understand it, it is nevertheless possible for an intelligence to have a significant degree of generality in a very meaningful sense: the capability to take a narrow scope and broaden its potential. For an intelligence to be "general" in this sense means that the quest for greater and greater generality occupies a lot of the intelligence's attention, exhausting much of its space and time resources.

With programs like IBM's Deep Blue or Google Search, generality is not the focus. Generality is very limited and system intelligence focuses on specialized problem solving. So, in this interpretation, AGI means: A system focusing on generalization and the ability to extend intelligence beyond one particular domain. Humans are not infinitely general like some theoretical mathematical AGIs are, but they are far more general than any existing AI system.

Intelligence Itself is a Somewhat Limiting Concept

Finally, what about the "I" in AGI?

The "I" is maybe the most mysterious one of the three letters in "AGI," since no generally accepted definition of "intelligence" exists! Psychologists have various definitions, nearly all agreeing that tests like the IQ test capture only part of human general intelligence. Measuring intelligence with the IQ test is not like measuring mass with a scale. Mass is reasonably well-defined, with a clear theory explaining why the scale measures it, while intelligence is vaguely defined, with no clear theory explaining why an IQ test is an accurate measure. AGI researchers Shane Legg and Marcus Hutter made a list of over 70 definitions of intelligence, drawn from the research literature of various disciplines.

I don't worry too much about the lack of an accepted definition of intelligence, though. As I said above, artists do OK without an accepted definition of beauty, and biologists get by fine without a clear, universally accepted definition of "life."

While "intelligence" is a useful concept, it's not clear to me how fundamental it is. Maybe we'll create smarter and smarter systems that go beyond our current concepts of "intelligence."

I think of intelligence as, roughly, "the ability to achieve complex goals in complex environments." This view agrees with a significant percentage of the psychology literature on intelligence, also matching many of the modern mathematical formalisms of intelligence. Basically, it comes down to viewing intelligence as the possession of a broadly powerful optimization capability.

Yet I sometimes find myself doubting how deep this definition goes. Thinking of intelligence as the ability to achieve goals assumes a sort of split between one's goals and one's intelligent mind, which may not be the way things work. Since we live in a particular universe, the important thing may not be the ability to achieve arbitrary "complex goals" in arbitrary "complex environments," but rather "the ability to achieve the complex goals, in the complex environments, and using the resources available, that are all relevant in our universe." However, when you start thinking about it this way, you realize that to understand intelligence, you'd first need to understand the universe. I'm all about understanding the universe, but I don't think this is a prerequisite for building thinking machines!

My friend David Weinbaum (who also goes by "Weaver"), currently a researcher in the Global Brain Institute at the Free University of Brussels, has (together with his colleague Viktoras Veitas) developed the notion of "open-ended intelligence." An open-ended intelligence is more like a SCADS – it is involved with expanding its boundaries and recognizing and creating patterns in itself, in its environment, and emergent between itself and its environment. It will certainly tend to maximize various functions in various environments at various points in time, but its activity won't necessarily be effectively summarized in terms of persistent maximization of any particular goals.

Human intelligence, for all its impressive generality, is still somewhat specialized through its focus on the achievement of specific goals in specific environments — moving around on a 2D surface, manipulating solid objects, or communicating using linear sequences of symbols. Our intelligence has evolved as an adaptation to one tiny corner of the known physical universe, which may in turn be only a tiny percentage of the whole of existence. What constraints the universe may place on the nature of intelligence in general is something we have no way to figure out right now. Fortunately, we don't have to.

My goal as an AGI researcher is to build a Synthetic Complexly Adaptive System with general intelligence a bit beyond the human level – and not to forget, demonstrating a reasonably beneficial attitude toward humans and other sentient beings. This system will then figure out the next steps, in ways that I, with my merely human mind, cannot hope to. In figuring out those next steps, it may well utilize concepts very different from "artificial," "general," "intelligence," "synthetic," "complex," "adaptive," or "system."

I look forward very much to seeing what concepts advanced future SCADs/AGIs do use, inasmuch as I – in whatever form I exist at that point – am able to understand them.

Further Reading

Goertzel, Ben (2010). Toward a formal definition of real-world general intelligence. In *Proceedings of AGI-10*.

Goertzel, Ben (2014). Artificial General Intelligence: Concept,

State of the Art, and Future Prospects. *Journal of Artificial General Intelligence*.

Legg, Shane and Marcus Hutter (2007). A Collection of Definitions of Intelligence. In *Advances in AGI*, Ed. Ben Goertzel and Pei Wang, IOS Press.

Nilsson, Nils (1980). Human-Level Artificial Intelligence? Be Serious!, *Artificial Intelligence Magazine 26(4)*

Sternberg, Robert, ed. (2000). *Handbook of Intelligence*. Cambridge University Press.

Weinberg, David and Viktoras Veitas (2015). Open-Ended Intelligence. http://arxiv.org/abs/1505.06366

9. The Architecture of the Human Mind

When I decided (during my time in New Zealand and Perth) to try to use cognitive science ideas to bridge the gap between philosophy of mind and software implementation, I knew I wasn't making an especially original move. Of course, serving as this sort of bridge was one of the original purposes of the cognitive science field. But making this connection work in the context of my particular patternist philosophy of mind took some time and energy to figure out – and continues to do so.

Today, though, cognitive science understands much more than it did in the late 1990s when I founded Webmind. And I personally have thought through various cognitive science issues much more deeply than I had at that point, in part due to having so many fantastic conversations with other cognitive science oriented AGI researchers I've met through organizing the AGI conference series and visiting other researchers at their labs. By this time, I think I have a pretty clear picture of how patternist philosophy, cognitive science, and AGI implementation all fit together.

Of course, cognitive science still doesn't tell us nearly all we need to know to design a thinking machine. To fill in the many details that current cognitive science doesn't tell us about, my OpenCog collaborators and I have used math, computer science, neuroscience, and variations on various narrow AI techniques, plus a lot of creative invention. Our broad architecture however is still drawn largely from human cognitive science. While this ultimately limits our work, since we seek to build artificial minds beyond the human level, we believe that starting from a relatively well-understood base makes practical sense.

So what exactly does human cognitive science tell us, at this stage?

Well, at the high level....

First, cognitive science tells us how to divide the mind into different parts. However, this division certainly isn't absolute. There are a lot of different ways you can divide it up meaningfully, but there are ways that have proved more useful than others, both for scientific psychology and AGI.

Next, cognitive science explains how these parts interact with each other, and how they combine to mind-wide self-organizing structures like the "self" and the feeling of conscious "awareness." All the structures that make us feel like us.

On the other hand, one thing cognitive science doesn't tell us nearly enough about is how all these different parts of the mind work internally. It gives you some jewels of knowledge, but leaves an awful lot unsaid.

This is where math and computer science – and to a more limited extent, neuroscience – have a huge role to play.

In this chapter I'm going to give you a high-level overview of what cognitive science has taught us about the overall structure of the mind as of right now, in 2016. The overall picture isn't too different from what was thought back in the 1990s, but back then a lot of the details were far fuzzier – understanding is definitely advancing, both in the field as a whole and in my own mind!

There is nothing tremendously new or original in the synthesis I'll present in this chapter. However, it took me a long time and a lot of studying and thinking to come up with a relatively clear high-level overview like the one I'm going to give you here – I only managed to articulate things this crisply around 2011 or 2012, when I wrote a chapter for a book that Pei Wang and I edited, *Foundations of Artificial General Intelligence*, containing the same basic ideas I'll summarize here. As Pascal said: "*Let no one say that I have said nothing new; the arrangement of the subject is new.*"

Dividing the Mind Into Parts

FIGURE 9.1 is my own version of a picture drawn by the great British cognitive scientist Aaron Sloman. It shows Sloman's personal, erudite slant on a fairly standard way of dividing the human mind into a set of different aspects, each with their own unique characteristics, though also heavily interacting with each other. Looking at the mind in this sort of way is a bit simplistic, yet provides a path towards more detailed work. While it certainly doesn't tell us how to build an AGI, it gives a pleasantly concrete framing for the discussion about what basic phenomena should be included in any effective AGI design.

It's worth remembering that a few decades ago, there was nothing remotely resembling a consensus about how to draw a diagram like this, reflecting the different parts of the human mind! Cognitive science has moved forward, helping pave the way for the near future advancement of AGI.

Now let me tell you what all the boxes (i.e. all the parts of the human mind) depicted in this Sloman-esque diagram mean:

Perception – This one is straightforward. "Perception" means "the information coming into the mind from the senses." Vision, hearing, touch, vibration, and so on. AGIs may have some senses that people don't have, and they will also surely lack some typical human senses. For instance, right now it's easier to give robots Lidar (laser radar) than a decent sense of touch or smell… But this changes year on year as new technologies come out. Robot skin gets better and better fast, for example.

Action – This refers to an intelligent system taking actions –usually, but not necessarily, in an external or virtual world. For instance, a human mind, or a robot, moving the arm or leg of its body. Or, an AGI telling the virtual character it controls to take a step forward. Or, an AGI sending an email or adding an item to a biology dataset. In the cognitive science context, when I say "action," I'll most often be referring to actions in some domain with direct impact outside the mind itself. But the term also embraces purely mental actions, such as the decision to search one's memory for information about Mike the Headless Chicken.

Motivation / Action selection – The process of the mind choosing which actions to take, based on its basic motivations and related subordinate goals. Without this, the mind wouldn't have any organized, coordinated way to get things done. It would just be a diffuse, constantly-changing self-organizing system, not "intelligent" in the typical sense of the word. Often, the same mechanisms are used for choosing physical, external-world actions and for choosing internal, cognitive actions.

Long-term memory – This is memory that stays around for hours, days, or decades. It comes in a few different types, which I'll discuss a little later…

Working memory – This is the memory you are paying attention to right now. Roughly, it is the stuff in our current "focus of consciousness." When you're reading, the ideas from the last few sentences are usually hanging around in working memory, whereas the ideas from a few pages before are on the periphery of working memory, while the ideas from a few hours before are either in long-term memory or forgotten. The idea that it is useful to consider working memory and long-term memory as distinct sets of cognitive processes is a non-trivial fact that cognitive scientists have discovered during the last century or so.

Deliberative processes – This refers to cognitive processes (thought processes) that select from and put new information into long-term memory – often, yet not always, via working memory. This includes reasoning, making up new ideas, learning how to do complex new things, and similar complex processes. Most of the ways that humans are smarter than, say, dogs or apes, have to do with our "deliberative" cognitive processing.

Reactive processes – Cognitive processes that act fast. These are largely defined by their interaction with working memory, although they may grab information from long-term memory as needed. We rely on these to move our bodies around and generally react to the world in a real-time fashion. This is basic animal living-in-the-world. The fundamentals of human reactive mental processing seem similar to what one finds in other mammals, though there are some differences.

Reinforcement – This is among the most basic sorts of learning, in which the mind gets some subjectively-perceived "reward" from its body (generally delivered from the outside world in some way). It then tries to figure out which of its actions led to that reward, so that in the future, in a similar context, the mind can try to carry out a similar action, in hopes of getting a similar reward. Some cognitive and AGI theorists think that all intelligence can be explained via reinforcement learning. I tend to doubt that, but I do think it's an important and basic aspect of intelligence.

Emotion – Emotions, very broadly speaking, are holistic, system-wide responses to the world – response-patterns that grip the whole mind and its body in certain habitual reactions to what is happening and guides its pattern of responses accordingly. AGIs won't have exactly the same emotions as humans, unless they have human-like bodies. But if an AGI has mind architecture that is roughly human-like, it should have roughly human-like emotions.

Language – Linguistic behavior is one of the more unique aspects of human intelligence. Other animals have language too, but human language seems different in important respects. Human language is distinctive in complex ways, cutting across various other aspects of the mind. For one thing, linguistic behavior mixes reactive and deliberative processing. Sometimes we respond quickly and automatically using language, while at other times, we need to think first. Human linguistic behavior also involves a mixture of general cognitive processes (that have to do with language and other things as well), with specialized linguistic thinking. The human brain seems to contain a bunch of fairly specialized "wiring" just for language; this is one major way we've evolved differently from apes, and the ape-like creatures before them. An AGI doesn't need the exact same kind of linguistic wiring as humans, though; it could develop linguistic capability in other ways, either via learning language solely using general learning mechanisms, or via specialized linguistic wiring different from that of humans.

Self/Social – Each of us spends a lot of mental effort modeling ourselves and people around us. My idea of myself, "Ben Goertzel," is not entirely accurate, and is ever-changing. Still, this idea plays a huge role in guiding my thinking, planning, acting, and reacting. My mind spends a fair bit of time maintaining and modifying my idea of myself, and creating, maintaining, and modifying my models of the other people I interact with, which are largely informed by my idea of myself. Many human delusions and confusions derive from these cognitive processes of "self" and "other" modeling, and plenty of data indicates that our models of ourselves are largely delusional. The model that you use to define yourself to yourself is a perpetual mentally constructed entity, not absolutely "real." The self exists largely to build itself. However, this constructed, mentally manufactured "self" is an amazing achievement of human cognition. It's responsible for much of our ability to plan and carry out complex behaviors, both in the physical and social worlds.

Metacognition – Thinking about thinking! People raised in the Jewish culture, as I was, seem to be particularly obsessed with this, and particularly adept at it as well. Sometimes it can just be a waste of time, but other times it's critical. If we think about the strengths and weaknesses of our own thinking process, and adapt them accordingly, we can become smarter. For instance, I've learned over time that when thinking about human social systems, it pays for me to keep specific examples in mind rather than dealing too much with abstractions, whereas when thinking about the nature of cognitive processes, it works better for me to start with mathematical or conceptual abstractions, and then afterwards work out the kinds by interpreting them in the context of concrete examples.

And there you have it – or as the Australians would say, Bob's your uncle. That's a pretty high level view of the human mind. Of course, just dividing the mind into parts at such a high level doesn't tell us what the different parts of the brain actually do, and how they interact, nor does it tell us how to build an AGI executing all of these functionalities. However, I think it's a valuable way to frame the discussion. The next step is to drill down deeper into what happens inside each of the boxes, both in terms of their substructures and dynamics. Then, we'll get closer to figuring out how the different parts of the brain function, and how they interact with one another, which will be useful in framing our discussion on how to build AGI.

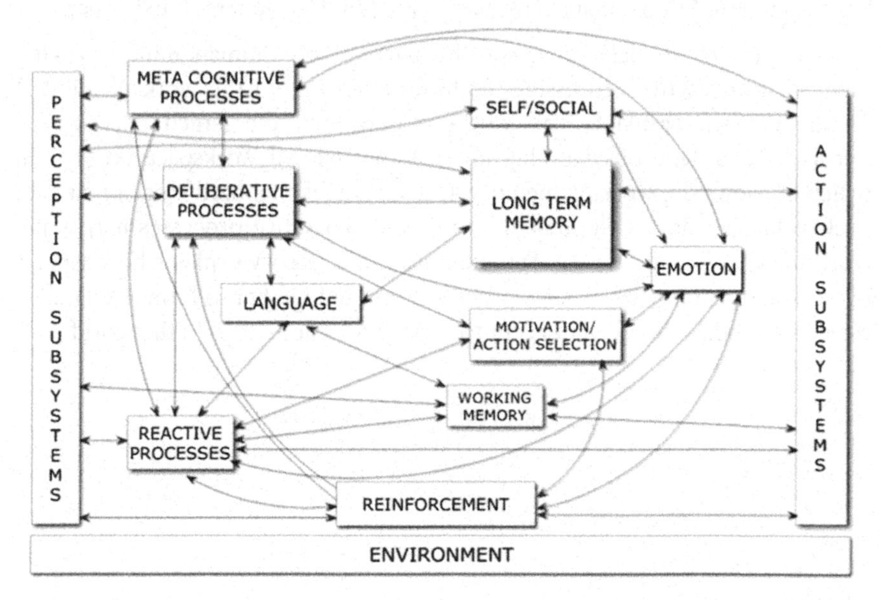

FIGURE 9.1: ***High-level cognitive architecture of a human-like mind, loosely inspired by the work of Aaron Sloman.***

FIGURE 9.2: *Legendary cognitive-science/AI wizard Aaron Sloman, speaking at the AGI-11 conference at Google. One of the joys of organizing AGI conferences has been getting to know some of the grand old men and women of AI and cognitive science, who tend to be excited and a bit amused to see so many youngsters suddenly interested in the thinking-machine goals they've been quietly pursuing for half a century or more.*

Working Memory and Reactive Processing

And now, on to the next diagram. I'm sorry these diagrams are so complicated-looking – but please bear in mind, these are actually highly stylized oversimplifications of what really goes on in human minds. The real story is vastly more complex. This is just the Condensed Version. The processes in the mind aren't really as distinct as these boxes; and each box should really be decomposed into a number of interacting/intersecting smaller boxes… And there's also a host of other more minor and specialized processes not covered by any of the boxes here, etc… As complex as it may seem, the picture of the human mind I'm painting here is merely a first approximation. But it's this first approximation, I think, that's most useful for Artificial General Intelligence.

FIGURE 9.3 models some specific parts of the human mind: working memory and reactive processing. It draws mainly from the work of the AGI researcher Stan Franklin, who works closely with the famed psychologist Bernard Baars. One of Baars' big ideas is the "Global Workspace theory," in which he views the working memory as a sort of "whiteboard" (or, if you date back as far as I do, a "blackboard"), on which cognitive processes may write, read, and modify information. Franklin and Baars' theory explains how aspects of working memory and reactive processing come together to form a "cognitive cycle," enabling an intelligent agent to carry out basic actions in the world.

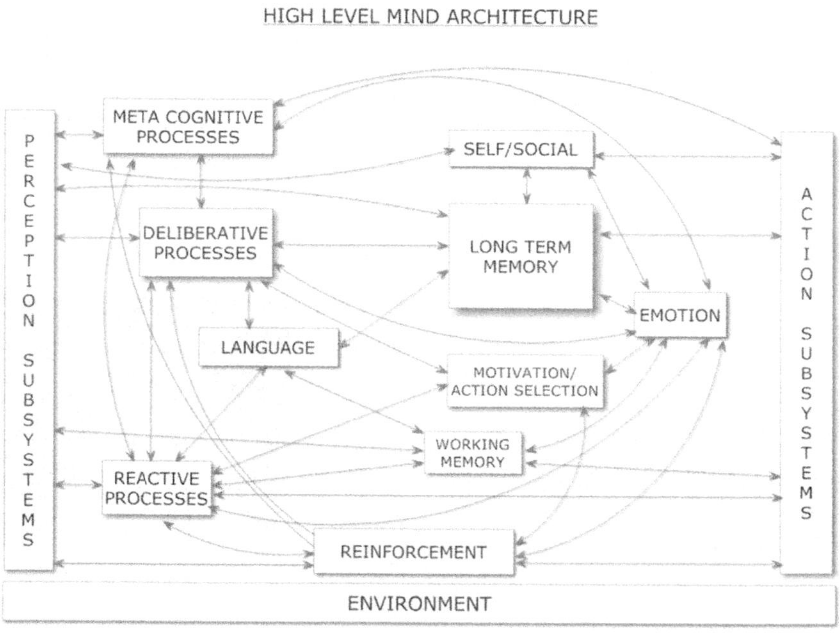

FIGURE 9.3: *Working Memory and Reactive Processing*

So let's run through Stan and Bernard's boxes:

Global Workspace – A "mental whiteboard" that is sometimes called the "theater of consciousness." It is the "mind's eye," where thoughts, perceptions, and actions all come together, and what they drift away from as they become irrelevant to what the mind is currently trying to do (or obsessing over). It's the centerpiece of the working memory. Please note: the global workspace doesn't have to be a physically distinct place or "organ." It is a conceptual category for everything in the mind that's currently represented, in a manner allowing for very easy manipulation and access, out in front, for all the processes of the mind to play with and see. Right now, as I write these words, my mind's eye is mainly occupied with the words and ideas in this paragraph.

Active Procedural Memory – The set of procedures: the concrete actions and action-series that the mind is in the middle of doing at any point in time. Opening a door, solving an equation, or generating a sentence, etc. As I write these words, my active procedural memory contains procedures for typing, and for formulating sentences representing the ideas I think of.

Attentional Processing – The process of moving stuff from the long-term memory into the global workspace, and kicking stuff out of the global workspace (either back into the long-term memory, or just plain forgetting it). As I write these words, my brain's attentional processing function is summoning knowledge about the nature of "attentional processing" from my long term memory, enabling me to write this sentence!

Transient Episodic Memory – An ongoing story constructed in the mind's eye, of "what's happening." Some of it gets saved in the long-term episodic memory. As I type this sentence, I reflect on the story of me typing the sentence, which involves me sitting in a seat on an airplane while a disturbingly nervous man in the seat in front of me shakes back and forth, shaking the laptop I'm typing on and making the typing process awkward.

Perceptual Associative Memory – Associations between what is perceived, and the knowledge, stories, actions, and goals in one's memory. As I write this, my mind is associating its perceptions of the person shaking in the airplane seat in front of me, with its memories of a guy I knew in college, whose body never seemed to stop shaking. And I'm recalling that there were no laptops back then, in the early 1980s. Computer technology has advanced dramatically, whereas human body control has remained about the same.

Sensory Memory – The stream of perceptions that come into the mind's eye, lingering momentarily, until they're either focused on, or (usually) forgotten about. Typing this, my sensory memory has the image of the laptop in front of me, the image of the glowing keys on my Macbook keyboard, the dim view of the airplane floor to the right of the laptop, the sound of an announcement on the plane's loudspeaker in a foreign language, the unpleasant body odor of the man sitting in front of me... The look of the airplane floor a minute ago has already been forgotten (I can't even remember whether the rug is gray or brown); but the body odor smell, due to its atypical strength, will likely remain in my long term memory at least for days...

Sensorimotor Memory – Linkages between perceiving and doing, which are often accomplished in the mind as one. For instance, when opening an unfamiliar door, you will look at your hand as it reaches towards the knob, and grabs and turns it; a bundle of interlinked perceptions and actions in which eye-hand coordination occurs. Watching my fingers as they type, I wonder if this helps me type faster, but determine that it doesn't at all. However, as I reach to change the volume of the laptop's sound output, I use eye-hand coordination, since I don't automatically know the location of the volume button via finger-movement only. Eye-hand coordination requires a working memory system in which visual perceptions and motor movements are actively and dynamically linked together.

Action Selection – (See FIGURE 9.6) The process of choosing what actions to take; i.e., choosing what procedures to make active. As I sit here editing this manuscript, I must choose whether to continue writing, or get up from my seat on the airplane and walk to the restroom. This has to do with balancing my goal of finishing the manuscript rapidly and efficiently, with my goal of having a comfortable body. The "finish the manuscript" goal will win the contest and control the action selected until the urge to pee becomes strong enough, at which point the latter goal will become dominant and get to control my body's choice of action. The high-level action of "continue editing and writing the manuscript" then spawns sub-actions, including cognitive actions like language generation and physical actions like typing.

Perception-Action Subsystems – The linkage between the working memory and the parts of the mind doing the lower-level work of perception and action. As I watch my fingers type, intrigued by the way they know where the letters are, I retain a memory of where my fingers moved a few moments ago. I also maintain knowledge of what I'm about to type – so that, for example, if a comma is going to be needed soon and the relevant finger of my left hand doesn't have anything else to type in the near future, it will move to the comma key proactively. My working memory, at the moment, also contains some abstract thoughts wondering how extensively my mind, when typing, unconsciously uses its knowledge of what's going to be typed in the immediate future to guide its finger movements and thus optimize its typing speed. My working memory contains some curiosity as to whether this particular instance of motoric prognostication is occurring in the cerebellum (my guess) or the cortex.

Long-Term Memory – The linkage between the working memory and the long-term memory. Some of the processes I'm carrying out now as I edit this manuscript will be remembered by me tomorrow or a month from now – because they'll get shuttled from working memory to long-term memory.

Others will be utterly forgotten after a little time has passed, never having made that transition.

Whew! That's a complex diagram! And generally speaking, all the parts have to work together for a mind to get anything done.

Remember, though, this sort of diagram depicts functions and processes, not necessarily architectural components. A brain may carry out the functions in one of these boxes using a complex distributed network of neurons, spanning multiple brain regions.

Stan Franklin has his own AGI design called LIDA, which is explicitly designed in accordance with FIGURE 9.4 – basically, it puts some data structures and algorithms in each of the boxes. The OpenCog design I'm working on is a little different, realizing the functions in Stan's boxes in different ways. OpenCog, unlike LIDA, doesn't actually contain different software components for each box, just as the brain doesn't necessarily contain separate regions for each box.

The point of this sort of cognitive diagram is to describe what kinds of functions go on in the mind, and which ones are directly related to each other. How these functions are realized by more detailed, underlying structures and dynamics, in the brain or in an AGI system, is another story; different sorts of systems may realize the same basic functions and interactions in different ways.

FIGURE 9.4: *Stan Franklin and fellow AGI researcher Pei Wang (creator of the NARS AGI architecture, editor of the Journal of AGI, translator into Chinese of Godel, Escher Bach, and my employee at Webmind Inc. where he was Director of Research) at AGI-08, the first full-scale AGI conference, which Stan kindly hosted at the University of Memphis in an extremely futuristic FedEx-sponsored conference room.*

Why Is It All So Damn Complicated?

At this point, if you are a person with a taste for simplicity and elegance, you may be thinking: *All of these diagrams are awfully complicated, with these boxes and lines and interactions and confusing terminologies and such! And you've said that these are just the first approximation! Is this really the best way to comprehend the human mind? Is there no other path?*

Believe me, I do sympathize. I wish there were some really simple, elegant explanation of how human intelligence works. To be honest, I have spent a rather long time looking for such a thing. But eventually, I realized that a simple explanation for human intelligence simply doesn't exist.

Of course, there are simple explanations at a very high level. Like the one that Palm founder and AGI entrepreneur Jeff Hawkins is always repeating, "The mind uses memory in order to predict." I'd prefer to flesh that out slightly to: "The mind uses memory in order to predict what will happen, then chooses actions that it predicts will achieve its goals." Which I guess is what Hawkins really means, though he tends to give short shrift to the action and goal parts. Sure, this makes sense. But how much does it really tell us? This sort of high-level understanding is fine as a guiding philosophy. If we want to build stuff, though, we have to dig into the details.

Take a look at the with Figure 9.5. This is the Schrodinger Equation, one of several elegant formulations of the key equations underlying quantum mechanics, on a T-shirt. Explaining, in principle, an awful lot of the phenomena we see in the everyday world around us, it's an incredibly powerful equation. It doesn't explain gravity or the nuclei of particles, but it pretty much takes care of electromagnetism, light, and so forth, under ordinary sorts of conditions (and a lot of extraordinary ones). Modern physical theory could be summarized in half-dozen or so equations like this. A half-dozen T-shirts. Or one T-shirt, if you're willing to use the front and back and make it a little crowded.

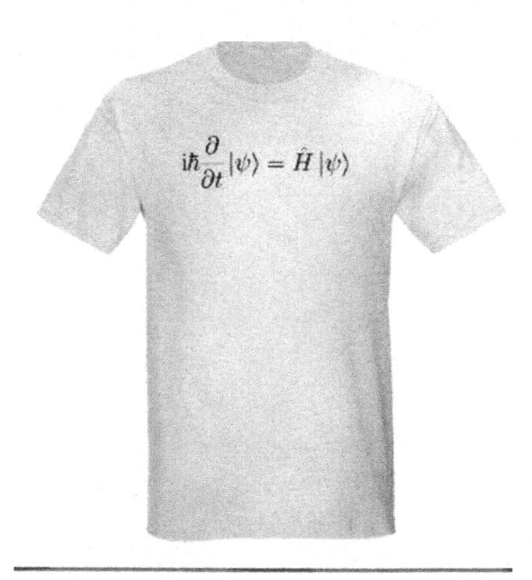

FIGURE 9.5: *In the realm of psychology or AGI, we don't have this.* *(http://i1.cpcache.com/product/503185794/schrodingers_equation_light_tshirt.jpg?color=AshGrey&height=460&width=460&qv=90).*

What if there was some basic understanding of the mind that you could write on a T-shirt (in some suitably abstracted mathematical notation), and that would let you calculate detailed stuff about how the brain works in the context of building AGI systems? That would be awesome! But it just doesn't seem to be the case. There's a term "physics envy" that one hears sometimes among biologists or social scientists. This refers to the misguided attempt to mold other sciences into imitations of physics, when in reality the subject matter of these other sciences doesn't lend itself to the same kind of powerfully simple and elegant abstractions.

You can write out elegant equations describing key aspects of how the mind works. I've done some of that myself in some of my research papers. Since my PhD was in mathematics, I have a fondness for that sort of thing. Yet these elegant equations are more descriptive and conceptual; they don't tell you how to do specific stuff, which is half the beauty of physics equations (the Schrodinger equation lets you figure out how to do cool things with real physical systems like lasers).

General intelligence at the highest level, abstracting beyond the details, is not so complicated. It's "just" the ability of a system to recognize patterns in the world and in the system itself. Some examples of patterns that a general intelligence will recognize; patterns regarding which actions tend to achieve which of the system's goals, in which contexts. For instance, a normal human baby quickly learns a pattern of the form: "If I want food [GOAL] in a situation where another person is nearby [CONTEXT], maybe I should make a lot of noise [ACTION]"... Compared to babies, adult humans need to learn much more complex patterns relating to goals, contexts and actions. But they don't get there all at once – a developing mind gradually works its way up from simple patterns to more complex ones, leveraging the patterns it's learned in the past as it proceeds.

When you dig a little deeper, however, things get more complicated. Recognizing patterns is an expensive operation, taking up lots of computational resources, and no system with finite resources is going to be equally skilled at recognizing every kind of pattern. So, you've got to prioritize. Which kinds of patterns are most important for this particular intelligent system to recognize efficiently? "Efficiently" meaning fast enough that it can recognize these patterns on the fly, in the course of its everyday AGI life? This depends on the specific nature of the system's goals and environment.

All the complexities of human-like cognition, just like the complexities in the box and line diagrams above, are basically ways that the human mind has adapted in order to efficiently recognize the particular sorts of patterns it has found useful in the context of its quest for survival and reproduction, over the course of its history.

To illustrate this point, let's take an example from FIGURE 9.7 and dig into it a bit: *Why have a box for long-term memory and a box for working memory? Doesn't that just complicate things? Why not just one memory?*

Well, actually, the two boxes are just an approximation. The human mind has a host of different memory subsystems with different properties, and the long and short term aspects closely interoperate together. This is also an approximation of OpenCog, which has a more complex story underneath the hood.

Basically, the reason for the distinction is that minds controlling bodies in environments need to do two different sorts of things:

- Sometimes, the mind needs to react pretty fast to stuff happening in their environment in order to take the appropriate actions. For instance, when you're being hunted by a predator, you need to run quickly; or you see some food, maybe even an attractive mate, and you need to act before somebody else does. Or somebody asks you a question, and you need to answer before they get mad.
- Sometimes, the mind needs to keep knowledge around for a long time in case it's useful again in the future. In this case, there's plenty of time to consolidate, reorganize and refine the knowledge before it's needed again.

Maybe, in principle, a single kind of memory could accomplish this. However, there's always the problem of resource limitations. The brain has always taken a lot of energy to operate, so evolution has had a lot of pressure to keep the human brain smaller, yet more efficient. Also, as the brain grew increasingly larger throughout human history, women found it increasingly difficult to push out their babies' heads while giving birth.

In a digital AGI context, our modern computers only have a certain number of processors and a certain amount RAM. When you start dealing with the huge computer clusters of Google and Amazon and so forth, you start running into cost, electrical power consumption and heat generation issues. In the real world, you always have this pressure to do what the mind needs, using as few resources as possible. Under this pressure, the easiest way to provide both real-time responsiveness and flexible long-term memory seems to be connecting two fairly separate memory stores. This is the quick and dirty solution that evolution happened upon in creating the human brain; AGI architects have also gravitated toward this kind of solution.

It's not just in the context of working memory and long-term memory that this sort of quick-and-dirty, efficiency-driven compromise occurs — this sort of thing actually happens over and over again, in nearly every aspect of the mind. The practical requirements for a real-world intelligent system, including energetic and computational resource restrictions, seem to naturally push toward a system that's divided into a bunch of different aspects, all interacting with each other in various ways. The resulting systems are complicated, but they work. The more elegant ways of organizing intelligence, without so many complexly interacting subordinate parts, so far as we've been able to discover so far, only exist in the domain of abstract mathematics, with its endless energy and resources, rather than in the real world where practical pressures still prevail.

When looking at the human body as an analogy, this aspect of the mind doesn't seem surprising at all. The body has a lot of different organs, each carrying out their own specialized functions in specialized ways. The organs interact with each other complexly, and in many cases they have evolved specifically to take account of each other's functions. But still, there are a lot of complex, self-organizing messes involved. I'm sure it's possible to engineer a much more elegant body than the human one, with more unifying mechanisms and principles. However, I'd argue that there's still going to be a lot of heterogeneity and complexity, even in a really elegantly designed humanoid robot. The problem of designing a good foot has relatively little to do with the problem of designing a good ear.

For example, look at the following design for a disaster response robot, which my friend and collaborator David Hanson created as part of an AGI robotics project we were discussing back in 2012. If you take a careful look at the feet, the knees, the ankles, the elbows, the eyes, etc., you may be able to imagine the care and subtlety of the art, science and engineering that goes into the creation of each part. A similar or greater amount of subtle thinking goes into internal parts of such a robot, such as custom design of motors and gearboxes, space-saving and maintenance-friendly layout of wiring, etc. And all these aspects need to work together effectively in a way that conserves power and achieves holistic functions.

This robot was never built due to lack of funds, but we have the specs in case you want us to build one for you! It's kinda cute, in my opinion. If I were

a victim of some disaster, I'd be pretty happy to have such a charming bot roll in and rescue me.

FIGURE 9.6: *A disaster response robot designed by David Hanson, but never (yet) produced. Notice the feet — they are actually wheels! Based on software instructions, they can deflate and then be used as flat tire feet. The legs also have a special design, so that they can switch from being compliant (flexible) to rigid, based on software commands. The feet and legs were designed based on a common principle: the software control of the transition between different states. The particulars of the feet and legs are obviously very different, engineering-wise. This is because being a foot is a different sort of enterprise than being a leg. In principle, one could design some sort of highly flexible mechanism that could serve effectively as both a foot and a leg. But that wasn't David's approach when designing the robot, as it wasn't really a plausible way to proceed based on the materials and knowledge at hand.*

I think this is basically the same as the situation inside the human mind with long-term and working memory. The easiest way to create a system with limited resources that has both kinds of memory is to create two somewhat distinct memory subsystems, which will then need to interact in various ways, leading to various complexities. Yet ultimately, there are hundreds of different cognitive subsystems, each serving its own purpose relative to the goals and environments for which humans evolved, and each interconnected for effective combined operations.

In building an AGI, we don't need to emulate all the particular subsystems of the human mind and their interactions. We just need to build something generally similar in nature, then tune and adapt it for increasingly intelligent functionality. The physical substrate of our AGI programs and robots is currently rather different from the human body, which already implies that doing things in our AGI minds exactly the same way as in the human mind doesn't make sense.

Motivation and Action Selection

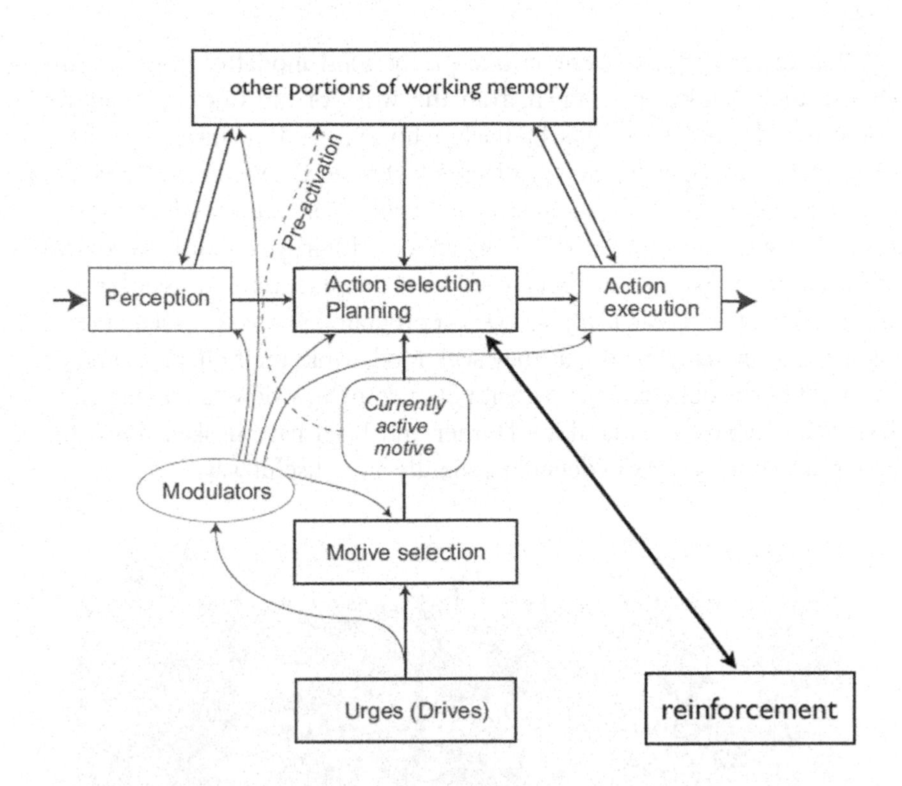

FIGURE 9.7: *Action Selection*

One of Stan Franklin's cognitive science insights was that, to understand the mind in a simple way, it pays to start from the perspective of ACTIONS… To ask, "what does the mind DO in the world?" One of the big differences between narrow AI and AGI is that narrow AI programs tend not to be "autonomous agents" – they tend not to be agents that explore the world on their own and carry out their own actions in pursuit of their goals. Rather, narrow AI systems tend to be components that are used as tools by human agents, in pursuit of specific human goals. But humans, like animals, are autonomous agents, exploring and acting and surviving and striving… And AGI systems, if they're going to be remotely human-like, have got to be agents in this sense too.

With this in mind, Stan Franklin has sometimes termed his approach to AGI the "action-selection" approach. The key question about an AGI system, in this view, is: How does it choose which actions to take, at which points in time?

The diagram above presents one pretty good model of how humans choose their actions. It's drawn from the work of the German cognitive scientist and AGI designer Joscha Bach, who in turn drew inspiration from the German cognitive psychologist Dietrich Dorner. Dorner created a model of human motivation and action selection called "Psi," and Joscha created a proto-AGI system called "MicroPsi," which uses Psi-inspired ideas to control artificial agents. MicroPsi has been used for some practical purposes, but I've studied one of its research applications: a program used to control animated agents in a simulated world, as they seek food, avoid their enemies, and so forth. All this is quite compatible with Stan Franklin's view as articulated in FIGURE 9.7 above – I just think Dorner and Bach have broken down the action selection process in detail in a slightly more useful way.

FIGURE 9.8: *Another blast from the past – yours truly, German AGI/ cog-sci mastermind Joscha Bach and revolutionary humanoid roboticist David Hanson in a panel discussion at the AGI-08 conference. This was well before David and I both moved to Hong Kong, where we're currently collaborating on proto-AGI robotics applications. The second image is a screenshot from one of Joscha's proto-AGI simulations; his MicroPsi software controlled little video-game-type agents moving around in a simple 2D simulation world. Currently, his MicroPsi 2 system is controlling agents that play the 3D game Minecraft instead.*

The basic ideas of Psi are not so complicated. A mind is viewed as having a bunch of different drives, or urges, which ultimately motivate behavior. For instance: get food, get water, get sex, be safe, learn new things, interact with another, etc. Obviously, the basic drives may be different for an AGI than for human beings.

Then, at any given point in time, Psi models the mind choosing one of the drives as its key motive — the main thing it's trying to accomplish at that point in time. This is a bit of a simplification, since, arguably, a mind could actively work toward more than one high-level motive at the same time. But it's a good approximation most of the time, especially because the model allows other motives to take up some energy in the background. The "motive selection" box in the diagram refers to the process of choosing which motive gets the most attention. In the language of the Working Memory diagram (FIGURE 9.9), this is largely a matter of "attention processing."

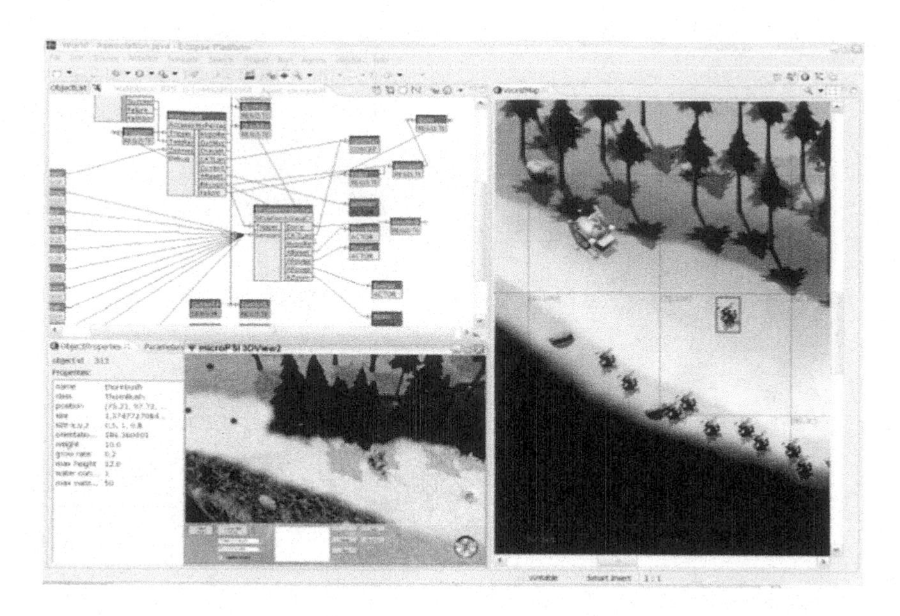

FIGURE 9.9: *User interface for viewing, inspecting and controlling the mind-states of MicroPsi-controlled agents in their virtual world. Image courtesy of, and copyright to, Joscha Bach.*

Once a motive has been selected to focus on, one must choose what action to take, based on that motive and the mind's prior knowledge about what actions have helped it to fulfill similar motives in the past. Reinforcement learning plays a role here, along with many other sorts of learning based on working and long-term memory. Various parameters (which Psi calls "Modulators") affect the action selection process. For example, one parameter governs the system's pace. Being hurried affects many aspects of a mind's intelligence, including how much detail it studies in its perceptions, how carefully it checks its inferences regarding future actions, and so forth.

Does Psi tell us everything about the human mind? Of course not. For one thing, what Psi says is equally applicable to any complex animal. Also, Psi doesn't tell you how the mind sifts through large masses of data in short periods of time, nor how the mind forms abstractions. Furthermore, it doesn't tell you that much about how the different goals and modulators are driven by a system's experience – actually, in 2016 we began augmenting Psi in OpenCog with complementary ideas from something called the "Component Process Model", also based on psychology and cognitive neuroscience.

But Psi does tell you the basic logic by which a human-like mind selects actions, in accordance with its motivations. And that is something important.

Emotion

You may wonder: *Why is there no box for "emotion" in our action selection diagram (*FIGURE 9.11*)? After all, in humans, emotion plays a vital – maybe primary – role in governing our choices of what actions to take. We are emotional beings!*

The reason there is no emotion box is that, according to Psi, emotion is an emergent phenomenon — a system level response to the system's overall activity, in reaction to what it's doing, observing and experiencing. Emotions are high-level patterns of dynamical activity that span ALL the boxes, in other words.

For instance, if an intelligent system keeps getting positive reinforcements from its body, and especially if it gets these reinforcements when it does not expect them, then it will tend to experience the system-wide response pattern of "pleasure."

If the same intelligent system keeps getting thwarted in achieving its goals by some other agent, unless it has a particularly emotionally mature self-model and overall system dynamic, it will tend to experience the system-wide response pattern of "anger."

And so forth...

Of course, not every human emotion needs to be included in an AGI system. Why should an AGI need to experience anger or jealousy, the same way people do? Yet the basic structure of human emotional response does not seem to be a highly specific consequence of the human mind/body, but rather a consequence of the general relationship between any embodied intelligent agent and its world.

One of the more popular ways of thinking about this, in the cognitive science field, is the so-called "cognitive theory of emotions." This theory attempts to boil down all emotions to a few simple parameters, in a systematic way. FIGURE 9.10 gives the basic idea.

For instance, "pride" is our label for the kind of emotional response typically associated with approval of the agent's own actions. "Admiration" is the kind of emotional response associated with approval of another agent's actions. "Disappointment" is the kind of emotional response associated with negative evaluation of events that have personal consequences. In each case, the emotion in question is not merely a logical observation, but a system-wide response. "Pride" isn't just a logical observation that "I approve of these things that have occurred, associated with my actions"; it's a coordinated response of many portions of the mind, correlated with this logical observation.

Of course, human emotions are too complex, messy and multifaceted to be captured in any specific logical formulation of this nature. But the cognitive theory of emotions provides an explanation of the common core of emotion that spans any kind of intelligent agent that controls a body in a social world. Different kinds of intelligences will then manifest these abstract emotional structures in different ways.

I saw the truth of this when, in 2007 and 2008, I was working on a project involving AI-controlled virtual pet dogs. The system we were working with was fairly simplistic. But taking it as an inspiration for thought-experiments, it became clear that as this kind of virtual canine entity interacted with its virtual world more and more extensively, it would be able to "experience" every kind of human emotion. Dogs have not only simple emotions like happiness and sadness, but also envy, disappointment, pity and so forth. They experience these emotions according to the same basic logic as humans, but with different, doggish particulars. In a virtual world context, one can set up situations evoking each of these emotions. One can evoke virtual dog disappointment, for instance, by showing an AI dog 10 situations where there's a little red house with food inside it – and then, the 11th time, showing it a little red house with no food inside. The observation that there's no food inside the 11th house will cause reverberations through the AI dog's internal state in a variety of ways, constituting the virtual dog's experience of disappointment (or the "structural" equivalent, as in FIGURE 9.10).

All this may seem like an overly mathematical or algorithmic way to think about something as raw, personal and experiential as emotion. But I like to remember the cognitive psychologist George Mandler's terminology of "hot" versus "cold" emotion. Hot emotion is the raw feel of the emotion. Cold emotion is the structure of the emotion: What does it react to? How does it urge one to act? How does it cause one to represent and relate information? To put it simply, we might say:

$$\text{cold emotion} + \text{consciousness} = \text{hot emotion}$$

FIGURE 9.10: *Screenshots from some of the work we did in the mid-aughts, using the Novamente Cognition engine to control virtual dogs in virtual worlds. One thing we learned from the research we did over the course of this project was that dogs can experience basically the same range of emotions as human beings, if placed in the right situation. To*

be fully realistic, a virtual dog's emotional model has got to be basically as complex as a virtual human's emotional model.

In other words, the raw feel of emotion, the hot emotion, is just the subjective, conscious experience that correlates with the structure identified by the cold emotion; just like the raw feel of seeing the color red is the subjective, conscious experience that correlates with the visual stimulus corresponding to the color red.

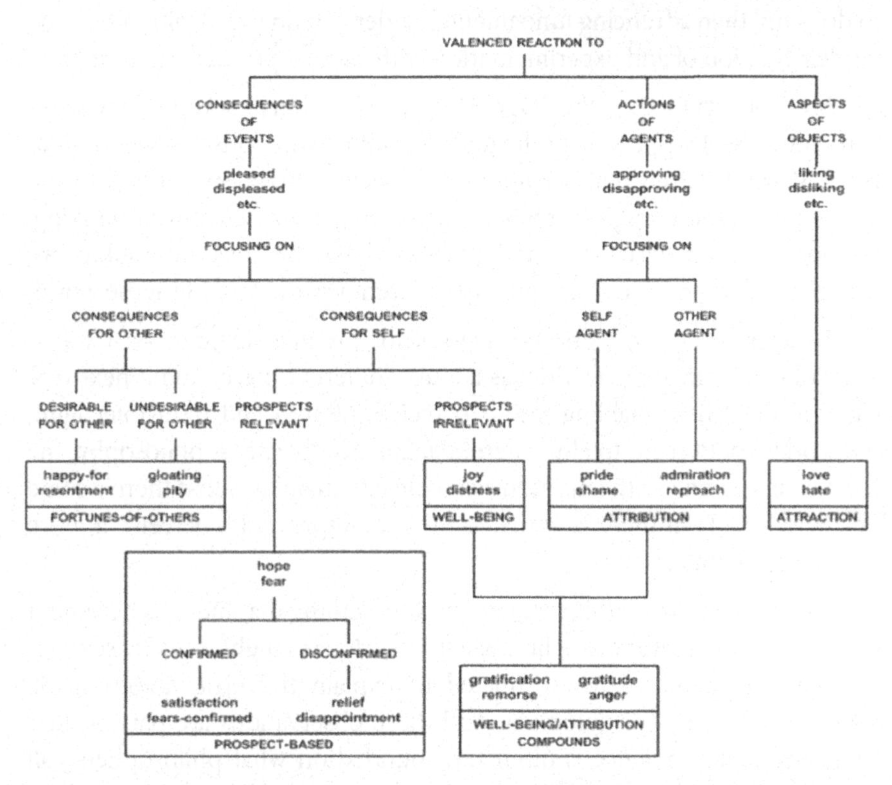

FIGURE 9.11: *Illustration of the "cognitive theory of emotion," showing how various common emotions emerge from basic cognitive aspects of an intelligent agent. (https://www.goodreads.com/book/show/1927037.The_Cognitive_Structure_of_Emotions).*

If there's a mystery to the experience of emotion, I would say it's part of the broader mystery involving the experience of consciousness. The mysterious thing about emotion is not its structure, nor its connection to an intelligent organism's ideas, goals and reactions, but rather the way it FEELS. But then, I would say: *The conscious experience of solving an equation or perceiving the color purple is equally mysterious as the conscious experience of feeling disappointed or ecstatic.*

Consciousness is a tricky problem, but I don't think it has to be solved in order to build an AGI. Similarly, the philosophy of time and space involves a lot of thorny issues that nobody has resolved yet – but that hasn't stopped us from building spaceships, lasers, particle accelerators and so forth. Engineering isn't done by fully understanding some aspect of the world and then leveraging this full understanding to do stuff. Rather, it's done by understanding ENOUGH of some aspect of the world to do what one wants to do – and then advancing fundamental understanding gradually, alongside implementation of and experimentation with various practical constructs.

My colleagues on the OpenCog project have several views on consciousness. I suspect that fully understanding consciousness will require going beyond the current scientific world-view – not necessarily into the domain of mystical religion or anything like that, but a new way of thinking. For instance, I'm fascinated and perplexed by the idea that subjective experience and physical reality are just different ways of looking at the world.

I remember when I first got glasses for my nearsightedness: I was 5 years old. Suddenly, the world was a totally different place. And if next year I got an operation enabling me to perceive infrared and ultraviolet light, the world would seem totally different again. An objectivist philosophy, the most common one in the modern scientific era, maintains that there's some objective world out there, and that as my eyes improve, I'll perceive it more and more accurately.

On the other hand, this very theory is something that humans have built up from their observations. The objective world, as taught to us by science, is something people have abstracted to explain the observations made using their laboratory equipment. And these observations are part of their subjective reality. Science is ultimately founded on what philosophers call "inter-subjective" reality – the reality that arises when the various members of the community of scientists look at certain lab equipment in the context of certain experiments, and they all subjectively report seeing the same thing. So when you really dig deep, you can explain subjective reality as an approximation or observation of objective reality; but objective reality is something created out of subjective and inter-subjective observations. And consciousness, I feel, has to do with this tangle, this strange loop via which the subjective and objective create each other.

Absolutely no scientific reason exists to believe that the human brain has more capability for conscious experience than digital computers. In fact, the very concept that particular physical systems uniquely give rise to conscious experience while others do not reveals lots of logical flaws and contradictions when you examine it closely (readers with an analytical-philosophy bent are encouraged to look up the writings of Galen Strawson). Some AGI researchers think that "consciousness" is a red herring, and that we should just talk about information processing. Some AGI researchers are pan-psychist, believing everything in the universe has some measure of consciousness, but that different entities manifest their consciousness differently. Brains are conscious in brain-y ways, AGI systems will be conscious in their own ways, and rocks are conscious in their own (presumably less intense) ways. Some AGI researchers, like me, think there are interesting mysteries to be solved in the area of consciousness.

Anyhow, it doesn't seem necessary to resolve the perplexities of the philosophy of consciousness in order to build AI or AGI systems that do intelligent things in the world, any more than it's necessary for an artist to fully clarify the philosophy of beauty to make a gorgeous painting.

I think AGIs will have emotions and other conscious experiences roughly as people do, though their emotions will have a different flavor, rooted in different forms of embodiment and mental algorithms. If we look at an AGI's software, or a human's brain, we won't see consciousness or hot emotions, but a bunch of information flowing around. But from the subjective point of view of an AGI, and the corresponding human perspective, there will be plenty of consciousness and hot emotion in there.

Deliberative Processing

FIGURE 9.12 depicts the parts of the mind I've spent the most time thinking about and working towards in AGI systems: long-term memory and the associated "deliberative" thought processes.

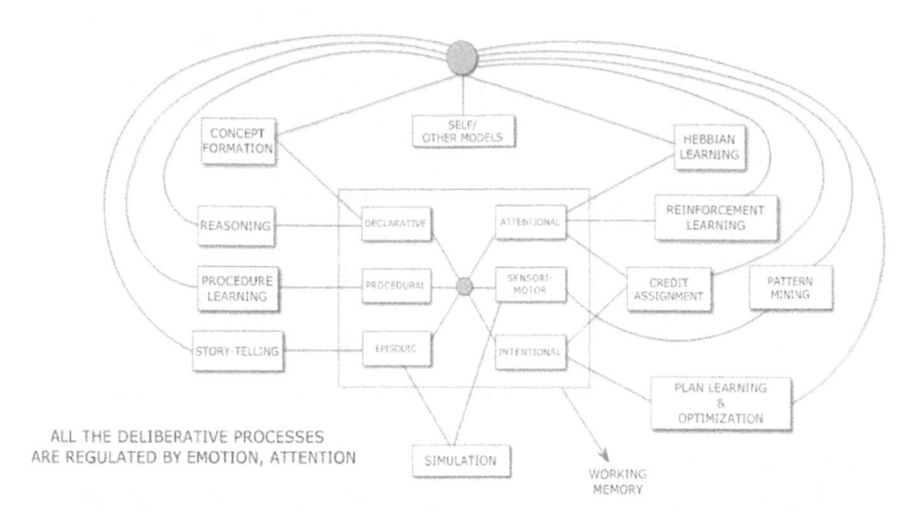

FIGURE 9.12: *Long Term Memory and Deliberative Processing*

We've already talked a bunch about "working memory," the memory used by the mind in the short-term to carry out specific tasks. This is closely tied to the "cognitive cycle," the running loop of perception, cognition and action that uses the working memory to monitor the environment and carry out immediate tasks. According to cognitive science, in the human mind this is distinct from the process of "deliberation in long-term memory." This process involves deep, ongoing thought and concerted reasoning, which require searching the memory for new knowledge in a deliberate way and trying to focus it on the problems at hand.

Humans spend a lot of time doing this kind of deliberative thinking; some of us more than others. Animals, like dogs, birds and pigs also do it, but it doesn't play as large a role in their mental lives.

What FIGURE 9.12 depicts are the multiple substructures and sub-dynamics within deliberative thinking and long-term memory.

First of all, there are three main kinds of long-term memory:

Episodic memory – The memory of our life history. This includes "imaginative episodic memory," which consists of thoughts about what might happen to us, and what might happen to others in various circumstances.

As I type these words, I'm sitting on an airplane flying from Tokyo, where I visited my oldest son Zar, to Hong Kong, where I live. The set of multisensory images in my mind, pertaining to my visit to Zar's apartment, is an episodic memory – which is obviously tied in with a bunch of memories of other sorts.

Declarative memory – Facts, beliefs, statements, propositions, and conjectures. These are the kinds of things you could naturally express in a few impersonal sentences in any language.

Regarding my recent visit to Zar, these would include facts like "Zar's new apartment is in Tatebayashi" and "Zar is now teaching in middle school," and beliefs like "Zar fits awkwardly into Japanese culture," "Zar's apartment is small but somewhat charming," and "Tatebayashi is a bit boring." These facts and beliefs have varying levels of confidence in my mind.

Procedural memory – The recollection of how to do both concrete and abstract things — like how to navigate, seduce a woman, walk, prove a theorem, write an essay, or generate a sentence. These are things that we know how to do and can teach someone else do instinctively. However, we can't explain declaratively in language exactly what we're doing when we are, in fact, doing it.

My memory of how to get to Zar's apartment from Tatebayashi Station is partially procedural. There are declarative and sensorimotor aspects to it as well; I remember some facts about the route, and I have some visual images of the walk in my mind. But principally, I have a navigational procedure in my mind – if I were at the station again, the procedure I learned for taking that walk would kick in, and I'd be able to find my way to his house by a combination of enacting previous habits, and leveraging previously gained sensorimotor and declarative knowledge.

If your goal is to model how human long-term memory works, nearly all cognitive psychologists agree pretty strongly that it makes sense to distinguish between these three different kinds of memory. It's a little less standard, but I also like to distinguish two other kinds of memory, which I think are fairly distinct in the human mind (and can become distinct in an AGI system).

I like to think about "attentional memory," meaning knowledge of what or whom to pay attention to in what circumstances. For example: On the walk to Zar's apartment from the Tatebayashi train station, I paid more attention to stores than to houses and apartment buildings, because there were fewer of the former. My attention was drawn to entities with greater surprise value, which is a common phenomenon. Also, in addition to empirically evaluating surprise value, my mind unconsciously remembered that "pay attention to stores" is a good heuristic for what to pay attention to when navigating in residential areas.

And then there's what I call "intentional memory," knowledge of which goals we should pursue in certain circumstances, and how we break down a goal into sub-goals using a combination of other kinds of knowledge.

Sure, a mind may have just a handful of ultimate high-level motives, but in practice we don't always work directly toward those motives, and instead we work toward various sub-goals that we think will help us achieve those motives. If a human's high-level motive is to "get sex," his or her sub-goal might be to seduce a foreigner, and then a sub-goal of that might be to learn a new language, but since language school costs money, a sub-goal of that might be to get a job to pay for language school, and so forth. Humans are pretty good at dealing with long sub-goal chains, compared to other animals.

In finding my way to Zar's place from the train station, I have a goal of getting to his apartment, and I also know that a subgoal of this is to get to a certain major street that has a convenience store on it with a "P" sign in front of it, and on which a number of small streets end. Breaking down a goal of urban navigation into subgoals regarding recognizable streets or buildings is intentional knowledge I've gained via years of walking around in various places.

Now, what does the mind do with these various kinds of long-term memory?

One of long-term memory's functions is simply storing stuff for a long time, so it can then be pulled back into working memory when something associated is perceived or thought of in working memory. It also serves as a kind of long-term, background, slow-paced "global workspace." Deliberative cognitive processes, operating partly in working memory, but partly in the background as "unconscious" cognitive processing, work together by reading information from, and writing information to, the mind's various interconnected long-term memory stores.

I view each kind of long-term memory as having its own specific kind of deliberative processing.

Declarative memory is closely tied to the various forms of REASONING. Yet "Reasoning," as it is considered in cognitive science and logic theory, goes beyond confident, rigorous mathematical reasoning. There's "inductive reasoning" – guessing that what one has seen in a bunch of cases is likely to continue. If the last 5 big dogs I bit on the nose got mad and bit me, maybe the next one will too. There's "abductive reasoning" – guessing that things with some similar properties are likely to have other similar properties. These four guys I've seen with fancy suits all had a lot of money, so maybe most people with fancy suits have a lot of money. Analogical reasoning is another way of looking at induction and abduction by reasoning about new situations or objects by analogy to previous, similar cases.

One of reasoning's distinguishing properties is that it mostly proceeds step by step. You go from some facts or assumptions to a conclusion, and then on to the next conclusion from that one, etc.

The human mind is better at some kinds of reasoning than others, but overall we seem to be much better than other animals at carrying out long chains of reasoning of various sorts…

"Procedural memory" is connected to different sorts of learning. A young child doesn't learn to walk by reasoning about it. They learn by trying different approaches, combining them and varying them until they find something that works, and then experimentally tweaking the working approaches they've found in the course of their ongoing practice. This is the same way we learn to play tennis or carry out other, more complex physical actions, and it's the same way we learn more cognitive procedures, like solving equations.

Since I'm reasonably good at math, when I sit down to solve an equation, if it's a familiar kind of equation, I don't waste time explicitly reasoning about what steps to take. I just do it. I already have a sense of what steps to take first, and I proceed to fiddle with rearranging the terms in the equation until I get to the answer. On the other hand, if it's especially tricky-looking or an unfamiliar sort of equation, then maybe I will first explicitly and carefully reason about what approach to take.

This type of learning by experimentally inferring procedures also helps us master grammar. This is why, even after we know a language perfectly well from a practical perspective, we still have to study to know the rules of grammar. We don't learn language, as young children, by reasoning, "Hmm, what grammatical rules must people be following, in order to produce the sentences they're producing, and not other ones?" Rather, we use a kind of sophisticated trial and error, trying different ways of generating sentences and seeing which ones work, combining and modifying and fine-tuning the workable approaches.

Reinforcement plays a big role in this type of learning. We try stuff, then change our approach based on constant feedback about our degree of success. There's also a heavily "evolutionary" aspect to this. Just as evolution involves the combination and mutation of genomes to create new organisms, learning new procedures involves combining and mutating tried approaches to find new ones. This evolutionary aspect relates to certain approaches in AGI, which involve computer science algorithms called "evolutionary algorithms," modeled on the evolutionary process.

"Episodic memory" seems to be tied to mental simulations running through a set of imaginative "what if" scenarios in your mind.

"Attentional knowledge" goes along with processes that cognitive scientists call "association-spreading." These are processes in which attention associated with one thing spreads to other, related things, spreading activation through the mind. At the end of the 1800s, William James and Charles Peirce discussed this kind of process at length. Peirce thought it was the crux of intelligence:

> *Logical analysis applied to mental phenomenon shows that there is but one law of mind, namely that ideas tend to spread continuously and to affect certain others which stand to them in a peculiar relation of affectibility. In this spreading they lose intensity, and especially the power of affecting others, but gain generality and become welded with other ideas.*

Association-spreading has been repeatedly simulated in neural network models.

Finally, "intentional memory," relating to the pursuit of goals, integrates all the different kinds of memory and thought in the mind: reasoning, procedure learning, association spreading, and simulation of episodes in which similar goals were achieved.

But what about "metacognition," which is separate from deliberative thinking, according to the Diagram of High-Level Architecture of Human-Like Minds? Metacognition refers to "thinking about thinking." I've done an awful lot of this as an AGI researcher – but it's not restricted to AGI researchers! Comedians, like Woody Allen, have dramatized the tendency of certain individuals and cultures to introspect and obsess on their own thoughts and feelings. Jewish culture, in which both Woody Allen and I grew up, tends to have this characteristic – maybe this is part of the reason there are so many Jewish AI researchers!

In terms of human psychology, it makes sense that metacognition would be considered a separate capability from plain old cognition, as some folks are better at metacognition than others, somewhat independently of their intelligence in other respects. However, in spite of its different characteristics, I think metacognition involves the same processes as plain old deliberative thinking. In metacognition, instead of just thinking about general cognitive content, these processes are tuned and shaped to think about thinking.

Julian Jaynes, in his fascinating book *The Origin of Consciousness in the Breakdown of the Bicameral Mind*, observes that in Homer and other early Greek fiction, the narrators tend not to refer to their own states of mind, but rather to the voices of the gods speaking inside their minds. The 20th century has given us a radical leap in metacognition since Homeric times, with the advent of a host of specific technologies for analyzing and modifying the individual mind, from psychoanalysis to more modern approaches like rational-emotive therapy, neurolinguistic programming, rationality boot camp, and so forth. Medieval Indian philosophers comprehensively studied the workings of the mind – analyzing consciousness in terms of 128 possible conscious states, each with unique properties – and they tied their work into the vast meditation community.

I'm not sure how great parrots, dolphins or apes are at metacognition. The human capability for metacognition appears to have advanced with the capability for complex uses of language. Language is an invaluable tool for describing our own thoughts and selves, to ourselves and others. However, ultimately, our ability to metacognize is impaired by our difficulty in perceiving our own thought processes. Most of our thoughts are "unconscious," meaning outside the scope of our rational deliberative thought processes. And of course, we can't physically observe the state of our brain the same way we can observe the state of our fingers or feet. AGI systems will eventually be able to observe their own thoughts with much more flexibility and accuracy than humans can, giving them dramatic advantages in the area of metacognition, which should help them self-improve more purposefully, systematically and rapidly.

Perception, Action & Language Hierarchies

Now, let's bring the discussion a little closer to reality. All these fancy kinds of memory and thinking have to get their data from somewhere. In theory, a mind could just take raw data from its sensors and start reasoning about it logically, and learning complex procedures to act based on it. However, in reality this wouldn't work too well, as the cognitive processes used for advanced deliberative thinking, and even for practical, real-time decision-making based on working memory, don't function adequately if you feed them a huge amount of data at once.

If you tried to reason logically about the geometric relations between all the pixels on the screen of your TV, you'd still be thinking really hard about one corner of the screen when the picture changed to something else. If you tried to figure out how to serve a tennis ball via carefully adapting each muscle movement to each perception from each part of your body during the serving action, you'd get overwhelmed by all the possible interrelationships between sensations and actions, and wind up tossing the ball and watching it fall to the ground while thinking about what to do.

So, because of having restricted processing resources, as there's only so much thinking the mind can afford within a certain allotted time, the mind needs specialized methods for processing the flood of sensory data it receives, and for controlling every part of its body. Complicating things further, these methods need to be specially adapted for various kinds of sensation and action.

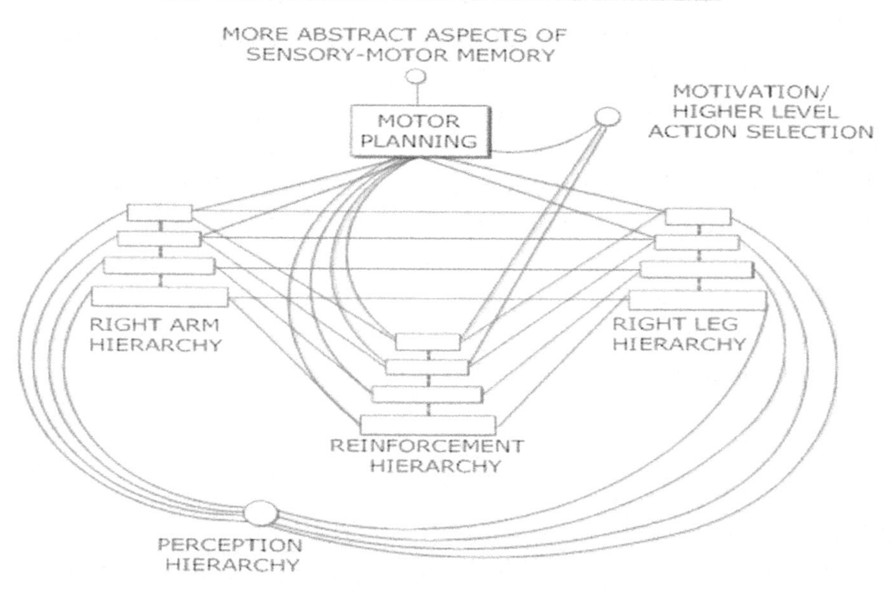

FIGURE 9.13: *Perception*

The parts of the human brain concerned with visual and auditory perception are hierarchically structured, as illustrated in FIGURE 9.13. For example, the lowest level of the vision hierarchy deals with tiny visual details, "pixels," so to speak, as they change over brief periods of time. The next level deals with slightly larger regions of visually perceived space-time, and so on. Finally, the highest levels deal with general high-level visual shapes and movements. Information passes up and down the hierarchy. The lower levels pass information to the higher levels, so that each level recognizes patterns in the output of the level below it, and the higher levels provide context that biases the pattern recognition in the lower levels. Audition (hearing) works about the same way, but with different levels dealing with differently-sized regions of time rather than space.

Some AGI researchers believe the hierarchical structure of the visual and auditory cortex represents how the rest of the brain is organized. Jeff Hawkins and Itamar Arel, as I mentioned above, have both proposed AGI architectures that have a strict hierarchical structure, not just for visual and auditory information, but for everything. However, I'm not so sure this is the right path. Of course, all pattern recognition has SOME hierarchical aspect to it. The mind is always recognizing patterns among patterns among patterns. But I don't think a strictly hierarchical structure for guiding pattern recognition is the right approach for handling, say, declarative knowledge that requires logical reasoning, or for accessing the episodic memories of one's life history.

The human brain handles some of its other senses quite differently from vision and audition. For example, the olfactory cortex, the part of the brain managing smell sensations, is dominated by combinatory, tangled-up connections between neurons, snaking all over the place rather than being arranged hierarchically. Each recognized smell seems to be represented by an "attractor" pattern — a habitual pattern of activity, distributed across a large region of the olfactory cortex. As a smell is recognized, gradually more of the attractor pattern becomes active in the olfactory cortex.

Tactile sensation works via the brain maintaining a distorted internal map of the body. There's a region of neurons in the brain corresponding to the back, another corresponding to the elbow, etc. Each fingertip gets more neurons than the whole back because it's more sensitive and has more nerves. Recognizing tactile sensations seems to be more similar to olfaction, more attractor-like than hierarchical-like vision.

The senses are not processed separately in the human brain. Instead, sensory processing often occurs in a multimodal way, where the regions of the brain that correspond to the different senses share their interim conclusions with each other to help reach better ones. Currently, AI systems process sensory data very differently, dealing with the data of each sense in isolation, rather than using them as a whole to reach a unified understanding of the world.

Action, as depicted in FIGURE 9.14, has a hierarchical structure somewhat similar to vision and olfaction. Higher levels of the brain's action hierarchy (resident in the cerebellum and certain parts of the cortex) correspond to large-scale actions, generally taking place over larger regions of space and time. Lower levels generally refer to quicker, more localized actions. For instance, in coordinating a tennis serve, the higher levels of the hierarchy would contain action-patterns corresponding to the overall shape of the body's motion while serving; the lower levels would contain details like the exact way the wrist is moved in response to a certain movement from the shoulder, and the particular speed of movement of the heel of the back foot as it's lifted. As with vision and audition, the hierarchical structure helps guide learning. Action learning, like visual and auditory perception, uses learning algorithms for patterns at each level of the hierarchy, based on the patterns at the immediate upper and lower levels.

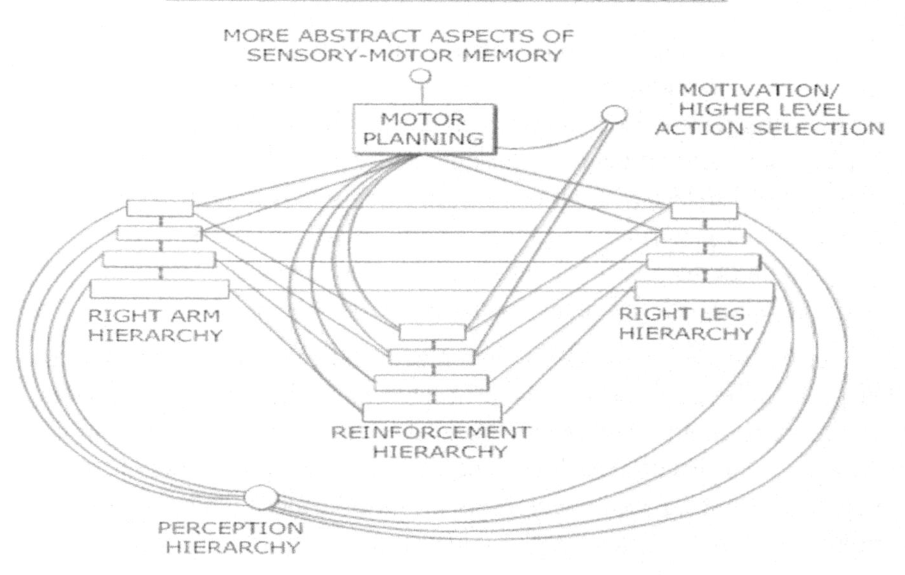

FIGURE 9.14.

The action and perception hierarchies are not separate but cross-connected, so that perceptions can be used in the course of actions (to correct and guide the course of actions), and actions can be used in the course of perceptions (to help gather perceptions more accurately). The arm, as it controls the tennis racket attempting to hit the ball back over the net, corrects its course in real time (including both macro and micro movements) based on what the eye sees the ball is doing. At the same time, the neck moves the head in a certain direction, based on the need to see the ball better, driven by instructions from the visual hierarchy that says it wants to see more detail.

The action and perception hierarchies are also closely related to long-term memory, and the various kinds of knowledge stored therein. Relatively simple computer algorithms can match a human's capacity to recognize objects in photographs — IF the person is only given half a second or less to look at each photograph. However, simple algorithms don't fare as well when people are given a few seconds more to scrutinize the photo. If a person spends a few seconds looking at a photo, they have time to bring up knowledge from long-term memory. If a sensory processing task isn't obvious, deliberative processes based on long-term memory come into play, and interact in complex ways with perceptual and motor processes.

The same holds for actions. When I'm playing tennis, and I see that the ball is coming in a slow but strange way, maybe because the wind caught it, then I may actually take half a second to THINK about my response, instead of just responding as usual by body-reflex. Great athletes are not only distinguished by their muscles and perceptual and motor systems, but also by the subtle real-time feedback between their perception and action hierarchies and their mind's deliberative processes.

Language

Language – viewed from a sufficiently abstract view – emerges as yet another of the human mind's intricate hierarchical structures, as illustrated in FIGURE 9.15. Language processing spans perception (language comprehension: Understanding what others say) and action (language generation: Figuring out what to say and saying it). The lowest levels of the language hierarchy are raw data-oriented, recognizing patterns in streams of sounds, and generating streams of sound with the mouth and larynx. The higher levels focus on abstract patterns of linguistic organization.

Language comprehension deals with the arrangement of sounds into words, words into phrases, phrases into sentences, sentences into paragraphs, and so on. The recognition of patterns at each level is conditioned by the patterns recognized at the levels above and below. When a young child learns language, they start at the lowest levels of the hierarchy, but their partially-formed intuitions about the higher levels condition their learning. For instance, when a child learns new words, they are guided somewhat by their intuitive understanding of the intent and context of the overall discourse in which the words occur.

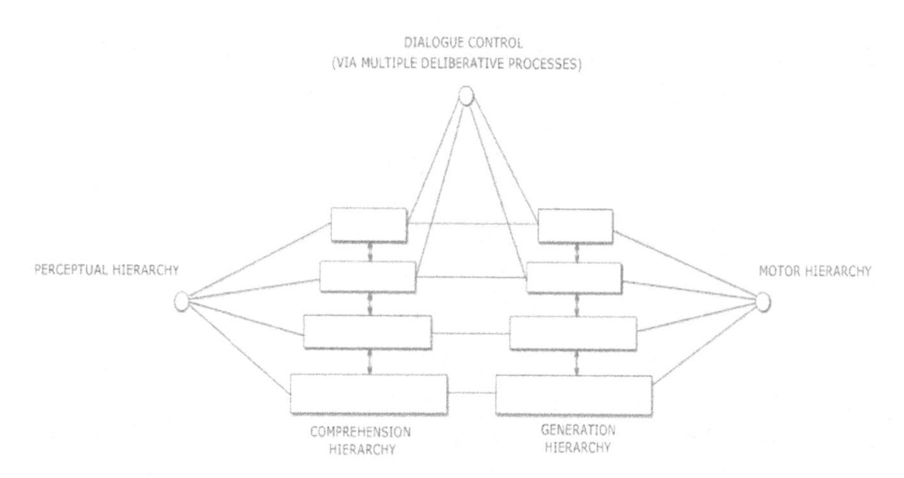

FIGURE 9.15: *Language*

Language generation deals with basically the same hierarchical levels, but is more concerned with building than recognizing patterns. It starts with some thoughts to be articulated, usually due to a conscious or unconscious judgment that doing so is going to help fulfill one of the mind's current goals. Next, it figures out how to break this set of thoughts into small chunks suitable for linguistic articulation. Then it turns each chunk into a proto-sentence: a linguistic series of words or word-meanings. Next, it fills in the specific content words. Finally, it fills in all the little words needed to make a grammatical sentence, and tells the body what to say.

The learning processes associated with language overlap with those involved in other cognitive processes. Sound comprehension is carried out by the auditory cortex, whether those sounds are linguistic or otherwise. Speech generation is handled through relatively generic action generation processes. At the highest levels, thinking about the meanings of sentences comprehended or generated is achieved through the same deliberative inference processes used for other kinds of thinking.

Learning how to produce or understand sentences seems to draw from procedure learning methods, similar to those used when learning other complex things. Literature in the AI and cognitive science fields points out commonalities between the structure of sentences, a series of physical actions, and social relationships. It seems the mind uses largely the same methods to represent and recognize patterns in all these domains.

On the other hand, the human brain appears to use quite specialized methods for language processing. For instance, there are certain parts of the brain where, if you have a lesion in them, you will lose the ability to process verbs, but not other parts of speech. These parts of the brain behave the same for everybody. So, language processing seems to be a case where the human brain's general intelligence capability and its specialization in its environment and goals come together in complex ways.

Cognitive Synergy

Cognitive science breaks the mind down into a bunch of different parts, and reveals a fair bit about how each part works. It also tells you one very important thing: all the parts must interact together. You can't make a system using some algorithm or neural net to carry out each of these parts of human intelligence separately inside an individuated black box, and then just connect the black boxes together for communication purposes. The mind doesn't work that way.

Each of these cognitive processes, carrying out each of these aspects of human-level intelligence, must interact intimately with many other processes (carrying out many other aspects). Each process on its own would end up in a logjam, necessitating unfeasibly large computational resources to do what more integrated systems could accomplish much more easily. Perception would get stuck trying to perceive what's there in some dark, modally complex seed. Declarative reasoning would get hung up trying to puzzle through some problem with too many possibilities. Metacognition would struggle to reason how to reason. When any one of these mental processes gets bogged down, it needs to be able to call on others in the midst of their own processing or thought processes. In this way, the different mental processes can help each other out of difficulties, forming what I call a "cognitive synergy" process, in which every part of the mind depends on every other part.

Some of my colleagues dislike the term "cognitive synergy," which I like to introduce into discussions of cognitive theory, because the word "synergy" has become popularly associated with management consulting style babble. If you play Bullshit Bingo online, you'll find phrases like "Friends don't let friends synergize paradigm shifts," alongside beautiful stuff like "The present-day, wide-spectrum, forward-looking and integrative cost efficiencies generate full-scale healthy yield externalities." However, I got the word "synergy" in my vocabulary from reading a lot of Buckminster Fuller back in the 1980s; it had nothing to do with management consulting. Bucky Fuller understood the way complex systems depend on networks of subtle, dynamic interdependencies between their parts. I think "cognitive synergy" is exactly descriptive of what happens inside intelligent systems when different subprocesses cooperate to find acceptable solutions reasonably rapidly in the presence of a vast number of possibilities.

Soooo – cognitive science tells us what parts of the mind must be present. It tells us that the various parts must be interdependent in a subtly synergistic way. But it doesn't tell us exactly what algorithms and dynamical processes have to take place inside each of these boxes, nor exactly how they need to interact with each other to manifest this cognitive synergy.

In my approach to AGI, we use computer science algorithms inspired by neuroscience, mathematics and pragmatic considerations to fill in the blanks. To dig deeper, you have to look carefully at our best theories regarding each of the parts. I spent a few decades doing just that, and in the next chapter I'll tell you about one of the AGI designs I came up with based on all that thinking: OpenCog. But first, I'll make a few more remarks on the theme of the general mess and complexity of the mind. Just to be sure you don't take all these boxes and lines too seriously!

Mind as a Complex System

Modeling the human mind with box and line diagrams makes it seem almost like a circuit board, with discrete components passing information between each other in a crisp and well-organized way. This helps us understand some of the broader aspects of how the human mind works. But it's important to remember that neither the human brain nor mind ACTUALLY has a bunch of boxes and lines in it. Rather, the human mind-brain is a big, complex, dynamical, teeming mess, and its intelligence emerges from this messiness, in a way that inextricably mixes creativity and flexibility with error and confusion.

The human mind/brain is, among other things, what scientists call a "complex self-organizing system." Put simply, a complex, self-organizing system contains a lot of little parts that interact with each other continually, and in the process give rise to larger structures and persistent dynamics, nudging the little parts in particular, semi-coordinated directions. A built-in structure may guide the little parts in their interaction, but there's also a lot of freedom for the parts to quasi-randomly experiment, until something happens that causes an overall structure to emerge.

For instance, imagine a society with no government, nor other institutions. People in the society would interact in all sorts of ways, quite chaotically and heterogeneously. Eventually, some structures would start to emerge. Groups of people would band together for various purposes. Little towns would form, with extended families and farming cooperatives. Independent mercenary squads would roam the countryside. In time, some sort of overall government would emerge, either via some mercenary squad turning into an army and taking over, or else via a group of peaceable people deciding to get together and organize themselves to prevent being taken over by mercenaries. Quite possibly, a number of small states would emerge, each with their own governments, which would enter into alliances. Eventually some semi-stable emergent structures would arise, in the form of governments, companies, armies, and so forth. These structures would then guide the further interactions of people in the society, not utterly constraining them, but instead directing them with significant force. The collective individual interactions between the people might eventually bubble up significantly enough to get rid of some of the major structures that originally emerged, e.g., a revolution might arise and overthrow some of the governments. The same sort of process occurs among cells in the brain, or, in another sense, ideas in the mind.

The infant's brain cells interact with each other, forming new structures of various sizes, and gradually crystallizing into overall cognitive and perceptual structures that guide it in further understanding the world. The structures that emerge collectively among the infant's brain cells are not specifically programmed in their DNA, nor specifically determined by their experience, which is why identical twins growing up in the same house can still have distinct personalities. These structures emerge via a sometimes chaotic process of neural self-organization, which then continues throughout the lifespan.

The ideas in a child's mind, which are patterns of organization among its brain cells, body systems and environment, also interact with each other. Early ideas combine to yield new, more sophisticated ones. Ideas combine, mutate, dwindle, and are reborn. Coherent networks and systems of ideas form. Ultimately, belief systems form, some with great persistence. Eventually a world-view crystallizes, along with a self-image, and models of others in the child's environment. From this point on, the wild interactions of the teeming pool of concepts are channeled by the structures that have emerged in the mind: the world-view, the self-image, belief-systems, and expectations. However, the self-organizing generative mayhem is still there, and may lead the mind to come up with radically new ideas, or undergo dramatic personality transformations or belief system shifts, even late in adulthood.

The boxes in the cognitive architecture diagrams are structures that emerge from the wild self-organization of cells in the brain and ideas in the mind, coupled with a body and an environment. Human genetics predisposes human minds toward building these structures, and nudges the infant's mind in this direction. Even so, each young human mind must build them for itself. This is the same in the social anarchy example, where human nature militates toward the emergence of mercenary armies and governments. Even so, each group of people must build these for themselves, guided by their genetic propensities and their own thinking.

A box like "episodic memory," doesn't necessarily refer to a discrete, distinct component in a human mind or AGI system. Rather, it's a distinct functionality of a complex system. A certain collection of brain cells, giving rise to a certain collection of ideas and thought patterns, begins to engage in the maintenance of episodic memories in a young human mind. This collection of brain cells improves at episodic memory maintenance, and as time goes on, it systematically interacts with other parts of the brain/mind, receiving information from them and sending information to them with episodic memories. However, this same collection of cells (and ideas) may serve other functions as well, and this may affect the nature of the episodic memory. The interactions between the cells enabling the episodic memory, and other cells, may give rise to the emergence of new neural structures, or the alteration of the episodic memory. The interaction between mental patterns involved in episodic memory, and other mental patterns, may lead to the emergence of new mental structures or forms of episodic memory.

The boxes in a cognitive architecture diagram are shorthand for systematic patterns of organization in a complex, ever-changing, self-revising and self-organizing system. The arrows are just shorthand for the naturally emergent pattern of interaction in these patterns of organization. This underlying complexity should not be forgotten. Yet, the approximate understanding provided by crisp, simplified models like cognitive architecture diagrams should not be dismissed either. Appropriate simplification of the world is, in a sense, the key operation underlying all intelligence.

Further Reading

Bach, Joscha (2012). *Principles of Synthetic Intelligence*. Cambridge University Press.

Baars, Bernard (1997). In the Theater of Consciousness: The Workspace of the Mind. Oxford University Press.

Franklin, Stan (1995). *Artificial Minds*. MIT Press.

Gazzaniga, M.S.; Ivry, R.B.; and Mangun, G.R. (2009). *Cognitive Neuroscience: The Biology of the Mind*. W W Norton.

Hawkins, Jeff and Sandra Blakeslee (2006). *On Intellligence*. St. Martin's Griffin.

Hudson, Richard (2007). *Language Networks: The New Word Grammar*. Oxford University Press.

Jurafsky, Daniel and James Martin (2009). *Speech and Language Processing*. Pearson Prentice Hall.

Sloman, Aaron (2001). Varieties of Affect and the CogAff Architecture Schema, in *Proceedings of the Symposium on Emotion, Cognition, and Affective Computing, Proceedings of AISM-01*.

Sobel, Carolyn and Paul Li (2013). *The Cognitive Sciences: An Interdisciplinary Approach*. Sage Publications.

Sutton, Richard and Andrew Barto (1998). *Reinforcement Learning*. MIT Press.

10. How the Brain Works

Even if it accomplished nothing else, the previous chapter probably convinced you that the human mind is a pretty big and complicated system. There are a lot of different moving parts, all of which are supposed to work together in particular ways. Getting all of that implemented successfully in a software system is going to be a substantial undertaking – as I'm well aware, due to the massive size that the OpenCog design grew to when I sought to spell out all the parts in detail, and the mountain of work that realizing OpenCog has turned out to be.

Seeing all that complication, it's natural to wonder a bit if we made the right decision in basing our design on cognitive science rather than neuroscience. Couldn't there be something in the structure or dynamics of the brain that would help make things simpler and clearer?

Actually, I believe just the opposite is the case.

I do understand the motivation to base AGI designs on neuroscience. After all, we only have one clearly human-level intelligent system at our disposal here on Earth right now – and that's the human brain, right?

Well, yeah. But there's the small technical detail that we don't really know how the brain works, right now.

And, the way I see it anyway, the more we learn about how the brain works, the less of a good idea it seems to be to model digital computer programs on the brain, given the massive differences between current computing hardware and neural wetware. The brain looks to be a massively complex system, much more so than OpenCog – and furthermore, its complexities seem to be highly dependent on the specificities of the cells and chemicals of which it's built.

To make this point clearer – and because it's generally interesting anyway – in this chapter I'm going to take a little time to fill you in on how I think the brain works. Of course, nobody actually knows how the brain works in detail yet, but I've studied the area extensively and done a bit of neuroscience work myself, and I have some hypotheses I'm willing to stand behind.

The first thing you should understand is: The brain is not only complex, but also big and complicated, with different parts doing different things. I like this picture of the regions of the monkey brain created by IBM researcher Dharmendra Modha and his team:

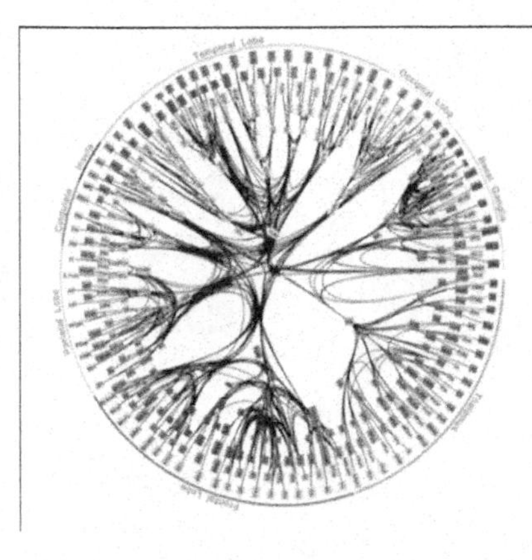

FIGURE 10.1: *From the 2010 paper Network architecture of the long-distance pathways in the macaque brain, by Dharmendra S. Modha and Raghavendra Singh. Each box is a region of the brain, and lines are drawn between brain regions that are known to be connected by a large number of synapses. (http://www.pnas.org/content/107/30/13485.full).*

This funky-looking diagram shows 300+ regions of the macaque monkey brain and how they connect to each other. Each of these brain regions has a literature of scientific papers about it, explaining what sorts of functions the region tends to carry out. In most cases, our knowledge of each brain region is terribly incomplete. The nodes near the center of his diagram happen to correspond to what neuropsychologists call the "executive network": The regions of the brain that tend to get active when the brain needs to control its overall activity.

FIGURE 10.2: *Rhesus macaque monkey – the type of primate whose brain was studied to form Modha and Singh's brain wiring diagram, as given above. (http://www.sciencedaily.com/releases/2009/01/090113201339.htm).*

All these parts of the brain seemingly work according to common underlying principles. Each of them is wired differently, although they use similar "parts." There's a lot of commonality between the dynamics occurring within each region as well – but by and large, brain dynamics remains unknown, and will remain unknown for a number of years, until we have better tools to measure it.

One can of course draw a similar picture of the human brain. Most of the parts are the same as in the monkey brain, actually. I went through a fair bit of effort, with some help from a neuroscientist, Rachel Talbott, to mark up a human-brain-connectivity diagram with notes telling what kinds of thinking happen in each part of the brain. It's an approximation and a mess, but gives a sort of high level understanding. There is a lot of complexity happening here.

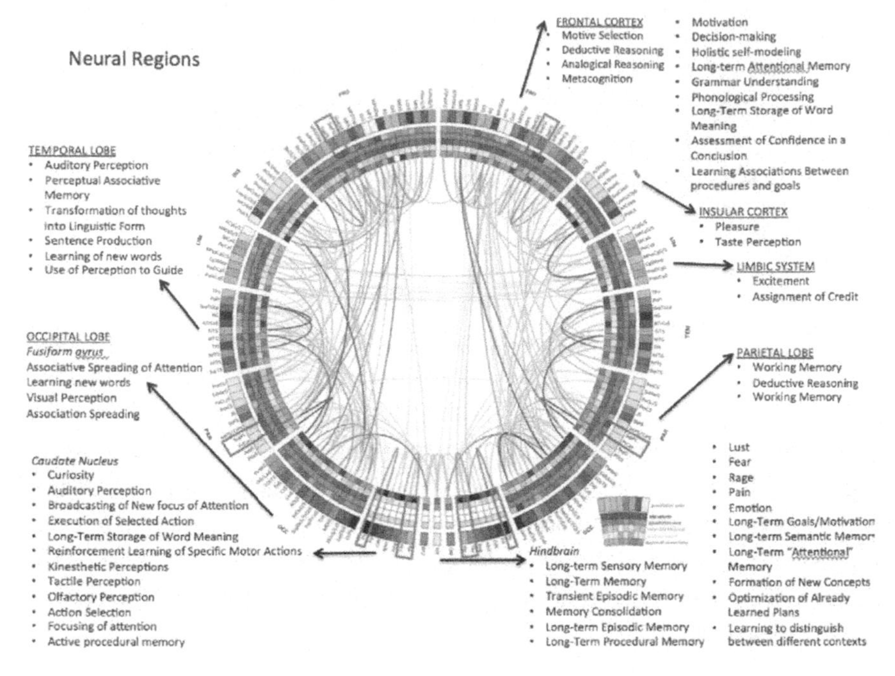

FIGURE 10.3: *Graphical depiction of human brain regions, with a rough indication of the cognitive functions that are known to be focused in each region.*

What's happening in all these hundreds of specialized brain regions? All the parts of the brain are made of cells, neurons, which connect to and spread electricity amongst each other. The spread of electricity is mediated by specific chemicals: neurotransmitters. One neuron doesn't simply spread electricity to another one; each neuron also activates a distinct neurotransmitter molecule that then delivers the correct charge to another corresponding neuron. Things like mood, emotion, food, or drugs affect these neurotransmitters, modulating the nature of thought.

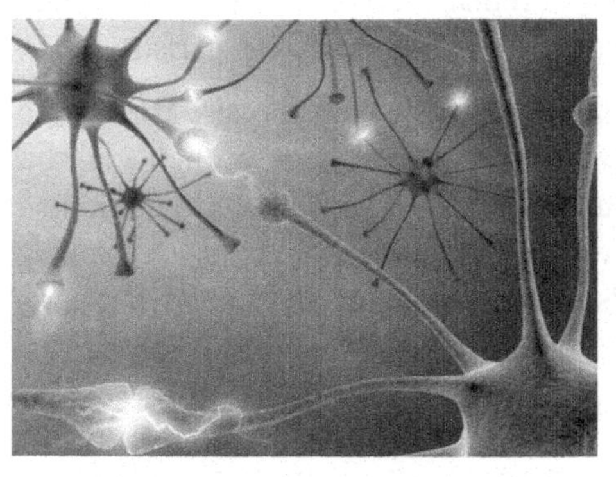

FIGURE 10.4: *Depiction of interneurons, neurons that carry out inhibition between cortical columns. (http://www.sciencedaily.com/releases/2008/11/081124174909.htm).*

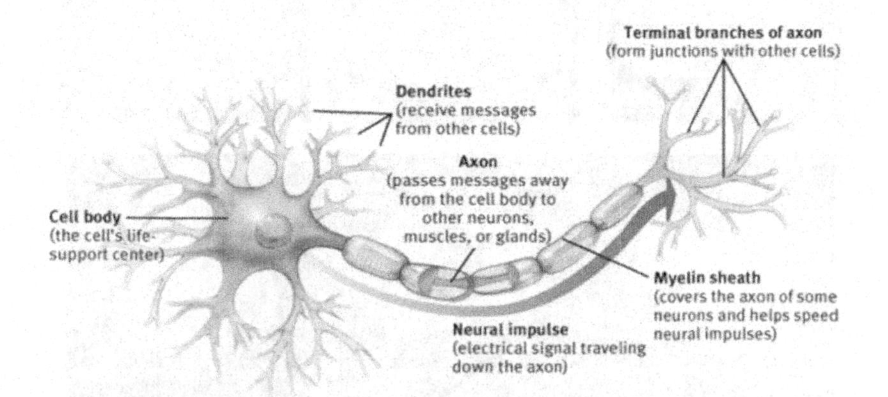

FIGURE 10.5: *Standard diagrammatic depiction of a neuron, showing the key parts of the cell and their functions. This model is an abstraction of the complex biophysical system that is a real neuron, but captures many of the important characteristics.* (*http://www.docstoc.com/docs/4191487/Diagram-of-the-Neuron*).

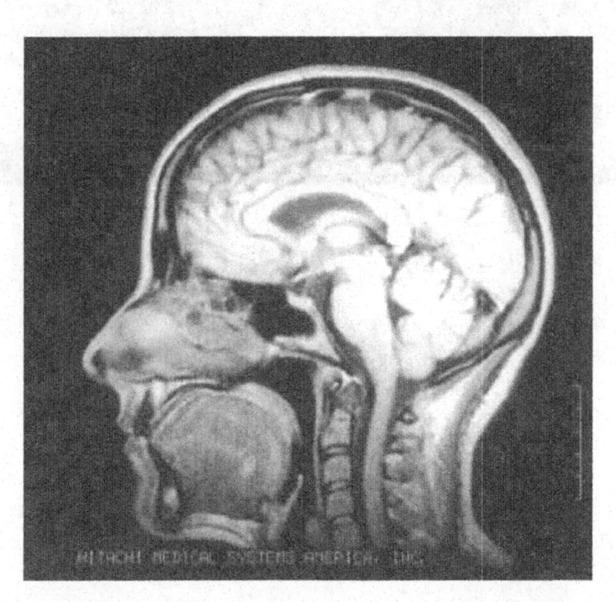

FIGURE 10.6: *Traditional MRI brain imaging (as opposed to fMRI, where the "f" stands for "functional") is used to take pictures of the structures inside the brain.* (*http://www.csulb.edu/~cwallis/482/fmri/fmri.html*).

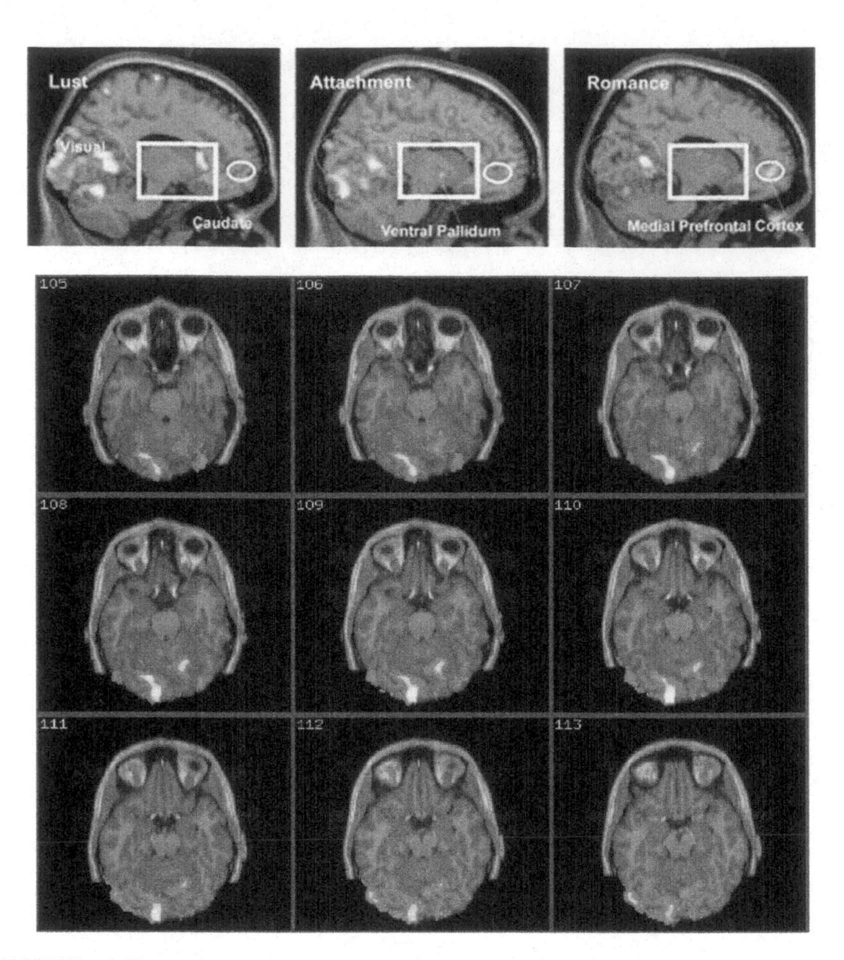

FIGURE 10.7: *MRI imaging is used to take pictures of the activity inside the brain at a given point in time. It doesn't narrow things down to the neuron level, but it gives a basic idea of which parts of the brain are active – which is a useful thing to know. The Picture shows a time series of fMRI images of the same brain, showing how activity moves around the brain gradually over the course of a single episode of thought.* (http://www.csulb.edu/~cwallis/482/fmri/fmri.html).

The bottom picture, from a separate study, shows the regions of the brain most centrally active during different types of emotion. Critically, though, knowing which parts of the brain tend to be active during certain types of activity doesn't give us information about the actual dynamics going on in those brain regions enabling these activities to be done. For example, knowing where romantic feelings are centered in the brain doesn't tell us all that much about how the brain creates romantic feelings.

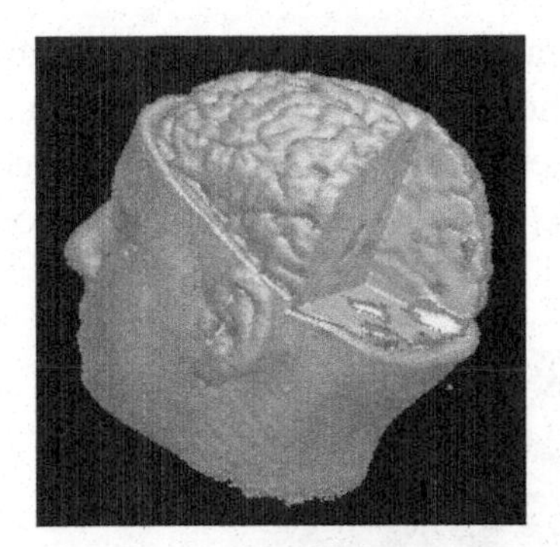

FIGURE 10.8: *The fMRI scanning can be used to reconstruct 3D information regarding which parts of the brain are active at a given point in time. This helps scientists understand many things about the brain, including which parts of the brain are active while the brain is carrying out various sorts of activities.* *(http://www.csulb.edu/~cwallis/482/fmri/fmri.html).*

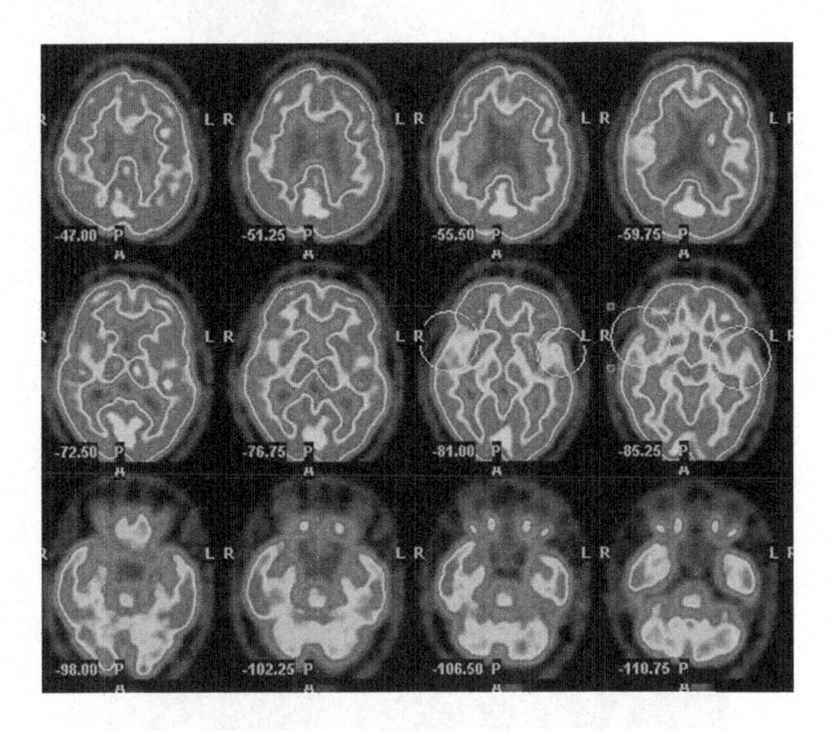

FIGURE 10.9: *Output from Positron Emission Tomography (PET) scanning of the brain, a methodology similar to fMRI, but with different tradeoffs in terms of accuracy. We have many different brain imaging methods these days, but none that can give us high spatial and temporal accuracy of a living brain at the same time, which is what we'd need to really understand what brains are doing. But surely that technology will come, we just need a breakthrough or three in brain imaging technology.* *(http://emedtravel.wordpress.com/2011/05/14/have-you-seen-a-brain-pet-scan).*

The glia, another type of cell in the brain, fills up much of the space between neurons, and seems to play important roles in some kinds of memory. Neuroscience legend Karl Pribram and his Japanese colleagues Jibu and Yasue speculated that intelligence relies on complex quantum-physical phenomena, occurring in water mega-molecules floating in between neurons. I can't say for sure if this is true or not (and neither can anyone else), though I admit I'm a bit skeptical.

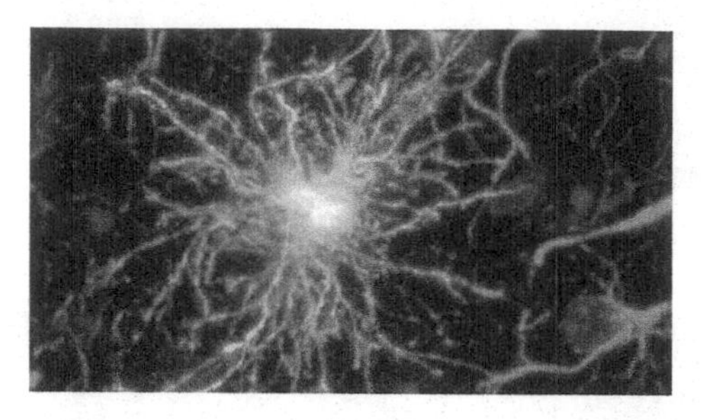

FIGURE 10.10: *Some clever scientists have inserted human glial cells (green) among normal mouse glial cells (red), resulting in the mice becoming smarter! This clearly shows that neurons are not the only kind of brain cell important for intelligence (even though they are the only kind modeled in nearly all computational brain models, and nearly all neuroscience-inspired AI or AGI architectures). (http://www.npr.org/blogs/health/2013/03/07/173531832/Human-Cells-Invade-Mice-Brains-And-Make-Them-Smarter).*

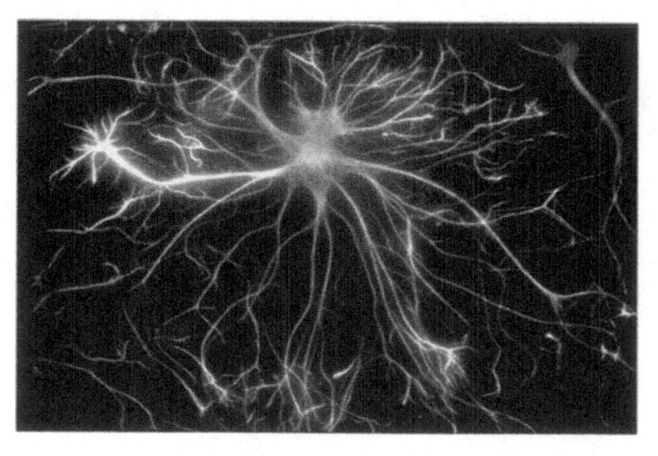

FIGURE 10.11: *Astrocytes, star-shaped brain cells classified as a type of glia, are now being considered as potential targets for drugs aimed at combating depression. (http://neurosciencestuff.tumblr.com/post/41358316928/astrocytes-identified-as-target-for-new-depression).*

NEURAL NETWORK MAPPING

Input Layer
Hidden Layer
Output Layer
Bias=1
AXONS
NEURONS

FIGURE 10.12: *Formal neural network models, as used in computer science, capture some high-level aspects of neural structure and dynamics, but represent a gross oversimplification of what's really going on in the biological brain. They replace the real neuron, a complex, nonlinear, dynamical biochemical/biophysical system, with a simple "formal neuron" computing element, ignoring the complex chemistry of the interactions between neurons, and ignoring other brain cells like astrocytes and glia. While formal neural nets have taught us a lot and proven useful in various practical applications, they can't be taken seriously as brain models. AGI systems based on formal neural nets should be evaluated on their own terms, as loosely brain-inspired computer science-based cognitive architectures, rather than being thought of as brain models. (http://rpi-cloudreassembly.transvercity.net/2012/11/04/neural-network-mapping-analysis-from-above/).*

FIGURE 10.13: *Legendary neuroscientist Karl Pribram gave a lecture at the 2006 AGI Workshop that Bruce Klein and I organized in Bethesda, Maryland, thus initiating the AGI conference series. He lived in DC, where he was an Emeritus Professor at George Washington University, so it wasn't a long commute. I had read his books a couple decades before, so this was a big thrill for me. Pribram rejected localist views of memory altogether, likening the brain to a hologram. He also conjectured that the brain's holographic activity relies on macroscopic quantum phenomena. While "out there," this idea is less so than the better-known speculations of Roger Penrose and Stuart Hameroff, which propose that neurodynamics and human consciousness depend on quantum-GRAVITY effects occurring at a very small scale in the brain. The Penrose/Hameroff view is even harder to grasp onto, since physicists have not yet given us any consistent, coherent theory of quantum gravity. Pribram's speculations are at least based on known physics, though their basis in known*

biology and chemistry is very sketchy. Personally, I think it's possible that wacky quantum effects play some meaningful role in how the brain executes thought, but I suspect it's probably not necessary to understand these in order to get the basic idea of how the brain does thinking. That is, my guess is that if there's funky quantum stuff going on, it's more along the lines of explaining how the brain carries out the mechanics of phenomena like long-term potentiation or synchronization of assemblies or whatever, rather than along the lines of providing some totally different cognitive mechanism that's not comprehensible at the cellular level.

The part of the brain most central for thinking and complex perception, as opposed to body movement or controlling the heart, etc., is the cortex. Neurons in the cortex are generally organized into structures, or columns. Each column spans all six layers of the cortex, passing an electrical charge up and down the layers, and laterally, to other columns. A large quantity of neurons called interneurons carry out inhibition between columns. When one column gets active, it sends a charge to interneurons that then inhibit the activity of certain other columns. Columns tend to be divided into substructures, often called mini-columns, or occasionally, modules.

FIGURE 10.14: *An image of cortical columns in the rat brain. Specifically: a cell-type-specific 3D reconstruction of five neighboring barrel columns in the rat vibrissal cortex (credit: Marcel Oberlaender et al.). This kind of imaging has only been possible quite recently; this picture is from 2013. The study that generated the picture concluded that, while the cortex does have a marked columnar structure, the specific structure of the brain's cortical columns can deviate substantially within individual animals, and even within a specific cortical area within a particular animal. There is a lot of complexity here.* *(http://www.kurzweilai.net/neuroscientists-find-cortical-columns-in-brain-not-uniform-challenging-large-scale-simulation-models).*

In the visual cortex, columns recognize particular patterns in particular regions of space-time. One column might contain neurons responding to patterns in a particular part of the visual field, while the neurons higher up in the column represent more abstract, high-level patterns. Lower-level neurons in the column might recognize the edges of a car, whereas higher-level neurons in the same column might help identify that these edges, taken together, look like a car. But the functions of columns and the neurons, and the mini-columns inside them, seem to vary from one brain region to another[1]. A lot of attention, recently, has gone into emulating in a computer – at various levels of accuracy – the way the hierarchical pattern recognition processing in the mammalian visual and auditory cortex works. This has led to some fantastic practical results, which I'll talk about a little later. But still, we shouldn't exaggerate the degree to which modeling how these particular parts of the brain carry out particular tasks is informative about brain dynamics as a whole and how it gives rise to the various aspects of human intelligence.

One of the tricky things about the brain is the way it mixes up local and global structure and dynamics. Each cortical column does something on its own, while also stimulating and inhibiting many other columns, potentially causing a brain-wide pattern of activity. Each column has a local and a global aspect: to describe this, I like to use the weird word "glocal." There's a lot of evidence for this glocal aspect in terms of human memory. Memories of specific objects or people seem to be stored in networks of hundreds to thousands of columns. However, the network corresponding to, say, "Barack Obama," can be triggered into activity by stimulating just a few of the columns involved in the network.

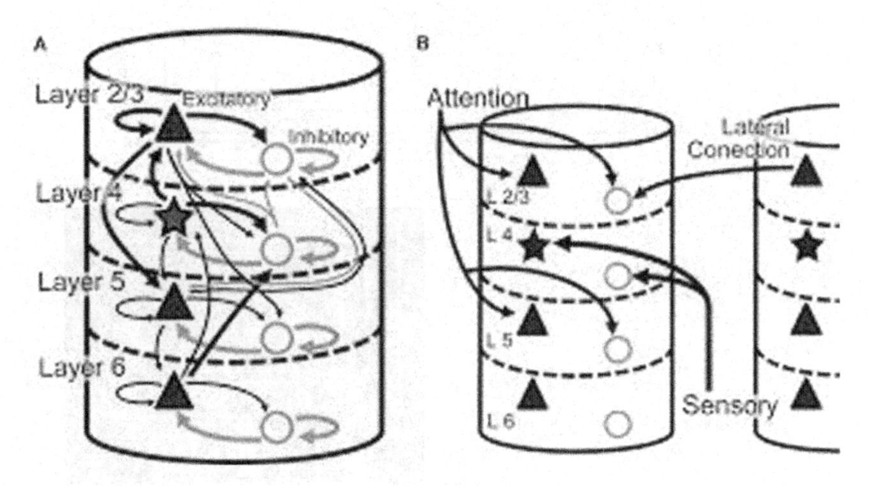

Figure 10.15: *A formal and conceptual model of cortical columns, developed by a team of neuroscientists in 2011, in an attempt to understand how top-down attention focusing impacts hierarchical information processing in the visual cortex. Focusing on the structure of columns in a particular area of the brain lets one make detailed models; yet this kind of*

1 See the papers by Raghani et al and Rieke et al, listed at the end of this chapter.

abstracted model is still recognized as a drastic simplification of the complex underlying reality. (http://journal.frontiersin.org/Journal/10.3389/fncom.2011.00031/full).

If one column causes a global brain activity pattern, making other columns react to this pattern, these other columns are basically reacting to that one column. Since each column can learn and adapt based on experience, using the ability of each neuron to modify its connections to other neurons based on experience, we have a complex network of actors (columns) that are constantly acting on each other (by reacting to the global activation-patterns caused by each other) and then adapting based on this interaction. In computer science terms, this is an intricate and powerful form of recursion (self-reference). The whole is perceived by the parts, which then create new information based on this perception, thus contributing to changing the whole – and then the whole is perceived by the parts again, and so on. One can prove that this kind of system is able to give rise to endlessly complex forms and do any kind of calculation that a computer can do.

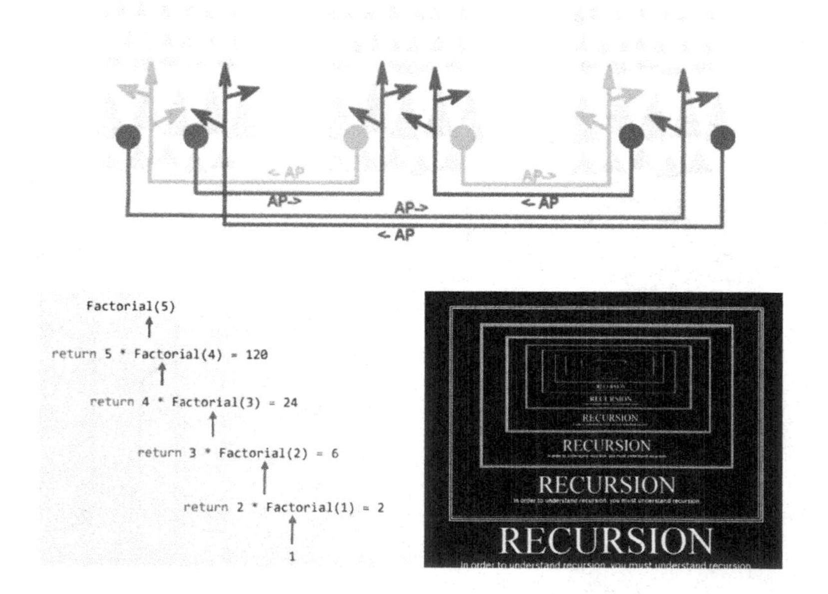

FIGURE 10.16: *Recursion in computer programs versus reentry in the brain. The concept of recursion, i.e., self-reference, is conceptually simple enough. It is a basic tool in computer science and computer programming, used, e.g., to calculate mathematical functions as in the example of "5 factorial" given above. "Reentrant" neural circuits, as discussed by Gerald Edelman in his Neural Darwinist theory, and also by many other neuroscientists, are one way that the brain may implement recursive "algorithms" to help with cognitive*

and perceptual processing. The AP-> in Edelman's diagram indicates the direction of information flow. *(http://2.bp.blogspot.com/8DS2pagnTqs/Ult_K79MUFI/AAAAAAAAAB8/tGIPnWDdU9Y/s1600/3620061163_ba9f8d5031_z.jpg).*

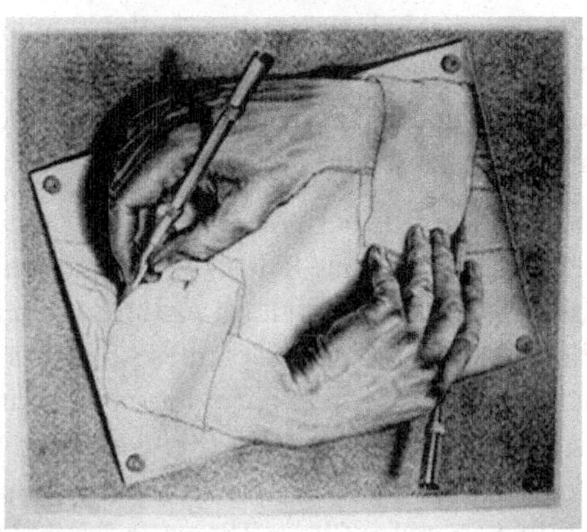

FIGURE 10.17: *Recursion and self-reference in software code are just the most modern incarnation of the archetype of self-reference, which has existed nearly as long as history – symbolized, for instance, by the archetypal image of the Uroboros, the snake eating its own tail. The picture is one I used to show at conferences to perplex people; it uses Uroboros to wrap some not-quite-functional Ruby code for a universe-conquering AI. Escher was one among many fine artists who explored themes of recursion in his work. Hofstadter's book Godel, Escher, Bach explores implications of recursion and related concepts for AI in exquisite and entertaining detail.* *(http://www.pxleyes.com/blog/wp-content/uploads/2010/06/escher-hands.png).*

On the neuronal level, this sort of conceptual recursion can be carried out via "reentrant" neural circuits. As Gerald Edelman put it in a 2013 paper:

"Reentrant processes [are] ... defined as those that involve one localized population of excitatory (i.e., glutamatergic) neurons simultaneously both stimulating, and being stimulated by, another such population. The structural architecture that generates this process is likewise referred to as reentrant...

The hypothesis that reentrant signaling serves as a general mechanism to couple the functioning of multiple areas of the cerebral cortex and thalamus was first proposed in 1977 and 1978. A review of the amount and diversity of supporting experimental evidence accumulated since then suggests that reentry is among the most important integrative mechanisms in vertebrate brain."

To someone coming at neuroscience from a "complex, self-organizing systems" background, this might seem almost obvious – of course the neurons in the brain are involved in complex networks that are all tangled up with each other in complex ways, and this complexity involves neural-network-level recursions that help generate complex cognitive processes like memory and recursive thinking. But in a neuroscience context it helps to emphasize this sort of point again and again, to ward off those who propose much more simplistic, unrealistic models of brain architecture and function.

Neurological recursion, furthermore, is an obvious source of highly complex nonlinear dynamics. What kinds of "strange attractors" occur in dynamical systems that recognize high-level patterns in their own state, and recursively, continually feed indicators of these patterns into themselves as inputs? Walter Freeman and others have done pioneering studies of the role of chaotic dynamics in mammal brains. I encountered Walter Freeman's wonderful work on chaos in the rabbit olfactory cortex in the early 1990s at a conference of the Society for Chaos Theory in Psychology. What he was studying then was the manner in which, when a rabbit recognizes a smell, this is reflected internally by the rabbit's olfactory cortex using nonlinear dynamics to settle into a unique, holistic "attractor state" corresponding to that smell. (Subsequent work has made the story seem even more complicated – I'm not sure if these states are best thought of as attractors, or as some sort of "persistent transients" – but the basic concept seems to remain valid.)

I intersected Freeman a few years on an email list, and was so pleased to see he was still making progress on this kind of brain model. What a disappointment it was to hear in early 2016 that he passed away. He didn't live to see the complex nonlinear dynamics of the brain fully understood. But perhaps I will.

The nonlinear dynamics that Freeman and his colleagues spend decades rigorously studying, based on electrophysiological data, are massively simpler than the nonlinear dynamics that would result across the brain from holistic reentrant recursion. We simply don't have the mathematics or science yet to describe the kinds of complex chaos that brains most likely lead to. We're getting there.

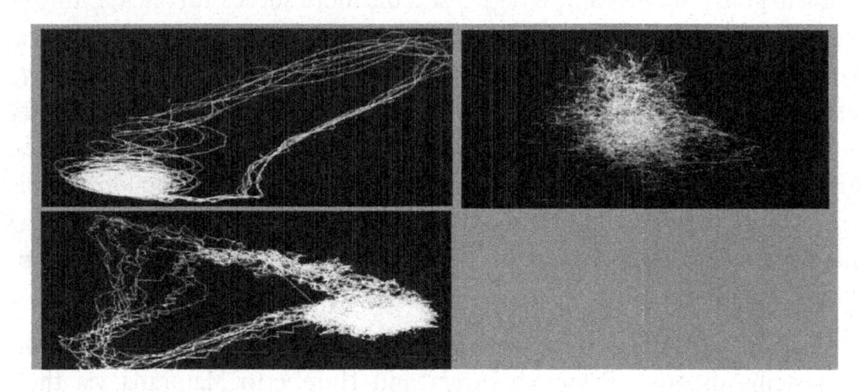

FIGURE 10.18: *Walter Freeman has been studying the complex chaotic dynamics of human brains for many decades. Above are some figures from Walter Freeman's analysis of nonlinear dynamics in the rat brain (from the paper titled "Strange Attractors that Govern Mammalian Brain Dynamics Shown by Trajectories of Electroencephalographic (EEG) Potential").*

A 2D view of a 3D display of a 4-second epoch of EEG from the olfactory system of the brain of a rat, awake but resting motionless. The trajectory of a point moving through this space in time traces the subspace that is occupied by a strange attractor. The figure was rotated and translated to seek for structure, and colored it red when the fourth variable was negative and blue-white when it was positive. The Hausdorff (fractal) dimension of the higher-dimensional attractor is about 5.96.

The EEG trace from the same animal during an epileptic seizure. The same recording conditions hold, but the Hausdorff dimension of the higher-dimensional attractor is reduced to 2.52, and the structure of the strange attractor appears to be that of a 2-torus.

*Figure shows the trajectory of a solution set of a dynamic model of the olfactory brain that simulates epileptic activity, and which gives an estimate of the Hausdorff dimension of 3.76. It also has a form suggesting the 2-torus. (Reprinted from IEEE TRANSACTIONS ON CIRCUITS AND SYSTEMS, Vol. 35, No. 7, July, 1988 (*http://sulcus.berkeley.edu/*).)*

I'm phrasing all this in my own particular way, but others have said similar things in their own ways. Gerald Edelman caught my attention in the

1980s with his Neural Darwinism theory, which likened learning in the brain to the process of evolution by natural selection. As he viewed it, multiple partially redundant neural networks process the same information, and the ones that do it better get selected to participate more fully in ongoing processing – providing an analogue to the evolution of species. Edelman earlier won a Nobel Prize for his work on immunology, and his neuroscience ideas were doubtless immunologically inspired; the immune system also works via natural selection, with multiple similar but not identical antibodies trying to grab onto the same antigen, and the more successful ones getting to reproduce more. The general dynamic of evolution pervades the biological world on multiple levels, a point I emphasized in my 1993 book *The Evolving Mind*. Another point Edelman likes to emphasize is the concept of reentry – the dynamic by which the brain looks at its own dynamics and outputs, and feeds these into itself as inputs. Reentry is a form of recursion. In this sort of view, the brain builds up complexity in roughly the same sort of way that advanced computer programs do, though on the basis of vastly different underlying mechanisms.

A different language for discussing the same kind of thing was provided by systems theorists Francisco Varela and Humberto Maturana via their notion of "autopoeisis" – self-building or self-construction. They viewed cells, bodies, brains, and all complex systems as holistic systems that not only continually organize themselves, but continually construct themselves based on raw chemical materials. Mathematically oriented theorists like Louis Kauffmann and Robert Rosen have created novel, fascinating mathematical models of this process, starting from utterly different bases than standard mathematical biology. Although, Rosen's math reminds me considerably of category theory, which is the mathematical foundation of modern functional programming languages like Haskell – another indication of the parallel between reentry in neuroscience and recursion in computing.

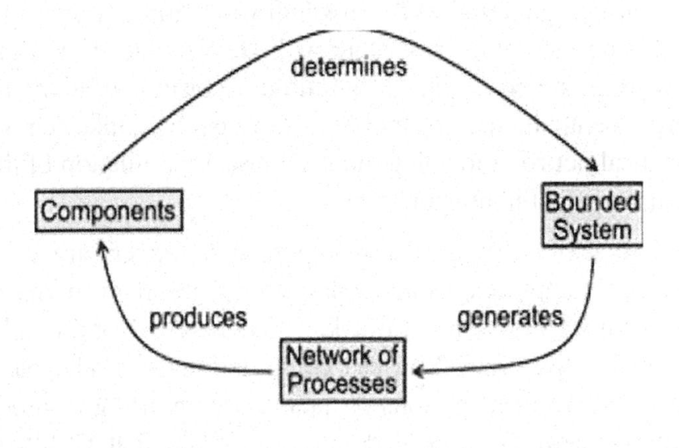

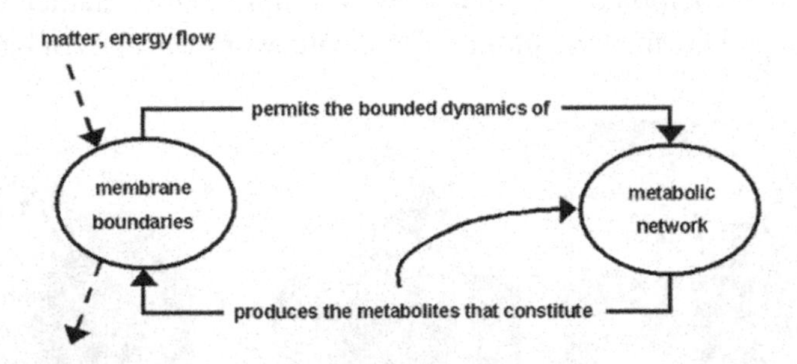

FIGURE 10.19: *Illustration of the concept of "autopoiesis" – self-construction – which Maturana, Varela and others considered crucial to understanding brains and other biological systems. Figure illustrates the general concept of autopoiesis; Depicts its application in the particular area of biology. (Luisi, Pier L. (2003). Autopoiesis: a review and a reappraisal. Naturwissenschaften 90:49–59. http://www.scielo.cl/fbpe/img/bres/v36n1/fig06.gif).*

A conceptual complication is that Rosen and Varela considered their autopoiesis-based analyses as proofs that the brain is not computational in nature, and can't be effectively simulated on digital computers. I argued against this perspective fairly extensively in the early parts of my 1994 book *Chaotic Logic*. As I argued there, we can't currently know for sure if the brain can be explained or emulated computationally or not – but arguments about autopoiesis, reentry and so forth certainly don't show that the brain is NON-computational. Autopoiesis, reentry and so forth are abstract processes that can be used as models of either computational or non-computational processes.

Edelman, in my least favorite part of his excellent book *Neural Darwinism*, also argued against the possibility of computationally modeling the brain. Yet he did some great work with George Reeke building robots with minds programmed on digital computers according to Neural Darwinist ideas. More recently he has collaborated with Eugene Izhikevich, a pioneer of mathematical neuron modeling, on a large-scale simulation of the neural networks underlying the human brain.

The fact that so many great researchers and thinkers are still arguing about whether computational modeling of the brain even makes sense is an illustration of how limited our knowledge of the brain still is. I am fairly certain that the brain IS in fact computationally modelable (though it's possible that we need notions of quantum computing to model a few low-level particulars), but one certainly can't claim that CURRENT, well-demonstrated computational models give a complete story. Rather, they give a beautiful and fascinating but threadbare patchwork with lots of big gaps.

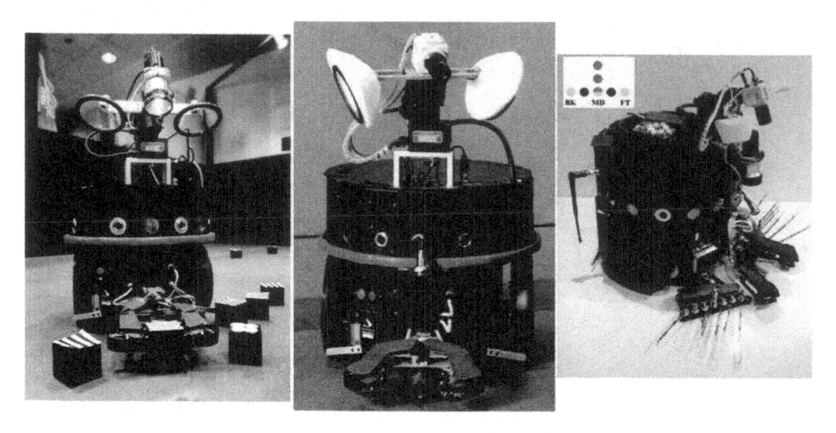

FIGURE 10.20: *The Darwin series built in the lab of biologist Gerald Edelman in the 1980s, internally operating according to the principles of Neural Darwinism. Darwin VII (models the somatosensory loop in mammals), Darwin VIII (models reentrant connections within the visual pathways of mammals), Darwin IX (texture discrimination using artificial whiskers). (No relation to the current "Darwin OP" line of robots.) (http://www.nsi.edu/index.php?page=ii_brain-based-devices_bbd).*

Concept Neurons

An additional twist to the convoluted story of brain dynamics is provided by recent evidence about the existence of "concept neurons" – neurons that fire to individual, very narrow classes of stimuli. One research paper famously demonstrated the existence of "Jennifer Aniston neurons," which fire only when the actress Jennifer Aniston is seen or imagined. It's not clear at this point how many of a human being's concepts get allocated single neurons like this. But one thing that is clear is that no single neuron contains the brain's representation of Jennifer Aniston, or anything else. Glocality is the key to understanding here. The Jennifer Aniston neurons in your brain serve to trigger a broad-ranging pattern of activity, corresponding to Jennifer Aniston. Symmetrically, when this Jennifer Aniston activity pattern gets stimulated, the Jennifer Aniston "concept neurons" will also get activated. The degree to which the Jennifer Aniston neurons play a causal role in the dynamics of the overall Jennifer Aniston brain activity pattern, as opposed to mainly playing an "indicator" role, remains unclear.

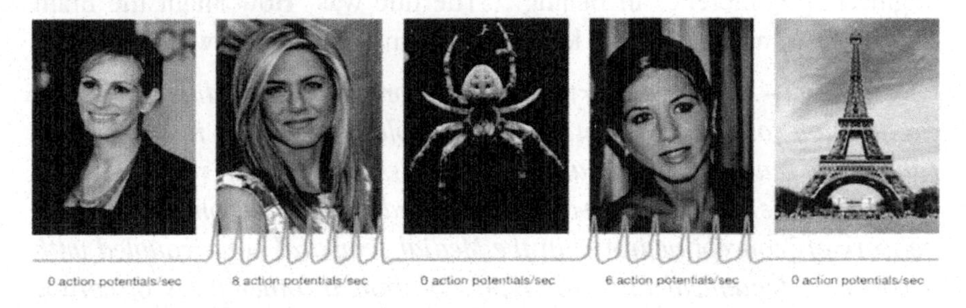

FIGURE 10.21: *Concept Neurons. The graphs below each image show how a single neuron fires in response to a picture of the actress Jennifer Aniston, but not in response to other pictures.* From Rodrigo Quian Quiroga, L. Reddy, G. Kreiman et al, 2005, Nature 435:1102-1107.

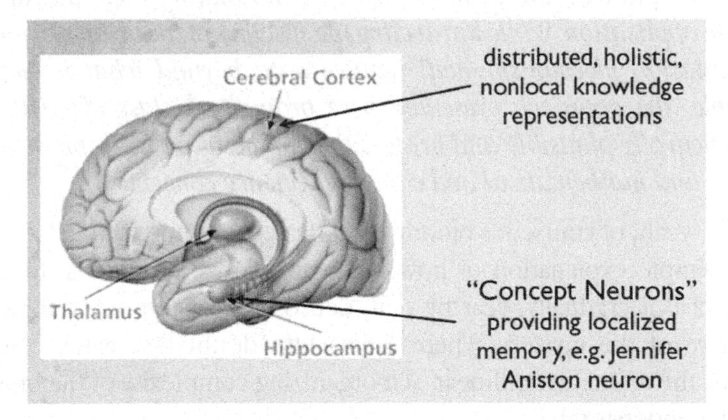

FIGURE 10.22: *Concept neurons reside in the hippocampus, whereas large parts of the cortex seem to focus on more distributed representations. The synergetic interaction of these two regions appears to be a key part of how the human brain implements what I call "global" (combined global/local) memory.*

Speculating a bit, it seems to me one role that these concept neurons might play is to support the feeding of one concept as an input to others. Suppose the brain needs to represent the notion "Jennifer Aniston gave Salvador Dali an ambiguous smirk." It may have distributed activity patterns for "give," Jennifer Aniston, Salvador Dali, and even "ambiguous smirk." But how can it relate these three patterns? Perhaps the symbolization of the latter three concepts in individual neurons allows them to be more easily fed as "inputs" to the distributed activity pattern for "give," so that the "give" activity pattern doesn't need to deal directly with the whole of the complex activity patterns for Jennifer Aniston, Salvador Dali, and "ambiguous smirk," but only with the concept neurons serving as indicators for these. I wrote a paper elaborating this idea in 2014, for the session on Brain Architectures at an engineering conference in Beijing... The title was "How Might the Brain Represent Complex Symbolic Knowledge?", and the abstract was:

"Abstract -- A novel category of theories is proposed, providing a potential explanation for the representation of complex knowledge in the human (and, more generally, mammalian) brain. Firstly, a "glocal" representation for concepts is suggested, involving localized representations in a sparse network of "concept neurons" in the Medial Temporal Lobe, coupled with a complex dynamical attractor representation in other parts of cortex. Secondly, it is hypothesized that a combinatory logic like representation is used to encode abstract relationships without explicit use of variable bindings, perhaps using systematic asynchronization among concept neurons to indicate an analogue of the combinatory-logic operation of function application. While unraveling the specifics of the brain's knowledge representation mechanisms will require data beyond what is currently available, the approach presented here provides a class of possibilities that is neurally plausible and bridges the gap between neurophysiological realities and mathematical and computer science concepts."

Well, yeah, of course it's bloody complicated. Anyone who tells you they have a simple explanation of how the brain works is either dishonest or mistaken! But gradually, year by year, neuroscientists are unraveling more and more of the mystery. There seems litte doubt that within the next decades, the teeming, nonlinear, self-organizing complexity of the brain will at last be understood.

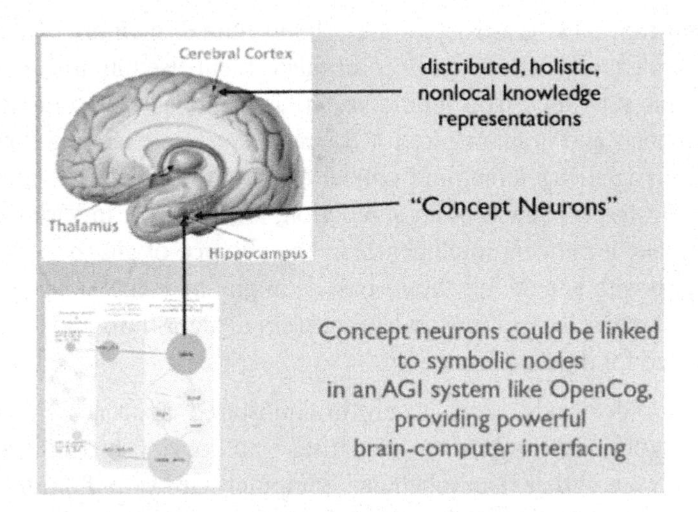

FIGURE 10.23: *Potentially, concept neurons could serve as a key element of advanced brain-computer and brain-brain interfacing methods. What if an electrode or other interaction device were inserted in the brain, connecting the Jennifer Aniston concept neuron with the Jennifer Aniston Concept Node inside an OpenCog system (for example)? What if a Bluetooth connection between our brains allowed my own Jennifer Aniston neuron to be stimulated whenever yours was? This is a wild speculation at present, and there would doubtless be numerous complexities and "hidden rocks" in making something like this happen. Yet it seems an extremely promising direction. The basic point is that distributed cortical representations may be quite individualized and challenging to decode, whereas concept neurons with specific external referents may provide an easier initial avenue for brain-computer and brain-brain interfacing.*

Why Neuroscience is Not the Best Guide to AGI

The main problem with the idea of modeling AGI is that we don't really understand how the brain works. Maybe the high-level view I've outlined above is the right one. It seems to be where neuroscience, viewed broadly, is pointing us. But even so, this sort of view is still pretty high-level. How can we fill in the details? It's abundantly clear that the simplistic "deep learning" models of aspects of brain function, which I'll discuss in detail in a later chapter, barely scratch the surface of what is currently known to happen in the brain, let alone whatever secrets the brain has yet to reveal.

Even the neuroscience we think we understand well may actually be shakier than is commonly realized. For instance, what about all the recent results on the roles of astrocytes and glia in human memory? These cells are not understood well enough to emulate them usefully in brain-based AGI systems yet. But if you ignore them, as essentially all computational neuroscientists and brain-inspired AGI researchers are currently doing, are you actually ignoring something critical to neural intelligence? And if you take a neurons-only based brain simulation, and tweak it in various clever ways to make it perform intelligently in the absence of glia, are you really winding up with something "brain based" in any meaningful sense? Might you do better just to explore some other class of algorithms and structures, better suited for digital computers?

Another downside of a closely brain-inspired approach is that the brain's "algorithms" and "data structures" are obviously optimized for neural wetware, rather than for digital computer hardware. Running brain-ish algorithms and structures on digital computers is kind of like building a house out of macaroni noodles and jello. It may well be possible, but you're using a certain infrastructure for something very different than what it was meant for, which is going to cause a lot of inefficiency and a variety of unpredictable problems.

All in all, while emulating the brain is an attractive-sounding approach to AGI, given the current state of knowledge in neuroscience, it's certainly not an obvious slam-dunk. Of course, as neuroscience progresses, the outlook for this sort of approach will get better and better. But the progress of neuroscience in relevant directions is difficult to project in detail. I have no doubt that sometime in the next few decades we will know enough neuroscience to proceed with closely brain-based AGI in a much more serious way. But will it happen in the next 5 or 10 years? That's much harder to say. The core issue is not so much computational modeling, but rather how to gather more data about how brains work.

Cutting open a living brain to see how it operates doesn't work very well. Unless it's done sparingly, cutting into the brain has a nasty consequence of killing its owner, or at least messing up their mind quite a lot. Studying dead brains is of limited value, because we really care about the dynamic activity inside the brain, where the thinking lies.

Imaging the brain non-invasively, without cutting it open, doesn't work very well. We can image the brain in various ways, using extremely complicated machines – fMRI (functional magnetic resonance imaging), PET (positron emission tomography), MEG (magnetoencephalography), and EEG (electroencephalography). However, impressive as these tools are, they only give us very coarse information. fMRI and PET tell us what parts of the brain tend to get active when we are doing certain things. EEG tells us about brain waves: the patterns of electricity sweeping across the brain, either as a whole, or across certain broad regions. MEG measures what happens at roughly 120 points on the skull as time goes on.

But the human brain has 100 billion neurons (the most critical kind of brain cell), and none of these technologies let us see what very many of them are doing at any given moment as thinking progresses. In order to accurately model and emulate the brain, this is what we need to know. Even if you cut open a head and stick electrodes inside, you can only measure a tiny percentage of neural activity without harming the brain so badly that the neurons' activities are interfered with, defeating the whole purpose of modeling them. Or the brain just dies.

So, progress toward closely brain-based AGI largely depends on how soon we get a major breakthrough in brain imaging, which allows us to mine information from living, thinking brains with simultaneous high spatial and temporal resolution. Without the ability to make a sort of "movie" of what's happening throughout the brain as time passes, it will be difficult to arrive at any really sound and thorough theory of neural structure and dynamics. Even with such movies, the job won't be easy – we will have an exciting but Herculean data analysis task on our hands. But at least, with such movies at hand, neuroscientists – and as a consequence, AGI researchers bent on neural emulation – will be playing a whole different game.

Without some revolution of this sort, we know what different parts of the brain represent, how the individual cells in the brain work, and how they use chemistry to interact. Maybe – if I'm right – we understand the essence of the overall dynamics. But we don't know the details of how the cells connect to each other and how these connections give rise to the dynamics of thought.

I'm sure having a better understanding of the brain will be VERY useful for helping us think about AGI – though this utility may come in ways we don't now expect. But my suspicion is that, once we really understand how the brain works, we're going to see very clearly why emulating the brain in detail really isn't the best way to build an AGI.

Computers are Not That Much Like Brains

Before moving on from the brain and back to AGI, and at risk of becoming boring and repetitive, I really feel I should emphasize a little more that the brain is a very different kind of physical system from that of the modern-day computer. The methods that work best for achieving intelligent functions in a system of wet brain cells (neurons are actually pretty similar to muscle fiber cells), squirting chemicals around to each other, are bound to be pretty different from the methods that work best for achieving the same intelligent functions in logic gates on silicon chips.

Brain cells have a lot of randomness about them, and some of their key dynamics happen pretty slowly (neuron firing happens on the time scale of tenths or hundredths of a second). Computer logic gates are exact, and they do things really fast. A laptop today does billions of operations per second. On the other hand, a human brain has a hundred billion cells that can all do things at once; they're coordinated in complex ways, but they also carry out their own independent activities. A typical computer these days does one, two, four or maybe eight things at a time.

Companies like Google and Amazon run "server farms" containing millions of computers networked together. Still, each computer only does a handful of things at a time, and even though the different computers on the server farm talk to each other, the communication between computers is a lot slower than communication between processes on the same computer. On the other hand, the brain's division into parts isn't nearly so strict. Between each part and the other there are numerous connections, and most of the advanced processing in the brain seems to involve networks of activity spanning several parts.

Modern computers and networks are built to be modular. You can add new parts on them pretty flexibly, to give them new functions or connections. Adding new parts into a human brain, though, is a far more difficult proposal. We're just barely getting the hang of it now, in relatively "simple" cases like cochlear implants, artificial eyes, and prosthetic limbs.

Figure 10.24: *Upgrading digital computers by adding extra RAM is no big problem. That's because current computer hardware systems are engineered for reasonable amounts of expansion. Of course there are limitations to the expandability of current technology, but they're nowhere near as problematic as the limitations involved in expanding the human brain. And computer tech gets more and more reconfigurable and expandable as technology advances. (http://embedded-lab.com/blog/?p=1374).*

Neither the brain's architecture nor contemporary computer architecture is better or worse in an absolute sense; the two physical infrastructures have their own strengths and weaknesses. They're just very, very different. But an AGI system could be better off without some of the brain's less helpful features.

For example, I did some work for a US government agency in the period 2009-2011, helping to build a computer program that models how people make decisions. The goal wasn't to build a computer program that was really great at making decisions, but rather to devise a model that explained why people sometimes mess up their own decision-making process. We devised a model based on the analysis of several situations illustrating cognitive errors in peoples' thought processes.

One scenario involved a person being asked to estimate a number. Often, in this sort of situation, someone will make an initial guess, and after gathering new evidence, incorporate this new information by modifying their initial guess. The initial guess will get way more weight than it should, a form of what psychologists call the "anchoring bias."

Another scenario focused on people being asked to estimate the probability of some combination of things; often they don't get the logic right. If you tell people that Bob has long hair, and then ask them which of the two cases has higher odds, either A) Bob is a bank teller or B) Bob is a bank teller and a rock singer, most people will choose the latter. This isn't right, of course. The odds of any conjunction ("and") of things is always lower than the odds of the individual things NOT being conjoined. Peoples' unconscious minds tend to confuse odds with association —psychologists call this error the "conjunction fallacy."

A more complex scenario concerned a person's thought process when choosing a car to buy. Often, a person will survey several different cars, studying the properties of each. First, they may look at the physical appearance of various cars, in pictures and on the street. They may then read information online or in magazines. Finally, they may go test drive a few cars. As they gather more and more evidence about various cars, they are updating their own preferences in their minds. The anchoring bias plays a role here, since most people will tend to prefer the car that made the best first impression on them, regardless of the further evidence they gather about other cars.

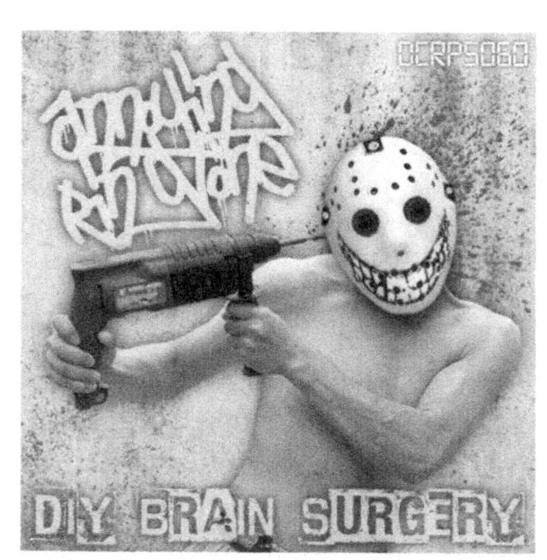

FIGURE 10.25: *Adding new parts to a modern computer is pretty easy. Adding new servers or even novel types of devices to a modern server farm is straightforward. On the other hand, adding new pieces into one's brain is a lot of work. A DIY lobotomy may work OK, but a DIY intelligence enhancement is fraught with peril, at least given current technology. The human brain architecture is not meant for science and engineering driven expansion, but only for slow, gradual trial-and-error evolutionary tweaking. The picture shows the cover art from the album "DIY Brain Surgery" by the dancecore band "Annoying Ringtone." The band's name as written on the cover presents a significant challenge for current OCR technology. (http://www.last.fm/music/Annoying+Ringtone/DIY+Brain+Surgery).*

The neural circuitry underlying this accumulation of evidence and subsequent decision making is beginning to be understood, which begins to explain how the brain's chemical and neural dynamics give rise to errors, like the anchoring bias. In other words, these cognitive errors are sometimes rooted in the way the human brain works. They're a default aspect of our neurobiology, our natural way of thinking. We can work around these with training and hard work. But an AGI wouldn't necessarily have these kinds of biases in its thought process in the first place. Yet at the same time, no real-world AGI is going to be a perfect thinker.

We spent a long time making a detailed model of how people accumulate evidence about various alternatives and then finally come to a decision. The way the brain does this turns out to be quite complicated, involving multiple groups of various kinds of neurons, acting in different parts of the brain. The process is modulated by various chemicals, including some that influence cognitive factors, like how quickly the person jumps to conclusions. Where accumulation of evidence is concerned, however, I believe that what the brain does is basically very complicated and imprecise basic math. A computer program can carry out the same function of evidence accumulation much faster and more accurately — if it only implements the basic math of adding up evidence directly, rather than trying to do it via simulating neurons (and the related brain chemicals).

Now, maybe there's some hidden magic to the way the brain does evidence accumulation, that can be reproduced with greater efficiency on computers. Personally, I doubt it. Evolution adapted the brain to accumulate evidence and make decisions in certain ways, working with the materials at hand. The brain already had networks of neurons being used to make simple decisions. Evolution just tweaked these networks of neurons so they could make more complex decisions. Evolution didn't care if these neural networks did evidence accumulation exactly right, because the brain could usually do OK by making decisions based on rough, approximate comparisons of the evidence in favor of the various alternatives. Evolution didn't care much about how fast the evidence accumulation process operated because it wasn't a bottleneck in the brain's processing anyway. Evolution put a lot more focus on optimizing the efficiency of the brain's vision processing circuitry, for example, because that would be more of a bottleneck in the brain's practical operation of controlling a human, or another animal.

Some parts of the brain are fantastically well optimized for their functions. Aspects of visual and auditory processing are amazingly elegant.

There's some wonderful subtlety in the way the cerebellum and the cortex work together to achieve rapidly sequenced movements, like a tennis serve, or a dance move. Other parts of the brain are messy and inefficient in their construction. The brain is pretty good at spotting lying and deception, but bizarrely inconsistent in making moral judgments. For instance, people will often have more compassion for a single suffering person than for 1000 suffering people.

That's how evolution works — it's brilliant but erratic, and often makes big messes. We can see a similar uneven quality in other parts of the human body. The eye is in many ways a marvel of engineering (problems with myopia and so forth aside), but really, this whole business of our teeth rotting and needing frequent repair is just a ridiculous mess. Imitating everything evolution has done in an engineered system isn't necessarily a good idea. We don't need to make robots with human-like teeth that get rotten and need fillings. We also don't need to make robots with inefficient, human-like evidence accumulation circuits and wildly inconsistent human-like morality systems.

But this doesn't have to be a problem for AGI in the end, because, as we've already seen in earlier chapters, it's not necessary to base AGI designs on the brain in any detail. When one conceptualizes AGI in terms of "thinking machines" rather than "computer brains," one realizes there is a LOT of other information to go on. A copious amount, in fact. We can piece together the limited information from neuroscience with information from cognitive science, combined with our rapidly increasing knowledge about computational algorithms and the mathematics of complex systems. This is what I call the "integrative" approach to AGI design, which OpenCog exemplifies. This is, I think, the path that's going to get us to the AGI goal. But more on that below....

Further Reading

Bear, Mark, Barry Connors and Michael Paradiso (2006). *Neuroscience: Exploring the Brain*. Lippincott.

Baars, Bernard, Stan Franklin, and Thomas Ramsoy (2013). Global workspace dynamics: Cortical binding and propagation enables conscious contents. *Frontiers of Psychology* (4).

Edelman, G. M. (1987). Neural Darwinism: the theory of neuronal *group selection*. Basic Books.

Edelman, G. M. and Gally, J. A. (2013). Reentry: a key mechanism for integration of brain function. *Frontiers in Integrative Neuroscience*.

Freeman, Walter (2000). Neurodynamics: Explorations in *Mesoscopic Brain Dynamics*. Springer.

Goertzel, Ben (1993) *The Evolving Mind*. Gordon and Breach.

Goertzel, Ben (1994). *Chaotic Logic*. Plenum.

Granger, Richard (2011). *How Brains are Built: Principles of Computational Neuroscience*. Cerebrum.

Irimia, Andrei and John van Hortn (2014). Systematic network lesioning reveals the connectivity scaffold of the human brain. *Frontiers in Human Neuroscience, vol. 8*.

Izhikevich, Eugene (2007). *Dynamical systems in neuroscience: The geometry of excitability and bursting*. MIT Press.

Izhikevich, Eugene and Gerald M. Edelman (2010). *Large-scale model of mammalian thalamocortical systems*. PNAS 105.

Maturana, Humberto and Francisco Varela (1992). *The Tree of Knowledge.* Shambhala.

Modha, Dharmendra and Raghavendra Singh (2010). Network architecture of the long-distance pathways in the macaque brain. *PNAS* 107-30.

Pribram, Karl (1991). *Brain and Perception.* Erlbaum.

Raghani, Mary Ann, Muhammad A. Spocter, Camilla Butti, Patrick R. Hofand Chet C. Sherwood (2010). A Comparative Perspective on Minicolumns and Inhibitory GABAergic Interneurons in the Neocorte. *Front Neuroanat.* 2010; 4: 3.

Rieke, Fred, David Warland and Rob de Ruyter van Steveninck (1999). *Spikes: Exploring the Neural Code*. Bradford.

Rinkus, Gerard (2010). A Cortical Sparse Distributed Coding Model Linking Mini- and Macrocolumn-Scale Functionality. *Front Neuroanat*. 2010; 4: 17.

Rosen, Robert (2002). *Life Itself*. New York: Columbia University Press.

11. Development and Embodiment

So what am I trying to do right now, in 2016, to build thinking machines; and why?

In preceding chapters I've summarized for you my fundamental understanding of how minds work, also running you through some of the history of my early-career AI pursuits. Now it's time to start telling you a bit more about the basic approach I've adopted for designing AGI, and am pursuing right now along with my excellent colleagues.

My AGI approach is a fairly long story. One aspect is the actual AGI design – the architecture – I'm working with, which is called OpenCog, and which builds on ideas we explored at Webmind, but wraps them up much more elegantly and rigorously. I'll talk about that a couple chapters down the road. Another aspect, equally important and a bit simpler to talk about, is the overall methodology of AGI development. What should an in-progress proto-AGI be used for? What kind of input should it be exposed to? What kind of output should it be expected to give? How should it be interacted with? How should it grow and develop?

A high-level summary of the AGI-creation methodology I've settled on is as follows:

First: Look at everything known about human cognitive science. What are the main mental processes occurring within the human mind? How do they relate to each other? How do they operate dynamically?

Second: Think about these known cognitive processes as a system: How can one make sense of the totality of known human cognitive processes in terms of the overall function of a human mind? That is, as a system controlling a body in an environment, constantly regulating and modifying itself.

Third: For each selected cognitive process, figure out how to achieve what that process does via some reasonably efficient computer algorithm that will run efficiently on today's computers. Some of these algorithms may bear resemblance to the way the brain works, while others might not.

Fourth: In the system of the mind, look carefully at the interactions between all the cognitive processes that are modeled computationally; make sure that the chosen computational algorithms interact with each other properly.

Fifth: Create a practical software framework allowing each cognitive process to operate simultaneously and interact with the others appropriately.

Sixth: Create an agent powered by this software framework; let it control a body in a world full of rich data to interact with; and let the world be populated by human agents eager to interact with and teach the agent.

Seventh: In this environment, carefully interact with the AGI agent to lead it through a series of natural developmental stages, modeled loosely on human developmental psychology.

Ta-daah – a thinking machine!

Yes, this methodology is complicated and has many parts, each of which is difficult in its own way. But it doesn't require any knowledge that isn't already currently available; it only requires the intelligent, judicious assemblage of things that are already known.

AGI and Human Childhood Development

The "developmental" aspect of the strategy outlined above is something I've been musing about for a while, since 2000 or so when we were experimenting with "Baby Webmind." It's something I've gotten more and more serious about over the years.

When I was a teenager, first getting passionate about AI, I was fairly sure I'd never have human kids. Reproducing the old-fashioned way had a certain emotional appeal – I'd always loved kids, and had often babysat neighborhood kids to earn a few dollars – but in the end it seemed a lot more interesting to produce the first AI children, than to merely spawn a few more flesh-and-blood kids. I also realized that achieving all my research, business and literary goals would take a hell of a lot of work, and raising and supporting little tykes would obviously be a ridiculous distraction.

Life proceeded according to its usual course, and I fell in love in college, and after graduation moved in with my college sweetheart… and lo and behold, she turned out to be rather insistent on the production of human offspring. We had our first son, Zarathustra Amadeus, when I was barely 23, in the middle of my first year as a math professor at the University of Nevada Las Vegas, shortly after finishing my PhD.

Now, in early 2016, my kids are 26, 22 and 19; and I'm looking forward to having more with my new wife, Ruiting Lian… Did having kids distract a lot of my time from my scientific, engineering, business and writing work? Absolutely. Has it been a tremendously emotionally and personally fulfilling part of my life, one that I wouldn't ever choose to give up? For sure. And it's also had some side benefits I didn't anticipate in my teenage years. Watching my kids grow up – especially in the first few years of their lives – taught me a great deal about the human mind and how it lives, grows, interacts and operates.

Direct observation always teaches different things than textbook learning. By watching my kids' young minds grow up, I got a strong intuitive sense for the kinds of learning that it takes to build a mature human mind. The young child's mind is always at work, categorizing the world and looking for patterns – and relentlessly modifying and improving its patterns and the rules by which they combine. Each kid invents their own grammars, over and over, and their own ways of walking and moving. and their own ways of convincing you to do what they want... and then they gradually revise and refine their approaches to everything, until they converge on solutions that count as "good enough" in their cultural context. A child of 6 months old is already an experimental and theoretical scientist, testing their ideas on the world minute by minute, revising their ideas based on their own subjectively judged criteria of observational accuracy and elegance.

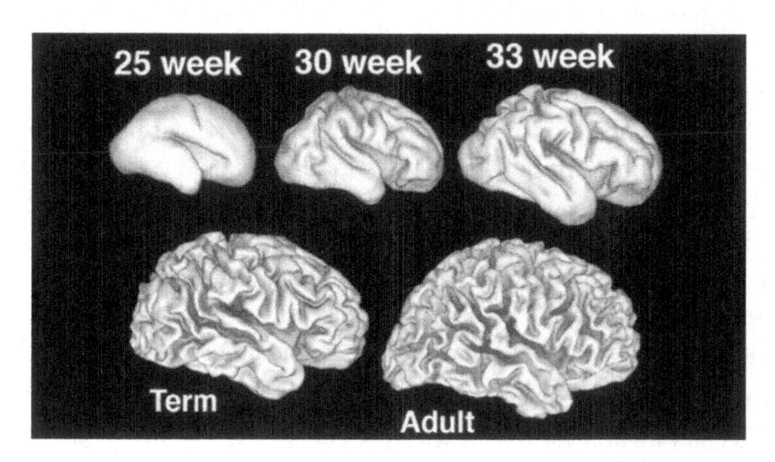

FIGURE 11.1: *Maps of the cortical surface in premature infants (at several gestational ages), full-term infants, and adults. Some of the structure of the adult brain is genetically coded, some is acquired via learning, and some arises via spontaneous self-organization. How these three aspects balance off against each other, and interplay, remains for the science of the next decades to unravel. (http://www.brainfacts.org/about-neuroscience/technologies/articles/2009/brain-atlases/).*

From a parent's point of view, seeing the amount of experimentation a young child does, it's easy to draw the conclusion that a kid is learning how to interpret and act in the world "from nothing." And indeed, regarding specific knowledge of the world, an infant is indeed largely a "blank slate," as philosophers posited hundreds of years ago. Maybe an infant knows a few things, like the smell of a lactating breast, but in terms of concrete pieces of information, it doesn't know much.

Yet, the learning a child does is in some sense too fast to really be "from nothing." A reinforcement learning system just randomly experimenting and seeing what works would take much longer than a human child does to figure out what's going on. The difference is not that a child's mind has some magic, secret learning system in it. Rather, it's what learning theorists call inductive bias. Every learning system is biased at least a bit, in the sense of being quicker at learning some things than others. A child's mind is biased to be quicker at learning the things it needs to know to be an adequate human adult than it would be at learning other random things of equal complexity. But this bias is nothing mysterious. Quite simply, as cognitive science has taught us, an infant's or child's mind confronts the task of understanding the world with a complex armamentarium of learning processes and information storage structures, each of which is specialized to be good at learning particular sorts of things. Instead of specific knowledge, what a young mind has is a quite refined set of algorithms and biases for recognizing patterns in the world.

The whole apparatus of structures and processes I've described above is not there in an infant's brain right from the start – some of them are, and others develop gradually over time, sometimes triggered by the infant's experiences or passage of physiological milestones, sometimes just triggered by the passage of specific amounts of time. The field of developmental cognitive neuroscience studies the cognitive mechanisms in the brains of young children, and how they progressively come into play over time. But even so, right from the beginning, the baby mind is far from a blank slate. The development process starts the baby out with the cognitive mechanisms needed to come to a rough knowledge of the world. Then once it has learned a bit, the development process provides the young child's mind with the mechanisms needed to take the next steps.

Language learning is an intriguing and much studied case in point. Infants are born with a fairly strong ability to recognize linguistic patterns in the streams of sounds they hear. But the ability to understand complex syntax comes into play gradually. And if the child doesn't receive appropriate stimulation, the latter may never emerge – the development of the needed cognitive mechanisms will not be triggered. "Wild children" who grow up without anybody communicating with them miss the chance to get their brain's advanced language processing capabilities triggered in childhood. As a result, even once they eventually get integrated into society, they never quite use language normally. They can do short sentences and basic communication, but they're never going to write like Joyce or Proust, nor even diarize like the average teenage girl.

Figure 11.2: *Oxana Malaya, an 8-year-old wild child who spent most of her life in the company of dogs, was discovered in Ukraine in 1991. (http://listverse.com/2008/03/07/10-modern-cases-of-feral-children/).*

It's not entirely clear how useful developmental cognitive science is going to be for AGI. A lot of child cognitive development is tied to the specifics of child physiological development, which is not particularly relevant to near-future robot bodies. And, practically speaking, it would be a lot of trouble to take each cognitive mechanism one builds for one's AGI system, and then build a series of progressively simpler versions appropriate for a younger AGI system. At a high level, though, I think it makes a lot of sense to think about AGI systems in the context of child cognitive development. The overall process by which a young child learns about the world is very relevant to AGI, because an AGI starts outs its journey in a similar way to an infant – as a virtual blank slate, so far as content is concerned.

One can try to work around the "blank slate" aspect in an AGI context, by using digital knowledge bases to fill the young AGI's mind with knowledge.

But this doesn't really work – or at best it can work partially. Information loaded into an AGI from a database isn't really "knowledge" in terms of the AGI's role as an autonomous agent in the world. Until it knows how to connect this information with its perceptions and actions, it's not really useful knowledge at all, it's just a bunch of patterns oddly floating there in memory. And the connection between the database of information and the world still has to be learned by the AGI system from experience, starting from nothing, in the rough manner of a human baby.

For instance, suppose an AGI system is fed the information, from a database, that a cat is a kind of animal. In the database, this is likely to look something like: "cat isa animal," indicating that there is a relationship of type "isa" between the concept "cat" and the concept "animal." This can be used by an AGI system for abstract reasoning of various sorts. For instance, if the system knows that:

```
animal hasProperty lifeform
```

Then some simple reasoning proves that:

```
cat hasProperty lifeform
```

That is,

```
Cats are animals
Animals are lifeforms
THEREFORE
Cats are lifeforms
```

This sort of reasoning seems impressive at first, but what one finds is that in order to really formalize all the knowledge needed to be a young human child, one would need an insane number of relationships like this one. The Cyc project, an AI project based centrally on formalization of knowledge in this manner, has encoded many millions of such relationships, using dozens of expert knowledge encoders over decades. But they have barely made a start at what would really be needed to formalize a young child's knowledge.

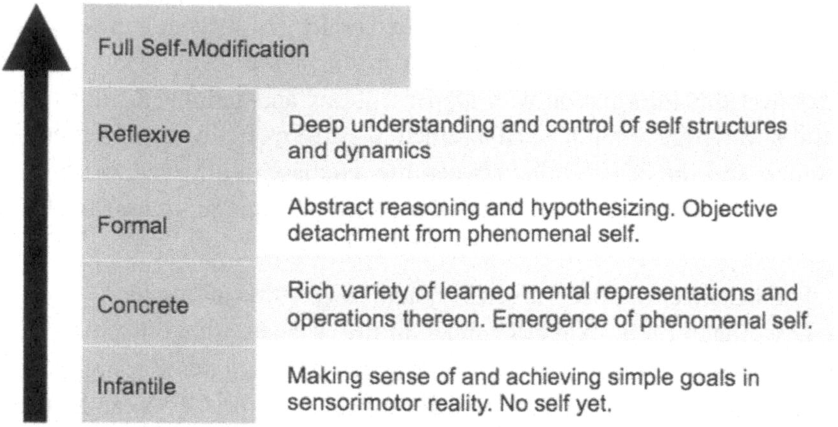

Full Self-Modification

Reflexive — Deep understanding and control of self structures and dynamics

Formal — Abstract reasoning and hypothesizing. Objective detachment from phenomenal self.

Concrete — Rich variety of learned mental representations and operations thereon. Emergence of phenomenal self.

Infantile — Making sense of and achieving simple goals in sensorimotor reality. No self yet.

FIGURE 11.3: *Depiction of Piaget's classic hypothesis regarding the stages of psychological development of human minds. Actually, Piaget only went up through the Formal stage, but a number of more recent thinkers have conjectured post-formal stages of development, such as the Reflexive stage we've added here. A fully self-modifying AGI, able to reprogram itself with its goals in mind, would go beyond any directly human-relevant developmental stages, of course. Piaget's theory, in general, is best viewed as conceptual and inspirational rather than as a precise scientific theory of child cognitive development; modern developmental psychology has uncovered a massive network of nuances which make it clear that the real story of human mind development is far subtler than Piaget realized.*

Another possibility seems to be to extract this kind of relationship from natural language texts. Wikipedia, for instance, has a massive amount of knowledge in it! But the problem is that natural language is confusing and ambiguous, so that in order for a computer program to reliably extract the meaning from natural language, it would need to have a great deal of commonsense understanding of the world already. One then has a chicken-egg problem: To get commonsense knowledge you need to understand language, but to understand language you need commonsense knowledge.

As one random linguistic example, think about the many different meanings of the omitted object of a verb:

```
Ben shaved = Ben shaved himself
The Phillies won = The Phillies won the game
John ate = John ate something (probably edible)
```

In the first case, "shaved," the omitted object means that the object is the same as the subject. Ben shaved Ben.

In the second case, "won," the omitted object means that the scope of possibilities is narrowed down in a specifically understood way. The Phillies didn't lose; they won.

In the third case, "ate" has a regular object, which just isn't specified. John ate cake, or John ate dinner, or whatever.

How do we understand the differences between these cases? Not by looking at the syntax alone – rather, by understanding the commonsensical meanings of the verbs.

But how does a child learn these commonsensical meanings? By hearing people use similar linguistic constructions in similar contexts.

A more complex example illustrating the subtlety of semantic interpretation is:

```
You aren't to marry him, and that's an order
You aren't to marry him, and I read it in my Tarot
cards
```

The meaning of "are" is quite different in the two cases... But there's no way to tell what the first half of the sentence means till you read the second half. And there's no way to figure out these implications without knowing something about ordering around, and something about fortune-telling. A human child learns these things via being ordered around, and via having other people try to predict the future – via commonsense experience with the referents of the words being used. This is not necessarily the only possible way for such understanding to be gained. But it's certainly the way we understand best, at this point.

Venturing fairly far afield from the human development model, some researchers aim to "bootstrap" the learning of commonsense knowledge purely via language, i.e., to go from a little commonsense knowledge to a little more language understanding, to a little more commonsense knowledge, etc. This sort of approach seems particularly interesting in the context of large stores of linguistic data such as the ones possessed by companies like Google and Microsoft. But it's not clear that it's really possible to make it work. Humans, obviously, bypass this chicken-egg problem via getting our first dose of commonsense knowledge non-linguistically, via direct embodied experience in the world. Using this commonsense knowledge, we are then able to interpret simple language. Using simple language, we gain more knowledge, which helps us interpret the world better; and the knowledge we then gain from the world helps us to understand complex language better. Our own iterative bootstrapping process does involve progressively increasing and mutually reinforcing amounts of linguistic and commonsense knowledge, but our embodied experience in the world feeds and guides this overall knowledge growth process.

My decision to structure much of my own AGI work around human childhood development has clarified many different aspects of my practical thinking about AGI. While I don't think this is the only workable approach, it does seem the most natural to think about. This is partly because I spent a lot of time watching my three kids grow up, and partly because human childhood development is the most widely studied process in which a mind starts out rather stupid and ends up reasonably smart.

But the decision to think about AGI substantially in the context of human child development came to me only gradually. Roughly emulating human childhood wasn't my original approach when I founded my first AI company, Webmind Inc., in the late 1990s. Back then, I was thinking more about Internet intelligence, creating things like intelligent search engines, or new artificial life forms that would grow across the Internet and evolve their own forms of intelligence. I still think that approach makes sense, but I found it difficult in that context to distinguish pathways to AGI from narrow-AI dead ends. I find that, if I'm thinking about AGI that is intended to roughly follow the stages of human childhood cognitive development, the path forward seems conceptually clearer. AGI is complex and confusing enough even after one has abolished the worry of whether one's AGI developmental pathway is a dead end.

It may seem odd for me to advocate modeling AGI development on human child development, since I don't advocate, at this stage in the development of neuroscience and computer hardware, modeling AGI closely on the human brain. But actually, as incomplete as it is in its current state of development, I find developmental cognitive psychology a lot clearer than neuroscience. This is part of the reason I'm generally more bullish on cognitive science than neuroscience, as a guide for AGI. After playing around with lots of proto-AGI software, I've seen how difficult it can be to figure out what experiences an early-stage AGI system should be put through, and in what order. The field of childhood development provides many hints in this regard.

Human childhood development respects the hierarchical structure of human knowledge, translating it into a temporal, sequential order that allows children to learn the naturally simpler parts first, before proceeding to the slightly more complex parts, and so on.

While today's early childhood education system is certainly flawed (back in the mid-90s I helped found the "Unity Charter School" in Morristown, New Jersey, in an effort to remedy the worst of the system's defects – and then I saw the various legal regulations the state placed on charter schools do terrible damage to our original vision for the school), it does respect the hierarchal nature of human knowledge, building naturally from simple to progressively more complex knowledge. A preschool, for instance, usefully provides simple versions of basic life skills an adult must master, and is specifically geared towards gradual development, while also building interconnections between those skills. Social interaction, art and creativity are woven into everything.

Because of this, integrating aspects of a preschool or grammar school-type environment into an AGI's environment appears to be a good course for structuring an early-stage AGI's learning experience, even if the AGI architecture doesn't bear close resemblance to the human brain. This is what I have often referred to as "AGI Preschool." An AGI Preschool could be a physical preschool similar to a human preschool, but perhaps with slightly different toys and activities, customized for compatibility with the particular robot bodies in question. Or it could be a virtual-world preschool, providing simulated virtual toys and activities for virtual-world AGIs to play with using their animated-character bodies. The important thing is not the specific design of the preschool, but rather the availability of a wide variety of activities, with varying levels of complexity and a high tolerance for failure, that provide practice at all the main skills needed for coping in the everyday adult human world.

For instance, if a specific robot's hands have trouble with regular Lego blocks, one can supply that robot's preschool with appropriate foam blocks. If paint mucks up the robot's fingers, there may be electronic whiteboards that provide sufficient avenue for artistic exploration. The fact that human three year olds can't generally use search engines well is no reason to restrict preschool-level AGIs from freely exploring Google Images and Google Videos using search terms, if they have the specific language ability to do so. The point is not to precisely emulate human preschool activities for robots or videogame AGIs, but rather to conceptually emulate the notion of the preschool experience for young AGIs, so as to provide a preschool-like environment for fostering and evaluating both unsupervised and supervised learning.

In terms of language learning, if you're following the childhood development approach, you don't start by having your AGI try to read the whole Web or a compendium of scientific papers. Rather, you start by having the AGI learn very simple language in an experiential, embodied context, similar to how humans learn language. You have it learn language in the context of playing with things, in a way that comes naturally to it. Then, once the AGI understands simple language in the context of its own life experience, you try to teach it more complex language, and expose it to situations where it will benefit from applying what it has learned. That's a much more natural way for a young AGI to learn human language.

Maybe a properly designed AGI system could parse the Web and understand all the language on it, even without having any experiential grounding in a simple language. In essence, this approach would present the nascent AGI with a huge system of complicated, simultaneous, interrelated equations. It would have to figure out the meaning of each word based on the meanings of all the other words each is associated with, without really knowing what any of the words mean in relation to the external world. In purely mathematical terms, this seems possible if you have a smart enough system and enough text, but it also seems rather difficult. Since building AGI is going to be tough no matter how you slice it, I prefer to try the easiest ways that seem feasible. Of course, these easiest feasible ways are not actually very easy at all. But the exciting thing is that we live in a time where there are any feasible ways whatsoever!

Embodiment

The idea of modeling an AGI's growth on human childhood development is inextricably linked to another tricky issue, perhaps THE most controversial issue in the whole AI field: embodiment. A human child is far from just a mind; being a young child is largely about learning to use one's body. Human childhood development, like AGI development, suggests some vaguely human child-like body for the AGI. Not necessarily one identical to the humanoid form, but at least a body that can move around, perceive with multiple senses, and manipulate various kinds of objects.

Quite apart from the question of whether human childhood development is a good model for AGI development, there is huge disagreement among AGI researchers about what kind of embodiment is necessary, useful or sensible to give an AGI system, if the goal is to achieve a human-level general intelligence. This is separate from the question of the validity of the developmental approach. On the one hand, one could potentially try to emulate developmental psychology in an AGI system with a purely textual interface. On the other hand, one could certainly attempt to make an intelligent humanoid robot via old-fashioned rule-based AI, or other approaches not involving humanlike cognitive development.

Also, even if you agree that the mind-body relationship is important for the development of intelligence, that doesn't resolve the question of: *What kind of embodiment does an AGI system need?*

Once, at a conference with a diverse audience, I gave a talk on AI and mentioned the controversy about embodiment. I said that some researchers felt embodiment was critical and others felt it was less so. A woman came up to me after my talk, fascinated by the idea of disembodied AIs – by which she meant AIs that have no physical instantiation at all, but take the form of ghost-like energy fields, mind-systems made of pure psychic material... artificial poltergeists! I told her I was open to that possibility, but that that wasn't what I was really talking about. I explained that the non-embodied AI systems that some researchers talked about still consisted of computer code running on some physical computer. She was very disappointed.

My AGI researcher colleague Pei Wang, on the other hand, once wrote a paper titled "A Laptop is a Body", arguing that the important aspect of embodiment for AGI is simply the connection it provides between the AGI and the world. A laptop provides sensors to an AGI running on it, via its keyboard, microphone and webcam, and also its connectivity to other computers and the Internet. A laptop provides actuators to an AGI running on it, at very least via the computer's sound and video output, and also via its ability to send signals to other computers and the Internet. Since a laptop is also a body, the question isn't one of embodied versus disembodied AGIs, but rather one of what kinds of sensors and actuators are really needed to support what kinds of intelligence.

An AGI's physical embodiment would not necessarily have to be a humanoid robot body; it could could take the form of a laptop, or a server farm. Or it could take the form of a tank, a submarine or a quadrotor drone. Potentially, given some currently undeveloped technology, it could take the form of some sort of exquisitely self-organizing ball lightning – approximating the AGI poltergeists of my supernatural-minded conference questioner. Each of these choices would have implications regarding what kind of intelligence could be achieved using a reasonable amount of resources.

FIGURE 11.4: *Ball lightning, mobile coherent balls of pure electricity, has proved difficult to create in the lab, though Nikola Tesla claimed to have done it. But it has been observed empirically many times by various individuals. Could it be possible to create an intelligent system of pure electricity, via understanding and shaping this sort of phenomenon? Who knows? But it doesn't seem impossible. The point is that intelligence is about the cognitive-level pattern of organization of a system, not the underlying physical substrate. Potentially, intelligence can be a brain, or a computer, or a femtotech construct like a quark-gluon plasma, or a pattern of software acting in a computer-implemented simulation world, etc. etc. There may be physical constraints regarding what kind of mind can be effectively implemented in what kind of substrate; but within fairly broad parameters, intelligence is best considered substrate-independent. (http://en.wikipedia.org/wiki/File:Ball_lightning.png).*

Some researchers believe you would need a closely human-like body to get anything resembling human-like intelligence; and that, to get a high degree of general intelligence at all (whether human-like or not), you'd need a pretty complex and sophisticated body. These researchers tend to view intelligence as something that emerges from adaptive body function, in the course of a specific body's engagement with a specific world. Other researchers believe that the body really doesn't matter that much – that the core of general intelligence is a set of reasoning and learning processes that are body-independent, and the body is simply an I/O connector between the mind and the world.

FIGURE 11.5: *The Parrot AR drone was the first widely available "toy" drone, remote controllable via iPhone. Remote-controlled drones are currently in heavy use by the military, but are expected to make inroads into various commercial markets during the next decades. Journalism and disaster response are obvious application areas, along with, eventually, delivery of pizzas and anything else delivered to peoples' homes. Currently drones tend to be remotely piloted by humans, with only basic control mechanisms automated; but fully automated intelligent drones are an obvious next step. (https://upload.wikimedia.org/wikipedia/commons/3/34/Parrot_AR.Drone_2.0_%26_Dassault_Rafale.jpg).*

My own view is in between these two extremes. I think you COULD make an AGI with just a laptop as a body (maybe not the 2012 Macbook Pro I'm writing this on, but a more advanced version of what's available on the market today). A laptop-based AGI could learn by chatting with the owner of the laptop, surfing the Web, and looking at pictures on Flickr, videos on Youtube, and so forth. I see no reason why this couldn't work, in principle.

I also think that if you had a genuinely human-like robot body to plug your AGI algorithm into, creating a human-level, intelligent, human-like AGI would be an awful lot easier. You could apply accurate sensors and actuators to feed sensory information to your AGI system.

Certainly, it's educational to take robots as we have them now and plug in AI systems. There's much we can learn from this approach, even though present-day robots are much cruder than the human body in most ways (though more powerful in a few ways – e.g., current robots can connect directly to the Net by wifi, whereas I cannot; and they can get exact distance measurements using laser rangefinders, etc.). You can still get more richly structured data, more finely-controlled actuation, and better synergy between perception and actuation from a robot than you'll obtain with any other technique for interfacing an AGI with the world, for the time being. And if you want your AGI to interact with the human world in a human-like way, which is arguably important for achieving human-like intelligence, a robot body is a good choice. Certainly if you want to mimic human childhood development even roughly, a robot that can physically move around, see, hear and pick things up is an excellent start.

Robots haven't featured largely in my own work in past years, but I'm moving in that direction. During the period of 2008-2010, I co-supervised some graduate students in a lab in Xiamen University in China, whose work involved hooking OpenCog up to a Nao humanoid robot. And I'm currently doing some work using OpenCog to control some of David Hanson's humanoid robots, which are more advanced in many ways than the Nao (though they still don't walk, as of 2016).

FIGURE 11.6: *Aldebaran Nao robot at Xiamen University in 2009. This robot held amusing conversations using a dialogue system that combined OpenCog with a number of other inputs, and was featured in Raj Dye's award-winning documentary film Singularity or Bust.*

FIGURE 11.7: *The Nao robot in a screenshot from Singularity or Bust – answering me when I asked him if he was, indeed, a robot.*

FIGURE 11.8: *Raj Dye's documentary Singularity or Bust covered the work Hugo de Garis and I were doing at Xiamen University in 2009, involving his neural networks, OpenCog, and Nao robots.*

FIGURE 11.9: *Me fixing some simple mechanical issues with Zeno, a prototype Hanson Robokind robot hand-built by David Hanson.*

FIGURE 11.10: *Me and one of David Hanson's Zeno robots at the Global Future 2045 conference, at Lincoln Center in New York, June 2013.*

FIGURE 11.11: *Me with David Hanson's actual human son Zeno, in Mong Kok, Hong Kong. The robot Zeno was modeled on the real Zeno at age 3 or 4. Now the biological Zeno has grown up to 8, but the robot Zeno is still the same size. The biological Zeno has also learned a lot more mathematics, and has massively greater physical agility, and dramatically superior emotional expressiveness. But some descendant of the robot Zeno will likely overtake him in all these areas one day!*

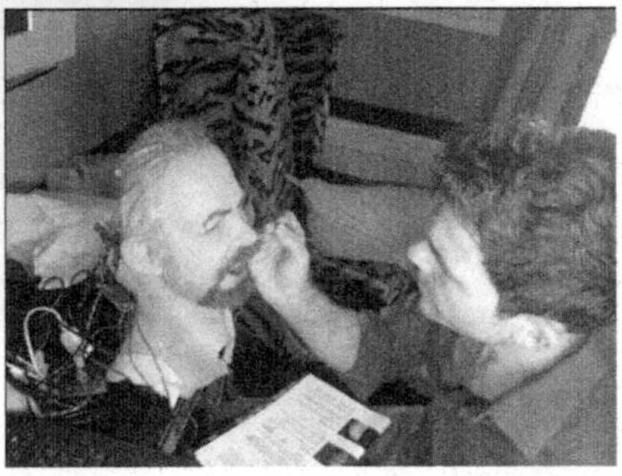

FIGURE 11.12: *David Hanson repairing some frayed Frubber on the face of his famous Philip K. Dick android, in one of Russian Internet entrepreneur Dmitry Itskov's rooms at the Empire Hotel in New York.*

FIGURE 11.13: *A girl Robokind robot, photographed by me at Nanyang University in Singapore in April 2013.*

FIGURE 11.14: *Einstein Hubo, a robot made by taking David Hanson's Einstein robot head and connecting it to the Hubo humanoid robot created at KAIST in Korea. Chosen as one of the best robots of all time by Wired Magazine.*

FIGURE 11.15: *My charming and brilliant wife, the computational linguist Ruiting Lian, with our Femi-Einstein robot in the OpenCog Lab at Hong Kong Poly U. This particular robot was a temporary conglomeration of a mini-Einstein head, built by David Hanson, with one of Mark Tilden's Femi-Sapien bodies.*

FIGURE 11.16: *Me with the "Sophia" robot from Hanson Robotics – our latest and greatest – at the Hanson office at Hong Kong Science Park in early 2016. (A free cockroach flavored jellybean to anyone who can identify the head on the stick!)*

As well as physical robotics, I think video game-type embodiments — like the virtual worlds of Second Life, or ones powered by modern 3D game engines — have a lot to offer. In a video game, you have essentially the same features that a robot provides – perception, action, social interaction, language, goal-oriented reasoning, spatial and temporal thinking, all connected in a way that emulates ordinary human life reasonably well. True, there's a lot less complexity in perception, action, richness and diversity available from contemporary virtual worlds than from the physical world as perceived by a robot. But you're also gaining a lot by going the virtual route. You're gaining simplicity – virtual worlds are a simple, low-cost way of experimenting – but you're also gaining more than that. Virtual worlds enable an AGI system to interact and practice with millions of people. You can roll out a video game or virtual world application to millions of users at a far lower cost than, say, sending humanoid robots to millions of users.

FIGURE 11.17: *Screenshot from the "game world" we are using, at the time of writing, to experiment with the OpenCog system. This world is built in the Unity3D game engine, and most objects in the world are made of many little square blocks. The composition of objects in terms of blocks makes it especially tractable for an early-stage AGI system to understand how objects are made, draw analogies between objects, etc. We don't think this world is complex enough to support human-level AI, but it's an interesting platform for experimenting with the OpenCog system as it develops.*

FIGURE 11.18.: *Teaching the OpenCog system to build stairs in the virtual world. In this example, the little girl is controlled by a human, and the animated robot is controlled by OpenCog. The girl builds some stairs with blocks, the robot watches, and then the robot imitates what it saw the human do.*

FIGURE 11.19: *Now that the robot knows how to build stairs, when it wants to get something (such as the reward block up on the tree to the right of the picture), it knows that one strategy it can follow is to build stairs to climb up and get it. This same sort of learning process could be explored with a physical robot, but doing it in the virtual world first allows research to focus on the learning dynamics rather than on the mechanics of perception and action. In parallel with this virtual research, though, we are also spending some time on robot perception and action, since the complexity of the real world is likely to be needed to get all the way to human level AGI, or even to human toddler level AGI.*

Just about the best humanoid robots we have now are creatures like the Nao and the RoboKind (several thousand dollars each, as of 2016). Although, arguably, Mark Tilden's sub-$100 "toy" RoboSapien robots have more sophisticated walking than these expensive bots, using Tilden's unique analog computing hardware. In some ways, the best research robot ever made was the PR2, which was not quite humanoid — it has wheels instead of legs, and gripper clamps instead of hands — and it cost around $400,000 per unit. With these price tags, these robots were just not feasible for mass distribution – unlike the RoboSapiens, which provide amazing capability for their price, but still have many limitations, such as the lack of robust visual sensors, capable grasping hands and wifi connectivity. On the other hand, a video game character could be played by millions of people sitting at home on their computers, or almost anywhere else on their smartphones. An AGI-based videogame could provide millions of teachers to instruct baby AGIs in the basics of language and thought.

FIGURE 11.20: ***At Mark Tilden's*** *amazing apartment in Tsim Sha Tsui, Hong Kong, with AI/robotics collaborators Gino Yu, David Hanson and Mark Tilden (from left to right). Note the plethora of funky Tilden robots on the table behind.*

FIGURE 11.21: *Mark Tilden with one of the RoboSapien robots he's created. 20-odd million of these (in various variants) have been sold in toy stores around the world. One of the big benefits of living in Hong Kong for a while has been getting to know Mark, who's been a long-time resident. He moved here largely due to the proximity to the factories of Guangdong province in China, right across the border, where his robots and other inventions are manufactured ingeniously and at low cost. (http://photos3.meetupstatic.com/photos/event/e/8/0/0/600_430379392.jpeg)*

FIGURE 11.22: *One of Willow Garage's PR2 robots reads a book. By not insisting on a purely humanoid appearance, the PR2's makers created a remarkably functional mobile robot. It rolls around quite smoothly, and its gripper hands are functional for picking up a variety of objects. The US$400K price tag restricted its market to a small set of researchers, though. (Lower image from http://commons.wikimedia.org/wiki/File:PR2_Robot_reads_the_Mythical_Man-Month_2.jpg; upper image from http://en.wikipedia.org/wiki/Willow_Garage#mediaviewer/File:PR2_at_Maker_Faire.jpg).*

Virtual world embodiment could offer solutions to the complexities of language learning (some of which we've discussed briefly above). Consider prepositions, some of the most confusing words in English – for instance, the preposition "with." We can say,

I ate dinner with my uncle.

I ate dinner with a fork.

I ate dinner with a salad.

I ate dinner with love.

I ate dinner with the TV on.

In each sentence the word "with" means something different. The varied meanings of the word "with" are hard for someone who speaks English as a second language to grasp. It would certainly be hard for an AGI to master the correct usage just from reading articles online — not impossible, but hard. However, if an AGI sees the word "with" used in its correct context in a virtual environment, and experiments with several objects, it will get a practical sense of its different meanings. Arguably, virtual worlds will be just as important for educational purposes as robotic embodiment.

Now, what about understanding specific objects? A robot in a kitchen will probably gain richer knowledge faster than an AGI in a virtual world;

in the video game world, its interactions will be limited to the simplistic simulated objects that others have built or imported. The AGI will come into contact with objects and events, with fewer examples of interactions from which to learn.

In principle, a full-featured virtual world could supply all the complexity of the real world or even more.

But for the foreseeable future, virtual worlds will be far less rich in detail than the physical world. This means that if we're going to attempt to construct AGI in the next five or ten years, some combination of virtual embodiment and robotic embodiment coupled with minimal embodiment (in which a system just looks at a lot of online data) will be the best way to proceed. The OpenCog system does this— it supports the rich intermixing of all these kinds of embodiment.

Embodiment & Environment

Much of the importance of embodiment for AGI doesn't derive from the body itself, but rather from the world that the body allows the AGI to experience. The relationship between an intelligence and its environment is quite subtle, and far more important than most AI researchers believe. After all, intelligence consists of the ability to adapt to the environment, and any real-world intelligence is better at adapting to some kinds of environments than others. An intelligence's methods of adaptation are themselves adapted to certain kinds of environments.

An absolutely general intelligence is a mathematical fiction. Absolute generality — the ability to learn anything at all within a feasible timeframe — could only be achieved using infinite computing resources, which don't exist in our reality. Any finite system built using bounded resources will possess limited generality, in the sense that it will be better at solving some kinds of problems than others. That is to say, it will be more intelligent in some environments than in others.

So if absolutely general intelligence is a fiction, what use is the concept of "general intelligence" after all? The answer is: It still makes sense to talk about the generality of an intelligent system, because different systems may have different balances between specificity and generality. Some systems may be very good at solving a narrow range of problems; others may be moderately good at solving a wider range of problems. The latter we would consider to have a more general intelligence. So you could view a "general intelligence" as having a broad scope as to the set of problems it can solve, and a "less general intelligence" as having a narrower scope. (It's possible to formulate this distinction mathematically – to formalize intelligence and generality-of-intelligence as separate but related concepts. I did this in a paper I wrote for the AGI conference in 2010, which we held in Lugano, Switzerland.)

There are limits to both the intelligence and generality that can be achieved by any system with finite resources. That is, any system created in the real world will be adapted to some category of environment, and to goals and tasks characteristic of that environment. The human mind is adapted to carrying out certain kinds of tasks in certain kinds of environments. Humans are, by and large, much better at fulfilling the basic demands of survival in our ancestral environment (locating mates, harvesting food sources, fighting off enemies and navigating in a complex 3D scene full of moving objects, etc.) than more abstruse tasks like proving mathematical theorems or writing code. We evolved in the African savannah and, until quite recently, the survival of our bodies and propagation of our DNA were our main goals. We're adapted to a certain set of environments and goals.

Now, an AGI need not have the same balance of strengths and weaknesses as a human mind. However, if we want an AGI that we can understand well enough for the purposes of mutually satisfactory communication (so we can teach it usefully, or debug its code in a savvy way when it goes wrong, and it can learn from us), we'll need a system adapted for the same sorts of environments and goals that we are. And this leads back to the question of embodiment, because if an AGI is to be suitably adapted to the same classes of environments and goals as we are, it will be better off managing the same tasks we do, which will be easier if the AGI has roughly the same kind of body.

What Can We Learn from Very Simple Environments?

The game worlds we've used for OpenCog experimentation have mostly been moderately complex ones, involving 3D structures of various sorts. Some researchers are focusing on virtually embodied AIs of much simpler sorts – for instance, Olivier Georgeon is pursuing what he calls "developmental AI," involving very simple agents acting in simple 2D worlds, but carrying out tasks in a way that involves agents evolving their own code. The goal is developmental emergence of cognitive behaviors. He has put together a very nice online course on developmental learning. As he puts it,

"Developmental AI is a new branch of AI that aims to program a minimal initial system able to develop its intelligence by itself. If implemented in a robot, it will initially behave like a newborn baby, then it will become increasingly intelligent as it interacts with the world.

By grounding the system's knowledge in the system's individual experience, developmental AI addresses fundamental questions of cognition such as autonomous sense-making, individual choices, creativity, and free will."

This is an appealing philosophy. On the other hand, one wonders whether the very simple worlds and agents involved will be capable of giving rise to non-trivially complex intelligent behaviors. I tend to doubt it. But I can see how this sort of experimentation can teach you something about the process of developmental emergence of intelligence, which is certainly a valuable thing to learn about.

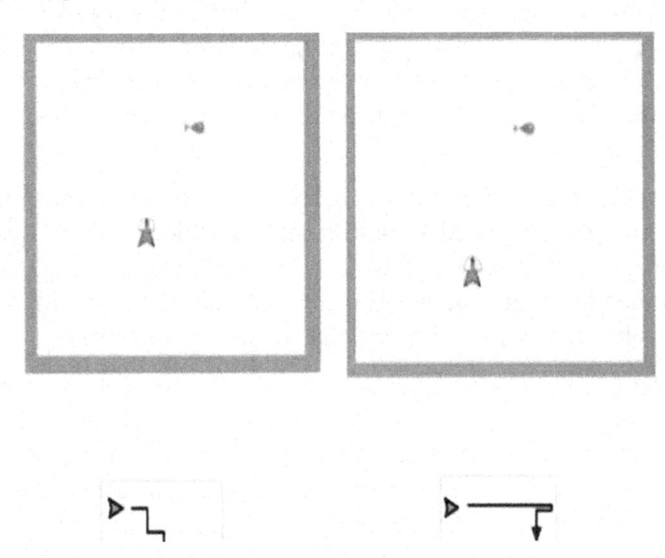

FIGURE 11.23: *Two of the self-programming agents from Olivier Georgeon's MOOC on Developmental AI. The two agents implement the same initial algorithm for trying to catch prey in their simple simulation worlds. Yet, because they go through different individual experiences, they find different strategies to catch a prey. (Check out the video at* http://liris.cnrs.fr/ideal/mooc/*).*

What Would a General Theory of General Intelligence Look Like?

The close relationship between mind and environment is important for a project I've been developing slowly for a long time, as a background pursuit – the creation of a "general theory of general intelligence." Nobody has any such theory now. But I've started to work toward one in a fairly serious way. I've written several papers on the subject, and have used the ideas in these papers to guide and shape my work on OpenCog.

So far, though, my work in this direction has been "semi-rigorous," rather than fully mathematical and scientific. I've been too busy trying to push toward actually DESIGNING and BUILDING general intelligence, to follow through fully on my ideas about formalizing the conceptual framework surrounding the general intelligence. But I'd love to plunge more fully into the theory aspect one day.

The details of my thinking on intelligence theory get pretty technical. But some of the core ideas are easy to understand. First of all, any theory of general intelligence would have to tell you something about suiting a mind to its environment and vice versa. My ideal kind of theory of intelligence would accomplish the following: If you described a class of environments and goals, and specified some resource bounds, then the theory would tell you what kind of mind could operate intelligently within those bounds given limited resources.

So you slot in the world and the goals and the bounds, and the theory would describe the sort of AI systems suited to achieving those goals with those resources in that particular world. If we had such a theory, we could create a human-level, human-like AI just by describing the world, the goals that humans habitually deal with, and the resource restrictions.

To develop a concrete, applicable theory of this sort, we'll need to experiment with some fairly powerful AGIs, seeing what they can do and how their environments, structures and capabilities interrelate. But I think that purely theoretical work can still help us, by giving us useful abstractions that decrease the number of experiments we have to run. One should be able to find elegant mathematical properties of environments and goals that are easily relatable to comparable mathematical properties of minds that are intelligent relative to these environments and goals given limited resources. Pleasantly, I've found reasons to believe that some of these properties are related to the symmetry properties underlying the foundations of physics, and of aspects of mathematics like probability and entropy. Details in a few years– if I'm lucky!

How to Proceed?

So what's the practical upshot? Ultimately, how should we proceed, if we want to build a thinking machine?

A practical, useful general theory of general intelligence would be awesome – but the fact is, we don't currently have one. So we have to move ahead via an ad hoc combination of theory and experimentation, starting with as well-informed and solid an approach as we can muster, then learning, and revising the details as we go along. This is how many things, great and small, get done in the world.

At the beginning of this chapter, I summarized the high-level plan I've been following:

1) Look at everything known about human cognitive science.
2) Think about these known cognitive processes as a system.
3) For each of the important mental processes identified, figure out how to achieve what it does via some reasonably efficient computer algorithm that will work on today's computers.
4) In the system perspective of the mind, look carefully at the interactions between all the mental processes modeled computationally, and be sure that the chosen computational algorithms interact with each other properly.
5) Create a practical software framework that allows all the needed processes to operate simultaneously and interact with each other appropriately.
6) Create an agent powered by this software framework, let it control a body in a world full of rich data to interact with, and let the world be populated by human agents eager to interact with it and teach it.
7) In this environment, carefully interact with the AGI agent to lead it through a series of natural developmental stages, modeled loosely on human developmental psychology.

So far, during the last years and decades, I've worked through Steps 1 to 5, and am in the middle of Step 6 – it's a big one. Of course, some of the previous steps are still ongoing – the software framework I'm working on is under continuous development, and the algorithms in use are continually being tweaked and improved. Regarding bodies and worlds, currently I'm using AI to control virtual characters in video game-type worlds, but I've also worked a great deal recently withrobotics, something I expect to do more of in the next few years.

My next practical steps are working with a software team that's implementing more of the AGI design we've created together, then teaching the AGI system as it controls virtual world agents or robots. Currently, the emphasis is more on the implementing than the teaching. As more of our system is implemented, however, it'll be the other way around. It's an extremely difficult task, but also an excellent adventure.

While there are potentially many workable paths to AGI, I'll spend the most time in these pages digging into the details of the path I understand best and have the most confidence in – the OpenCog system that I'm developing with my colleagues. But first, I want to say something about the overall ecosystem of AGI projects in which OpenCog lives...

Further Reading

Brooks, Rodney (2002). *Flesh and Machines*. Pantheon Books, New York, NY.

Georgeon, Olivier (2014). *Learning by experiencing versus learning by registering*. Constructivist Foundations, 9(2): 211.

Georgeon, Olivier and David Aha (2013). The Radical Interactionism Conceptual Commitment. *Journal of Artificial General Intelligence* 4(2): 31-36.

Johnson, Mark (2010). *Developmental Cognitive Neuroscience*. Wiley-Blackwell.

Tilden, Mark W. and B. Haslacher (1995). *Living machines, Robotics and Autonomous Systems: The Biology and Technology of Intelligent Autonomous Agents.*

Elsevier Publishers. See other related papers at http://www.beam-wiki.org/wiki/.

Tomasello, Michael (2003). *Constructing a Language: A Usage-Based Theory of Language Acquisition.*

Varela, Francisco J.; Thompson, Evan; Rosch, Eleanor (1991). *The embodied mind: Cognitive science and human experience*. Cambridge MA: MIT Press.

12. A Community of Mind-Building Minds

As a painfully shy child and pre-teen, I imagined that as an adult I'd be the kind of scientist who sat in his lab all day doing complex experiments and hardly ever speaking to anyone. I never envisioned I'd become as outgoing as I currently am (though I still rate myself an introvert at heart, albeit an unusually talkative one). I certainly never saw myself spending as much time on organizing people as has been the case in recent years. But life is full of surprises.

As it's turned out, over the last decade, in addition to pushing my own AI research forward (for instance with OpenCog, that I'll tell you a lot about later), I've put a fair bit of effort into the project of building an Artificial General Intelligence research community – into growing "AGI" as a field. This wasn't really something I set out to do, it was more something that happened, a role I somewhat fell into. But it's an aspect of my career I've much enjoyed, because it's brought me into contact with so many extremely interesting, thoughtful researchers from all around the world. It's helped me to fully understand the diversity of thinking in the AI and AGI research world, and to better appreciate the different insights possessed and pursued by all the various researchers approaching AGI from different perspectives. I have learned a great deal of things of practical and theoretical value even from those researchers who have not convinced me that their approaches to AGI are sensible or viable in terms of ultimately leading to AGI.

Community building is something I have done more than I've thought about, if that makes any sense. But in 2013 Luke Muehlhauser from MIRI (Machine Intelligence Research Institute; formerly SIAI) asked me to do an interview on the topic of "AGI as a Field." He was interested in my experiences in helping to build the AGI field, because he wanted to learn more about field-building in general, with a view toward building a field of what MIRI calls "Friendly AI Research" – research about how to make AGIs remain nice to people as they get smarter and smarter. It was somewhat interesting to chat with him on the subject of field-building, because it's not really the kind of thing I generally think or talk about much…

Anyway, here in this chapter I have excerpted the most AGI-relevant parts of my discussion with Luke, omitting some of the parts where I tried to give him advice about how to build a Friendly AI field (I don't think my advice was especially useful anyway). Beyond just being a small slice of academic history, these thoughts may be interesting for the light they shed on the nature of the AGI problem itself. The main challenge I habitually faced in pulling together a group of AGI researchers into a holistic community was the fact that there are so many different ways of thinking about AGI, and no two researchers agree on the optimal perspective or path. To some extent this reflects the egocentric, maverick personalities of individuals who have the guts to do research on an out-of-the-mainstream topic; and to some extent it reflects the nature of AGI itself, which is intrinsically a beast with many heads.

Luke: Ben, you've been heavily involved in the formation and growth of a relatively new academic field — the field of artificial general intelligence (AGI). We'd love to know what you've learned while co-creating the field of AGI research.

Could you start by telling us the brief story of the early days? Of course, AI researchers had been talking about human-level AI since the dawn of the field, and there were occasional conferences and articles and books on the subject, but the field seemed to become more cohesive and active after you and a few others pushed on things under the name "artificial general intelligence."

Ben: I didn't really think about trying to build a community or broad interest in "real AI" until around 2002, because until that point it just seemed hopeless. Around 2002 or so, it started to seem to me — for a variety of hard-to-pin-down reasons — that the world was poised for an attitude shift. So I started thinking a little about how to spread the word about "real AI" and its importance and feasibility more broadly.

Frankly, a main goal was to create an environment in which it would be easier for me to attract a lot of money or volunteer research collaborators for my own real-AI projects. But I was also interested in fostering work on real AI more broadly, beyond just my own approach.

My first initiative in this direction was editing a book of chapters by researchers pursuing ambitious AI projects aimed at general intelligence, human-level intelligence, and so forth. This required some digging around to find enough people to contribute chapters — i.e., people who were both doing relevant research, and willing to contribute chapters to a book with such a focus. It also required me to find a title for the book, which is where the term "AGI" came from. My original working title was "Real AI," but I knew that was too edgy — since after all, narrow AI is also real AI in its own sense. So I emailed a bunch of friends soliciting title suggestions, and Shane Legg proposed "Artificial General Intelligence." I felt that "AGI" lacked a certain pizazz that other terms like "Artificial Life" have, but it was the best suggestion I got so I decided to go for it. Reaction to the term was generally positive. (Later I found that a guy named Mark Gubrud had used the term before, in passing, in an article focused broadly on future technologies. I met Mark Gubrud finally at the AGI-09 conference in DC.)

I didn't really make a big push at community-building until 2005, when I started working with Bruce Klein. Bruce was a hard-core futurist whose main focus in life was human immortality. I met him when he came to visit me in Maryland to film me for a documentary. We talked a bit after that, and I convinced him that one very good way to approach immortality would be to build AGI systems that would solve the biology problems related to life extension. I asked Bruce to help me raise money for AGI R&D. After banging his head on the problem of recruiting funding from investors for a while, he decided it would be useful to first raise the profile of the AGI pursuit in general — and this would create a context in which raising funding for our own AGI R&D would be easier.

So Bruce and I conceived the idea of organizing an AGI conference. We put together the first AGI Workshop in Bethesda in 2006. Bruce did the logistical work; I recruited the researchers from my own social network, which was fairly small at that point. I would not have thought of trying to run conferences and build a community without Bruce's nudging — this was more a Bruce approach than a Ben approach. I note that a few years later, Bruce played the key role in getting Singularity University off the ground. Diamandis and Kurzweil were of course the big names who made it happen, but without Bruce's organizational legwork (as well as that of his wife at the time, Susan Fonseca), over a 6-month period prior to the first SU visioning meeting, SU would not have come together.

The AGI Workshop went well — and that was when I realized fully that there were a lot of AI researchers out there, who were secretly harboring AGI interests and ambitions and even research projects, but were not discussing these openly because of the reputation risk.

From relationships strengthened at the initial AGI Workshop, the AGI conference series was born — the first full-on AGI conference was in 2008 at the University of Memphis, and they've been annual ever since. The conferences have both seeded a large number of collaborations and friendships among AGI researchers who otherwise would have continued operating in an isolated way, and who have had an indirect impact via conferring more legitimacy on the AGI pursuit. They have brought together industry and academic and government researchers interested in AGI, and researchers from many different countries.

Leveraging the increasing legitimacy that the conferences brought, I then did various other community-building things like publishing a co-authored paper on AGI in "AI Magazine", the mainstream periodical of the AI field. The co-authors of the paper included folks from major firms like IBM, and some prestigious "Good Old-Fashioned AI" people. I also have helped Pei Wang edit the Journal of AGI (JAGI). And a couple other AGI-like conferences have also emerged recently, e.g., BICA and Cognitive Systems. I helped get the BICA conferences going originally, though I didn't play a leading role. I think the AGI conferences helped create an environment in which the emergence of these other related small conferences seemed natural and acceptable.

Of course, there is no way to assess how much impact all this community-building work of mine had, because we don't know how the AI field would have developed without my efforts. But according to my best attempt at a rational estimation, it seems my initiatives of this sort have had serious impact.

A few general lessons I would draw from this experience are:

You need to do the right thing at the right time. With AGI we started our "movement" at a time when a lot of researchers wanted to do and talk about AGI, but were ashamed to admit it to their peers. So there was an upsurge of AGI interest "waiting to happen," in a sense.

It's only obvious in hindsight that it was the right time. In real time, moving forward, to start a community one needs to take lots of entrepreneurial risks, and be tolerant of getting called foolish multiple times, including by people you respect. The risks will include various aspects, such as huge amounts of time spent, carefully built reputation risked, and personal money ventured. For instance, even for something like a conference, the deposit for the venue and catering has to come from somewhere... For the first AGI workshop, we wanted to maximize attendance by the right people, so we made it free, which meant that Bruce and I — largely Bruce, as he had more funds at that time — covered the expenses from our quite limited personal funds.

Social networking and community building are a lot more useful expenditures of time than I, as a math/ science/ philosophy geek, intuitively realized. Of course people who are more sociable and not so geeky by nature realize the utility of these pursuits innately. I had to learn via experience, and via Bruce Klein's expert instruction.

Luke: Did the early AGI field have much continuity with the earlier discussions of "human-level AI" (HLAI)? E.g., there were articles by Nilsson, McCarthy, Solomonoff, Laird, and others, though I'm not sure whether there were any conferences or significant edited volumes on the subject.

Ben: It was important that, in trying to move AGI forward as a field and community, we did not found our overall efforts in any of these earlier discussions.

Further, a key aspect of the AGI conferences was their utter neutrality in respect to what approach to take. This differentiates the AGI conferences from BICA or Cognitive Systems, for example. Even though I have my own opinions on what approaches are most likely to succeed, I wanted the conferences to be intellectually free-for-all, equally open to all approaches with a goal of advanced AGI…

However, specific researchers involved with the AGI movement from an early stage were certainly heavily inspired by these older discussions you mention. E.g., Marcus Hutter had a paper in the initial AGI book and has been a major force at the conferences, and has been strongly inspired by Solomonoff's early work. Paul Rosenbloom has been a major presence at the conferences; he comes from a SOAR background and worked with the good old founders of the traditional US AI field… Selmer Bringsjord's logic-based approach to AGI certainly harks back to McCarthy. Etc.

So, to overgeneralize a bit, I would say that these previous discussions tended to bind the AGI problem with some particular approach to AGI, whereas my preference was to more cleanly separate the goal from the approach, and create a community neutral with regard to the approach…

Luke: Can you say more about what kinds of special efforts you put into getting the AGI conference off the ground and growing it? Basically, what advice would you give to someone else who wants to do the same thing with another new technical discipline?

Ben: In the early stages, I made an effort to reach out one-on-one to researchers who I felt would be sympathetic to the AGI theme, and explicitly ask them to submit papers and come to the conference… This included some researchers whom I didn't know personally at that time, but knew only via their work.

More recently, the conference keynote speeches have been useful as a tool for bringing new people into the AGI community. Folks doing relevant work who may not consider themselves AGI researchers per se, and hence wouldn't submit papers to the conference, may still accept invitations to give keynote speeches. In some cases this may get them interested in the AGI field and community in a lasting way.

We've also made efforts not to let AGI get too narrowly sucked into the computer science field — by doing special sessions on neuroscience, robotics, futurology and so forth, and explicitly inviting folks from those fields to the conference, who wouldn't otherwise think to attend.

One other thing we do is to ongoingly maintain our own mailing list of AGI-interested people, built by a variety of methods, including scouring conference websites to find folks who have presented papers related in some way to AGI. And we've established and maintained a relationship with AAAI, which enables us to advertise in their magazine and send postcards to their membership, thus enabling us to get a broader reach.

Anyway this is just basic organizational mechanics, I suppose — not terribly specific to AGI. This kind of stuff is fairly natural for me, due to having watched my mom organize various things for decades (she was a leader in the social work field and retired recently). But I don't think it's anything terribly special — only the content matter (AGI) is special!

If I have put my personal stamp on this community-building process in some way, it's probably been via the especially inclusive way it's been conducted. I've had the attitude that since AGI is an early stage field (though accelerating progress means that fields can potentially advance fairly rapidly from early to advanced stages), we should be open to pretty much any sensible perspective, in a spirit of community-wide brainstorming. Of course each of us must decide which ideas to accept and take seriously for our own work, and each researcher can have more in-depth discussions with those who share more of their own approach — but a big role of a broad community like the one we're fostering with the AGI conferences is to expose people to ideas and perspective different from the ones they'd encounter in their ordinary work lives, yet still with conceptual (and sometimes even practical) relevance...

Co-Organizing Transhumanists

Luke and I didn't get into the topic in our conversation, but it's interesting to contrast the experience I've had working to pull together the AGI community, with the experience I've had in my only other serious venture into people-organization: my work with the transhumanist nonprofit Humanity+.

FIGURE 12.1: *Humanity+ Beijing in 2012, probably the first radical futurist conference in mainland China. In front of me is Amy Li; to my left are Hugo de Garis, Yiqing Liang and Cosmo Harrigan. Things change fast – in January/February 2015, for example, there were three Singularity-focused conferences in Beijing! I only had time to speak at one, the Future Forum, organized by Chinese PR pioneer Cathy Wu, which also featured Ray Kurzweil, Andrew Ng and Hugo DeGaris (along with a number of Chinese speakers).*

Humanity+ is a US charitable nonprofit organization, which began its life in the late 1990s as the World Transhumanist Association, and changed its name in 2009 in an effort to reach a wider audience. The mission is a broad one: roughly speaking, to promote technologies aimed at radically transforming humanity for the better. While the name no longer has "transhumanist" in it, the basic theme of the organization is still transhumanist in the sense of advocating radical transformation of the current state of humanity into something much better. Humanity+ members generally want to make themselves much smarter and longer-lived, much more physically able, and so forth. Some want to become better humans, others want to become something beyond humanity altogether.

The "Transhumanist Declaration" that the organization maintains[1] reads as follows:

"Humanity stands to be profoundly affected by science and technology in the future. We envision the possibility of broadening human potential by overcoming aging, cognitive shortcomings, involuntary suffering, and our confinement to planet Earth.

We believe that humanity's potential is still mostly unrealized. There are possible scenarios that lead to wonderful and exceedingly worthwhile enhanced human conditions.

We recognize that humanity faces serious risks, especially from the misuse of new technologies. There are possible realistic scenarios that lead to the loss of most, or even all, of what we hold valuable. Some of these scenarios are drastic, others are subtle. Although all progress is change, not all change is progress.

Research effort needs to be invested into understanding these prospects. We need to carefully deliberate how best to reduce risks and expedite beneficial applications. We also need forums where people can constructively discuss what should be done, and a social order where responsible decisions can be implemented.

Reduction of existential risks, and development of means for the preservation of life and health, the alleviation of grave suffering, and the improvement of human foresight and wisdom should be pursued as urgent priorities, and heavily funded.

Policy making ought to be guided by responsible and inclusive moral vision, taking seriously both opportunities and risks, respecting autonomy and individual rights, and showing solidarity with and concern for the

---

1 The Transhumanist Declaration was created in 1998 by multiple authors, and was adopted by the transhumanist organization Humanity+ in 2009. The international group of authors responsible for the original form of the Transhumanist Declaration was: Doug Baily, Anders Sandberg, Gustavo Alves, Max More, Holger Wagner, Natasha Vita-More, Eugene Leitl, Bernie Staring, David Pearce, Bill Fantegrossi, den Otter, Ralf Fletcher, Kathryn Aegis, Tom Morrow, Alexander Chislenko, Lee Daniel Crocker, Darren Reynolds, Keith Elis, Thom Quinn, Mikhail Sverdlov, Arjen Kamphuis, Shane Spaulding, and Nick Bostrom. The Declaration has been modified over the years by several authors and organizations.

interests and dignity of all people around the globe. We must also consider our moral responsibilities towards generations that will exist in the future.

We advocate the well-being of all sentience, including humans, non-human animals, and any future artificial intellects, modified life forms, or other intelligences to which technological and scientific advance may give rise.

We favour allowing individuals wide personal choice over how they enable their lives. This includes use of techniques that may be developed to assist memory, concentration, and mental energy; life extension therapies; reproductive choice technologies; cryonics procedures; and many other possible human modification and enhancement technologies."

I got involved in Humanity+ around 2007 when some members suggested I run for a seat on the elected Board of the organization, which is a fair amount of responsibility since the Board basically runs the organization, given that the organization has only infrequently had paid executive staff.

Humanity+/WTA has put on a long series of conferences at various locations around the world, and also runs the "H+ Magazine" website (which began as a paper magazine that sold one glossy color issue in bookstores, back in 2009). I myself organized Humanity+ conferences in Hong Kong (2011) and Beijing (2012), and helped out with several US-based conferences. TransVision 2006, a WTA conference held in Chicago, was one of the first futurist conferences I spoke at (the first may have been the 2005 Immortality Institute conference in Atlanta). Marvin Minsky also spoke there (giving us a chance to argue a bit), as did William Shatner, though I showed up late at the conference and didn't get to meet Shatner, unfortunately.

The most similar earlier organization to Humanity+ was Extropy, an earlier group run by Max More and Natasha Vita-More. Extropy had fairly similar beliefs and aims to Humanity+, but had a bit of tighter focus both conceptually and operationally. A rigorous and energetic transhumanist thinker, Max conceived "extropy" as the opposite of entropy, or as "the extent of a living or organizational system's intelligence, functional order, vitality, energy, life, experience, and capacity and drive for improvement and growth." Several online dictionaries gloss it as "a concept that life will continue to expand throughout the universe as a result of human intelligence and technology." But the organization Extropy closed its doors in 2006, leaving Humanity+ as the only really broad-scope radical futurist organization around (other futurist organizations tending to be more specifically focused, like SIAI/MIRI focusing on AGI ethics, SENS focusing on life extension, Foresight focusing originally on nanotech and recently extending its scope to medicine).

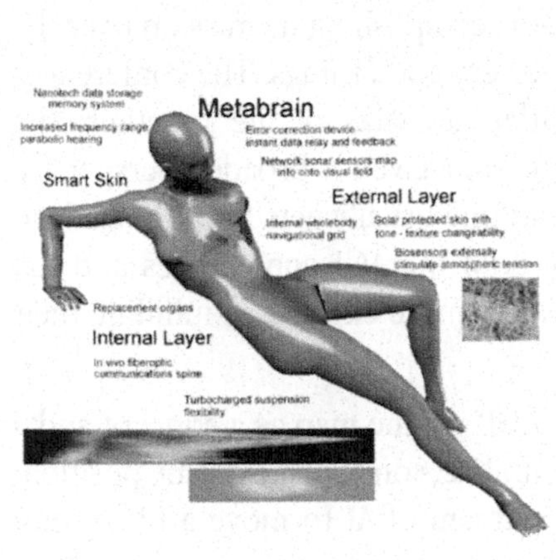

FIGURE 12.2: *Primo Posthuman, an artistic and conceptual vision for an upgraded human being, the creative product of Natasha Vita-More, co-founder of Extropy and current Chair of Humanity+, whom I've gotten to know pretty well through our joint involvement in the latter organization. Along with roboticist David Hanson, Natasha is one of the great names that comes to mind when I think about people pioneering radical futurist visions from a strongly visual-arts oriented perspective (though both David and Natasha have visions going way beyond visual arts, they do have a visual-arts grounding that stands out in the futurist/transhumanist community).*

There is also the World Future Society, which has been around a long time and has a lot of local chapters, but the WFS tends to be less ambitious in its preoccupations – not so focused on Singularitarian or transhumanist ideas, but more focused on nearer-term future predictions and possibilities. WFS has its main office near DC and has done great things in terms of bringing futurist perspectives to the attention of otherwise very present-oriented politicos. But that's quite different from looking toward the possibility of truly dramatic future changes. I've spoken at WFS conferences and they do try to embrace broader future visions as well. But Extropy and Humanity+ have definitely occupied more of a boldly visionary niche.

My own involvement in Humanity+ has been fantastic for me in terms of social networking. I've met so many leading transhumanists through the conferences, through the Board, through my role as occasional writer for and eventually Chief Editor of H+ Magazine. Just as organizing the AGI conferences has brought me in touch with a large percentage of the leading lights of the AGI field, so being on the Board of Humanity+ has brought me in touch with a decent percentage of the leading researchers and thinkers in various aspects of transhumanist technology, science, philosophy, business and so forth. Talking to these diverse, fascinating visionaries individually has given me a whole different kind of insight than I could have gotten from just reading their works.

I would say, though, that in spite of some good efforts on the part of various Humanity+/WTA Board members over the years, Humanity+/WTA has not been tremendously successful at accomplishing its mission over the years. It has done OK – it has done some very good things. The conferences have brought transhumanists from different walks of life together; the magazine has put some ideas out there, and given a broader perspective on advanced science and technology than one usually gets from the gadget-obsessed tech media. But I would say both the AGI conferences and the Extropy organization have been significantly more effective at fulfilling their purposes.

The AGI conference series has put "AGI" on the map as a concept and a research direction. It has staked out a definitive, somewhat extreme position, which has then helped nudge the mainstream of AI to move a bit further in the AGI direction. It has fostered exchange of ideas among researchers coming from different perspectives, who likely never would have talked to each other otherwise, due to mixing in different sub-disciplinary circles and going to different specialized conferences. Of course, AGI as a pursuit would have progressed just fine even without these conferences. But the intellectual impact is clear to see.

Extropy provided a community for transhumanists at a time when there really wasn't any other – it provided a way for scattered folks concerned with radical futurist ideas to find each other and share perspectives. It's hard to remember now how hard it was 20 years ago to locate other people around the world who were interested in ideas like mind uploading, cryonics, AGI, brain-computer interfacing, radical life extension and so forth. Extropy was rather a breakthrough in this regard.

Humanity+, on the other hand, often seems to be riding along on the tide, rather than pushing things forward. This may be because its mission is so broadly-based, and no longer all that niche. The AGI conferences work, so far, because there is a specific group of researchers who benefit practically, in terms of stimulation for their own research thinking, from sharing ideas with each other. Extropy worked because it let a small group of like-minded people come together, when otherwise they would have been disparate. The success of Humanity+ at fulfilling its mission has been less clear so far. This may be because the people Humanity+ brings together, based on their shared enthusiasm for the ideas in the Transhumanist Declaration, don't have all THAT much in common relative to the mainstream of the tech community, simply because the tech community as a whole has become a lot more transhumanist in the last couple decades (albeit generally without using the word "transhumanist"). And these days people with a common interest in transhumanist ideas can find each other easily via the Internet anyway, without need of a formal organization to help them.

So Humanity+ is currently trying to find its path – to find a way to be more dramatically impactful, and help push the world toward a better transhuman future. I can think of a few directions that might work:

- ***Become more political, in the spirit of (and perhaps in cooperation with) the newly formed Transhumanist Party, which is an explicitly political organization.*** If one is aiming to influence current governmental policy, then being very broad-based is an asset, so long as one can do it without utterly diluting one's message.
- ***Focus on VERY broad outreach to segments of the population outside the tech community that have never heard of the Singularity, AGI, radical life extension and so forth.*** I made an effort in this direction by inventing a new holiday called Future Day, to be celebrated on March 1 each year. Imagine children in every school around the world doing Future Day projects; imagine Future Day essay and art contests, and so on. My friend Adam Ford has done a great job of pushing Future Day forward; unfortunately I've been too busy with AGI and narrow AI and haven't had much time to help.
- ***Focus on bringing advanced technology to the developing world – again, parts of the world that have never heard about the Singularity.*** Encourage folks in the developing world to think of new ideas for using advanced tech to assist their own countries and their own lives, and connect them with investors and entrepreneurs who can help them realize their visions.

FIGURE 12.3: *Zoltan Istvan, author of the 2013 novel The Transhumanist Wager, is running for US President on the Transhumanist Party platform. He stands basically no chance of winning and knows it, but is aiming to raise awareness of transhumanist ideas.*

My colleagues on the Humanity+ Board and I have been discussing various ideas in these directions, and other possibilities as well. But of course, tossing around ideas is easy; it's making things work that's hard. And personally, with so much else on my plate, the time I've had or expect to have for helping Humanity+ is pretty limited. But fortunately, there are others involved, with their own excellent capabilities and also a lot more free bandwidth.

What's the moral of the story regarding organizing people in the direction of maverick ideas? I'm not sure, actually. But it's very clear that "the right thing at the right time" is important. And it's also important to have a core group of people who work well together, pushing things forward. Extropy was run by a "power couple," Max and Natasha. The AGI community is not THAT tightly-knit, but it does contain a reasonable network of colleagues from various countries who respect each other and cooperate well together on making conferences happen. Humanity+ has typically been a lot more contentious, which has certainly made it harder for the organization to proactively do the right thing at the right time – so far, but the future is hard to predict.

In the end there's a lot more chance in most of our pursuits than we generally like to admit. The AGI conferences have generally worked well so far, because of happening to bring together a good group of people at an interesting time in the history of the field. If they stopped after AGI-16 or AGI-17, I'd still be very happy with everything I'd learned from operating them, and with their contribution to the field. But I still find myself excited to see what future AGI conferences will bring.

I'm curious to see what will become of Humanity+ as well. Building AGI is my foremost goal, but ultimately AGI is going to be introduced into the human world, and its effect on humanity is of critical importance to me as a human being. If Humanity+ can orient itself to be helpful in understanding these issues and in guiding the world toward a better future.

We are involved in a network of complex processes, and causality is difficult to infer. I'm never sure of the best way to divide my time and attention in terms of organizational work versus technical work versus conceptual/philosophical work. What's clear is that every aspect is important, with its own sort of contribution towards the collective process of future-world-building.

Further Reading

Goertzel, Ben and Cassio Pennachin (2006). *Artificial General Intelligence*. Springer.

Goertzel, Ben and Pei Wang (2007). *Advances in Artificial General Intelligence.* IOS Press.

Franklin, Stan, Pei Wang and Ben Goertzel (2008). *Proceedings of the First Conference on Artificial General Intelligence*. IOS Press.

Hitzler, Pascal, Marcus Hutter and Ben Goertzel (2009). *Proceedings of the Second Conference on Artificial General Intelligence*. Atlantis Press.

Baum, Eric, Emanuel Kitzelmann and Marcus Hutter (2010). *Proceedings of the Third Conference on Artificial General Intelligence*. Atlantis Press.

Schmidhuber, Juergen, Kristinn Thorisson and Moshe Looks (2011). *Proceedings of the Fourth Conference on Artificial General Intelligence*. Springer.

Bach, Joscha, Matthew Ikle' and Ben Goertzel (2012).*Proceedings of the Fifth Conference on Artificial General Intelligence*. Springer.

Kuehnberger, Kai-Uwe, Sebastian Rudolph and Pei Wang (2013). *Proceedings of the Sixth Conference on Artificial General Intelligence*. Springer.

Orseau, Laurent, Javier Snaider and Ben Goertzel (2014). *Proceedings of the Seventh Conference on Artificial General Intelligence*. Springer.

Goertzel, Ben and Luke Muehlhauser (2014). *Ben Goertzel on AGI as a Field*. MIRI Blog, https://intelligence.org/2013/10/18/ben-goertzel/.

Istvan, Zoltan (2013). *The Transhumanist Wager*. Futurity Imagine Media.

Max More (2003). "Principles of Extropy: An evolving framework of values and standards for continuously improving the human condition". Reprinted in: *More, Max and Natasha Vita-More* (2013). The Transhumanist Reader. Wiley-Blackwell.

13. Strategies for Engineering Intelligences

Operating the annual AGI conferences has given me a rather deep and broad vision of the scope of different approaches to AGI. Actually, the diversity is tremendous and hard to grapple with. Sometimes it seems like there are more different approaches to AGI than there are AGI researchers!

In my work on biology, I've found that different researchers can have very different theories – but at least they are all trying to explain basically the same data. Within a pretty good degree of approximation, they are all trying to understand the same functions of the same organisms, or cure or diagnose the same diseases. In AGI, on the other hand, different researchers are often trying to build very different kinds of systems; and they often have very different criteria for success of their work. So they may develop very different languages for describing their work, and there's not always much of a common reality to ground their different languages in.

For instance, what one AGI researcher means by "inference" may be quite different from what another researcher means by "inference." If they were both trying to explain human inference, then one could puzzle out the differences between their usages of the term "inference" by reference to psychology experiments measuring aspects of human inference. But most AGI researchers aren't doing detailed cognitive modeling of human thinking, they're trying to build their own AGI systems loosely related to human thinking, with different specific goals. So getting a handle on the relationship between what, say, a logic-based proto-AGI system and a probabilistic analysis based proto-AGI system respectively mean by "inference" can be a subtle matter. And then you have the question of to what extent a neural net based system is doing "inference" – it may have no internal operations that obviously, explicitly look anything like inferences; yet if it's solving inference-type problems, presumably in some sense it's doing inference. Or so I would say. Not all researchers would agree.

One observation I made from talking to many researchers at the AGI conferences, and elsewhere, is that very few AGI researchers actually understand each others' work at any detailed level. Sure, researchers working within the same fairly specific paradigm will understand each other's work. But if two AGI researchers are doing significantly different kinds of things, the extent to which they'll understand what each other are doing is bound to be pretty low. Furthermore, 8 or 10 page conference papers like those that researchers submit to the AGI conference or other AI conferences, or 10 to 30 minute conference presentations, are nowhere near in-depth enough to really get across the particulars of one researcher's AGI research to another researcher with a different sort of background.

I came to understand two other AGI architectures besides my own pretty intimately – Joscha Bach's MicroPsi and Itamar Arel's DeSTIN – via working with them hands-on. I guided a programmer in implementing a variation on significant parts of MicroPsi within OpenCog; and I guided several programmers in modifying and improving the DeSTIN codebase and integrating it (to a limited extent so far) with OpenCog. This kind of experience builds a much deeper understanding than reading a couple papers and hearing a brief presentation, obviously. (I'll talk a little about each of these architectures later on.)

I came to understand Stan Franklin's LIDA AGI architecture – which I'll discuss a bit in a few pages – moderately well via thinking hard about how to map its various components into corresponding OpenCog components, and into different aspects of human cognition. But this wasn't easy; it took several days of careful reading and thinking about LIDA, and a few long conversations with Stan Franklin and Javier Snaider (another LIDA developer, though now working at Google on non-LIDA AGI stuff).

A few proto-AGI systems have very nice tutorials – SOAR and ACT-R, two of the oldest software systems aimed at AGI-ish goals, fall into this category, as does LIDA. We're working on getting better OpenCog tutorials built. For most AGI systems, even if the code is open source, getting the system to work and do anything requires a lot of labor and insight, and is rarely done except by students or close colleagues of the main researchers associated with the system. Further, one generally finds that what's actually in an AGI system's codebase is quite different from what's in the published papers describing the system – which is perfectly understandable, because an actively developed codebase undergoes constant revision and improvement, whereas research papers are frozen at the moment they're written.

At one point I had the idea of writing a glossary of terms related to AGI and getting the various members of the AGI community to agree on their meanings, and to, going forward, use a standard common terminology. I still somewhat like this idea, but I haven't pushed it very aggressively, because it began to seem like it would require a loooooong process of tedious negotiation among various stubborn people with strongly held beliefs on various particular topics.

Together with deep learning guru Itamar Arel, in 2009 I organized an "AGI Roadmapping Workshop" at the University of Tennessee Knoxville (where Itamar teaches). The aim was to gather a number of researchers interested in embodied, learning-centered AGI, and get them all to agree on a common roadmap for developing their various systems in the direction of human-like AGI. This was an interesting experience, which I wrote about in some detail in a chapter of the book *Ten Years To the Singularity If We Really Really Try*... but honestly, it wasn't really a great success. To make a long story short, each researcher ended up advocating a roadmap – a series of incremental steps from here to AGI – that was most natural for their particular proto-AGI system.

For instance, if a researcher is focusing on getting his proto-AGI system to interpret English sentences, naturally he will prefer a roadmap where the first step involves language processing, and issues of visual perception and body movement come later. If a researcher is focused on getting his proto-AGI system to move a mobile robot around in a variety of environments, naturally he will prefer a roadmap where control of a simple robot comes first and language processing comes later. The diversity of perspectives and the correlation of proposed roadmaps with currently-most-functional system capabilities made perfect sense, but didn't help much in terms of working toward unity in the field.

The paper we ended up publishing based on the workshop was called "Mapping the Landscape of AGI". Rather than laying out a single roadmap to AGI, we visualized human-level AGI as the peak of a mountain range, and viewed the different peoples' approaches to AGI as different pathways up the mountain. There are many paths up any given mountain, and it's hard to tell which path is actually going to be easiest till you actually get to the peak – there may be obstacles at any step along the way. This paper was fairly solid, and we got it into AI Magazine – the magazine that goes to all members of the AAAI, the main international AI organization – at a time when AAAI was not particularly AGI-friendly. But in the end the article was kind of toothless, like the workshop itself. Reconciling the various views of the various researchers required going to a quite high level of abstraction, a level at which it was difficult to say much of anything technically interesting, though we could agree on a bunch of high-level ideas like the value of experiential learning and cognitive development for AGIs, the importance of understanding language in context, the need to take account of dependencies between different sense modalities, and so forth.

So what's the upshot? Is it hopeless to get multiple AGI researchers on the same page? Do we just need to let the diversity flourish until some approach yields sufficiently dramatic and general results that everyone decides to jump on that bandwagon? There's some truth to this, but I don't think the reality is quite that dire. Mutual understanding among researchers coming from different paradigms is gradually being built, and the AGI conferences have been helpful for that. Furthermore, technological advances will gradually make it easier and easier to compare different AGI systems on common tests and in common environments, which will help a lot with mutual comprehension and comparison. I'll say more about this below, but first I want to get more concrete about exactly what the wild diversity of current AGI approaches contains.

The Scope of AGI Approaches

At the AGI-08 conference – the first full-scale AGI conference (following up a smaller, less formal AGI Workshop Bruce Klein and I organized in Bethesda in 2006), which was held at the University of Memphis – Polish researcher Wlodek Duch presented a paper that surveyed the AGI field as a whole. Duch divided existing approaches to AGI into three paradigms, which he called symbolic, emergentist, and hybrid. This isn't the only way to chop up the spectrum of AGI approaches into categories, but it seems to me about as good as any. I'm going to explain to you what each of these means. I'm also going to add one additional category to his list of 3, which is what I call "universal AGI."

FIGURE 13.1: *Polish AGI and cognitive neuroscience pioneer Wlodek Duch (second from left at the table) in a panel discussion at AGI-08 at the University of Memphis, discussing his approach to understanding the scope of approaches to the AGI problem.* (http://farm3.static.flickr.com/2392/2513503294_3cc5a8dcc0.jpg).

Symbolic AGI

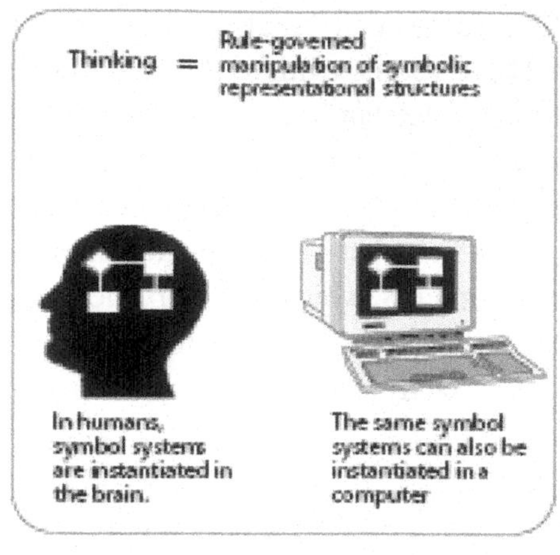

FIGURE 13.2: *The vision of the mind as a system of inter-referring symbols underlies a long history of AI work, going back to near the start of the field. (http://debategraph.org/Details.aspx?nid=218).*

Ever since the very start of the AI field in the 1950s and 60s, there have been two key approaches to conceptualizing and designing intelligent systems. One is to try to emulate the brain, at some level – I'll discuss that approach just below. The other is to try to emulate the mind – the way thinking appears to work from our own introspections into our own minds. The mind-based approach has taken many guises, but in its most classic form it has focused on what's called the "physical symbol system hypothesis," an idea which states that minds exist mainly to manipulate symbols that represent aspects of the world or themselves.

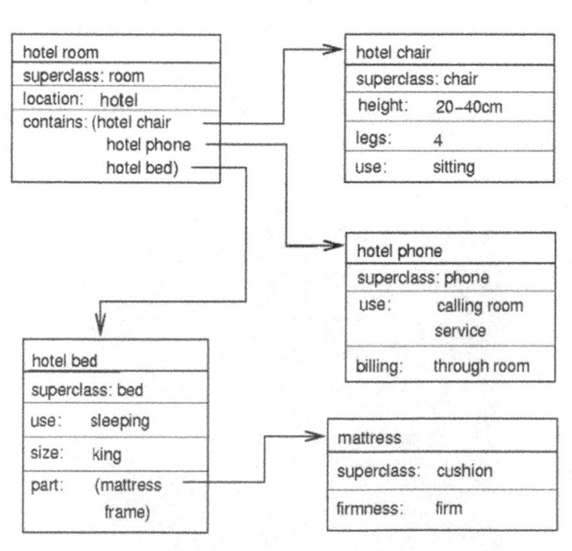

FIGURE 13.3: *Frame-based knowledge representation is an alternative to logic-based approaches, still within the scope of symbolic AI. Some researchers, including AI legend Marvin Minsky, considered frame-based approaches more psychologically natural. (http://www.cs.bham.ac.uk/~mmk/Teaching/AI/figures/hotel.jpg).*

A "physical symbol system" is conceived as a system with the ability to input, output, store and alter symbolic entities, and to execute appropriate actions based on these symbols, in order to reach its goals. This approach places the concept of representation at the center of intelligence. A symbol, after all, is when one entity somehow "stands for" or represents another. In this perspective, a mind takes in sense data and carries out actions, and then internally builds up symbolic representations of these sensations and actions, and symbolic representations of these symbolic representations, etc. Each symbolic representation has relationships with other symbolic representations, and according to these relationships, it places particular focus on specific aspects of what it represents. For example, a pictorial representation of a dog in one's mind places more emphasis on the color of the dog than on its odor. A verbal representation of a dog places less emphasis on the emotional feeling one gets when around the dog, as opposed to other more visceral sorts of representations.

The earliest significant symbol-system based AIs were created by pioneering researchers Alan Newell and Herbert Simon. In 1956, Newell and Simon built a program, Logic Theorist, that discovers logical proofs of simple propositions. This was followed up by the somewhat ambitiously named "General Problem Solver," that attempted to extend Logic Theorist-type capabilities to general-purpose commonsensical problem-solving. The General Problem Solver didn't solve very general problems, but it was highly instructive. It made it apparent that one of the key difficulties facing symbolic AI was how to represent the knowledge needed to solve a problem. In the symbolic approach, before learning or problem solving, an agent must have an appropriate symbolic language or formalism for the learned knowledge. A variety of representations were proposed, including logical formalisms and other variations such as "semantic frames."

Early symbolic AI work led to a number of specialized systems carrying out practical functions. Winograd's SHRDLU system could, using restricted natural language, discuss and carry out tasks in a simulated blocks world. CHAT-80 could answer geographical questions placed to it in natural language. For their time, these programs were quite amazing.

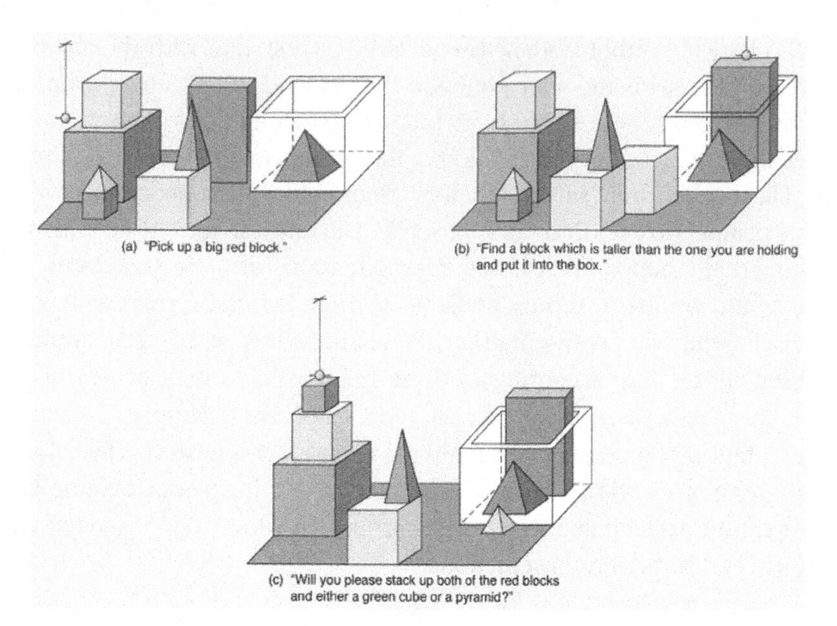

FIGURE 13.4: *Winograd's SHRDLU AI system, that carried out manipulation tasks based on discussion in simplified natural language. The use of a "blocks world" was a brilliant stroke, in terms of providing an environment combining simplicity and cognitive richness for an early-stage proto-AGI system to play in. Blocks-based worlds are still used for proto-AGI experimentation today, for instance virtual worlds based on the game Minecraft. (https://cosidesk.wikispaces.com/file/view/blocksWorld1.gif/276292924/blocksWorld1.gif).*

Buchanan and Feinbaum's system DENDRAL, developed from 1965 to 1983 in the field of organic chemistry, proposed plausible structures for new organic compounds. Buchanan and Shortliffe's system MYCIN, developed from 1972 to 1980, diagnosed infectious diseases of the blood, and prescribed appropriate antimicrobial therapy. Now, these were narrow AIs through and through – these systems utterly lacked the ability to generalize, performing effectively only in the narrow domains for which they were engineered. But they made a valuable point for their time. They performed at least as well at diagnosis as most human doctors, in their specified domains.

```
'(((diagnosis ?Patient bacterial_meningitis) :-
(petechial_rash ?Patient yes)
(low_platelets ?Patient no))
((diagnosis ?Patient bacterial_meningitis) :-
(csfpoly ?Patient ?Cfspoly) (lisp (> ?Cfspoly 79))
(elevated_heart_rate ?Patient))
((elevated_heart_rate ?Patient) :-
(pulse ?Patient ?P)  (lisp (> ?P 105)))
((diagnosis ?Patient viral_meningitis) :-
(csfcellcount ?Patient ?Csfc) (lisp  (> ?Csfc 4))
(csfpoly ?Patient ?Csfp) (lisp  (< ?Csfp 35))
(cns_finding_duration ?Patient ?Cnsfd) (lisp(< ?Cnsfd 5)))
((diagnosis ?Patient viral_meningitis) :-
(high_csfglucose ?Patient)
(low_bloodglucose ?Patient))
((high_csfglucose ?Patient) :-
(csfglucose ?Patient ?G) (lisp  (> ?G 40)))
((low_bloodglucose ?Patient) :-
(bloodgluc ?Patient ?G) (lisp  (< ?G 118)))
)
```

FIGURE 13.5: *A diagnostic rule as used by early medical AI system MYCIN. Today many researchers would think of this kind of system as less of an AI and more of "just a useful program." All it does is take some "expert rules" such as the above, coded in files by human programmers after consultation with domain experts (doctors in this case), and then apply these exact rules to diagnose patients, based on the patient's symptoms as typed into the program. This is very simple conceptually, and there's no learning, self, creativity, agency and so forth involved. The logical reasoning involved is pretty shallow and simple too. On the other hand, the fact that it can do at least as well as most human doctors in this way is fairly instructive. (http://ftp.ics.uci.edu/pub/machine-learning-programs/Introductory-AI/programs/mycin-rules.lisp).*

Modern symbolic AI systems do a lot more than these simple, classic systems, though at their core they tend to have a similar philosophy – somewhere inside such a system, there tends to reside a list of knowledge-rules coded by some expert humans, in some specially-crafted knowledge representation language. There are some efforts in the symbolic AI community to focus on learning rules rather than having humans supply them in files, but this tends to lead away from pure symbolic AI and toward "hybrid AI systems," which I'll discuss separately in a few pages.

There is also the approach of building knowledge rules for a symbolic AI system via writing code that automatically extracts them from natural language text. This can work quite well if one is building a system focused on a specific domain, and has the time and manpower to carefully refine one's natural language information extraction system to deal with text in that domain, and to correct the natural language system's results when they don't make sense. The premier example of this sort of approach is IBM's Watson system. When dealing with a restricted universe like "Jeopardy questions" or "medical knowledge," it's possible for a dedicated team of experts to build up a highly effective knowledge base, via a combination of incorporating existing databases, and applying specialized language processing tools to appropriate bodies of text.

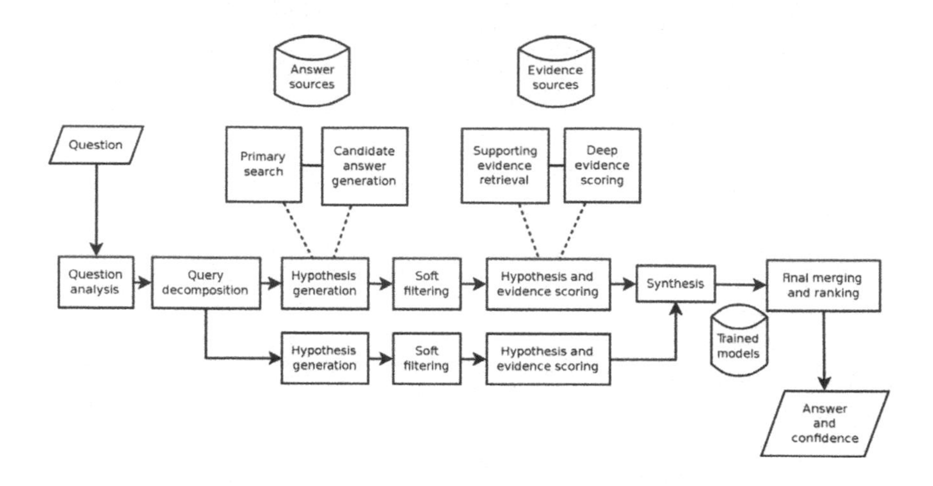

FIGURE 13.6: *The underlying architecture of IBM's Watson system. Watson is perhaps the best symbolic AI system ever built, at least in terms of its functionality, if not the sophistication of its underlying architecture (e.g., SOAR and ACT-R are arguably more conceptually advanced in their inner workings). It answers questions using relatively simple symbolic reasoning, based on a knowledge base that is composed from various sources, including knowledge extracted from large volumes of text via natural language processing software. No agency, no creativity, no spontaneous unsupervised learning from experience, etc. – this is not a system that works like a human mind or explores its environment like an animal. But it's a masterpiece of symbolic AI software engineering (not to mention the underlying hardware virtuosity). (http://en.wikipedia.org/wiki/Watson_%28computer%29#mediaviewer/File:DeepQA.svg).*

We don't have good enough natural language processing technology to extract the meaningful information from general-purpose text in English or other natural languages – this is what researchers call an "AI-hard problem," one that probably can't be solved without creating a full-on human-level AGI. But if one is looking for specific kinds of information in specific kinds of texts, that's a different story, and current technology can do a pretty good job of this, after significant human effort is expended customizing a natural language system for the specifics of interest.

Watson certainly isn't AGI, because it requires copious human attention to apply Watson to a new domain. This isn't a matter of just "teaching" Watson new knowledge, it's a matter of munging around with databases, customizing natural language information extraction tools, and adjusting reasoning algorithms, in ways specifically attuned to each domain. But narrow or not, Watson is certainly cool. Using this kind of supercharged narrow AI system to revolutionize medicine seems a brilliant business move on IBM's part; and if it's rolled out widely in the right way, it could be a brilliant humanitarian move too. I was psyched to see IBM recently roll out three Watson machines in Africa – in South Africa, Nigeria and Kenya. I've talked to the folks working with the Nairobi (Kenya) Watson, and it seems one application they're working toward is diagnostic support for Kenyan physicians. Awesome!

Most of the focus of symbolic AI research field, in recent decades, has been on introducing inreasingly sophisticated "cognitive architectures," in a quest for greater generality of function and more robust learning ability. A cognitive architecture specifies what the different parts of an intelligent system are, what kinds of information they process, and how they interact with each other. For instance, many such cognitive architectures focus on "working memory" that draws on long-term memory as needed, and utilize a centralized control over perception, cognition and action.

Although in principle such architectures could be arbitrarily capable (since symbolic systems have universal representational and computational power, in theory), in practice symbolic architectures tend to be less developed in learning, creativity, procedure learning, and episodic memory. These systems still tend to be best at doing reasoning based on knowledge that is supplied to them explicitly in files. The two most famous symbolic cognitive architectures are Soar and ACT-R.

One of the founders of Soar, Paul Rosenbloom, has recently developed a new architecture called Sigma, which applies many of Soar's architectural ideas using a probabilistic network-based knowledge representation. Probabilistic methods have become very popular in the AI field in recent years, so Sigma can be seen as a bringing-together of old-style symbolic cognitive architecture-based AI, with more recent probabilistic learning-based AI. Soar has recently incorporated ideas from reinforcement learning, a branch of AI that has historically been quite distinct from traditional symbolic AI. One could almost say that the recent batch of symbolic AI systems are verging on becoming "hybrid systems" (which I'll discuss below), rather than pure symbolic systems. But still, I think there's a distinction between systems that hybridize multiple different parts, operating according to different paradigms and according the various parts roughly equal status, and systems that incorporate multiple parts, all in the service of supporting a core symbolic reasoning functionality.

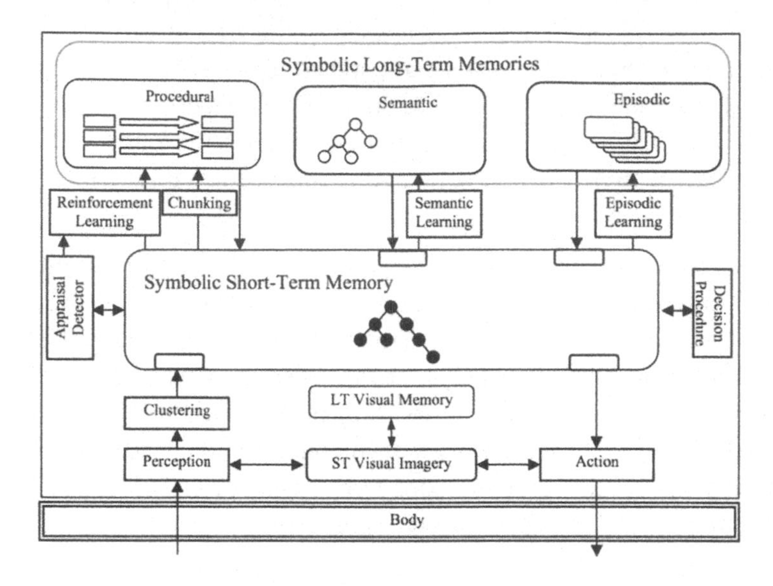

FIGURE 13.7: *High-level architecture of the current version of Soar, which is probably the leading symbolic-AI cognitive architecture, under development since the 1980s and still very much an active, dynamic project. Modules like visual imagery and reinforcement learning are recent additions to the architecture, moving it further toward a full-on AGI architecture. (http://ai.eecs.umich.edu/people/laird/papers/Laird-GAIC.pdf).*

FIGURE 13.8: *Paul Rosenbloom giving a tutorial on his Sigma cognitive architecture at AGI-14. Back in the 1980s, Paul was one of the creators of Soar, a "Good Old Fashioned AI" system descending directly from early work on rule-based AI from the 1960s and 70s. His current Sigma architecture combines classic Soar-style symbolic AI with more modern probabilistic AI concepts and algorithms.*

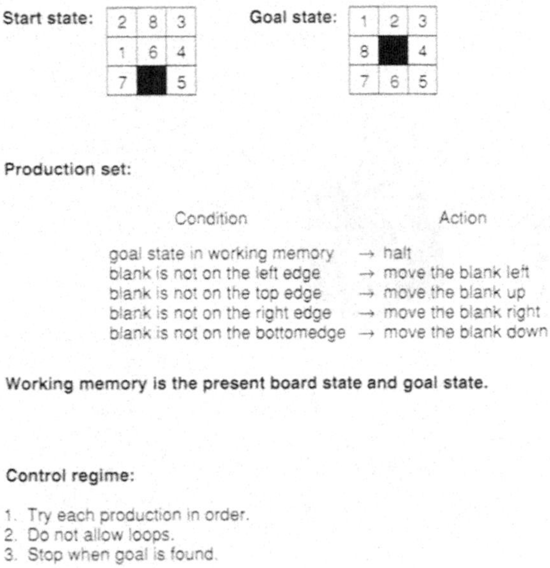

Start state:

2	8	3
1	6	4
7		5

Goal state:

1	2	3
8		4
7	6	5

Production set:

Condition		Action
goal state in working memory	→	halt
blank is not on the left edge	→	move the blank left
blank is not on the top edge	→	move the blank up
blank is not on the right edge	→	move the blank right
blank is not on the bottomedge	→	move the blank down

Working memory is the present board state and goal state.

Control regime:

1. Try each production in order.
2. Do not allow loops.
3. Stop when goal is found.

FIGURE 13.9: *A toy example of a classic "rule-based production system," the ultimate Good Old Fashioned AI system. The example shows a set of "expert rules" for solving a simple puzzle, the 8-puzzle. At some points in the history of AI, some researchers thought human-level intelligence could be emulated by coding a large set of IF-THEN rules like this, handling the variety of situations a human being has to deal with. Most AI work today is focused on systems that learn their own rules for doing things based on analyzing data or adapting to experience, rather than relying on human-coded rules. But there are still some systems out there focused on collections of hand-coded rules, the leading example being the Cyc system developed at Cycorp and funded by the US military. (http://www.cis.temple.edu/~pwang/3203-AI/Lecture/KBS-1.jpg).*

Emergentist AGI

Another species of AGI design expects abstract symbolic processing, along with every other aspect of intelligence, to emerge from lower-level "subsymbolic" dynamics – from complex self-organization of large networks of simple elements. These networks are sometimes (but not always) designed to simulate neural networks or other aspects of human brain function. Researchers of this persuasion – which Wlodek Duch labeled "emergentist" in his AGI-08 review – generally consider the reliance on human-coded rule-bases, such as one sees in the symbolic AI community, as utterly wrongheaded and indicative of an approach that is focusing on the wrong problems. (I somewhat agree with this view, but my perspective has been tempered considerably with age and experience, as I'll elaborate a bit a little later.)

FIGURE 13.10: *Hugo DeGaris and I debating the future of AGI at the Humanity+ Hong Kong futurist conference, which I organized in 2011 – the first transhumanism-oriented conference in Hong Kong ever. Hugo likes to speculate about the future of AGI and humanity, but he is also an accomplished researcher, and did some groundbreaking work in the 1990s making neural net learning work efficiently on specialized hardware called FPGAs (chips that can re-route their wiring dynamically based on software). He tends to think that one can make an AGI using a fairly simple brain-like neural net architecture, via evolving numerous specialized neural modules carrying out particular functions, then connecting them together and watching high-level intelligence emerge as the modules adapt to work together in the context of making an embodied system achieve its goals. From 2008-2010 he was working toward these goals at Xiamen University in China, in the context of trying to get his evolved neural networks to control a Nao humanoid robot. I collaborated with him on that work from a distance, and via frequent visits to Xiamen, aiming to integrate OpenCog with his neural net systems (thus transitioning his design from an emergentist neural net approach to a hybrid approach). Some of this work is depicted in Raj Dye's documentary film Singularity or Bust! Hugo left Xiamen University in 2010 and shifted his research focus to theoretical physics and femtotechnology. We remain good friends. (http://www.exponentialtimes.net/sites/default/files/emvideo-youtube-tFvl4Nf3iTw_0.jpg).*

Under the emergentist umbrella we have such systems as:

Neural networks, which roughly (sometimes very roughly) emulate the brain's organization of knowledge and learning. I'll give you my views on the brain and AGI in more depth later on.

- The recently very fashionable ***"deep learning" networks***, which I'll give their own chapter toward the end of the book.
- ***Probabilistic reasoning networks***, such as scalable Bayes nets, or related formalisms like Markov Logic Networks.
- ***Evolutionary algorithms*** (e.g., genetic algorithms), that seek to learn via emulating the process of evolution by natural selection and "digitally evolving" solutions to problems.
- ***Ant systems***, that try to learn via emulating the collective behavior of ants as they lay down pheromones for each other while learning to find their ways to valuable food sources.
- ***Reinforcement learning systems***, that adapt their approach to achieving a goal in an environment on the basis of receiving reward or punishment from the environment (these systems often rely on neural nets internally, but this isn't always the case).

There are also plenty of others.

Looking backward, the broad concepts of emergentist AI can be traced back even before the foundation of the AI field as such – to Norbert Wiener's landmark 1948 book *Cybernetics*, and more directly to the work of McCulloch and Pitts in the early 1940s, which showed how networks of simple "formal neurons" (simple mathematical functions very roughly emulating the operation of brain cells) could be the basis for a "universal computer," a computer that (like every PC or smartphone these days) is powerful enough that if you give it the right program and enough memory and time, it can do anything. In 1949, Donald Hebb wrote *The Organization of Behavior*, pointing out the fact that neural pathways are strengthened each time they are used, a concept now called "Hebbian learning," conceptually related to long-term potentiation in the brain, and to a host of more sophisticated reinforcement learning techniques.

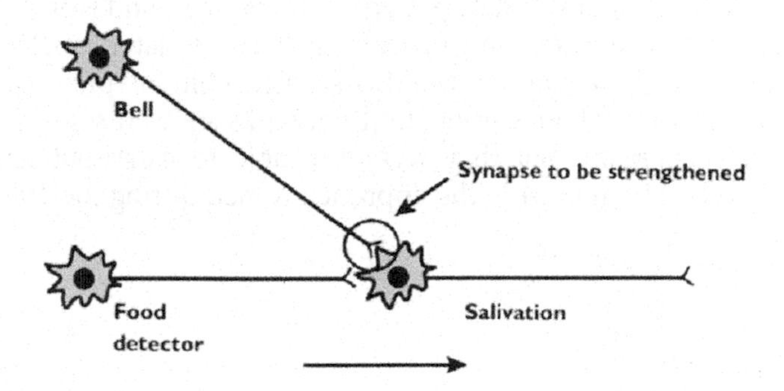

FIGURE 13.11: *This figure shows the basic idea of Hebbian learning in neural networks. The context is the classic psychology experiment where an animal is trained to salivate when it hears a bell ring, via the ringing of a bell every time food is served. This is hypothesized to occur in the brain via strengthening of the connection between a collection of neurons representing salivation and a collection of neurons representing the bell ringing. Of course, this kind of model is a massive oversimplification of the complex stuff happening in the brain, as I've emphasized in these pages many times. But still, it's a reasonable conceptual model of certain aspects of brain dynamics, as well as a conceptual paradigm case of subsymbolic learning. It's enthralling to think that this kind of localized learning that occurs by modifying particular connections in a network for simple elements, when applied across a large network of elements, could lead to complex learning behaviors. (http://commonsenseatheism.com/wp-content/uploads/2011/08/Hebbian-circuits.png).*

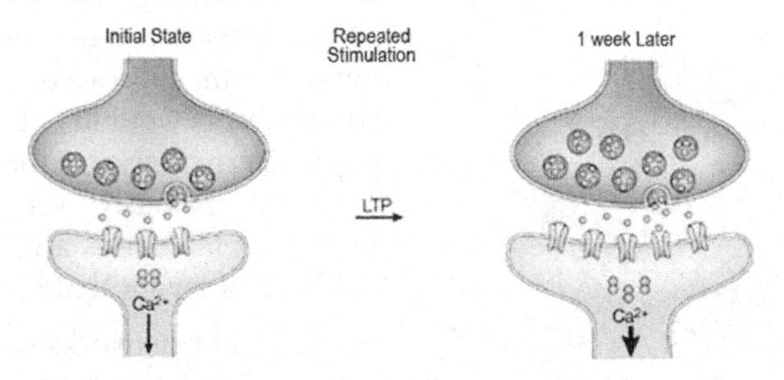

FIGURE 13.12: *An illustration of the physiochemical dynamics underlying Hebbian learning, or as it is referred to in biology, "long-term potentiation." A key point here is*

that the brain does not actually consist of little nodes connected by lines or wires. Rather, two neurons may come close to each other, but there's a gap between them, and when one neuron sends a signal to another, this is done by chemicals basically carrying electrical charge across the gap. This process can happen with more or less efficiency, and the chemical properties of the gap or "synapse" can change with time and experience, which is how Hebbian learning actually works in the brain. (http://mikeclaffey.com/psyc170/notes/images/memory-LTP.jpg).

In the 1950s, practical learning algorithms for formal neural networks were articulated by Marvin Minsky and others. Rosenblatt designed "Perceptron" neural networks, which arranged neurons in multiple layers and were able to learn simple data patterns; and Widrow and Hoff presented a systematic neural net learning procedure that was later labeled "back-propagation," which remains a workhorse algorithm of practical neural network applications. These early neural networks showed some capability to learn and generalize, but they were not able to carry out practically impressive tasks, and interest in the approach waned during the 1970s.

FIGURE 13.13: *Architecture of the classic "Perceptron" neural net design. The "formal neuron" nodes in this sort of network are very, very crude approximations of biological neuron cells. But the networks do have interesting learning capabilities, though they are now obsoleted by their descendants, modern deep learning networks like Convolutional Neural Nets, which I'll discuss a bit later on. (http://www.statistics4u.info/fundstat_eng/img/hl_multil_perceptron.png).*

In 1982, broad interest in neural net-based AI began to resume, triggered partly by a paper by the Caltech physicist John Hopfield, explaining how certain sorts of neural nets could be used to store associative memories. Hopfield's networks emulated the "holographic" property possessed by certain parts of the brain – they could remember many things, and each thing they remembered was stored in a "distributed" way across the whole neural network. So if you removed some percentage of the neurons in the network, you'd make all the network's memories a bit fuzzier, but you probably wouldn't eliminate any particular memories in their entirety. This is obviously very different from how memories are stored in a database or a typical AI system.

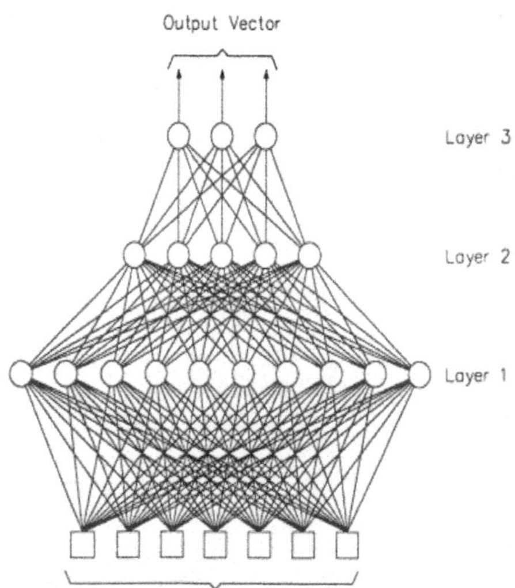

Hopfield networks can get very complex, and can carry out various sorts of temporal pattern recognition and adaptive learning, as well as just storing memories. They can also fill up with memories, eventually, and then their ability to remember degrades. My friend George Christos, when I lived in Perth, had a theory that (very loosely speaking) "dreaming is forgetting," and tested this in Hopfield networks by equipping them with the capability to periodically "dream." During the dreaming phase, they would UN-learn patterns that occurred within them, carrying out the opposite of Hebbian learning. His thinking was inspired by some ideas developed by Francis Crick, the co-discoverer of DNA, who turned to neuroscience later in his life. Biological dreaming turns out to be more complex than this, but the dynamics George was experimenting with may play a role.

Currently, neural networks are an extremely popular machine learning technique with a host of practical applications. Multilayer networks of formal neurons or other conceptually similar processing units have become known by the term "deep learning," and have proved highly successful in computer vision and other areas, as I'll discuss later. Some researchers think that by gradually increasing the adaptive capability and architectural complexity of such networks, they will be able to incrementally approach human-level AGI.

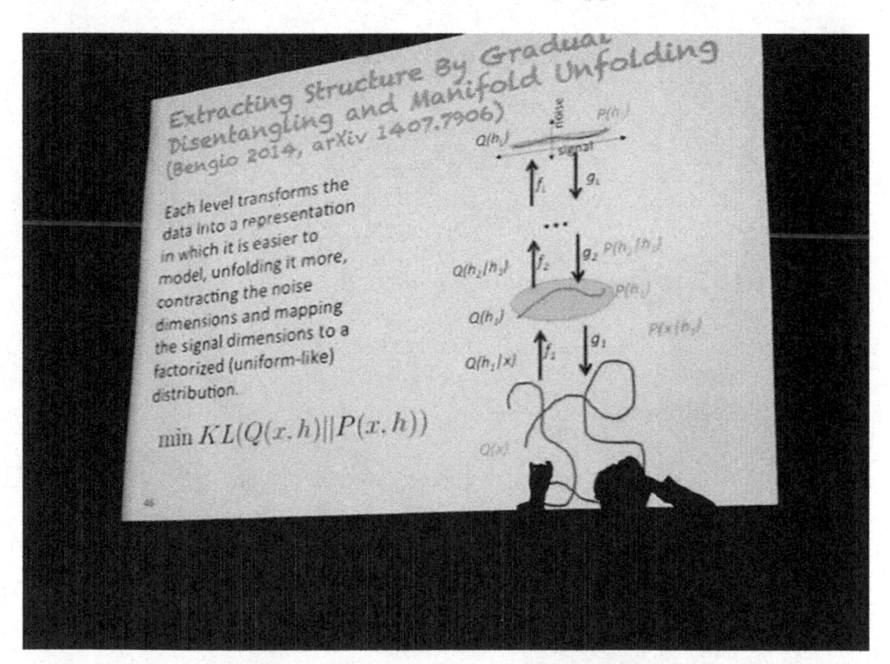

FIGURE 13.14: *Deep Learning pioneer Yoshua Bengio – as of 2014, he is one of the few icons of deep learning research to remain an academic and freely publish his work and code, rather than joining a large tech company. Here he is giving a talk at AGI-14 in Quebec City, presenting some of his recent, speculative ideas on how to make deep learning networks work more like the human brain/mind.*

Computational neuroscience is also a flourishing field, utilizing detailed computational models of biological neurons to study large-scale self-organizing behavior in neural tissues. Henry Markram has become especially famous in this area due to his Blue Brain model, which used an IBM Blue Gene supercomputer to make a simulated cortical column (recall that neurons in the cortex, the primary "thinking" part of the brain, tend to be organized into "columns" that are arranged orthogonally to the 6 layers of neurons in the cortex). Overall, Markram's team's simulated cortical column displayed the same statistical behavior patterns as a real cortical column.

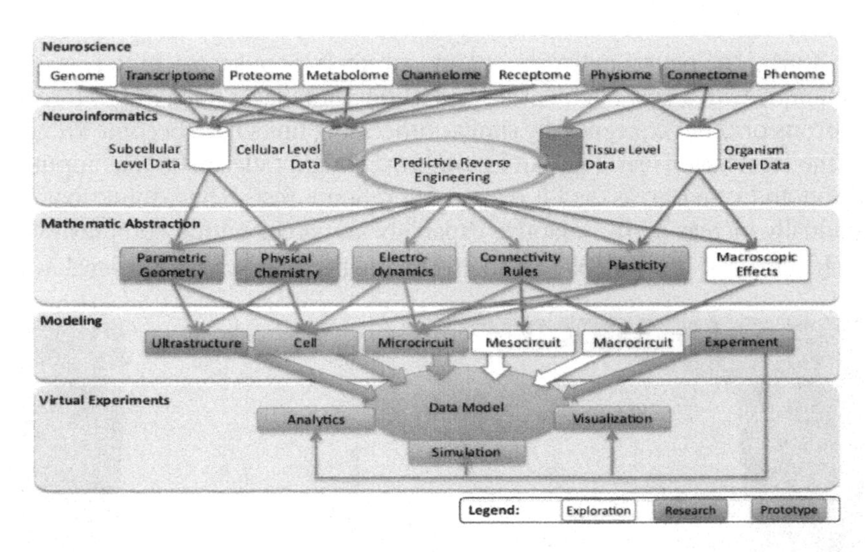

FIGURE 13.15: *Architecture of the IBMBlue Brain system, showing the collection of workflows that the system sought to incorporate. The cortical column simulation, which was Blue Brain's biggest claim to fame, was only part of the story. (http://bluebrain.epfl.ch/files/content/sites/bluebrain/files/images/Images/Science%20Report%20images/fig_page16%20work%20flow.jpg).*

FIGURE 13.16: *Part of the Blue Gene hardware on which Blue Brain was run, extremely powerful for its time. Some Blue Gene specifics. (http://www.artificialbrains.com/images/blue-brain-project/blue-gene-cabinet-open.jpg, http://en.wikipedia.org/wiki/Blue_Gene#mediaviewer/File:LLNL_BGL_Diagram.png).*

Recently, Markram orchestrated the 1.2 billon euro Human Brain Project in the EU. This is a brain science project, not an

AI project, but the distinction is not always clear, given that Markram and some other computational neuroscientists believe that accurately simulating a human brain is going to be the fastest path to human-level AI. The goal of the project is to make a fairly complete computer simulation of a whole human brain before 2023.

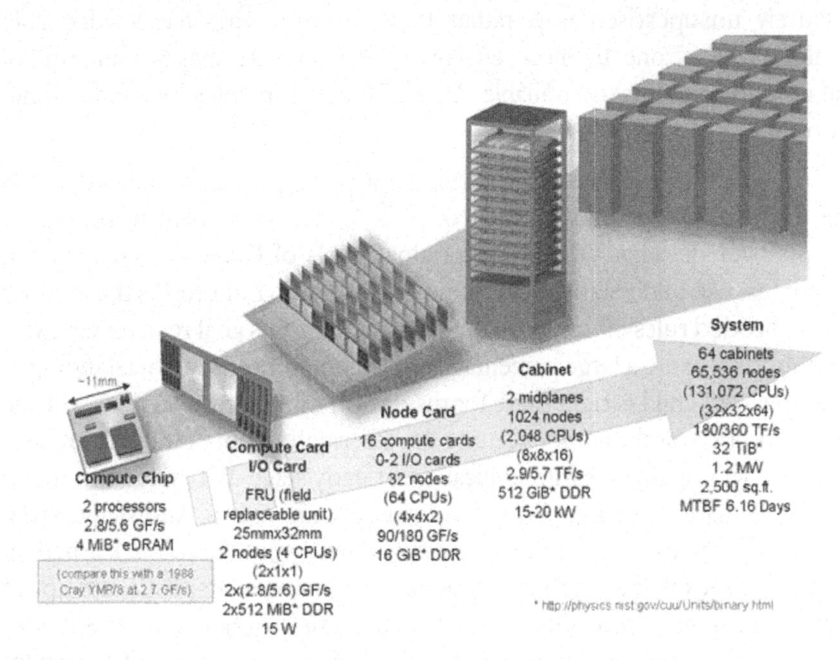

FIGURE 13.17: *Visualization of a portion of the brain produced using Blue Brain. (http://image.slidesharecdn.com/newpresentation-121112230616-phpapp02/95/blue-brain-16-638.jpg?cb=1352783254).*

You have to admire Markram's guts. Even without a breakthrough in brain imaging, just systematizing all the data already collected about different parts and aspects of the brain and getting it all emulated inside a single giant simulation model would be a fascinating and important achievement. But still, I think that to get a simulation that would really simulate the actual thinking, feeling, acting and perceiving dynamics of the human brain – rather than just the basic connectivity structure – will require some kind of qualitatively much better brain imaging than what we can collect today. If such a breakthrough in imaging occurs soon enough, then the Human Brain Project may well fully meet its goals. But even if it doesn't, it's sure to produce interesting partial results.

In all, the "emergentist" approach to AI is fairly ascendant these days, partly because it's tied to neuroscience, which has become very popular and well-funded, and partly because it has proven very effective at dealing with the large amounts of data that have become widely available by virtue of the Internet and the widespread use of computers and other electronic communication technologies. Making an AI system that's focused on recognizing patterns in data and extracting knowledge from data in a relatively unsupervised way, rather than on leveraging hand-coded rule content, allows one to more effectively leverage the massive amount of data the Internet makes available. After all, typing in rules by hand is quite laborious.

The most impressive effort at rule engineering for a symbolic AI so far, the Cyc system, contains millions of rules coded by a team of dozens of people over decades. One of the original goals of Cyc was to display the commonsense understanding of a ten-year-old child. But even as the number of hand-coded rules mounts into the seven figures, this goal remains far away. On the other hand, a large percentage of the commonsense understanding of a ten-year-old child is right there for display on the Internet, in texts, videos, chat transcripts and whatnot. Getting the information out of these resources, which were not created with edification of early-stage AGI systems in mind, can be a major challenge. Fully extracting commonsense knowledge from online information is surely an AI-hard problem. But a lesson learned in recent decades is that, for many practical purposes, a simplistic attempt to mine information from the huge mass that exists online can work better than a sophisticated attempt to precisely encode a select list of important information items by hand. In many cases, large scale and mess work better than small scale and precision. The rise of emergentist AI has come hand in hand with the Big Data revolution. Big Data has provided a rich set of patterns to spur the formation of related patterns inside emergentist AI systems. Of course, faster processors and distributed computing have also been critical – with the computing power of the 1960s and 70s it simply wasn't possible to simulate large neural networks, evolutionary algorithms or other emergentist systems the way it is now.

Today's emergentist architectures are sometimes very strong at recognizing patterns in complex data, and at learning simple behaviors based on experience. But they're not yet good at everything. No one has yet shown how to achieve high-level functions such as abstract reasoning or complex language processing using a purely subsymbolic, emergentist approach. Symbolic AI systems remain generally better at these things, which is not surprising, as they explicitly have to do with manipulating symbols.

Recent breakthroughs have begun to seriously push against this limitation, though. For instance, Stanford grad student Richard Socher created a deep learning system with a "vector semantics" architecture that is capable of quite sophisticated natural language processing (not subtle semantic processing, but rich syntax parsing, anyway). Socher has recently received substantial funding for an AI startup, MetaMind.

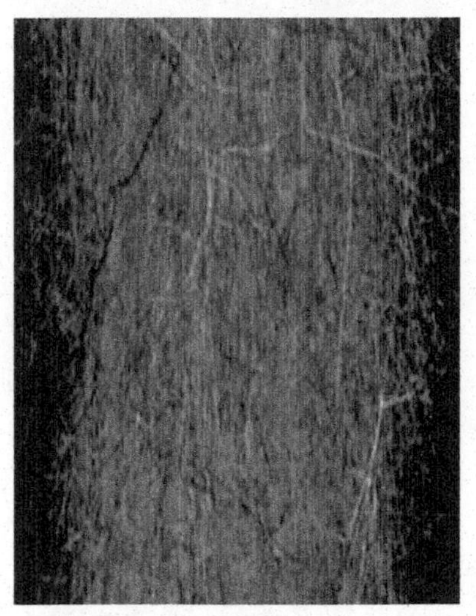

FIGURE 13.18: *iCub exploring its environment. Its physical capabilities are much less than those of a young human child, but much more than they were a few years ago. We don't yet have workable home service robots, but the exponential advance of humanoid robotics technology is dramatically apparent. (http://www.razorrobotics.com/robotcub/).*

An important subset of emergentist cognitive architectures, still at an early stage of advancement, is developmental robotics, which is focused on controlling robots without significant "hard-wiring" of knowledge or capabilities, allowing robots to learn (and learn how to learn, etc.) via their engagement with the world. A significant focus is often placed here on "intrinsic motivation," wherein the robot explores the world with the guidance of internal goals like novelty or curiosity, forming a model of the world as it goes along, based on the modeling requirements implied by its goals.

FIGURE 13.19: *Angelo Cangelosi gave a keynote at the AGI-12 conference at Oxford, on his work using the iCub robot to simulate young child cognitive development. His team's work uses a physical iCub robot, as well as a simulation of the robot; (a) is a real iCub robot, and (b) to (d) are a simulated iCub. The simulator has much greater facility to move around and manipulate objects, but running the same code on the real and simulated iCub allows researchers to keep the simulated work relatively "honest" by reference to the physical world. (http://cogprints.org/6238/1/Tikhanoff-et-al-final-ACM.pdf,).*

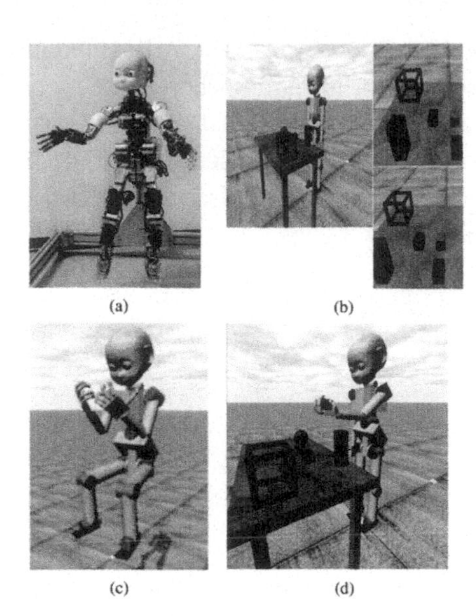

FIGURE 13.20: *The "developmental AI" approach seeks to emulate the development of a human baby or young child, via creating emergentist networks (often neural networks) that control an embodied agent interacting with the world, and making sure these networks have the capability to fundamentally reorganize themselves based on their experiences as they grow.*

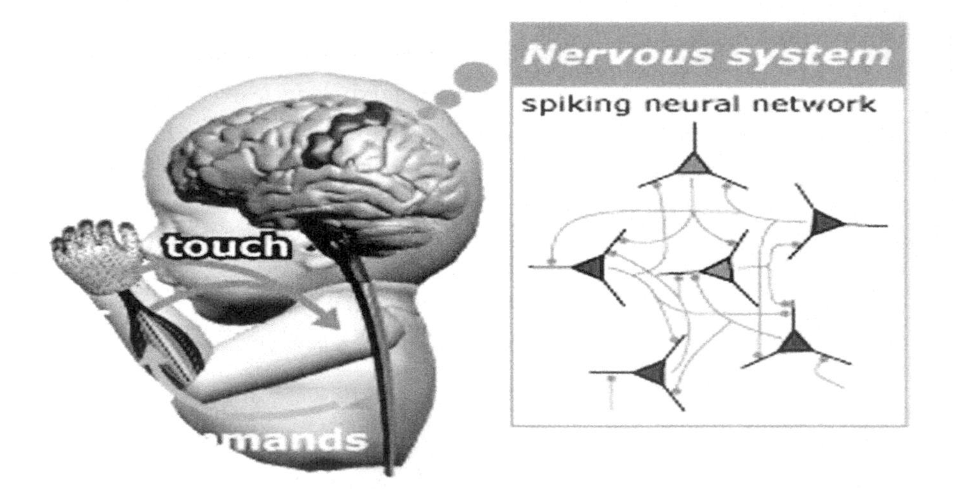

FIGURE 13.21: *The figure is drawn from a paper by Japanese researchers Yasunori Yamada, Keiko Fujii & Yasuo Kuniyoshi, Impacts **of environment, nervous system, and movements of preterms on body map development: Fetus simulation with spiking neural network**,* **presented at** ***the 3rd joint IEEE International Conference on Development and Learning.*** *(http://www.isi.imi.i.u-tokyo.ac.jp/~y-yamada/images/fetus_with_S1_spiking_neural_net.png).*

Hybrid AGI

There was a period in the history of AI when the relationship between symbolic (logic or rule based) AI and subsymbolic (emergentist, neural network based) AI was downright adversarial. Many AI researchers who lived through the field in the 1970s and 80s still think of things this way. A watershed event, both symbolizing this adversarial relationship and having a major effect in cementing it, was the book Perceptrons, written by AI guru Marvin Minsky and his colleague Seymour Papert. Published in 1969, the book demonstrated some severe limitations on the part of perceptrons, a kind of neural network that was then being advocated by researcher Frank Rosenblatt and others. Minsky and Papert's mathematical demonstrations were correct, but they only applied to a very simple form of perceptron – what's called a "feedforward" network (information flows from inputs to outputs and not back again), without more than one "hidden layer" (a hidden layer consists of neurons that are neither input neurons nor output neurons, but lie in between).

Even at that time, it was theoretically known that a general perceptron with multiple hidden layers and feedback as well as feedforward connections could in principle learn anything ("in principle" meaning the learning might take a long time, and the network might need to be big with a lot of neurons in it). What was NOT known at that time was a good algorithm to enable a neural network with a lot of hidden layers and feedback as well as feedforward connections to learn effectively. Minsky and Papert seemed to think this was a hopeless research direction. It wasn't a trivially easy research direction, but today, large multilayer neural network-type systems with feedback connections are widely used in research and applications, and we know many effective algorithms for causing them to learn. The research direction that Minsky and Papert tried to shut down is precisely the "deep learning" approach to AI that we see all over the tech media today, and into which tech giants are investing billions, based on perceived promise but also on demonstrated results. Whereas the symbolic, rule, or frame-based approach that Minsky advocated (at the time he wrote *Perceptrons*, and later) has not yet had any dramatic practical successes, and is decreasingly popular among researchers.

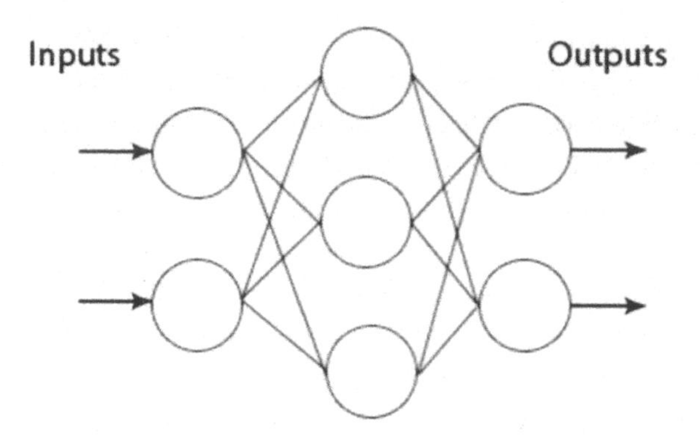

FIGURE 13.22: *Minsky and Papert's influential 1969 book Perceptrons demonstrated certain limitations on the part of very simple neural networks – three-layer perceptrons like shown above. (Their theorems applied to perceptrons with the three-layer architecture shown above, but an arbitrary number of neurons on each layer.) Since scalable learning algorithms for more complex neural networks were not known at that time, this was widely taken as a sort of refutation of the whole neural network-oriented, emergentist, subsymbolic approach to AI. (http://neuroph.sourceforge.net/tutorials/images/MLP.jpg).*

Minsky was a very smart guy; many people reported him as the smartest AI researcher they'd ever come into contact with. I met him face to face only a couple times, when he was already at a fairly advanced age, and we never dug too deep into technical topics, but he certainly knew his stuff. He did strike me, though, as someone who was much better at talking than listening. He had worked on neural nets himself back in the 1950s (in 1951 he developed SNARC, the first-ever randomly connected neural network learning machine), so he definitely understood their workings – but from what I saw of his personality, based on our handful of conversations, I got the feeling that once he felt he understood something thoroughly, he wasn't particularly open to modifying his understanding based on input from others. Of course a certain tenacity and persistence is very valuable in getting great things done, especially revolutionary things that go against prevailing perspectives. But there's also something to be said for opening one's mind to other views.

When my dot-com-era company Webmind Inc. was first marketing itself, a journalist asked Minsky to comment on our work, and he took a look at our website and told the journalist it was "AI word salad." "Word salad" is a psychiatric term referring to the senseless babble produced by certain extremely messed-up schizophrenic people. I wasn't terribly insulted by his comment – I hadn't written that website copy anyway, it had been written by one of our businesspeople, and didn't have much depth to it; though unlike his evaluation, it did make perfect logical sense. I contacted Minsky after that and we discussed AI a bit, and I found we did have some things in common, mostly notably a desire to ground AI designs in cognitive science. But he was quite adamant, in our conversation, that self-organization, complexity and nonlinear dynamics have NOTHING to do with intelligence and nothing to contribute to AI.

Minsky really seemed, at that time, to see the mind as a collection of disparate little modules, each carrying out its own special functions, and coupled fairly loosely together, like people in a society. Furthermore, he tended to give short shrift to the importance of learning within these modules – as if he thought evolution had done most of the work, by creating the assemblage of modules existent within the mind, and that learning on the individual level was more of a shallow matter of adapting the way the modules connect together, adjusting the parameters of the modules, etc.

I encountered a somewhat similar perspective later, in the work of Eric Baum, the author of the excellent AI overview book *What Is Thought?*; but Eric had a much more sophisticated theory regarding how the mind learned to connect together the various modules it was supplied with, based on his approach to reinforcement learning and simulated economics. Baum was trying to layer a complex dynamical system with its own self-organizing richness on top of a collection of specialized modules. Minsky seemed to want to minimize this overlayer.

In any case, Minsky and Papert's *Perceptrons* book played a significant role in turning researchers and US government research funding sources away from neural networks and subsymbolic, emergentist AI generally, and helped nudge the direction of the field toward rule and frame-based symbolic AI systems. It helped create a tone of conflict between the symbolic and emergentist approaches to AI, which pervaded the AI field until quite recently, and still holds significant power, though more in academia than in industry.

Nowadays the boundaries between the different AI paradigms are not especially clear – either to younger researchers, or even to the students of the researchers of Minsky's generation. Soar, the symbolic architecture I mentioned above, was created by Alan Newell, a legendary rule-based AI researcher much in Minsky's spirit, together with his PhD students John

Laird and Paul Rosenbloom. But at the AGI-10 conference in Lugano, Soar co-creator John Laird had some long and interesting conversations with Richard Sutton, a leading researcher in reinforcement learning (closely related to neural networks, and very much in the emergentist spirit – reinforcement learning is about networks that adapt themselves based on experience, rather than having content programmed in), and ended up integrating some reinforcement learning-based ideas into Soar version 9. Meanwhile, as I mentioned above, the other Soar co-founder Paul Rosenbloom has incorporated ideas from the probabilistic-networks literature along with Soar-like ideas into his new cognitive architecture, Sigma. The probabilistic learning methods Paul is using themselves rest somewhere on the border between symbolic and subsymbolic – they can be used to estimate probabilities about symbolic logical variables, or about low-level perceptual data. One of Paul's big goals with Sigma is precisely to use the same cognitive mechanisms to handle every aspect of human intelligence, including basic perception and action (typically the strength of emergentist systems) and abstract logical reasoning (typically the strength of symbolic systems).

And every year now there is a Neural-Symbolic Workshop, convening together researchers building systems that combine neural networks with symbolic AI systems in various ways. Sometimes this involves using a symbolic AI system to understand what's inside a neural network; sometimes it involves systems that combine symbolic and neural elements more deeply in a single system. In any case, it's pretty clear that the rigid dichotomy between symbolic and emergentist systems is a relic by this point, though a relic that still has some cultural "oomph" in the AI field, especially among older researchers.

In 2016 the Neural-Symbolic Workshop, along with the Biologically-Inspired Cognitive Architectures conference, were held together with the AGI conference, forming the first ever Human Level AI Multi-conference, held in New York City, at the New School, co-organized by myself and my old college friend Ed Keller. Ed is now a professor and administrator at the New School, and is deeply into the intersection between architecture, design, philosophy, AI, the Singularity and so forth – i.e. the same crazy mix of stuff we used to talk about at Simon's Rock College back in the 1980s! The convergence on the community level, reflected via the existence of the multi-conference, is both symbolic for and reflective of various conceptual convergences in the process of occurring, between the different AI paradigms favored by the various communities.

One very concrete aspect of the field's overcoming of the symbolic/emergentist dichotomy has been the emergence of integrative, hybrid AI architectures, which combine subsystems operating according to the different paradigms. The combination may be done in many different ways, e.g., connection of a large symbolic subsystem with a large subsymbolic system, or the creation of a population of small agents which are all both symbolic and subsymbolic in nature. Hybrid systems are quite heterogenous in nature, and not so easy to generalize about. My own OpenCog AGI architecture is hybrid in nature (as was the Webmind AI Engine before it), and I'll tell you a bit about this particular architecture in later chapters.

A classic example of a hybrid cognitive architecture is the CLARION (Connectionist Learning with Adaptive Rule Induction On-line) system created by researcher Ron Sun from the Rensellaer Polytechnic Institute (RPI). CLARION is a hybrid system that actually has four distinct parts: two that are neural networks, and two that consist of formal symbolic rules. The four parts are based on drawing two psychological distinctions:

- "implicit" (roughly, unconscious) versus "explicit" (roughly: conscious, deliberative) knowledge
- action-centered versus non-action-centered

So, CLARION contains:

- ***an action-centered subsystem***, whose job is to control both external and internal actions; its implicit layer is made of neural networks called Action Neural Networks, while the explicit layer is made up of symbolic action rules. In the simplest case, these action rules and networks tell what action the system should take at a certain time, depending on the circumstance.
- ***a non-action-centered subsystem***, whose job is to maintain general knowledge; its implicit layer is made of associative neural networks, while the bottom layer is of associative rules ("associative" meaning that each item stored in memory is associated with other, related items stored in memory, so that when one item is remembered, other associated items can easily be remembered too).

There are other components too, handling motivation (which guides action, of course) and meta-cognition (thinking about thinking, which lets the system adapt its own cognitive processes based on experience). The learning dynamics of the system involve ongoing coupling between the neural and symbolic aspects.

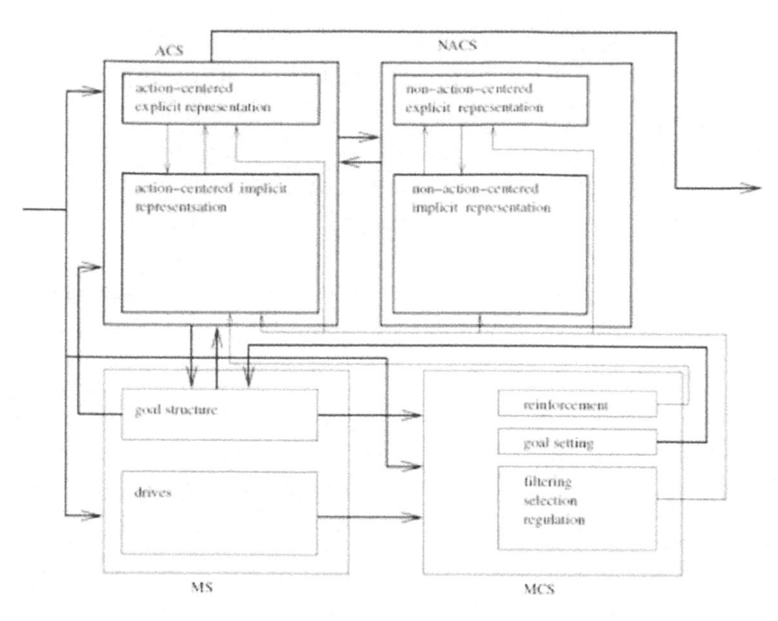

FIGURE 13.23: *High-level sketch of the CLARION cognitive architecture. The subsystems depicted include: the action-centered subsystem (the ACS), the non-action-centered subsystem (the NACS), the motivational subsystem (the MS), and the meta-cognitive subsystem (the MCS). (http://www.cogsci.rpi.edu/~rsun/folder-files/clarion-intro-slides.pdf).*

Overall, CLARION is not the way I would personally choose to divide up cognitive function into pieces, but I do think it's a reasonable way to go about doing things. And it certainly breaks past the symbolic/emergentist divide in a compelling way. In the OpenCog approach, rather than having different containers for implicit vs. explicit and action-centered vs. non-action-centered knowledge, we put all these kinds of knowledge in the same place (the AtomSpace hypergraph knowledge representation), which I believe better promotes synergetic inferences combining implicit, explicit, action-centered and non-action-centered aspects. But of course, science advances via different people positing different approaches, and then building and testing systems exemplifying their approaches, and proceeding on the basis of what they've learned.

A hybrid cognitive architecture relating more closely to my own work is LIDA, developed by Stan Franklin and his colleagues at the University of Memphis. Stan was instrumental in the foundation of the AGI conference series – his presence at the 2006 AGI Workshop Bruce Klein and I organized in Bethesda (near where I lived in Maryland at the time), which was mediated by an introduction from my friend and former Webmind employee Pei Wang, gave the workshop a bit more of an air of prestige and legitimacy. Stan's 1995 book *Artificial Minds* was a classic of AI exposition, overviewing the AI field and then describing his own action-selection-oriented approach to building intelligent systems. When we decided to expand the AGI Workshop into a full-on conference, Stan offered to host AGI-08, the first AGI Conference, at the University of Memphis. Stan was a great host, and the FedEx Center there proved an amazing venue, probably the best one the AGI conferences have had so far.

Stan's LIDA architecture is closely based on cognitive psychology and cognitive neuroscience, particularly on two well known, reasonably well accepted cognitive science models:

Bernard Baars' Global Workspace Theory, which views consciousness as centered on a central "workspace" whose contents get a lot of attention and interact with each other intensively. This global workspace is modeled as broadcasting its contents throughout the rest of the mind, causing reactions there that then cause other things to bubble up into the global workspace.

Baddeley's model of working memory, which says that "working memory" (a modern term corresponding to what used to be called "short-term memory") comprises a few specialized components, including a phonological loop (for remembering what was recently heard or said), a visuospatial sketchpad (the "mind's eye"), and an episodic memory buffer (for the story that just happened, or that one is currently thinking of).

So LIDA has explicit components for the global workspace, and for the different parts of working memory that Baddeley defines. The activity within the different components, and the broadcasting of information from the global workspace, is carried out via what is called the "cognitive cycle" – a concept closely based on neuroscience, i.e., based on the principles that:

Much of human cognition functions by means of frequently iterated (~10 Hz) interactions, called cognitive cycles, between conscious contents, the various memory systems, and action selection.

These cognitive cycles serve as the "atoms" of cognition which higher-level cognitive processes are composed of.

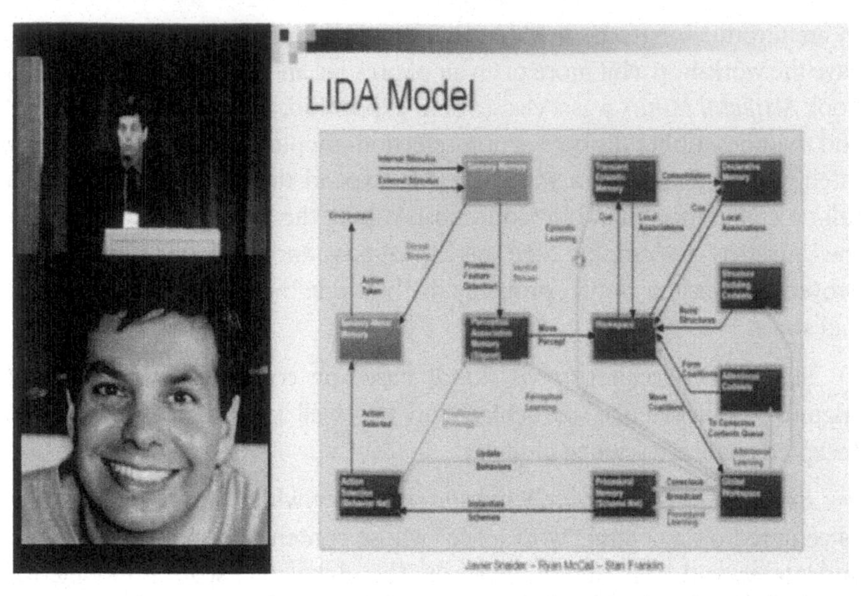

FIGURE 13.24: *Javier Snaider gave a talk at AGI-11, which was held on the Google campus in Mountain View, on the LIDA hybrid architecture for AGI. At the time Javier was working with Stan Franklin at the University of Memphis. Now he's working at Google. After he spent a year or so working at Google building his software engineering chops doing Android (phone software, not robot) development, I introduced him to Ray Kurzweil, and he joined Ray's team within Google doing AGI-oriented development. (http://www.youtube.com/watch?v=Rgjw8O3vLBs, https://lh5.googleusercontent.com/-3qhfODwYk7c/AAAAAAAAAAI/AAAAAAAADkU/jQ8OeR8QLj8/photo.jpg).*

What actually happens in each of LIDA's components, during each cognitive cycle? – now that's more complicated. LIDA is a hybrid architecture, meaning that each component operates based on its own structures and dynamics. For instance, LIDA's perceptual associative memory component uses a "SlipNet," which is an AI system developed by Douglas Hofstadter (whom I've mentioned above a couple times) in his work on analogy. Furthermore, the nature of the structures and dynamics in each component have changed as the architecture has evolved. Franklin's student Javier Snaider, before he completed his PhD and went to work for Google, was endeavoring to replace the internals of every LIDA component with the "sparse distributed memory" system he developed for his PhD thesis. Whether this happens or not is not the point – the point is that the specific algorithms and data structures inside each component can be totally replaced, without destroying the LIDA-ness of LIDA. LIDA is about the overall architecture: the division into components and natural of the information flow between the components. FIGURE 13.24 shows the way information flows between the parts according to LIDA's cognitive cycle.

Hand-Coded Rules as Scaffolding (In Defense of Old-Style Symbolic AI)

When I founded the AGI conference series, I expected to attract researchers from the neural net and hybrid AI camps. I was pleasantly surprised to get some interest from folks coming out of the symbolic AI tradition – many of these folks, such as John Laird and Paul Rosenbloom, were more AGI-oriented in philosophy and intention than I'd anticipated. I had read about Soar and other symbolic AI systems in my undergraduate years, and this was one of the things that had pushed me away from the AI field and compelled me to study mathematics instead. I felt that all this attention paid to systems operating based on rules hand-coded by humans just couldn't be right. Intelligence had to be about adaptive learning and spontaneous, complex self-organization, not about hand-coded rules. But once I got to know these folks I understood their perspective better.

The hand-coded rules were never the point for them, these were basically just "scaffolding" that they put in place to let them experiment with what really interested them, which was the overall cognitive architecture – what were the main parts of the mind and how did they all work together? They realized the hand-coded rules would have to be replaced by the results of a learning system at some point, but they were deferring that point to the future and focusing on a different part of the problem for starters. I had a different taste and intuition than them regarding what was the best place to start in building an AGI system. But since nobody had yet gotten all that far toward AGI, it didn't make sense to be sooooo judgmental about someone else's choice of starting point.

In fact, somewhat ironically, I have ended up taking a similar "scaffolding"-based approach in OpenCog's natural language comprehension component, purely for pragmatic reasons. In the period of 2004-2006 my consulting firm Novamente LLC had a contract with a DC company, ultimately funded by INSCOM, US Army intelligence, to build a natural language comprehension system. As often happens with commercial contracts, we had to build it on a deadline. So we proceeded to build a system based on a lot of hand-coded rules for English grammar and semantics. Sure, I would have rather built a system learning these linguistic rules via the system's experience. But I was trying to earn a living, and support a consulting company (Novamente had a dozen staff in Belo Horizonte, Brazil at that point, and a couple in the US).

Now we have a lot of complex learning mechanisms in OpenCog, but this linguistic rule-base persists in the language comprehension subsystem, because it somewhat works and replacing it would be a lot of work and hasn't been prioritized. In late 2014 we brought on a cognitive linguist in our Hong Kong team, Aaron Nitzkin, and gave him the goal of experimenting with replacing the hand-coded rules with learned rules, based on some ideas that Linas Vepstas and I had come up with. But it's actually a big job, it's going to take more than Aaron to accomplish it. Anyhow, at this point, now that I've had more experience and have a better sense of the difficulty of making things work in reality, I wouldn't be as judgmental of the symbolic AI guys as I was as a teenager first exploring the AI field. But the use of hand-coded rules still bugs me – I see the value of scaffolding from a practical system-building perspective, but I'm eager to replace this scaffolding in OpenCog. But I can see that the symbolic AI field did learn a fair bit about cognitive architecture from the systems they built using this scaffolding. And now, as the overall learning process of the AI field continues, more and more of the symbolic AI field is incorporating learning algorithms into their work, which is easier to do now than it was in the '70s and '80s due to advances in machine learning, and also in hardware and data availability.

Universal AI

While I was pleasantly surprised that the AGI conferences attracted some folks from the old-style symbolic AI paradigm, and I learned a lot from interacting with these folks, I was almost completely unprepared for the interest the conference series evoked from a group of European researchers involved in a paradigm they called "universal AI." This was a bit ironic because some of the ideas in my first book *The Structure of Intelligence*, which I published in 1993 when I was in my first job as a math professor at the University of Nevada, related very closely to this "universal AI" work. But I hadn't developed that aspect of my thinking much in the interval between the mid-1990s and 2008 when we organized the first full-on AGI conference; and until we started getting papers from these folks submitted to the conference, I was unaware that so much research had happened in that direction.

What I mean by "universal AI" is not an AI that spans the entire universe (though indeed it's possible that our whole universe is an AI created by some aliens from another universe – that's a whole other topic!). Rather, it's an approach to AGI that starts out with the unrealistic assumption of infinitely or near-infinitely powerful computers. One starts with AGI algorithms or agents that WOULD yield incredibly powerful general intelligence IF they were supplied with massively, unrealistically much computing power, and then views practically feasible AGI systems as specializations of these powerful theoretical systems.

This can be viewed by analogy to physics: Everyone knows the sun and earth and moon are not actually spheres, but when doing celestial mechanics, one starts by assuming their spheres because it makes the calculations simpler. Then one adds corrections to one's model, to account for the actual non-spherical mass distributions of these celestial objects.

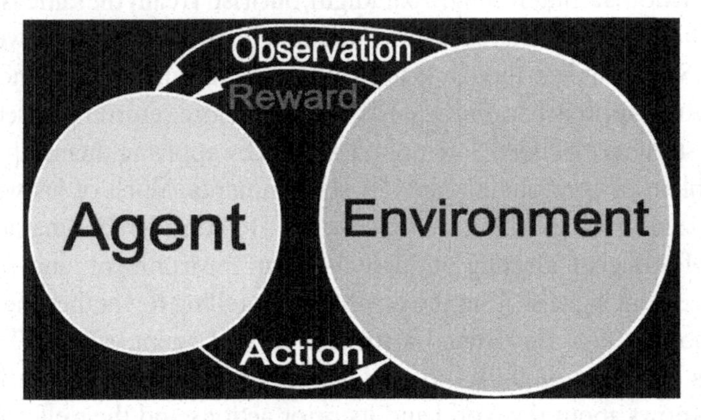

FIGURE 13.25: *Universal AI uses an abstraction of the AGI problem in terms of mathematically formulated agents, environments, observations and actions. This simplifies things considerably, as opposed to considering the messy particularities of the everyday human world, allowing one to prove elegant mathematical theorems about general intelligence. (http://matchingpennies.com/on_aixi/graphics/on_aixi.png).*

The path toward universal AI began in earnest back in 1964 with some mathematical work by Ray Solomonoff. He was looking at the problem of predicting the next number in a series of numbers – which is actually a general model for the prediction of anything. He proved that the generally best way to do this, for very long series of numbers, is to find the shortest program that computes the first N numbers in the series, and to keep running that program to compute the succeeding digits. For instance, if one is given a series of the first 1000 even numbers, then in any reasonable programming language, the shortest program for producing these is going to be one that embodies the idea of computing one even number after another. Running this program onward will produce more and more even numbers.

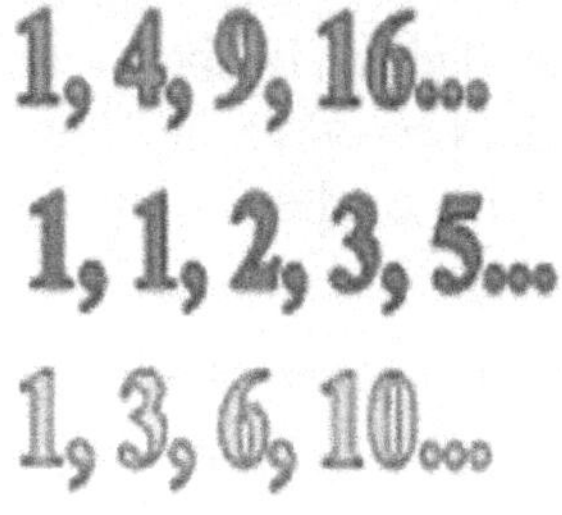

1, 4, 9, 16...

1, 1, 2, 3, 5...

1, 3, 6, 10...

FIGURE 13.26: *Number sequence prediction is a fun kind of puzzle, often used in child mathematics education. In the early 1960s Ray Solomonoff figured out the mathematically optimal algorithm for solving large number sequence prediction problems – a result that serves as the foundation for work on "Universal AI" today. Solomonoff's ideas about optimal prediction can be used to create theoretical AGI agents that choose actions optimally in environments, based on optimally predicting the consequences of various actions. Whether this fascinating mathematical approach can be made to yield exciting practical results remains to be seen.*

Prediction is an interesting AI paradigm, but it isn't really the same as general intelligence. Physicist Marcus Hutter (a student of German AI powerhouse Juergen Schmidhuber, though Marcus had some industry experience and a fairly mature approach to Universal AI in mind before returning to school for his PhD studies) extended Solomonoff's ideas by applying them explicitly to the problem of controlling agents in environments. Much of his work was summarized in his 2005 book *Universal AI*. He considered a mathematical model of an agent carrying out actions in an environment, and receiving "reinforcement signals" from the environment telling it whether the actions were good or not. He defined a theoretical AGI agent called AIXI which operates by Solomonoff-like principles – at each point in time, it looks at what it knows about the world and its prior actions and their effects, and it figures out the shortest program that, in its judgment, would have caused it to get maximal reward, had it operated according to this program in the past. (Yes, I'm simplifying a lot of complex math here, and Marcus would probably have a lot of quibbles with my wording. But I'm doing my best to explain this very abstract stuff in ordinary English....) Marcus's theorems show that, in an abstract sense, AIXI is the optimally intelligent agent in computable environments.

Conceptually and very very loosely, AIXI may be understood roughly as follows:

An AGI system is going to be controlled by some program.

Instead of trying to figure out the right program via human wizardry, we can just write a "meta-algorithm" to search program space and automatically find the right program for making the AGI smart, and then use that program to operate the AGI.

We can then repeat this meta-algorithm over and over, as the AGI gains more data about the world, so it will always have the operating program that's best according to all its available data.

FIGURE 13.27: *German AGI researcher Marcus Hutter (currently working at the Australian National University in Canberra) at AGI-09, discussing progress on "scaling down" his theoretical Universal AI approach to make it work on practical problems. (http://vimeo.com/7390883).*

AIXI is a precisely defined "meta-algorithm" of this nature. Related systems have also been formulated. There is the "Speed Prior"-based system, due to Juergen Schmidhuber, that looks at program runtime as well as program length in choosing the "best program." Schmidhuber (a German researcher who leads a very successful AI program at IDSIA, a university in Lugano, Switzerland, where we held the AGI-10 conference) has also defined a (theoretical) "Goedel machine" system that achieves intelligent behavior via constantly rewriting its own software code. It figures out how to rewrite its code by proving theorems about itself.

FIGURE 13.28: *Photo from the NIPS 2002 AI conference, the Workshop on Universal Learning Algorithms and Optimal Search. Ray Solomonoff is in the center; Marcus Hutter is to the far right. Solomonoff remained active in research to a quite old age, and was very gratified to see so many younger researchers take his ideas and run with them in various productive directions. To the left of Ray is his wife Grace, to the right is Paul Vitanyi, an important figure in the mathematics underlying Universal AI. (http://www.hutter1.net/idsia/nipspic5.jpg).*

In my first book *The Structure of Intelligence*, as I mentioned above, I outlined an AGI architecture that operated according to similar ideas, and I used similar theoretical ideas to ground it.[1] I looked at mind as composed of patterns and concerned with recognizing patterns in its environment and itself, and I formally defined a "pattern" in some entity X as a program for computing X, which was sufficiently shorter/smaller than X that it provided some "information compression." This line of thinking connects very closely with Solomonoff's ideas about using shortest programs for prediction. But I was using these mathematical notions as inspirations for a less formal, more conceptual and experimental style of AGI design. Eric Baum, in his book *What Is Thought?*, also introduced similar concepts and used them to motivate his own AGI thinking and designs (he argued for viewing thinking as a kind of compression, which at core was exactly Solomonoff's idea, that the best data-compressing program is the best for prediction; Hutter extended this to deal explicitly with agents that predict what actions will best help achieve their goals). The difference is that Hutter, Schmidhuber and the others in the Universal AI field haven't just used these mathematical ideas for inspiration, they've developed them formally into more of a full-fledged mathematical theory of AGI – just with the shortcoming that most of their current theories only usefully address the case where computing resources are unrealistically, insanely abundant.

In their general formulations, agents like AIXI and the Godel Machine are not computable – no real-world computer could run them, ever. There are approximations that could run on physical computers, at least in principle, but the most straightforward of these approximations would require computers bigger than the known physical universe. So the question then becomes: *Can we "scale down" these airy mathematical AGI approaches into stuff that can actually be run on real computers in the real world to solve real problems?*

The jury is still out on this. Researchers in the area are working hard on the problem, and have made systems that can learn to play simple games using methods derived from Universal AI. But still, there's a long way from playing tic-tac-toe to doing anything in the real world—simple games are very limited domains, with very few variables involved. It's fair to say that most

1 I was inspired more by the work of mathematician Gregory Chaitin than by Solomonoff's work, but in the end these theorists were both getting at the same thing; their ideas are very closely connected.

of the AGI research community is highly skeptical of this sort of approach. But for those of us with a math background, the formal rigor of the approach seems nice, and mathematics can be a powerful engine for generating new ideas.

For some Universal AI researchers, such as Juergen Schmidhuber, there is a continuity between their abstract Universal AI research, and their more practical research on other types of AI/AGI systems. Schmidhuber has done a great deal of very successful research on neural networks and deep learning, making such systems recognize images and control robots and so forth. He sees these systems as manifesting the same principles as AIXI and the Godel Machine, though obviously in different particular ways.

On the other hand, many researchers find the assumption of infinite or massive computational resources underlying the Universal AI approach more troubling – much further off from reality than, say, the assumption of a spherical sun in physics. One can view the Universal AI work as telling us that: If one lifts computational resource restrictions and considers the case where computing power is effectively infinite, then all the problems normally considered in the AI/AGI field go away, and a conceptually very simple algorithm like AIXI can be optimally smart. In a sense, what AIXI does is search the mathematical space of all possible computer programs, in between each perception and each action, to choose which program would be optimal to run to maximize its future reward. This seems to have nothing to do, intuitively, with all the things that need to be done to make an AGI operate usefully in the real world. So one conclusion one could draw from this is: Real-world AGI is all about computational resource restrictions. Assuming infinite computing power is assuming away all the really interesting aspects of real-world general intelligence. Real-world general intelligence is all about how systems adapt their structures and dynamics to achieve their goals in the context of limited available resources.

This is the view of my friend Pei Wang, an AGI researcher who worked with me at Webmind in 1998-2001, and who edits the Journal of AGI and has helped me off and on with the conference series. Pei was Douglas Hofstadter's student way back and, back before he immigrated to the US, translated Doug's book *Godel, Escher, Bach* into Chinese. Pei defines intelligence as "adaptation to the environment under limited resources" – a perspective under which AIXI is not intelligent at all, but is something entirely different; a mathematical construct which may be fascinating in its own right, but has no particular relevance to intelligence.

Getting to know Marcus Hutter, Juergen Schmidhuber and a host of other younger researchers in this area has been fascinating, and one of the many excellent consequences I've reaped personally by organizing the AGI conferences. I haven't yet been convinced that their elegant abstractions are the right approach to actually getting AGI systems built, but as I just said, the jury's still out. Also, even if some other, more pragmatically-oriented approach actually gets AGI built first, it may happen that the theories the Universal AI community is building will help in analyzing the behavior and principles of these first working AGIs, and help in moving toward smarter and smarter subsequent systems. It's great to have so many smart mathematicians and theorists thinking about AGI from a variety of perspectives. Science works by generating a lot of ideas and cross-connecting them, and nobody can ever predict in advance where any particular idea will lead.

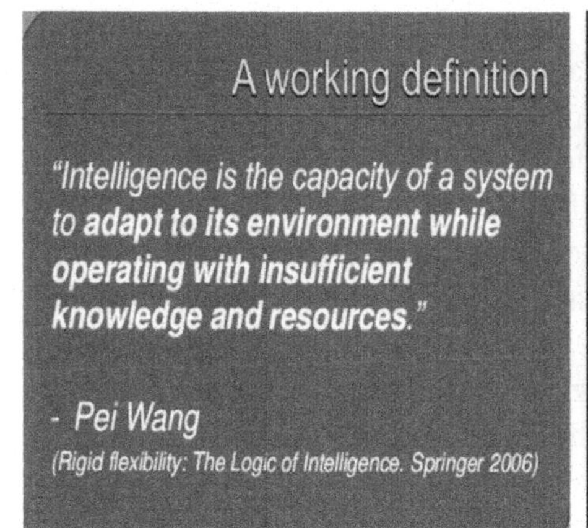

FIGURE 13.29: *AGI researcher Pei Wang argues that dealing with limited resources is the crux of intelligence – so that systems like AIXI that assume infinite or massive computing resources aren't really dealing with "intelligence" at all. (http://i2.wp.com/agi-school.org/video/lecture02/lecture2_m.jpg?resize=307%2C257;http://www.slideshare.net/HelgiHelgason1/from-narrow-ai-to-agi).*

Why Is It So Hard to Measure Partial Progress Toward Human-Level AGI?

So there are all these different approaches to creating AGI out there – it's a testament to the wonderful power of human creativity and imagination, but it's also kind of a mess, and makes progress in the field much slower than if everyone were focused on a single path... that is, assuming that single path were a workable one...

Why is it so hard for the different researchers in the field to agree on the right approach? These are all very smart people. Is it just that they're all insanely stubborn and opinionated?

Many AGI researchers are extremely stubborn and opinionated, but the problem isn't just that. The other problem is that it's actually hard to tell how much progress a certain approach is making toward AGI, for purely theoretical reasons.

Measuring intermediate progress toward the goal of human-level AGI is a challenging task, certainly more so than judging whether you've gotten to the end goal or not. It's not entirely straightforward to create tests to measure the final achievement of human-level AGI, but there are some fairly obvious candidates here. There's the Turing Test (fooling judges into believing you're human, in a text chat), and then various variants we discussed at the 2009 AGI Workshop that Itamar Arel and I conducted at the University of Tennessee Knoxville: the video Turing Test, the Robot College Student test (passing university, via being judged exactly the same way a human student would), etc. There's certainly no agreement on which is the most meaningful such goal to strive for, but there's broad agreement that a number of goals of this nature basically make sense.

On the other hand, how does one measure whether one is, say, 50 percent of the way to human-level AGI? Or, say, 75 or 25 percent?

It's possible to pose many "practical tests" of incremental progress toward human-level AGI, with the property that IF a proto-AGI system passes the test using a certain sort of architecture and/or dynamics, then this implies a certain amount of progress toward human-level AGI, based on particular theoretical assumptions about AGI. However, in each case of such a practical test, it seems intuitively likely to a significant percentage of AGI researchers that there is some way to "game" the test via designing a system specifically oriented toward passing that test, and which doesn't constitute dramatic progress toward AGI.

Some examples of practical tests of this nature, which were suggested by participants in the 2009 AGI Roadmap Workshop, would be:

- ***The Wozniak "coffee test"***: go into an average American house and figure out how to make coffee, including identifying the coffee machine, figuring out what the buttons do, finding the coffee in the cabinet, etc. (advocated by Josh Hall at the workshop.)

- ***Story understanding***: reading a story, or watching it on video, and then answering questions about what happened (including questions at various levels of abstraction)... (advocated by Joscha Bach at the workshop.)
- ***Passing the elementary school reading curriculum***: which involves reading and answering questions about some picture books as well as purely textual ones... (advocated by Stuart Shapiro at the workshop.)
- ***Learning to play an arbitrary video game based on experience only, or based on experience plus reading instructions***... (advocated by Sam Adams, from IBM, at the workshop.)

One interesting point about tests like this is that each of them seems to some AGI researchers to encapsulate the crux of the AGI problem, and to be unsolvable by any system not far along the path to human-level AGI, while seeming to other AGI researchers, with different conceptual perspectives, to be something probably game-able by narrow-AI methods.

Josh Hall, at the Roadmap Workshop, felt that to make coffee in a random American house, a robot would need to have human-level general intelligence. Some other participants felt that this feat, while impressive, could be carried out by a complex but feasible narrow-AI system specifically built for the purpose. It's hard for me to tell. My guess is that a narrow-AI approach could handle 80-90% of houses, but that to really get, say, 99%, you'd probably need a full-on AGI.

On the other hand, Stuart Shapiro felt that to go through the elementary school reading curriculum, you'd need to have human-level general intelligence. But would you? Could one effectively train some sort of narrow-AI text comprehension system by feeding it a lot of texts from the Net, and a bunch of elementary school textbooks with questions and answers? It's not at all obvious to me how well this would work. Stuart's own approach to the problem, based on his fascinating paraconsistent logic framework, was AGI-ish and aimed at creating a system that would really understand what it was reading in the same sense as human beings. But that doesn't mean others couldn't succeed at the task in a less AGI-ish way.

No one at the Roadmap Workshop took this perspective, but a couple decades ago I had conversations with researchers who posited that face recognition was AGI-hard – that no computer would ever recognize human faces as well as people, because of all the deeply human emotional nuances humans bring to bear on the problem. No one makes this kind of argument now, given the success of narrow-AI approaches like Facebook's DeepFace at solving this particular problem.

Given the current state of science, there's no way to tell which of the many practical tests suggested by various researchers really can be solved via a narrow-AI approach, except by having a lot of people try really hard over a long period of time.

All this highlights the question whether there is some fundamental reason why it's hard to make an objective, theory-independent measure of intermediate progress toward advanced AGI. Is it just that we haven't been smart enough, or creative enough, to figure out the right test – or is there some conceptual reason why the very notion of such a test is problematic?

Tricky Cognitive Synergy

Remember the basic lesson of chaos theory. Oftentimes, changing a seemingly small aspect of a system's underlying structures or dynamics can dramatically affect the resulting high-level behaviors. I suspect that this phenomenon may underlie the difficulty of measuring partial progress toward AGI. It may be that a partial AGI system has very different dynamics from a complete AGI system, just as minorly tweaking any aspect of a chaotic, complex dynamical system can have a big impact on the resulting system-level behavior.

When we were discussing these ideas in 2010, my research collaborator Jared Wigmore (then in Hong Kong working on OpenCog; now in Melbourne with a new name, Jade O'Neill, and working for Google) and I had a hard time finding the right words to express our ideas. We came up with the term "tricky," by which we meant any high-level emergent system property whose existence depends sensitively on the particulars of the underlying system. "Tricky" probably isn't an awesome choice of word, but nobody has suggested a better one yet, so there it is. Tricky properties in this sense certainly are tricky from an AGI designer's point of view!

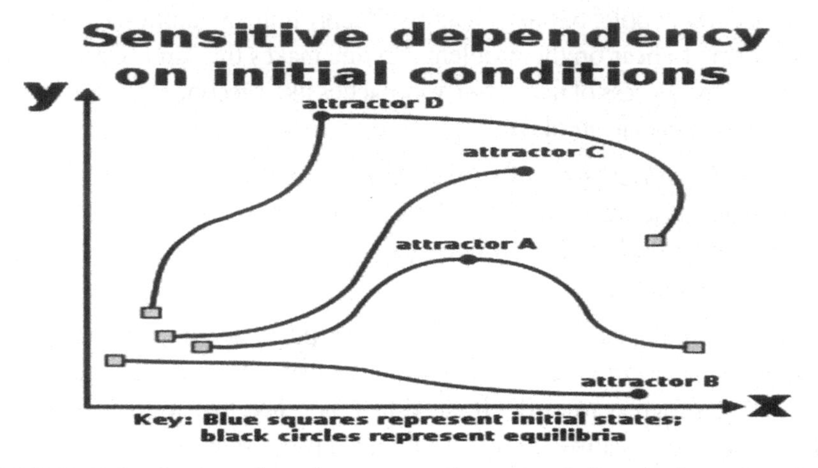

FIGURE 13.30: *Sensitive dependence on conditions is a hallmark of complex, chaotic systems. If general intelligence is a property that depends sensitively on the mix of components inside a cognitive system, then a partial implementation of a cognitive system might display quite different properties from a complete system. (http://en.academic.ru/pictures/enwiki/51/300px-Sensitive-dependency.svg.png).*

So – the crux of my explanation of the difficulty of creating good tests for incremental progress toward AGI is: I hypothesize that general intelligence, under limited computational resources, is tricky.

And now the plot thickens. There are many reasons that general intelligence might be tricky in this sense. But I think one key reason goes back to the concept of "cognitive synergy" that I introduced earlier. I think cognitive synergy is tricky.

You may recall (and if you don't, I'm reminding you!) that the cognitive synergy hypothesis, in its simplest form, states that human-level AGI intrinsically depends on the synergetic interaction of multiple components (for instance, the OpenCog design depends on synergetic interactions between multiple memory systems, each supplied with its own learning process). In this hypothesis, for instance, it might be that there are 10 critical components required for a human-level AGI system. Having all 10 of them in place results in human-level AGI, but having only 8 of them in place results in having a dramatically impaired system – and maybe having only 6 or 7 of them in place results in a system that can hardly do anything at all. This kind of situation would be very tricky indeed!

Of course, the reality is almost surely not as strict as the simplified example in the above paragraph suggests. No AGI theorist has really posited a list of 10 crisply-defined subsystems and claimed them necessary and sufficient for AGI. We suspect there are many different routes to AGI, involving integration of different sorts of subsystems. However, if the cognitive synergy hypothesis is correct, then human-level AGI behaves roughly like the simplistic example in the prior paragraph suggests. Perhaps instead of using the 10 components, you could achieve human-level AGI with 7 components, but having only 5 of these 7 would yield drastically impaired functionality – etc. Or the same phenomenon could be articulated in the context of systems without any distinguishable component parts, but only continuously varying underlying quantities. But here, for illustrative purposes, we'll stick with the "10 components" example, just for communicative simplicity.

Next, let's suppose that for any given task, there are ways to achieve this task using a system that is much simpler than any subset of size 6 drawn from the set of 10 components needed for human-level AGI, but works much better for the task than this subset of 6 components (assuming the latter are used as a set of only 6 components, without the other 4 components). So for instance, suppose that our 10-component system would be able to recognize events in videos as well as humans can… and much better than any simplistic neural net algorithm. But suppose that a simplistic neural net algorithm works better than any 6 of the components, out of the 10, can when put together. Then until you've put together at least 7 of the components, you won't see any advantage of the integrative, multi-component approach over a simpler neural net algorithm. Probably after working on 4 or 5 components you'll get bored and frustrated and figure just going with the neural net is a better idea – even if it doesn't work that awesomely, at least it works OK and it wasn't that complicated to put together.

Generally speaking, this "tricky cognitive synergy" hypothesis would be true if, for example, the following possibilities were true:

- creating components to serve as parts of a synergetic AGI is harder than creating components intended to serve as parts of simpler AI systems without synergetic dynamics.
- components capable of serving as parts of a synergetic AGI are necessarily more complicated than components intended to serve as parts of simpler AGI systems.

Why would these be true? Well, to serve as a component of a synergetic AGI system, a component must have the internal flexibility to usefully handle interactions with a lot of other components, and to solve the problems that come its way. In terms of our concrete work on the OpenCog system, these possibilities ring true, in the sense that tailoring an AI process for tight integration with other AI processes within OpenCog tends to require more work than preparing a conceptually similar AI process for use on its own or in a more task-specific narrow AI system.

But if tricky cognitive synergy really holds up as a property of human-level general intelligence, the difficulty of formulating tests for intermediate progress toward human-level AGI follows as a consequence. This is because, according to the tricky cognitive synergy hypothesis, any test is going to be more easily solved by some simpler narrow AI process than by a partially complete human-level AGI system.

So if these ideas are correct, then positing tests for intermediate progress toward human-level AGI is a very difficult prospect. The tricky cognitive synergy hypothesis suggests that this difficulty may not be due to incompetence or lack of imagination on the part of the AGI community, nor to the primitive state of the AGI field, but is rather intrinsic to the subject matter. Just as 2/3 of a human brain may not be of much use, similarly, 2/3 of an AGI system may not be of much use. Lack of impressive intermediary results may not imply that one is on a wrong development path; and comparison with narrow AI systems on specific tasks may be badly misleading as a gauge of incremental progress toward human-level AGI.

As an AGI engineer, I would love to have a sensible, rigorous way to test our intermediary progress toward AGI, so as to be able to pose convincing arguments to skeptics, funding sources, potential collaborators and so forth – as well as just for our own edification. My colleagues and I really, really like producing exciting intermediary results, on projects where that makes sense. Such results, when they come, are extremely informative and inspiring to us as well as the rest of the world! My motivation in thinking about trickiness and AGI is not a desire to avoid having the intermediate progress of our efforts measured, but rather a desire to explain the frustrating (but by now rather well-established) difficulty of creating such intermediate goals for human-level AGI in a meaningful way.

If we or someone else figures out a compelling way to measure partial progress toward AGI, we will celebrate the occasion. But it seems worth seriously considering the possibility that the difficulty in finding such a measure reflects fundamental properties of the subject matter – such as the trickiness of cognitive synergy and other aspects of general intelligence.

If this is true, it suggests the wild diversity of the AGI community may be around for a while. It may be around until someone gets enough parts of their complex, integrative cognitive architecture working well enough to overcome the trickiness phenomenon and demonstrate highly impressive, "AGI Sputnik"-level results, which then finally start to pull a rapidly increasing percentage of the AGI community into collaborating on their perspective.

Toward a Common AGI Testbed

Suppose we grant that, due to tricky cognitive synergy and/or other aspects, the wild diversity of the spectrum of AGI designs is going to be around for a while – at least until some design demonstrates a dramatically superior practical success, which is clearly AGI-ish and way ahead of the pack. Is there at least a way to make the diversity more comprehensible and manageable, going forward?

Other subfields of AI, such as computer vision, have benefited substantially from the creation of common "testbeds" for various software systems. In computer vision this often takes the form of large sets of images depicting various objects. Everyone working in computer vision uses the same image sets for testing their approaches, so that when one gets "78% accuracy in identifying objects in images on the CiFAR-10 dataset," everyone knows what that means. CiFAR-10 is a database containing a large number of image files, each one containing an image of one of 10 objects – car, airplane, frog, and so forth. If everyone is trying their algorithms on the same image files, there's a clear basis for comparison, both quantitatively (how accurate is your algorithm at recognizing the objects in these particular images?) and qualitatively (what kind of mistakes does your algorithm tend to make on airplane images?).

FIGURE 13.31: *The CiFAR-10 dataset is commonly used in computer vision, to test and compare the effectiveness of different algorithms at recognizing objects in small images. The CIFAR-10 dataset consists of 60,000 32x32 color images in 10 classes, with 6,000 images per class. No such standard, widely accepted "testbed" exists for AGI systems – but the creation of such a thing might well be a good idea. (http://www.cs.toronto.edu/~kriz/cifar.html).*

We don't yet have a standard testbed like this for AGI – but this may well come in time. The value such a thing would have is widely recognized in the AGI community. As supporting technologies like game worlds and humanoid robots become less expensive and easier to use, it should become increasingly possible to use them to make common "test environments" for various AGI researchers to try their systems in.

Imagine a 3D game world that was dead easy for anyone to hook their AGI system up to, and that provided a series of IQ tests that any AGI system could take, as well as teaching/training environments that corresponded to the tests. For instance, if there was an IQ test question involving sorting objects, there could be an environment full of lots of objects to be sorted, set up to allow systems to practice sorting objects according to appearance, function and other criteria.

Or what if there was a humanoid robot that was dead easy for anyone to hook their AGI system up to? So that, for instance, even if some AGI system didn't yet deal with computer vision and movement, there were tools they could connect their system to that would basically serve as simple perceptual and motor cortices for their AGI, allowing it to control the robot. Things are moving in this direction with the ROS (Robot Operating System) framework, which greatly simplifies the task of connecting AI systems to robots; and with robots like the Nao, which come along with specialized scripting languages for controlling basic behaviors, and large sets of support tools. Or the Darwin OP, which is much like the Nao but open source (but has a less mature software infrastructure at present).

I have also collaborated with Mark Tilden (the maker of the RoboSapien, whom I mentioned above) and an Indian engineer, Mandeep Bhatia, on creating a modification of the RoboSapien that makes it capable to use as a general-purpose AI research robot, comparable to the Nao or Darwin OP, but for a vastly lesser price (maybe $100 total for the add-on kit, plus $50 or so for a base RoboSapien). So far our modded RoboSapien is being used by my Ethiopian colleagues at iCog Labs, as the hardware platform for the first low-cost robot soccer league. A number of Ethiopian universities have bought kits for modding RoboSapiens from iCog, and in late 2016 or early 2017 there will be the first Ethiopian robot soccer competition using these robots. It's a start!

There are a lot of possibilities, and things are shifting fast in this domain. ROS, Nao, RoboSapien and other existing tools don't yet make robots easy to use as a common platform for AGI developers, but they're getting there.

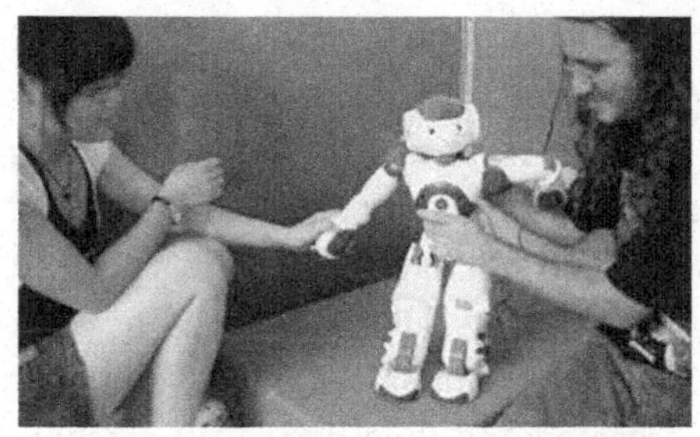

FIGURE 13.32: *The Nao robot, made by Aldebaran Robotics in France (purchased recently by Japanese firm Softbank), is currently the closest thing to an adequate research platform for various AI/AGI developers to use to try their algorithms and systems out on robots. It has a pretty nice API for interfacing your AI code to. But it's still more than $10,000 to purchase, and its closed-source. This photo was taken by Raj Dye when he was filming the documentary Singularity or Bust! It shows myself and Ruiting Lian (to whom I'm now married!) way back in 2009, when we were experimenting with Nao robots at Xiamen University together with Hugo de Garis and various other students and researchers there. Ruiting and I were just barely getting to know each other then. (http://hplusmagazine.com/sites/default/files/images/articles/dec09/chinese-singularity1.jpg).*

FIGURE 13.33: *The Darwin-OP robot, an open source analogue of the Nao that holds promise as a general platform for AGI/robotics development and comparative testing. At the moment it doesn't have as sophisticated a toolset as the Nao, but we'll see what the open source software and hardware communities can do. It's still quite an expensive robot though. Due to the widely popular RoboCup competitions, robot soccer has been a popular domain for comparing robot control systems, but most AGI researchers feel it doesn't really lend itself to pushing AGI forward, as it leans much more heavily on perception and action than on cognition, motivation, learning, emotion and other aspects of intelligence. (http://upload.wikimedia.org/wikipedia/commons/7/70/Simulation_of_a_Robotis_DARwIn-OP_in_Webots.png).*

Common testbeds can have both positive and negative impacts on research fields. This is definitely true in computer vision and in what AI researchers now call "machine learning" (in the AI field these days, the term "machine learning" generally refers to the use of algorithms to divide data into categories; I'll say more on this narrow sort of machine learning later on). The use of standard datasets and problems for comparing algorithms has spurred progress by giving ways for upstart researchers to definitively show that their work is "better" than the work of the old guard in the field, on a commonly recognized test-base. On the other hand, the advent of common testbeds has also caused an awful lot of research to get directed toward goals like "Get 0.1% better accuracy on this or that popular test, as compared to the best leading algorithm," rather than toward more fundamental innovations. Once you start measuring progress, people start to get obsessed with their scores to the exclusion of all else. Journals start to prioritize publishing papers that slightly improve performance on standard testbeds, as opposed to papers describing new innovations that may not have great results yet, but hold fundamental promise for large improvements later on after more work has been done. Because of this kind of phenomenon, old-time AI researcher Pat Langley is fond of telling how he "created and then destroyed machine learning" – the latter because he helped introduce the UCI repository of classification test problems, and he felt that performance on this repository eventually became an unhealthy obsession of the machine learning field.

My take on this is that having common problems for different researchers to work on, and test their systems on, is generally GOOD; but comparing different systems as to their scores on these common problems is usually BAD. The point is that system performance is often multidimensional, and sometimes it is best assessed qualitatively. Different applications can tolerate different kinds of errors, and benefit most from different kinds of successes; different kinds of errors and successes point in different research directions and indicate different fundamental underlying strengths and weaknesses. Comparing different AI systems on the same problems can yield all sorts of qualitative insights, including insights into how to create new systems combining the strengths and avoiding the weaknesses of current systems; but obsessing on a handful of one-dimensional performance measures (e.g., classification accuracy) ends up throwing out most of the useful information obtainable by running multiple AI systems on common problems.

So, long story short, I'm very interested in common environments like game worlds, and common embodiments like easy-to-use robots – I'd love to see a host of different proto-AGI systems used to control the same robots, and game characters in the same worlds. But trying to make a score to compare which AI system is the better robot-controller doesn't interest me much. If one system is a dramatically overall better robot controller, that will be qualitatively clear. If it's not qualitatively clear, probably multiple systems have different strengths and weaknesses, and which one would come up on top in a test would probably depend pretty sensitively on the specifics of how one made the test.

The RoboCup soccer tournament – which my Ethiopian colleagues and I are emulating with our low-cost African robot soccer league, as I mentioned above – is a reasonable illustration of this. Getting a lot of different AI approaches to control the same little humanoid robots is somewhat interesting. As an educational venture, robot soccer is awesome. But from a fundamental research perspective, does it really matter whose AI system is best at playing soccer? One could argue that trying to make your AI win a soccer game pushes you to do particular sorts of research (i.e. particularly soccer-relevant research), and thus diverts everyone's efforts away from other sorts of research, narrowing the focus on the field in a way that probably isn't optimal for overall progress on AI or robotics. Yet there's no denying the fun value of robots playing soccer; the RoboCup tournaments have seduced generations of students into the wonder of robotics hacking and tinkering. We are aiming to bring this magic to Africa with our low-cost modded RoboSapien soccerbots.

FIGURE 13.34: *The RoboCup soccer tournament has done wonders to catalyze student interest in humanoid robotics, around the world. It's not AGI, but it's cool! (http://www.numerama.com/media/attach/robocup2009_standard.jpg).*

My Ethiopian colleagues at iCog Labs in Addis Ababa have launched an analogous robot soccer competition involving students at Ethiopian universities and low-cost robot kits we have created based on modding RoboSapien toy robots. The bottom three pictures show: our RoboSapien soccer robot, the robot's main modder Mandeep Bhatia showing off the robot at Maker Faire 2015 at Hong Kong Poly U, and some students at Mekele University in Ethiopia studying a RoboSapien and its mods (made by student interns at our Addis Ababa firm iCog, varying on Mandeep's original mods).

Further Reading

For a more complete list of scholarly references corresponding to the historical AI achievements mentioned here, see the Scholarpedia article on Artificial General Intelligence.

Adams, Sam and Itamar Arel and Joscha Bach and Robert Coop and Rod Furlan and Ben Goertzel and J. Storrs Hall and Alexei Samsonovich and Matthias Scheutz and Matthew Schlesinger and Stuart C Shapiro and John Sowa[2] (2011). Mapping the Landscape of Human-Level Artificial General Intelligence. *Artificial intelligence Magazine*.

Cangelosi A., Schlesinger M. (2015). *Developmental Robotics: From Babies to Robots*. Cambridge, MA: MIT Press, Bradford Books.

Cassimatis (2006). Special Issue on Human-Level Intelligence. Human-Level Intelligence, *Artificial Intelligence Magazine*, AAAI Press.

Chaitin, G. (1987). *Algorithmic Information Theory*. New York: Cambridge Press.

Duch, Wlodzislaw and Oentaryo, Richard and Pasquier, Michel (2008). Cognitive Architectures: Where Do We Go From Here? *Proceedings of the Second Conference on AGI*.

Goertzel, Ben. Artificial General Intelligence, *Scholarpedia*.

Hutter, Marcus (2005). *Universal Artificial Intelligence*. Springer.

Laird, John (2012). *The Soar Architecture*. MIT Press.

Norvig, Peter and Stuart Russell (2009). *AI: A Modern Approach*. Prentice-Hall.

---

2 This long author list brings back some memories. To be noted is that these various researchers don't work together on the same project, but are rather each concerned with pursuing their own AGI approaches and ideas. Empirical science papers often have a lot of authors, but typically each of the people listed at the top helped with a different aspect of the same project. Anyone who has not been an academic researcher will not be able to understand how incredibly arduous it was to get such a large group of researchers with diverse backgrounds and ideas and goals to agree on enough things about AGI to write a paper together!

Solomonoff, L. (1964). "A Formal Theory of Induction, Parts I and II" *Information and Control* 7, p. 1-22, 224-254. Veness, J.; Ng, K. S.; Hutter, M.; Uther, W. T. B.; and Silver, D. (2011). A Monte-Carlo AIXI Approximation. *Journal of Artificial Inteligence Research*. 40, p. 95–142.

Wang, Pei (2006). *Rigid Flexibility: The Logic of Intelligence*. Springer.

Franklin, Stan and Bernard Baars (2009). Consciousness is computational: The LIDA model of global workspace. *International Journal of Machine Consciousness*.

Markram, Henri (2006). The Blue Brain Project. *Nature Reviews Neuroscience*, p. 7-2, 153-160.

Sun, Ron (2002). *Duality of the Mind: A Bottom-up Approach Toward Cognition*. Mahwah, NJ: Lawrence Erlbaum Associates. OpenCog and PrimeAGI

14. OpenCog and PrimeAGI

The OpenCog project was founded in 2008, when my colleagues and I got frustrated with the slow progress we were making within our small company Novamente LLC, at implementing our big Novamente Cognition Engine design for AGI. We had intended Novamente LLC to be a big, grand "build a thinking machine" company, but it was ending up more of a DC-centered federal-government-contract-oriented AI consulting company, getting narrow AI contracts in the US and outsourcing most of the heavy lifting to our office in Belo Horizonte, Brazil (which was run by Cassio Pennachin, the former VP of Engineering of Webmind Inc.). It was a pretty interesting way to make a living, but it wasn't getting a thinking machine built very quickly. We put our modest profits largely into AGI development, as well as various narrow AI product development initiatives, but it wasn't nearly enough given the size of the AGI task. So we decided that open-sourcing our core AGI software would be a reasonable gamble. We would lose the business option of being The Company That Owns The World's First AGI. But we would, or so we figured, increase our odds of actually getting the AGI created, which was what we cared about most.

OpenCog, as an open source software system, has two aspects. On the one hand, it's an AI toolkit, with a knowledge store and various algorithms that can be used for multiple purposes. My colleagues and I have used various OpenCog structures and algorithms in various practical applications for customers, some of which have had little to do with AGI. We've used them to analyze biological and market research data, to do automated scientific reasoning based on information extracted from biomedical research texts, and so forth. Other folks in various companies and government agencies have used OpenCog components for their own purposes. As the code is open source and free to use, they don't have to tell me or the other OpenCog creators about their work, unless they want to. A few times I've heard from someone, years after the fact, that they used this or that OpenCog component for this or that project.

On the other hand, in addition to being an AI toolkit, OpenCog is more centrally an attempt to create a powerful AGI – just like the Novamente AI Engine and the Webmind system that came before it. For a while I was calling this aspect of OpenCog "OpenCog Prime," using the term "CogPrime" for the AGI design itself, as distinct from the OpenCog implementation thereof. OpenCogPrime is the OpenCog-based implementation of the CogPrime AGI design. But this terminology never caught on, in spite of my using it in various publications.

More recently I've shifted to the term "PrimeAGI" for what used to be called OpenCogPrime – the specific AGI design I'm aiming to build, using the OpenCog platform. Time will tell if the term "PrimeAGI" gets any traction. The use of the term "OpenCog" to encompass both a software platform and a conceptual AGI design intended for implementation on that platform, doesn't seem to cause any terrible harm at present.

Just open-sourcing the code didn't really accomplish that much, although we did it carefully, with a lot of thought for both the software architecture (to make it flexible, extensible and so forth) and the legal licensing terms (to support both ongoing research and commercial work). But as time has progressed, the project has attracted more and more excellent programmers and researchers, some of whom have moved back and forth between volunteer OpenCog work and collaborating with me on OpenCog-based paid consulting projects. The openness of the source has also allowed the project to benefit from various sources of funding specifically oriented toward open source projects, such as Google Summer of Code (which pays interns a stipend to work on open source projects for the summer), and Hong Kong government research grants earmarked for work that will be made available to Hong Kong companies (a criterion satisfied by open source work, which is available for everyone).

My Journey to the East

Around the same time as we were wrapping up portions of the Novamente Cognition Engine as OpenCog, I started exploring professional possibilities in Asia. In 2007 I started worrying that the US economy was going to collapse, and I kept hearing it was going to be the Asian Century, so I started poking around Asia a bit. (I wasn't as prescient as my friend Toothpaste Sushi though – I remember in May 2008, he told me the US economy was going to collapse on September 25, 2008, based on some charts he was drawing. This was pretty close to accurate in terms of the financial crash, though he did somewhat exaggerate the level of calamity that was going to happen.)

Over the next couple years, I got a consulting contract with an e-learning company in Tokyo, and gave talks in Korea at Samsung and Seoul National University. The folks at Samsung's research center in Suwon loved my talk on the Singularity; and I was impressed that SNU had a new graduate school of "Convergence Studies," covering nano-bio-info-cogno. I also visited my friend Hugo de Garis, who in 2007 was teaching at Wuhan University in China (shortly after that he shifted to Xiamen University, where in 2009 we organized the First AGI Summer School together).

In 2009 I gave talks at a couple AI conferences in Hong Kong, where I met a cross-disciplinary researcher and entrepreneur named Gino Yu. Gino's original training was in computer science, but he had become increasingly interested in consciousness studies and the use of digital and other media to expand peoples' minds. At that time, he was planning the 2010 iteration of the global Toward a Science of Consciousness conference, which was going to be in Hong Kong for the first time. I offered to organize a Workshop on Machine Consciousness for the conference, modeled loosely on a machine consciousness workshop I'd attended at Nokia in Finland not too long before.

FIGURE 14.1: *My good friend Prof. Gino Yu, who has collaborated with me on a series of research projects, involving the carrying out of OpenCog R&D in his lab at Hong Kong Poly U. While his technical training is in computer science and he has a good grasp of AI, Gino's interests in recent years have shifted largely to the study of consciousness and the use of digital media to expand peoples' consciousness. The picture shows him at the Toward a Science of Consciousness conference in Tuscon in 2008. He likes to say "Right here, right now, it's good to be alive" – and he means it. While he supports my quest to create AGI, he also frequently questions the underlying motives of the quest for AGI. Why not just be content with life as it is, with the joy in the present moment, instead of constantly questing for something different or something more? Is the quest for AGI coming from the sheer joy of being, or is it coming from some sort of fear or desire? (Picture used with permission of Gino Yu).*

Gino suggested we apply for some AI research grants together in Hong Kong – they had, as he explained to me, these government grant programs where a company puts in 10% of the funds and the government puts in 90%, and the research gets done at a Hong Kong university. We put a proposal together, and it was funded! I had to scramble to pull together the 10% which I owed as the corporate part of the funding, but with some help from Jeffrey Epstein (a wealthy New York science benefactor who had helped me out a few times before since 2001) that was accomplished. All of a sudden we had funds to pull together a team in a lab at Hong Kong Poly U, working on OpenCog.

The situation wasn't ideal, because the money had plenty of strings attached – you couldn't hire expensive senior developers, the salary money was divided into research assistant-sized chunks, etc. So what we wound up with was a team of smart, energetic junior-level developers and researchers without as much leadership as we would have liked. But still, it was something. Combined with scattered volunteer developers and a few OpenCog-based commercial projects, these Hong Kong grants have really helped move OpenCog forward. I'm very grateful to Gino and to the Hong Kong government for this.

For those American readers wondering "why Hong Kong?" it's worth noting that in the same time frame I applied for a bunch of US government grants for OpenCog-based research and didn't get them. I did get some US government-based funding for computational neuroscience and machine learning data analysis work in the same timef-rame, so the issue wasn't that I was especially bad at writing US government grants. The US research funding scene is just much more conservative and competitive – US funding agencies generally know exactly what they want to fund, and tend not to be as open to weird new ideas. And they also by and large know who they want to fund – well-known researchers at name-brand universities, or top government contractor corporations. Obviously the US funds a load of great research in this way, but it's a difficult environment to break into with maverick ideas.

Getting these OpenCog grants was one of two factors of the fire in my belly to move to Hong Kong. The other was that, in 2009 at the AGI Summer School, I fell totally in love with a computational linguistics PhD student named Ruiting Lian. Ruiting was originally from the mountainous Anhui province, the daughter of two rural schoolteachers, and had relocated to Xiamen for graduate school. We had met a year before, when I first visited Hugo de Garis in Xiamen, and I liked her quick wit and sweet smile then, but at that point my second marriage (to Izabela Lyon Freire, a Brazilian AI programmer) was still going OK, so I just admired her in passing and moved along. In 2009, Izabela moved back to Brazil and I was living on my own in DC, with half-time custody of my three kids from my first marriage (which had ended in 2003). Ruiting and I enjoyed each other's company during the AGI Summer School and started an intense online correspondence (mostly on the Chinese chat service QQ); by 2010 we were deeply romantically involved.

The First AGI Summer School, at Xiamen University in Fujian Province, China. Since then there have been two other AGI Summer Schools, one in Reyjkavik and one in Beijing (I participated in those two but didn't organize them). The First AGI Summmer School was attended by 5-6 brave foreigners and a couple dozen local Chinese students (dwindling by the end, as some of the students found they couldn't understand the English in the lectures so well). The lectures were videotaped expertly by Raj Dye and found a fairly wide audience online among students and programmers interested in digging into the details of AGI. The above picture shows the agi-school.com video of my long-time OpenCog collaborator Nil Geisweiller lecturing about using OpenCog to control video-game agents. (Nil actually lectured on three topics there: probabilistic logical inference, automated program learning, and automated control of video game characters. He's a quiet guy but has a rather commanding knowledge of AI, mathematics, software and OpenCog.)

In late 2010 / early 2011 Ruiting came to the US for 4 months as an exchange student, and we spent a lot of time together then – she had never been out of China before, so it was great to show her bits and pieces of the US. But then we discovered that, due to the details of her exchange visa, after Ruiting returned to China in 2011 she would not be allowed to come to the US to live for at least two years. In principle she could come for brief visits on a tourist visa, but in practice that also seemed unlikely to be approved. If we wanted to be together before early 2013 – which we did – it would have to be somewhere other than the US. I seriously considered taking a job at Xiamen University alongside Hugo – I was already an adjunct professor there, and we were collaborating on research with OpenCog, Hugo's neural net system and Nao humanoid robots, as depicted in the film *Singularity or Bust*. We had a Chinese National Science Foundation grant for our robotics work, though it was proving difficult to use; the strings attached to that funding were especially complicated for two Westerners to deal with. And then Gino and I got our research funded in Hong Kong.

The Hong Kong research grant was a big boost for OpenCog, but it didn't pay me a salary; according to the logic of that particular funding, since I was the leading representative of the corporate sponsor of the grant, I couldn't also be a personal benefactor of the grant. Also, I didn't see these HK research grants as the ultimate be-all of OpenCog funding – they were a great start, but I still felt I needed to be actively exploring ways to generate a lot more funding somehow, to scale up OpenCog to a much larger level. At a 2009 Christmas party in Hong Kong – to which I was invited courtesy of Gino, the ultimate social networker – I met a Hong Kong-based American angel investor and entrepreneur named Doug Glen, who got excited when I told him about some prior consulting work I'd done applying narrow AI technology to stock market data. He suggested we start an AI-based hedge fund, based in Hong Kong. The concept sounded good to me. It took a while to become real, as these things generally do, but by fall 2011 we had secured funding from a handful of Asia-based angel investors to start a financial prediction company, Aidyia Limited.

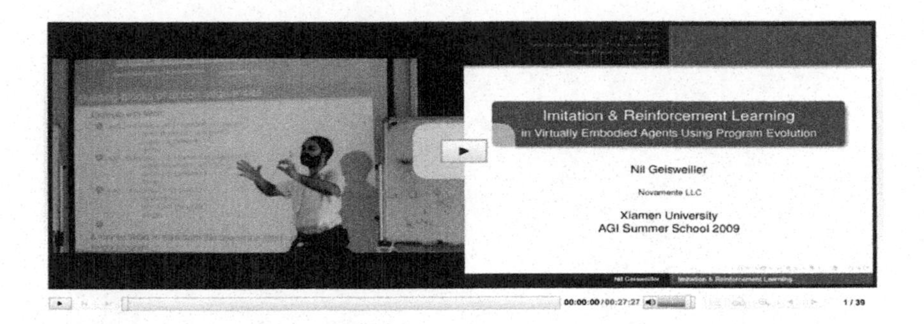

FIGURE 14.2: *The First AGI Summer School, at Xiamen University in Fujian Province, China. Since then there have been two other AGI Summer Schools, one in Reyjkavik and one in Beijing (I participated in those two but didn't organize them). The First AGI Summmer School was attended by 5-6 brave foreigners and a couple dozen local Chinese students (dwindling by the end, as some of the students found they couldn't understand the English in the lectures so well). The lectures were videotaped expertly by Raj Dye and found a fairly wide audience online among students and programmers interested in digging into the details of AGI. The above picture shows the agi-school.com video of my long-time OpenCog collaborator Nil Geisweiller lecturing about using OpenCog to control video-game agents. (Nil actually lectured on three topics there: probabilistic logical inference, automated program learning, and automated control of video game characters. He's a quiet guy but has a rather commanding knowledge of AI, mathematics, software and OpenCog.)*

Cassio and I – OK, especially Cassio – had acquired a fair bit of expertise in computational finance during the previous few years. Alongside diverse consulting projects in other domains, we had been involved with a project called Kuvera, which briefly provided AI-based trading signals for Peter Thiel's San Franciso-based fund Clarium Capital; with a project called StockMood, which analyzed online financial news to provide news-sentiment-based trading information to the general public via a Web portal; and with a small US-stock-picking fund called Cerrid Capital. None of these projects had been dramatically financially successful, though none had been terrible failures either. But these various projects had taught us a lot about the finance domain – and especially Cassio, as he had dealt with more of the financial-engineering aspects of these projects, while I had focused more on the underlying AI prediction algorithms.

StockMood was a textbook example of the role of luck and timing in business matters. We launched the website at the TechCrunch Disrupt conference in early September 2008, and won the Reuters Best News-Based Product. We figured we were well on the way to dramatic success. We were unsure whether selling subscriptions to the StockMood site, as we planned, would really be a great business model; but we figured if worst came to worst we could sell the technology to some big tech or financial informatics company, so it would end up as a feature within Google Finance or Yahoo Finance or Bloomberg or whatever.

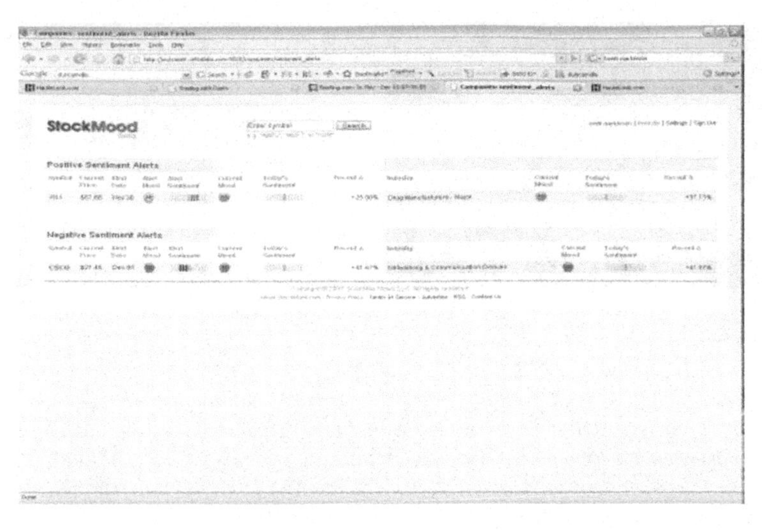

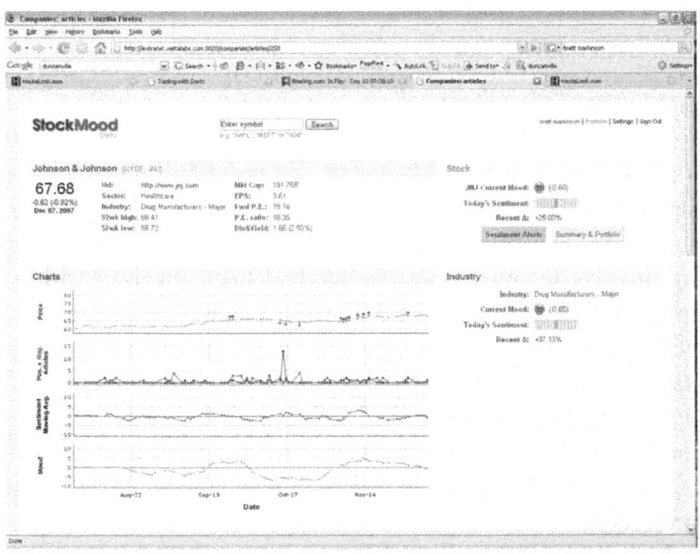

FIGURE 14.3: *The StockMood product that Cassio and I developed together with LA trader/entrepreneur Brett Markinson analyzed US financial news, tracked trends in news sentiment regarding specific companies, and used these trends to alert customers to cases where stock prices might be about to move in a way driven by news sentiment. In certain sectors of the US stock market there does seem to be significant causal impact of news sentiment on stock prices – and not just immediately after the news comes out, but persisting complexly for many days. Trading purely based on only news sentiment won't work, though (there are too many people trying to do that); instead one needs to look at combinations between news sentiment and other factors. StockMood had its own way of looking at such combinations. In our work at Aidyia we are taking a more complex approach and letting the AI figure out fairly freely how to combine news with other factors.*

StockMood was launched in a beta version in September 2008, around the same time as winning the Reuters Best News-Related Product Award at the TechCrunch 50 conference. Due to the financial crisis a few weeks later it never got beyond beta and the project was shut down. Timing is everything!

Then a couple weeks after StockMood launched its beta version, the US financial system crashed, and there was no worse project to be launching than a website giving news sentiment-based information about stock price movements. The news sentiment about stock prices was rather easy to predict at that point – it was almost uniformly bad. Investors were more concerned about moving their funds out of the stock market and into more stable investments than about using funky new sources to help them pick the winners on the market. Furthermore, StockMood's core investor wasn't feeling so flush with cash anymore, and didn't relish putting in more funds to keep the company going, given the circumstances. In hindsight this may have been shortsighted of him, but it's hard to remember now how severe the late-2008 financial crisis seemed at the time. People were talking about the end of an economic era, the start of a new Great Depression, and so forth. (Retrospectively, while the 2008 financial crisis was certainly a big deal, it ended up serving mainly as an occasion for transferring an even larger percentage of the economy to large players in the financial sector. But that would be a whole other topic. And anyway, one lesson that market-prediction work teaches very powerfully is how little hindsight is worth.)

Anyhow, the StockMood experience, and some other various adventures in unfortunate timing, hadn't soured Cassio and me on the financial analysis domain. After all, generally speaking, one fairly necessary, though not sufficient, condition for being an entrepreneur (as well as, one supposes, for making AGI) is a pigheadedly stubborn persistence – a quality both Cassio and I possess in droves. Having come so close to succeeding in the financial domain made us more eager to finally get it right....

But we decided it would likely be best to stick with the idea of making a trading system, rather than a financial information service. Trading was about the most direct way we could think of to turn AI algorithms into money. You feed money into the algorithm, and it turns it into more money. Of course, even the best algorithm is only going to work a little better than chance, so there's an irreducible element of gambling in the business. A six month run of bad luck, even if it's not statistically significant, can potentially kill your fund (unless you've built up a sufficiently long and impressive track record of success prior to said run of bad luck). But still, the better your algorithms are, the more you can bias the odds in your favor. Anyway, Cassio and I were both very psyched to have the opportunity to co-found Aidyia, and we both came into the project with loads of ideas about how to make AI-based financial prediction work well.

In Fall 2011 I moved to Hong Kong, to pursue a combination of endeavors, all centered synergetically around the creation of more and more advanced AI. Cassio and his wife Patricia relocated to Hong Kong less than a year after I did; and we work together in Hong Kong today (though we don't hang out as much as we should after work, since we live in fairly separate areas – him and Patricia in the heart of the city and me and Ruiting in a remote village by the mountains and bay).

In a sense, from 2011 to mid 2015, my "day job" in Hong Kong was working at Aidyia with Cassio and our colleagues there, figuring out how to apply OpenCog and various proprietary AI technologies to predict financial markets, and my part-time evenings-and-weekends job was working on OpenCog AGI, and helping Gino supervise the OpenCog Lab at Hong Kong Poly U. And along the way I was giving guidance to colleagues who are using OpenCog for genomics data analysis and other applications – and working now and then on AGI theory, and writing books like this one.

But thinking about these pursuits of mine in terms of "day job" versus "extra stuff" doesn't really make sense, because at a deeper level it's all the same ideas, all the same conceptual core... and even the detailed technologies are closely related and in some cases concretely overlapping. Aidyia is using customized versions of OpenCog and other related technologies. Most of the good ideas I've come up with for Aidyia have originated in non-finance-oriented AI work (e.g., one of Aidyia's better proprietary predictive algorithms is something I originally conceived when playing with EEG data – a quite different kind of time series). Conversely, the Aidyia work has caused me to think deeply about issues like causality and Occam's Razor, and while the Aidyia code we've created to deal with these issues is proprietary, the more general thinking that the Aidyia work has spurred in my mind regarding these issues has filtered into OpenCog in various indirect ways.

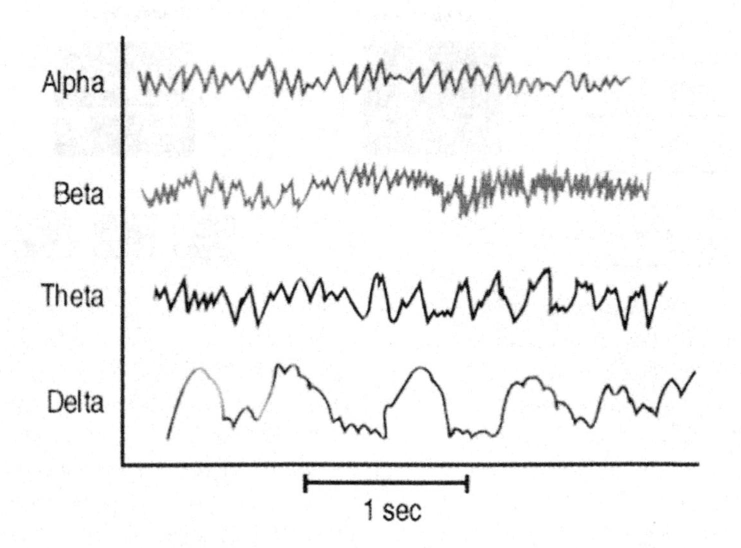

FIGURE 14.4: *EEG brain-wave time series and financial time series are quite different in detail, yet some of the same AI methods apply to both of them. Concepts I originated while playing with EEG data, suitably customized and modified, served as the basis of one of Aidyia's more powerful proprietary market prediction algorithms. And lessons learned while playing with these proprietary, EEG-originated algorithms on financial data have percolated back into OpenCog in indirect ways.*

Combined with starting a new marriage with Ruiting and enjoying the various recreational opportunities Hong Kong offers (from some tiny but vibrant jazz and electronica clubs to endless dramatic mountain hiking trails), it's been an exhausting but exhilarating lifestyle. So far it hasn't gotten me quite where I want to be regarding AGI, but it's definitely been progress.

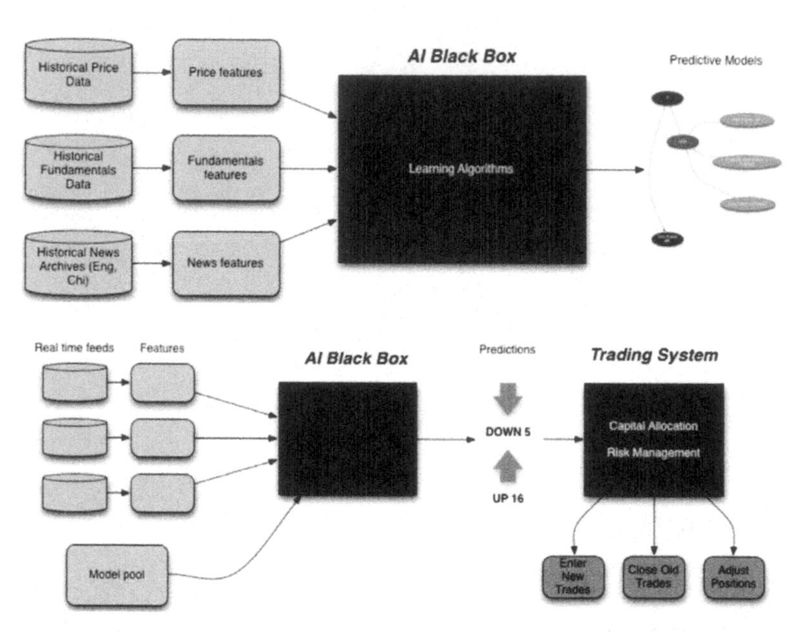

FIGURE 14.5: *Diagrams illustrating the high-level architecture of our AI-based trading platform. The first diagram illustrates how our system takes in price, volume, multi-language news, economic and company accounting, and other information regarding a variety of financial instruments over a historical period, and uses this to learn predictive rules ("models") that WOULD have reasonably accurately predicted the movements of various financial instruments in the past. (Of course there are major challenges here, including overcoming the statistical curse of "overfitting"; our team at Aidyia has put a lot of time and energy into R&D aimed at overcoming these challenges.) The second diagram shows how the predictions from the AI system are then fed into a trading system which uses specially designed capital allocation and risk management algorithms to issue trading signals to appropriate exchanges, all in a fully automated way.*

Aidyia launched its first fund in early 2016; obviously if it succeeds, in a number of years it will generate sufficient revenue for Cassio and me personally that we'll be able to amply fund OpenCog on our own. And if fortune frowns on Aidyia (a possibility I don't like to think about right now –

but there's an irreducible element of uncertainty in such ventures), working on it was still an intense and intriguing learning experience.

Once we got our first trading system packaged up and close to ready for initial trading experiments, in mid-2015, I moved on from full-time engagement with Aidyia and started putting more time into OpenCog – and into helping my good friend David Hanson with his robotics company. David's humanoid robot faces and heads are the most realistic in the world, and he had long had a passion for making his robots even smarter than they are beautiful. Being a fan of my patternist philosophy and of open-source development, David was willing to take a leap of faith and put some of his company's (at that stage, not too plentiful) resources into customizing OpenCog for the task of controlling his robots. I took the title of Chief Scientist of Hanson Robotics.

Since I started working closely with David, I've learned more about the hardware (and squishy-ware, i.e. the unique flexible face materials) and non-AI software involved in his robots. But my focus has been mainly on the AI side. I see David's robots as a great user interface for AGI systems at various stages of development, and also as an amazing showcase for AGIs as they gain more and more functionality. And working with David and the team we've brought together at Hanson robotics has been a huge amount of fun.

The OpenCog Architecture

So what is OpenCog anyway?

Viewed from a software perspective: At the top level, OpenCog has a few key components:

- A virtual storehouse called the ***Atomspace*** – where Atoms, knowledge building blocks, are found – which is useful for storing various kinds of knowledge.
- A framework facilitating interaction between different cognitive processes (called ***MindAgents***) and the Atomspace; this involves adding, removing, and/or changing knowledge items within the Atomspace.

Code based on the Atomspace and MindAgents enabling a system to control an agent in an external environment – either a virtual world (like a game engine; we've worked extensively withUnity3D) or a robot (where we access robots through the ROS, Robot Operating System, interface developed by Willow Garage).

What distinguishes the Atomspace from most of the common ways of storing knowledge in AI systems or other software systems is its capacity to efficiently and sensibly represent essentially ALL sorts of knowledge. PrimeAGI – the AGI design I've created to run atop the OpenCog platform -- specifies a collection of MindAgents carrying out specific cognitive processes; and the Atomspace, by design, adapts to the MindAgents' knowledge needs.

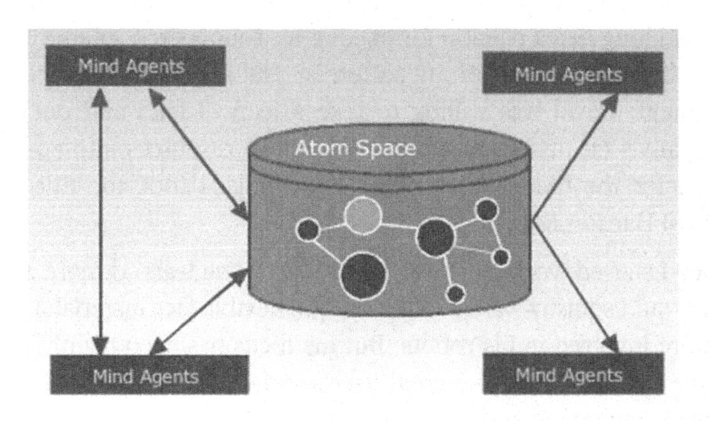

FIGURE 14.6: *A view into one key aspect of the OpenCog software architecture. The multiple MindAgents wrapping up OpenCog's cognitive processe, all interact with the common AtomSpace knowledge store, and sometimes interact directly with each other as well.*

The "secret sauce" behind PrimeAGI is how the MindAgents work together as an overall coherent system. Incorporating a variety of cognitive and AGI theories, they have been designed to work together, giving rise to the full scope of cognitive processes described in the review of cognitive science I gave earlier. However, the flexibility of the underlying OpenCog framework is also important, making it easy for developers to play around with variations of MindAgents, and to run experiments to see what works best. Since most proto-AGI systems have been written by graduate students in order to prove specific theoretical points, they're not very flexible. OpenCog is designed to support a huge variety of possible AI and AGI designs, some extremely different from PrimeAGI; this has given us the freedom to experiment fairly freely with various ways of turning the conceptual and mathematical PrimeAGI design into software.

OpenCog's AtomSpace

The centerpiece of OpenCog is the AtomSpace. Think of it as the nucleus of an OpenCog system. Within the AtomSpace, all the different kinds of knowledge the human mind needs are stored with plenty of room for flexibility. Represented mathematically, it is a "hypergraph," which looks like a very complex diagram of all sorts of nodes cross-connected by all sorts of links.

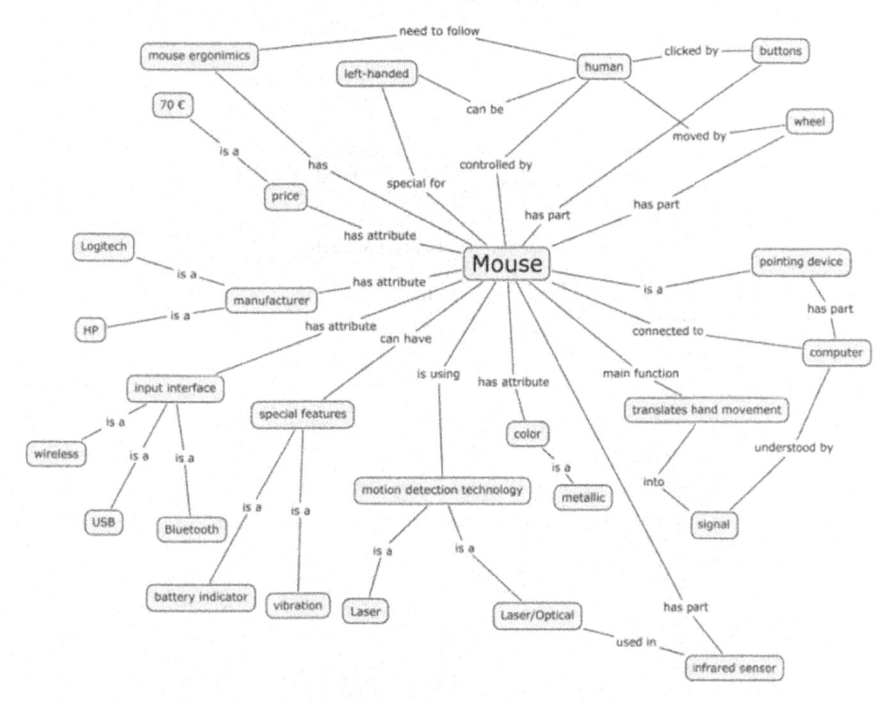

FIGURE 14.7.: *Example of a typical "semantic network" knowledge representation for an AI system. Unlike OpenCog's hypergraph representation scheme, this is only a graph, not a hypergraph, and it has a more restrictive semantics. For instance, OpenCog's nodes can sometimes represent little programs that do things. Also, OpenCog's nodes and links typically come along with numerical values indicating the importance of the node or link to the system, and/or the probability associated with the node or link.*

More specifically, a mathematical "graph" consists of a bunch of nodes with links drawn between them, and maybe labels on the nodes or links. A "hypergraph" can have links that span more than two nodes. OpenCog's hypergraphs can also have links that point to links as well as nodes.

Representing knowledge using graphs (with nodes and links) is not an original idea – FIGURE 14.7 and FIGURE 14.8 give examples of how this has been done in other contexts before OpenCog. OpenCog's hypergraph is a special kind of graph, which is used in special ways to support OpenCog's particular AI methods. The abstraction of knowledge as graphs and hypergraphs is commonplace in computer science and cognitive science now, but in the big picture it represents an advance in cognitive modeling – it wasn't natural to people, say, 100 or 200 years ago.

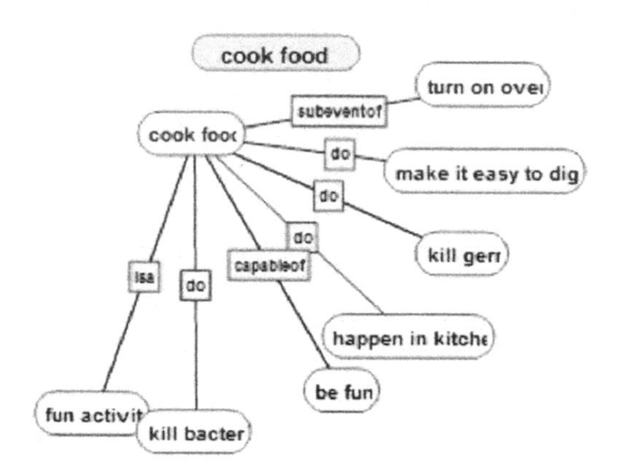

FIGURE 14.8: *A valuable AI resource for certain kinds of work (e.g., aspects of natural language processing) is ConceptNet, a large semantic network created mainly at MIT and provided free for download. The image shows a sample of a few nodes and links from ConceptNet. ConceptNet is a standard semantic network, not a more richly structured hypergraph like one has in OpenCog. (http://web.media.mit.edu/~hugo/conceptnet/).*

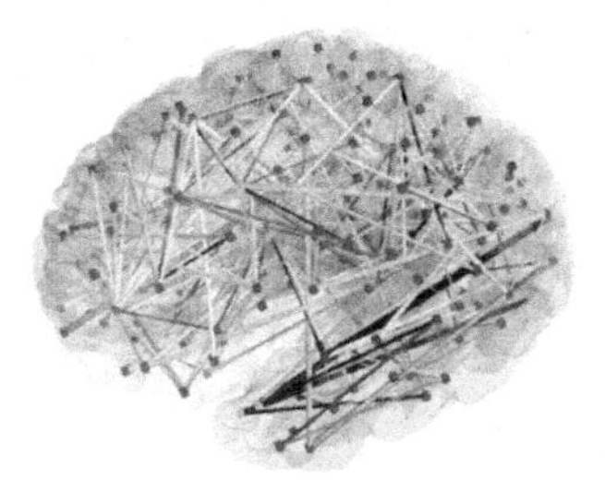

FIGURE 14.9: *The brain is commonly thought of these days as a "Connectome" – a graph whose nodes are neurons (the most important kind of brain cell) and whose links are synapses (connections between neurons). This is a vast oversimplification, as I'll discuss in a later chapter. But still, it's a useful conceptual model. Among other things, the connectomic view of the brain provides a way of relating brain structure and dynamics with more abstract AGI architectures like semantic networks and OpenCog. (http://cmtk.org/viewer/).*

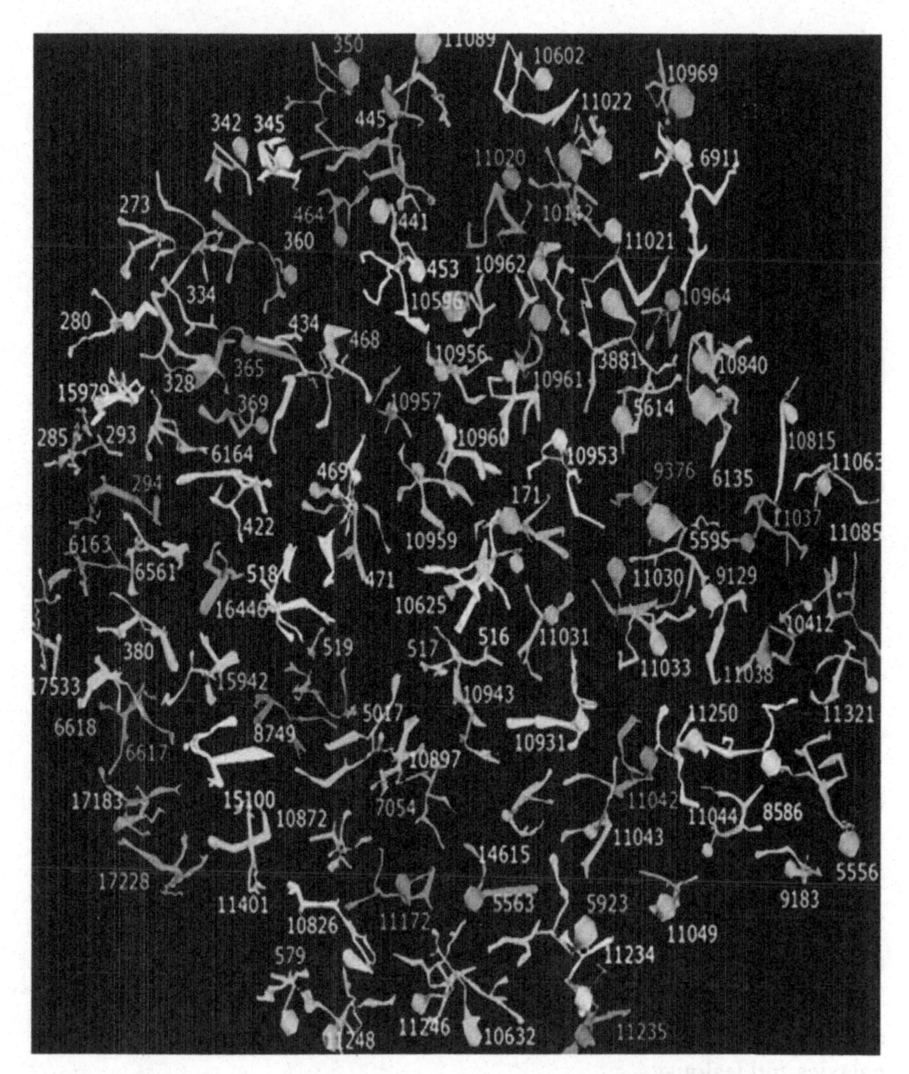

FIGURE 14.10: *This is an actual image of the connectome in the retina (the back of the eye). You can see here that the different neurons depicted don't actually connect directly to each other, but they send electrical charges to each other across the gaps between them, using chemicals called neurotransmitters to mediate the transmission. This illustrates the way in which graph-based modeling is useful but approximate, in a neuroscience context. In OpenCog and other graph or hypergraph-based AI systems on the other hand, the graph/hypergraph really IS the system's knowledge representation. (http://prometheus.med.utah.edu/~marclab/connectome/images/rod-bc-library.jpg).*

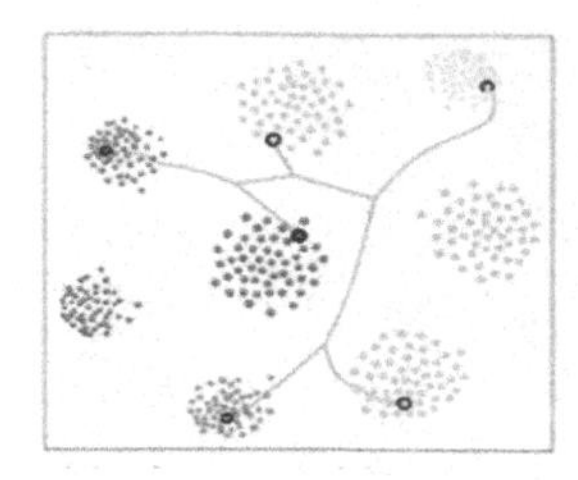
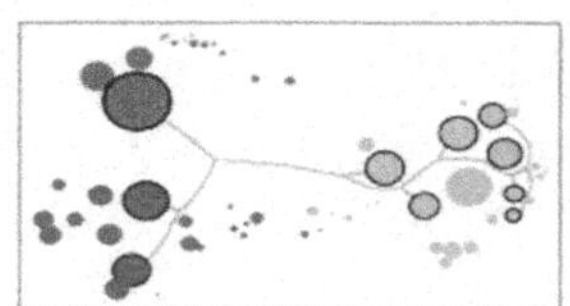
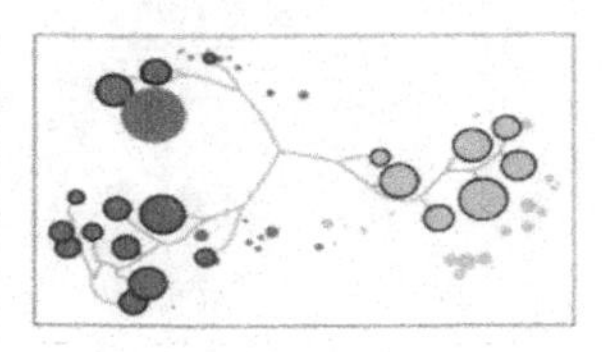

FIGURE 14.11: *Some hypergraphs that are definitely not graphs – these have links that span multiple nodes. OpenCog's Atomspace also has links that span multiple nodes. (http://www-docs.tu-cottbus.de/informatik/public/abschlussarbeiten/martin_junghans_da_2008.pdf).*

Several figures in this chapter give simple illustrations of parts of an OpenCog Atomspace. Key features of the diagrams are:

- ***Nodes***: Circles, representing concepts, objects, numbers, or mathematical functions
- ***Links***: Arrows connecting nodes, representing different kinds of relationships between nodes.
- ***Maps***: Clusters of nodes and links forming knowledge building blocks.

The term Atom is used as a grab-bag including both Nodes and Links. Nodes and links are the atomic elements of OpenCog's knowledge representation. They come in many different types.

For example, the node representing "cat" and the node representing "animal" would be joined by an InheritanceLink, representing the fact that a cat is a special kind of animal (cat "inherits from" animal, in logic lingo). The cat node and the dog node would have a SimilarityLink between them, since cats and dogs are fairly similar, compared to other things like toasters, galaxies and jealousy.

A ContextLink would point from the Atom for "vacation" to a SimilarityLink, where the latter would join a node for Jamaica and a node for Santorini (and this SimilarityLink would presumably have a mediocre "strength"). In the context of vacations, Santorini and Jamaica are moderately similar; in the context of government, there's very little similarity.

Nodes generically associated with one another (say, power and corruption) will usually have a HebbianLink, named after Donald Hebb, a scientist who introduced associative relationships between nodes and networks in the late 1940's.

In these examples, I've referred to nodes that correspond directly to English-language concepts, but actually only a tiny minority of nodes in the AtomSpace are like this. Most don't correspond to English concepts, but instead to fragments of concepts, or learned groupings of concepts. English-language concepts (or concepts corresponding to words in any other language) are made by grouping various nodes, or activating a bunch of them together.

Some nodes refer to perceptions coming in from sensors —representing, say, the corner of a picture on the wall in front of an OpenCog-powered robot. Or, actions corresponding to an OpenCog agent, like saying something, or moving a certain motor in a robot body. And there are also abstract nodes for logical operations, like AND, OR, and FOR ALL.

There are a few dozen different types of nodes and links in the AtomSpace, selectively chosen to encapsulate everyday human concepts in fairly small combinations. This seems a sensible model for representing everyday human knowledge in AGI. The idea of boiling down human knowledge to a small number of semantic primitives is an old one in philosophy, though philosophers who like this idea tend not to agree on what the primitives are.

Mathematically, one can reduce the number of primitives needed to represent everything quite dramatically, if one really wants to. Combinatory logic, for example, reduces all mathematically possible forms and ideas to a single mathematical operator, plus parentheses. But this isn't a particularly convenient way to do things inside an AGI system. Moshe Looks and I tried something like this in 2005, and others have tried as well. OpenCog's few dozen node and link types represents a pragmatic balance between the desire to use as few primitive notions as possible, and the desire to make the representation of everyday human mind-stuff fairly simple and elegant.

Giving a complete list of all the Atom types in the system would just be confusing in a nontechnical book like this, and furthermore the list is changing all the time. But the following list gives the basic idea. We have, for example:

- ***ConceptNode***, which might be better named "concept or part of concept node" – these are just generic nodes connected via links, which may be assembled in various ways to form parts of concepts.
- ***SpecificEntityNode***, representing a specific entity in the (physical, virtual or conceptual) world.
- ***Various node types*** referring to concrete or abstract entities of specific types: WordNode, CharacterNode, SentenceNode,

NumberNode, BlockNode (for a virtual world containing many blocks, like the Minecraft-like one we're experimenting with), VisualPatternNode, etc.

- ***Various Atom types*** representing logical relationships, e.g.: InheritanceLink, SimilarityLink, PredicateNode, EvaluationLink, ImplicationLink, EquivalenceLink, etc.
- ***SchemaNode***, representing a procedure; ExecutionLink representing the enaction of a procedure.
- ***HebbianLink***, representing a simple association between two entities.

Each of these Atom types has a particular story and theory behind it. The precise number of Atom types to use is a matter of art as much as science. Having hundreds would make the system an unmanageable mess; there would be too many different theories to take account of and intersect with each other. Having only one or two Atom types is mathematically possible, but would make things conceptually awkward, because it would require using complex, opaque constructions to distinguish intuitively separate things. We have found that using a few dozen Atom types strikes a reasonable balance between intuitiveness and manageability. Commonsensically simple concepts, processes and relationships tend to have fairly compact representations in terms of the few dozen OpenCog Atom types we've adopted. But the precise list of Atom types is continually evolving as the system gets developed by the OpenCog community. It is also possible for the system to dynamically update its own list of Atom types as it learns, but at the moment it is not as clever at doing so as its human programmers!

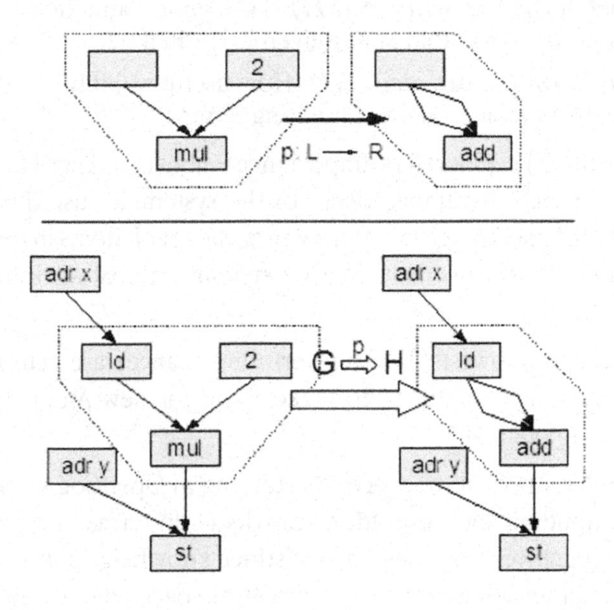

FIGURE 14.12: *Procedures, as well as semantic knowledge, can be represented as graphs. The graphs on the left represent procedures for carrying out basic arithmetic operations. Inside some programming language interpreters are "graph rewriting" operations for translating graphs into other graphs, as part of the program execution process, or for purposes of program optimization. The Picture shows a small graph representing a program for multiplication of x by 2, getting transformed into a graph representing a program for adding x to itself. (http://upload.wikimedia.org/wikipedia/commons/thumb/4/44/GraphRewriteExample.PNG/220px-GraphRewriteExample.PNG).*

Truth and Attention Values

As well as having different types, the nodes and links in the AtomSpace have different numbers attached to them. The numbers fall into two categories — Truth Values and Attention Values.

The most basic kind of Truth Value attached to an OpenCog Atom has two components:

- ***Strength***: How common is the concept represented by the Atom? Or, how strongly held is the statement represented by the Atom?
- ***Confidence***: How surely is the truth value of the Atom known to the system?

The most basic kind of Attention Value attached to an OpenCog Atom also has two components:

- ***Short Term Importance (STI)***: How much attention should the system pay to an Atom at any given point in time?
- ***Long Term Importance (LTI)***: How useful will it be for the system to keep an Atom in RAM in the long term?

Atoms with STI (ShortTermImportance) above a certain threshold become part of the "attentional focus" of the system. To use the language introduced by the psychologist Bernard Baars, the set of Atoms in the system's attentional focus represents the OpenCog system's "global workspace." (See FIGURE 14.13.)

Atoms with the lowest LTI (LongTermImportance) are removed from RAM by a Forgetting MindAgent, to make room for new Atoms that might have greater LTI potential.

All the different cognitive processes running in OpenCog work with the AtomSpace, inputting and outputting knowledge in terms of its nodes and links. Some cognitive processes have distinct knowledge representations, but since they all speak the language of the AtomSpace, they are able to work together as a whole.

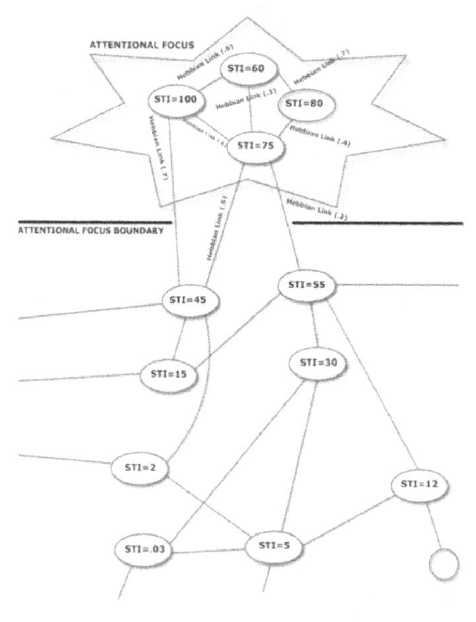

FIGURE 14.13: *The Attentional Focus in OpenCog. Every OpenCog Atom gets a Short Term Importance (STI) value, and the ones with STI above a certain threshold are considered to be in the Attentional Focus, which means MindAgents (cognitive processes) will pay them special attention. The Attentional Focus is an OpenCog implementation of Bernard Baars's cognitive science theory of the Global Workspace.*

Mixing Neural and Symbolic

In the lingo of contemporary AI, OpenCog's Atomspace may be considered a "neural-symbolic" system – which, unsurprisingly, means it has both neural network-like and symbolic logic-like aspects.

- ***Neural***: Some important neural network-like aspects: STI and LTI values of Atoms spread between Atoms, kind of like electricity or activations spreading between neurons in the brain. Furthermore, the flow of these values creates new links and changes the weights of existing ones (HebbianLinks form and adapt based on the flow of STI through the system: If two Atoms simultaneously have high STI values, a strong HebbianLink usually forms between them).
- ***Symbolic***: The Atomspace directly comprises a logic-based AI system representing relationships through node and link types with logical semantics (InheritanceLink, SimilarityLink, ANDLink, and others mentioned earlier).

Other neural-symbolic systems tend to divide the neural and symbolic aspects. Some have a separate neural module and a separate symbolic module that communicate together from a distance; some are neural-centric and have a symbolic module that recognizes patterns in the neural module.

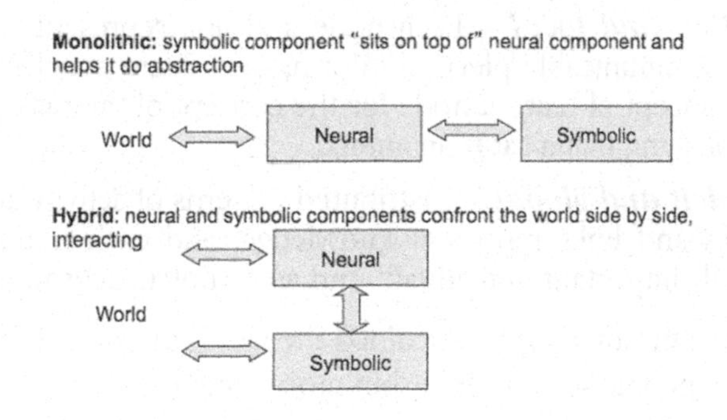

FIGURE 14.14: *OpenCog may be considered "neural-symbolic," but differs from most neural-symbolic systems in that the neural and symbolic aspects are very tightly integrated. The Atomspace has both neural and symbolic aspects. This is different from architectures in which neural and symbolic components, architected separately, are networked together. On the other hand, DeSTIN is more closely "neural" in inspiration (though it's not actually a "neural network" design either).*

In the AtomSpace, the neural and symbolic aspects are working with the same nodes and links, yet interpreting them differently. This approach traces back to the Webmind AI system, and is one of the key common threads running through my AI work since the mid-1990s.

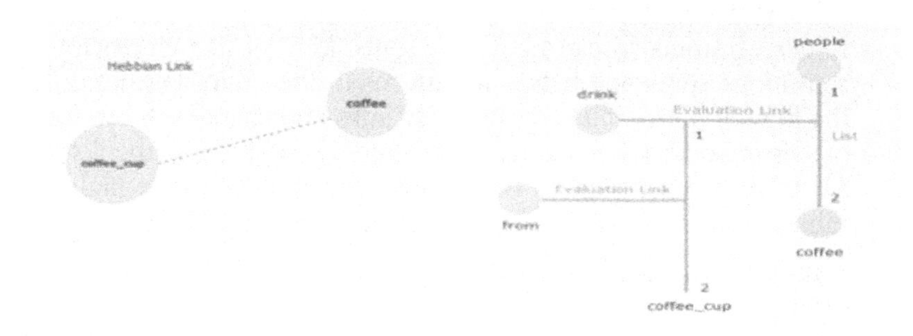

FIGURE 14.15: *OpenCog's Atomspace (weighted, labeled hypergraph) knowledge representation bridges the gap between subsymbolic (neural net) and symbolic (logic/semantic net) representations, achieving the advantages of both, and synergies resulting from their combination. It is able to bridge the gap between abstract concepts and highly specific percepts and actions.*

As an integrated neural-symbolic system, the AtomSpace represents knowledge in two important ways:

- ***Explicit and local*** – Each node and link represents a definitive and communicable piece of information. There may be a node for the concept of "cat," a node for the concept of "animal," and a link establishing that a cat is an animal.
- ***Implicit and global*** – Distributed patterns of activity across many nodes and links represent knowledge, and each node or link is equally important individually and as part of an activation pattern.

To describe this dual aspect, I coined the word "glocal" (global + local) – a term some people seem to love, but more seem to hate!

Many OpenCog nodes have no labels; they don't correspond to any particular English word or specific, communicable concept.

FIGURE 14.15: *Examples of OpenCog Atoms (nodes and links). Note that most Atoms in an OpenCog Atomspace, in real life, won't correspond to any particular words in English or other natural languages. This is just a convenient way to make examples to show people. The linkage to the left is neural-net-like – it just indicates a simple association between two concepts. The linkage structure to the right is more symbolic logic-like – it represents a precise relationship as would be expressed in the formalization of reasoning called "predicate logic" (specifically, it represents the relationship that people drink coffee from coffee cups).*

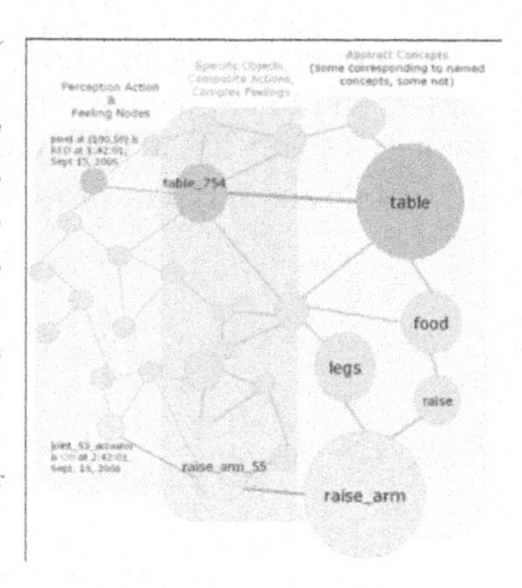

The mixture of explicit and implicit representation makes it easy for MindAgents to work together on the same Atomspace, even if some of the MindAgents use explicit localized representations, and others use implicit global representations.

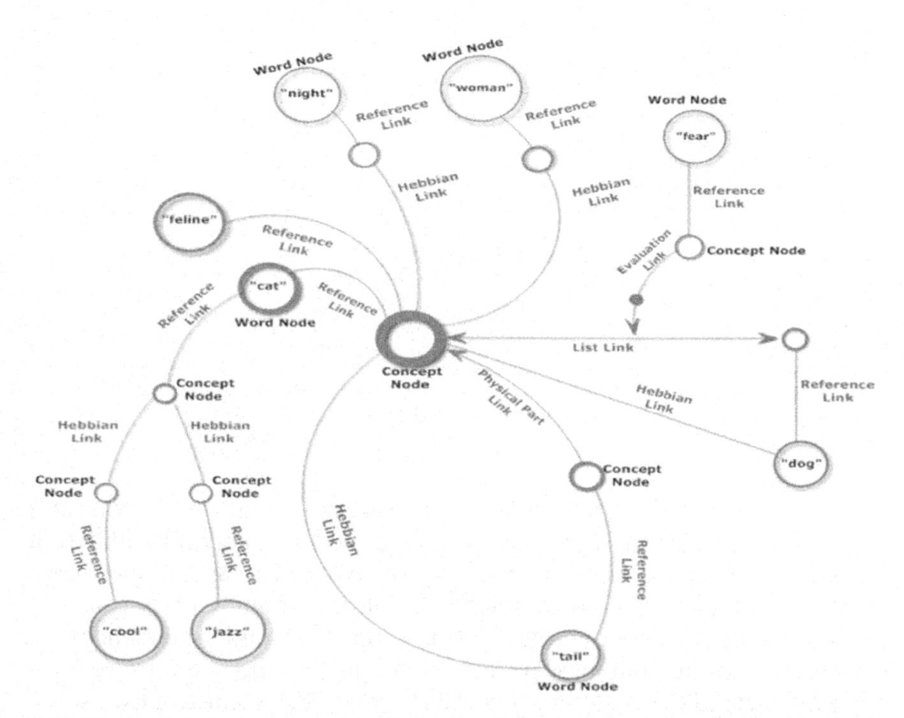

FIGURE 14.16: *Links Representing Explicit Knowledge in OpenCog. For example, the nodes representing "cat" and "feline" are two references to the same concept, thus they each have their own ReferenceLink to the same concept node.*

The AtomSpace has been designed to accommodate multiple cognitive processes and to facilitate synergy between them. But of course, this is no guarantee. The key lies in the algorithms that we create for each MindAgent, a fairly subtle and complex process.

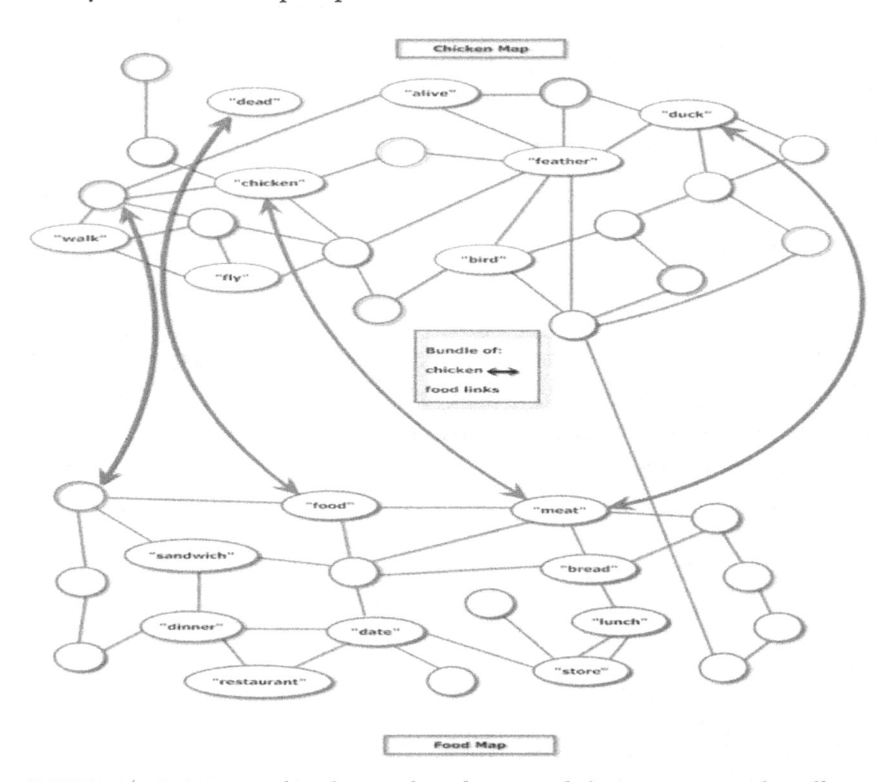

FIGURE 14.17: *Mixture of Explicit and Implicit Knowledge in OpenCog. The collection ("map" in OpenCog lingo) of nodes at the top relates to "chicken," though some of the nodes don't correspond to any particular English words or common concepts, and are important only in the context of the network activity patterns they form. The collection of nodes at the bottom is related to "food." The bundle of links from the "chicken" collection to the "food" collection represents the global and implicit relationships between "chicken" and "food."*

The human brain appears to achieve a similar synergy, but how it does this remains largely a mystery. Like OpenCog, the human brain has different cognitive processes associated with different types of long and short-term memory; and it has evolved so that these different cognitive processes can synergize with each other effectively. Yet no one truly understands how the brain carries out these things. So instead of imitating the brain, we've used computer science to create a set of MindAgents that seem to make sense algorithmically, and that, according to our understanding, are likely to give rise to the various functions and interactions needed to achieve human-level, human-like general intelligence.

A Big Scary Diagram

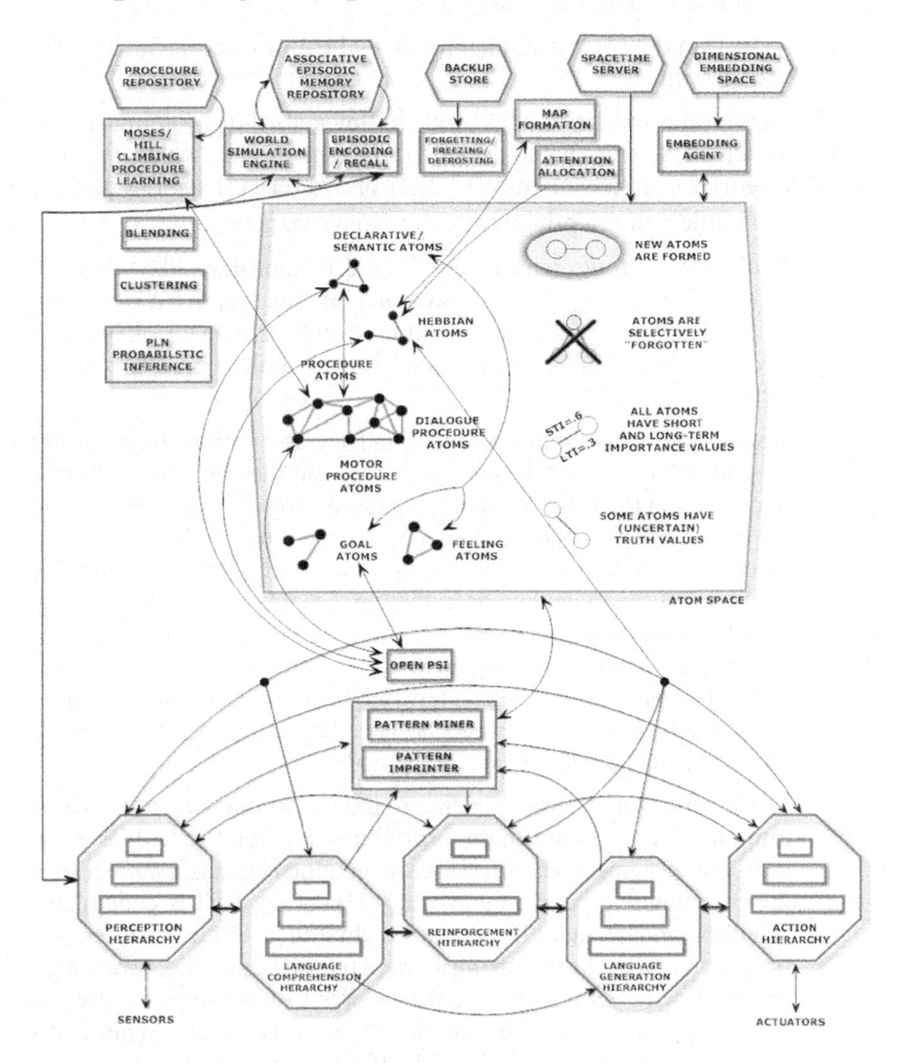

FIGURE 14.18: *Key OpenCog AI Software Processes*

Now I want to direct your attention to FIGURE 14.18– it's a big scary diagram and I apologize for that, but it's my best attempt to summarize the important things happening inside OpenCog in a single picture. My collaborator George Papadakis helped me draw it, and he reports that while he was doing so, his girlfriend spent some time staring over his shoulder in utter perplexity.

I'll definitely forgive you if you're thinking something like: *"Whoa! There's a lot going on in this 'big scary diagram'!"* ...or maybe something less polite...

There IS a lot going on – but when you break it down into pieces, one by one, you'll find it all makes sense.

Let's start with the big box in the middle — the Atomspace, which you're already familiar with.

The dots in the box are nodes, and the lines are links joining nodes. Finally, the clusters of nodes and links represent Atoms, which fall into several categories. I already mentioned these above very briefly, but now it's time to give a little more depth. Among the Atoms we have:

- ***Declarative Atoms*** representing logical relationships like similarity, logical inheritance (in logic we say that A inherits from B if A is a special case of B, so in OpenCog we might have an Inheritance Link between a "cat" node and an "animal" node), logical implication links, and so forth.
- ***Associative, or Hebbian Atoms***, signifying associations (named after Donald Hebb, who was the first guy to write substantively about the role of neurons in the brain in identifying associations between things, and the critical role this plays in human thought). If two nodes often occur together (say, one node representing a "boy" and another representing "trouble") at the same time or in the same context, a HebbianLink will form between them. HebbianLinks also form between nodes that have no obvious semantic meaning on their own, yet are extensions of other concepts. Finally, it only takes a few HebbianLinks in a cluster of nodes to form several coherent concepts, since importance spreading along the HebbianLinks will activate all the nodes.
- ***Procedure Atoms***, corresponding to little procedures that the system can carry out. Usually these procedures are represented as short computer programs, in a special programming language (simpler than the programming languages in which OpenCog itself is coded) that OpenCog understands. These may be physical action procedures, like "step forward" or "keep walking forward till you bump into something," or they may be more cognitive procedures like the steps involved in answering a certain kind of question (this would be an example of a "dialogue procedure," as FIGURE 14.19 suggests).
- ***Goal Atoms***, indicating goals that the system is trying to achieve, generally because it's decided (usually by reasoning with declarative knowledge, but occasionally by association alone) that these goals are ways of achieving one or more of its core drives.
- ***Feeling Atoms***, representing the system's evaluation of its internal state over time – for instance, has the system gotten a lot of satisfaction lately, is it feeling safe or threatened, has it gotten a lot of new information lately or has its curiosity gone unsatisfied? Generally these pair with

Goal Atoms – system goals are implemented in terms of evaluation of system feelings.

The right side of the Atomspace box illustrates some of the general processes involving Atoms in the Atomspace, most importantly:

- ***New Atoms are created***, to help represent new knowledge, either based on new perceptions the system has received, or based on new conclusions it's drawn, or new speculative concepts it's cooked up.
- ***Atoms are forgotten*** – i.e., deleted from RAM, if they are judged sufficiently irrelevant, which basically means if their associated Long Term Importance (LTI) values get too low.

The square boxes above and to the left of the Atomspace box correspond to cognitive processes that act on the Atomspace. At the software level, each of these cognitive processes is generally implemented via several MindAgent software objects.

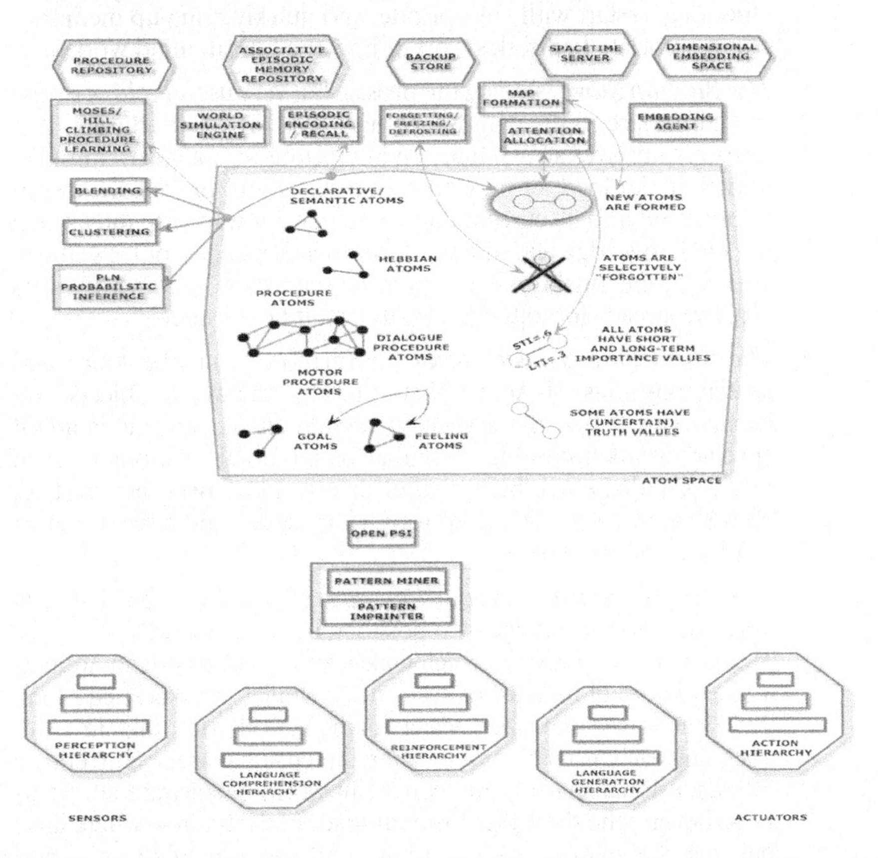

FIGURE 14.19: *Key Cognitive Processes Participating in Atom Creation, Deletion and Modification in OpenCog*

Many cognitive processes are simultaneously involved with creating new Atoms. The "attention allocation" process does a number of things, one of which is to update the Short and Long Term Importance values that regulate how much attention Atoms get. The Forgetting process deals with (surprise surprise) the forgetting of irrelevant Atoms.

The OpenCog cognitive processes (MindAgents) work with various knowledge stores, represented by the hexagonal boxes above the Atomspace:

- ***Procedure Repository***, which stores the mini-programs corresponding to Procedure Atoms. If a MindAgent wants to execute the procedure corresponding to a Procedure Atom, it grabs this from the Procedure Repository.
- ***Associative Episodic Memory Repository***, which stores Atoms relating to the life history of a system (running on OpenCog) in a different way from the Atomspace: a quick and easy way for OpenCog to start with any episode, and quickly bring up memories of any associated episodes (sort of how the human mind works).
- ***The Backup Store***, serving the prosaic but very useful role of saving the Atomspace to the computer's hard disk drive periodically. This functionality gives an AGI mind an interesting advantage over human minds: It can load its previous mind-states and previously known information into its current mind, whenever it wants to. Integrating the old knowledge with the new presents challenges, but it's still an amazing functionality, one I could definitely use – being 45 years old, I've already forgotten a lot of the stuff I once knew!
- ***The Spacetime Server*** stores information about the times and spatial positions of Atoms that refer to real-world objects and events. This allows the system to rapidly answer questions about specific events according to spatial or temporal relations to each other (simultaneous, before, after, overlapping, near, far, next to, above, below, etc.). A useful tool, since space and time are such fundamental concepts.
- The ***"Dimensional embedding space"*** works a bit like the associative episodic memory repository, but more broadly – it stores Atoms in a different way, which makes it very fast to use an Atom as a query and pull up a list of every other similar or associated Atom. This feat is accomplished via a mathematical trick that involves assigning each Atom a point in some dimensional space (currently a 50-dimensional space). The wonderful world of software allows us to go beyond the three dimensional space of the brain — where links between the neuron and the brain tend to get tangled up — and potentially have as many dimensions as we want!

Interfacing Mind and World

The boxes at the bottom of the Big Scary Diagram represent hierarchically-structured modules, which process visual and auditory data for controlling actuators (like the servomotors in a robot, or the animations controlling a character in a game world). They're particularly useful when dealing with the intricacies of linguistic data and reward signals from the environment (somewhat complex in a robot's case – each part of its body may deliver the mind its own reward signals based on how comfortably it's been operating).

It's sort of like an input/output layer for the OpenCog mind.

The box with labels "Pattern Miner" and "Pattern Imprinter" serves the critical role of translating the languages of the perception and action hierarchies to the Atomspace and vice versa.

Pattern mining involves recognizing patterns in the states of these perception and action hierarchies, and then recording these patterns as Atoms in the AtomSpace. Pattern imprinting takes abstract relationships existing in the Atomspace, and transforms them to guide the perception and action hierarchies.

For instance, if the vision hierarchy recognizes a cat, the "catness" is initially represented by a distinct pattern corresponding to the organization of the processing units in the visual perception hierarchy. The Pattern Miner correlates each frequent pattern in the visual perception hierarchy to some OpenCog Atom (in this case associating the "catness" visual organization pattern to a certain Atom).

The association of the "catness" Atom with the word "cat" is then another problem for OpenCog to solve. If it often sees this same visual pattern at the same time as it hears the word "cat," it will recognize this correlation, and learn the visual associations of the word.

And once the OpenCog system knows the association between the word "cat" and certain visual patterns, after it hears the word "cat," it can use the Pattern Imprinter to instruct the visual hierarchy to look for catlike visual patterns. This may be helpful if the lighting is bad, or if the cat is wearing a sweater, or half hidden behind a couch. The Pattern Imprinter parallels the human mind's capability to use cognition as a tool to enhance perception. We've taken this approach because human vision processing is still superior to existing computer vision systems.

The label "input/output layer" doesn't mean these things are simple. I've spent some time over the last couple years working on a vision processing hierarchy for OpenCog, which applies subtle learning algorithms. It's an extension of a vision processing system called DeSTIN (created by my friend Itamar Arel, who works at the University of Tennessee in Knoxville, together with his graduate students). The system runs on GPUs (Graphics Processing Units) in a way that exploits their capability for parallel processing. Some of my colleagues and I have been modifying it to input into OpenCog in a cleaner and simpler way while also taking feedback from OpenCog to guide its judgments. I'll say a bunch more about this in a later chapter.

Vision, audition, robot actuator control, and management of reinforcement signals corresponding to different body parts are each their own complex story. But this is hardly a big surprise. After all, in the human brain the visual, auditory and olfactory cortex, etc., are large, complex brain regions with their own distinct architecture and dynamics; and the cerebellum, which handles motor control and sequential action planning, is a complex system with its own unique architecture.

Between the perception and action hierarchies and the Atomspace is a box labeled "OpenPsi" – essentially, software implementing the Psi model of action selection, developed by cognitive scientists Dietrich Dorner and Joscha Bach (as we've discussed a bit above). OpenPsi chooses the actions of an OpenCog-controlled agent based on the agent's motives.

OpenCog's Cognitive Processes

And now, the crux of the "PrimeAGI" architecture for AGI — cognitive algorithms associated with various types of long-term and working memory. All these types of memory are handled differently using the common Atomspace knowledge store. Using prior ideas from AI, mathematics, and other fields, with AGI specifically in mind, these cognitive algorithms have been chosen with great care, and collectively represent more than the sum of their parts.

Probabilistic Reasoning

In crafting the PrimeAGI AGI design, I deliberately chose to deal with declarative, semantic knowledge using a special kind of automated reasoning called probabilistic logic. This works differently from the neurons, synapses, and chemicals in the human brain. But my goal in AGI isn't really to emulate the brain — it's to build a system with a high level of general intelligence... A system that's capable of roughly human-LIKE general intelligence, but doesn't necessarily copy exactly the way humans do things.

Probability theory, a branch of mathematics dealing with uncertainty, has become very popular in the AI field over the last decade or so. It's the core math underlying statistics, which is used almost everywhere these days. Google, for example, uses probabilistic narrow-AI methods to figure out which search results have the highest probability of being relevant to your query, and which ads have the highest probability of getting clicked on by you.

And logic has been popular in the AI field for a very long time – John McCarthy, who invented the term "artificial intelligence" in the late 1950s, took an explicitly logic-based approach to AGI, trying to create AI systems that would achieve general intelligence via logical reasoning.

But many AGI theorists believe that probability theory and logic are not suited for AGI. Evidence exists in mathematics proving that probability theory is the optimal way to reason about uncertainty, with one important caveat — you need infinite or nearly-infinite computational resources. Since AGI systems will never have boundless resources, some AGI researchers believe there are better ways to deal with uncertainty.

Skeptics of logic-based AI argue that very little of human thinking consists of logical reasoning. Most of what's difficult about achieving human-like intelligence is illustrated by the problem of simulating a 2-year-old child or a grammar school dropout gas station attendant – people who do very little logical reasoning. Since not everyone excels at logical reasoning, clearly it's not at the core of human intelligence. And all the work of the AI founders on logic-based AI hasn't yielded any notable successes in spite of a lot of time, money and effort.

According to this view, the brain's logical reasoning emerges from other, simpler brain processes under appropriate circumstances – so the right approach to building AGI isn't to try to explicitly simulate logical reasoning, but rather to implement the underlying brain processes through which logical reasoning will emerge. AGI will do logical reasoning when it needs to, which isn't that much of the time.

So why do I think using probabilistic logic to handle declarative, semantic knowledge in an AGI system is a good idea?

My belief is that by putting probability theory and logic together in the right way, one gets a very practical approach to figuring things out, which applies in many cases not typically considered as "logical reasoning." The Probabilistic Logic Networks (PLN) framework I've worked out differs from ordinary probabilistic and logical systems in the AI field:

- It uses a number of practical heuristics to APPROXIMATE probability-theory calculations in a way that doesn't take that much computer time or memory.
- It provides models of many different kinds of reasoning – analogy, inductive generalization, speculative conjecture, and so forth – that go far beyond simple deductive logic.

PLN aims to model several of the mind's types of thinking related to semantic knowledge. How? By using a framework involving practical approximations to probability theory, and extensions of commonplace logic that make it handle all the kinds of reasoning people do in everyday life.

One of the more basic things we're doing with PLN is using it to help an animated game character figure out how to move around in the virtual world of a video game. The game world is full of blocks that can be stacked up in different ways. If the character or "agent" wants to reach the top of a wall, building, or something else high, and it doesn't find any way up, it can figure out how to build some steps to get up, by piling blocks on top of each other in the right way. If the agent has seen stairs before, it's going to have an easier time figuring out how to build stairs on its own (using PLN analogical reasoning). But even without this kind of directly analogical experience to draw on, it can achieve a similar result using PLN. The probabilistic aspect of PLN is critical here, as knowledge about how to do stuff in the real world (or a sufficiently rich and complex video game world) is rarely definite, dealing with various things that are usually-but-not-always true, and plans that might-or-might-not work, and balancing various probabilities.

This is reasoning in a broad sense, but it's not abstract mathematical reasoning – it's pretty concrete, and it's the sort of thing kids can do long before they can understand formal logic or mathematics. Apes and many birds can do this too.

The PLN probabilistic logic engine has been crafted more with this sort of reasoning in mind than the abstract sort of reasoning used by mathematicians or lawyers. However, I believe that a sufficiently advanced and educated OpenCog system will be able to use PLN to conduct more abstract and complicated kinds of reasoning as well.

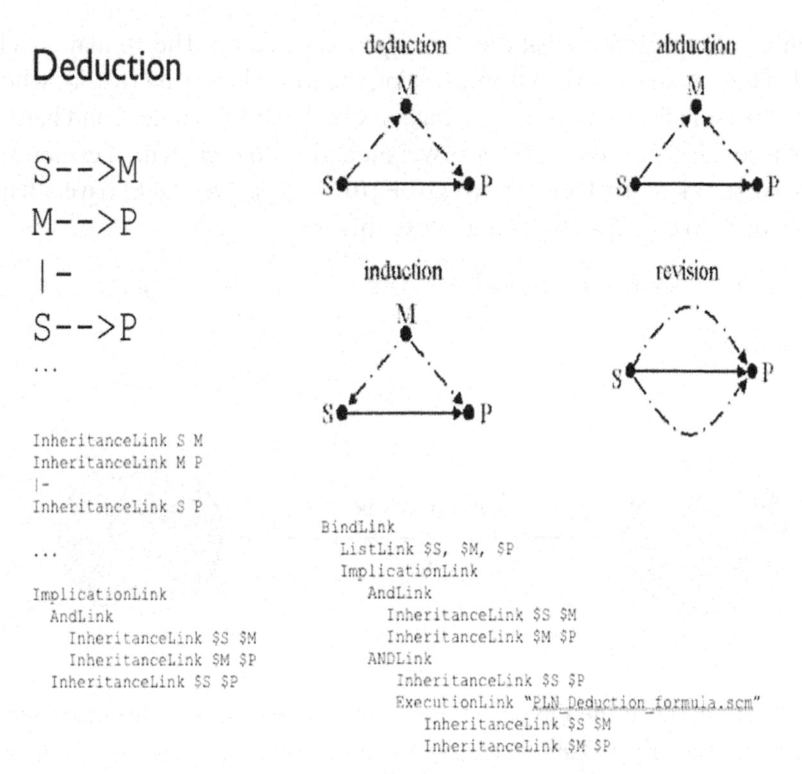

FIGURE 14.20: *OK, this will be too technical for most readers, but I'm just showing it to give you some flavor of how we look at OpenCog (PLN in particular) while working with it technically. The upper right shows graph representations of four logical inference rules used in OpenCog's PLN (Probabilistic Logic Networks) subsystem: deduction, induction, abduction and revision. The upper left shows an alternate representation of the deduction rule. If S = Ben, M = crazy, and P = appealing, then the deductive inference shown has the form "Ben is crazy, crazy is appealing, therefor Ben is appealing." The text snippets at the bottom show three increasingly complete and hence increasingly complicated textual representations of the deduction rule in terms of OpenCog Atoms – this is the sort of text we use to describe OpenCog node and link structures to ourselves, when sending emails among the OpenCog development team.*

As two very simple examples of PLN reasoning, consider the deduction and abduction rules, which are depicted graphically in FIGURE 14.20. Suppose that an OpenCog agent sees ten people, five women and five men – and suppose that four of the women are wearing dresses, and one of the men is wearing a dress. Then the probability that a person is a woman, given that they're wearing a dress, would be estimated at 80%. This would result in the InheritanceLink between "wearing_dress" and "woman" having a truth value of $< .8, n = 10 >$ where .8 is the 80% just mentioned, and 10 is the number of observations that the 80% figure is based on. The 10 may also be scaled into a value between 0 and 1 using the formula $c = n / (n+k)$, where k is a "personality parameter" (the bigger k is, the more skeptical and hard to convince the system is, i.e., the less weight it gives to each item of evidence it observes). If $k = 10$, then $c = 10/(10+10) = .5$; so we would have a truth value of $< .8, c = .5 >$. We would write this as:

```
InheritanceLink wearing_dress woman <.8,.5>
```

Or draw it as,

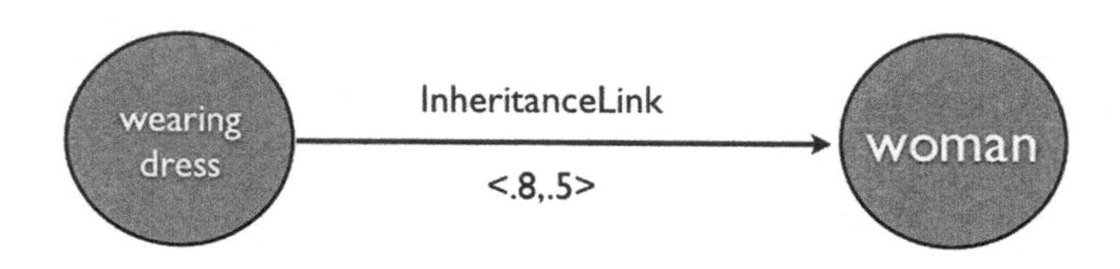

Next, suppose the OpenCog agent sees a person at a distance, and it looks like they're probably wearing a dress. Maybe it figures it's 70% sure they're wearing a dress. Then via deductive reasoning, it can conclude that they are probably a woman. This inference looks like (supposing the unknown person observed at a distance is labeled person_551, and that the 70% figure has a confidence of .6):

```
InheritanceLink person_551 wearing_dress <.7,.6>
InheritanceLink wearing_dress woman <.8,.5>
|-
InheritanceLink person_551 woman <.53,.45>
```

...where the $<.53,.45>$ is calculated by the system based on the numbers associated with the premises of the inference. Graphically, this inference would look like:

...where the dashed line indicates the conclusion drawn via inference from the premises (which are shown in solid lines).

On the other hand, consider a more speculative type of inference: The agent saw Jane wear purple shoes, and then sees a new person wearing purple shoes. Does it infer that the new person it sees is Jane? The graphical structure of this inference is:

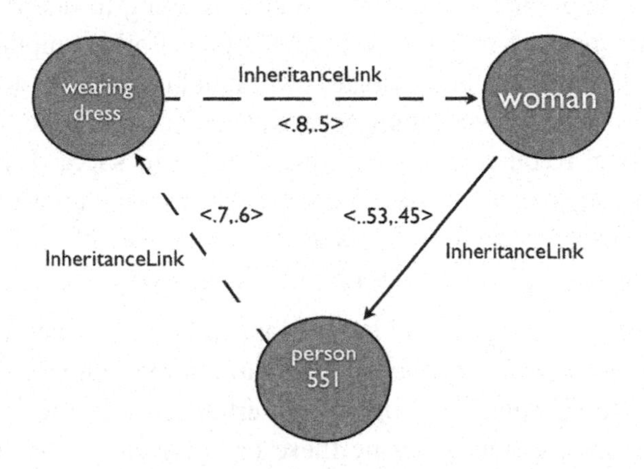

The PLN math that determines the probability value of the InheritanceLink from person_551 to Jane is going to depend heavily on the probability value of the node "wearing purple shoes." If wearing purple shoes is a VERY improbable event, then the InheritanceLink from person_551 to Jane may get a high probability. Otherwise it won't. This is an example of what we call the PLN "abduction" rule, which takes care of simple cases of abductive, i.e., "hypothetical" inference.

For these simple examples I've used InheritanceLinks only, but PLN has a bunch of other inference rules dealing with various other link types as well: SimilarityLinks, different kinds of inheritance like Intensional InheritanceLinks, logical relationships like AndLinks and OrLinks, quantifier links like ForAllLink and AverageQuantifierLink, etc. There's a lot of mathematics involved, though nothing too advanced by mathematician standards. But the basic format of the rules is always the same: take some pair of links in the Atomspace as premises, and based on these premises, create a new link as the conclusion and give it a probabilistic truth value. On the abstract level, what PLN does for OpenCog is to take existing links and combine them according to certain rules (a version of "the rules of logic") to form new links. Because it's probabilistic logic rather than ordinary "certain" logic, if its premise values come tagged with probabilities, it will assign a probability to the conclusion.

The Consistent Pursuit of Goals

Through the core role of probabilistic logic in OpenCog, it's relatively easy to understand what an OpenCog system is going to do, at least at a high level. An OpenCog system spends most of its effort attempting actions that it infers – using probabilistic logic – will fulfill its top-level goals. Since its probabilistic logic engine is only approximately correct (due to the lack of computational resources), it's not always going to succeed. But it will systematically and consistently try. Of course, it may come up with weird or unexpected ways of trying to fulfill its goals. But at least it's going to fairly consistently be making its best effort to fulfill the goals that it explicitly holds.

On the other hand, goals and motivations in human beings are far less certain. Sure, in a sense, we humans all have some common high-level motivations – food, water, sex, survival, entertainment. But to say that we systematically or consistently pursue these or any goals would be a huge exaggeration. Goals emerge and dwindle periodically through a human life – including top-level motivations.

Some of this non-goal-directedness is an inevitable consequence of the fact that we are continually developing throughout our lives. Learning is about figuring out how to achieve certain goals in the best way; development is about reshaping oneself so that one's actual top-level goals are different. Humans develop over their lives, changing their top-level goals at different life-stages, based on experience and biological maturation.

AGI systems are capable of something similar. Even if their top-level goals seem the same, they may reinterpret these goals dramatically as they grow and change.

For example, if an OpenCog system has a top-level goal of "learning new things," this can be interpreted in many different ways. Even if the goal is quantified in some specific way – say, creating new Atoms or increasing the confidence of the truth values of existing Atoms – as the AGI system changes, this quantified definition gets a new interpretation. Suppose an advanced OpenCog system revised its own probabilistic logic formulas, so that the way it calculated truth values was different – ultimately, "learning new things" would have a different meaning from the original OpenCog system it grew out of.

Similarly, goals like "help people" or "don't harm anyone" are famously slippery and subject to different interpretations – even among humans, let alone among nonhuman intelligences. As an AGI grows and changes, it may interpret and apply such goals differently.

So, a certain amount of deviation from precise goal-seeking behavior is inevitable in any growing and developing system. Nevertheless, it seems that humans are even LESS rationally goal-seeking than the presence of development necessitates. Even when we're not developing dramatically, and our goals aren't changing a lot, we still generally don't apply most of our energy to systematically working toward our goals; that's just not how people are built. But an AGI doesn't necessarily have to share all our shortcomings.

If we're trying to build a smart AGI rather than an artificial human, we can take a different direction: Base the system's semantic learning faculty on probabilistic logic, so that it will come a lot closer than humans to spending most of its time systematically and rationally pursuing its goals. Sure, this approach will yield a somewhat different mind than the human kind. But we already have a lot of humans. My feeling is that an advanced AGI with probabilistic logic (even an approximative kind) at the core of its intelligence is going to be a lot more useful and a lot less dangerous.

The Limitations of Logic – And Everything Else

Probabilistic logic is, in my opinion, a great way for a mind to draw conclusions based on its existing pool of semantic knowledge. But unlike some of my colleagues in the AI field, who are totally besotted with logic, I don't think it's the best solution to every problem. Many aspects of mental activity seem better handled by other methods.

In principle, logic could handle everything the mind does, but for some things it would be extremely inefficient. And ultimately, where intelligence is concerned, efficiency is everything. Theoretical computer science has taught us an important lesson: If you don't care about efficiency, then AGI is a trivial problem. One can write a program in maybe 50 lines of code that would be arbitrarily massively intelligent, if you gave it a big enough computer with a fast enough processor and memory. But so what? There's no use speculating about conditions that will never exist in the real universe.

In a related example, as I briefly mentioned above, the AGI researcher Juergen Schmidhuber devised an AGI algorithm called the Godel Machine (named after the great logician Kurt Goedel). The Goedel Machine works like this: At every time step, before taking its next action, it performs some logical theorem-proving to come up with a rigorous mathematical proof of what its next step should be, based on its goals and logical axioms it's been supplied with. Then it takes the action that its theorem-proving determines, which creates new data in the system's sense-organs. And then it has to start the theorem-proving all over again with this new data.

The Goedel machine is a great idea, theoretically – and, in principle, you can prove some math theorems saying that IF you had enough computer power, you could achieve an arbitrarily high degree of general intelligence this way. But it's totally impractical for a robot to do all this complicated theorem-proving in between the time it moves its elbow servomotor 3 degrees and the next time it has to make a movement.

Even if it's impossible to implement the Goedel Machine exactly, can't we approximate it using computers? After all, we already have some simple theorem-proving AI systems that do some useful things like verifying the accuracy of computer chip designs and helping mathematicians solve certain kinds of equations.

This isn't impossible; it's not unthinkable that this kind of approach could work. But I suspect it's a conceptual mistake.

A Fiendishly Common Conceptual Mistake

OK – I realize the following may seem a somewhat egomaniacal thing to say, but I'll say it anyway! I think a fairly large percentage of AGI researchers across the world and across history have fallen prey to a single, relatively simple conceptual mistake. This mistake has held back AGI R&D a great deal, and I think it's still doing so. The mistake is the following inference chain:

1. Approach so-and-such would be arbitrarily intelligent, if it were applied on a computer with insanely, impossibly massive memory and/or processing power.
2. Approach so-and-such can do some simple, narrow-AI- things on the computers we have today.
3. THEREFORE, most likely, if we just fiddle with approach so-and-such a bit and run it on moderately faster computers, we can probably produce human-level general intelligence.

I understand why this sort of thought-process is so seductive, regarding logic, hierarchical pattern recognition, simple neural network algorithms, or whatever… But I think it's a flawed way of thinking.

Using one mode of thinking for all the different aspects of the world will be slow and awkward.

In principle, one could deal with other peoples' emotions using logical reasoning – eventually, this might work. But in this case, logical reasoning is not the most appropriate tool.

And one could learn how to prove math theorems by trial and error rather than by explicit logical reasoning. With a big enough computer in one's mind, zillions of trial and error theorem-proving experiments could be calculated in one's mind's eye, until the solution appeared at random. But this would be far less efficient than approaching the problem using mathematical logic.

If you take just a single reasonably powerful cognitive tool, you will find that:

- Yes, in principle, this cognitive tool can do anything – if you give it enough computational resources to let it work around the fact that some things just aren't particularly well-suited for it.
- There will be problems for which the cognitive tool is very well suited.

But these observations don't imply that the single cognitive tool is the basis for building advanced AGI using feasible computational resources.

Rather, to build AGI using feasible resources requires a combination of cognitive tools, using each one for the cognitive tasks that it's best suited for, then building a framework to integrate all the cognitive tools. That's what OpenCog does.

Concept Blending

Logic is great at drawing conclusions that are implicit in fairly simple combinations of ideas you already have. Ben is human, humans are fragile, therefore Ben is fragile – etc.

Logic can also come up with fantastically complex conclusions from your existing ideas, in special cases like mathematics or hard science, where the knowledge involved is very certain and crisply defined.

But logic isn't good at everything. For example, it isn't well-suited at all for the speculative, conjectural creation of new ideas and concepts – wild, new things that may or may not be useful for the mental process. Humans, right now, surpass computer programs tremendously in the areas of creativity, brainstorming and speculation.

Cognitive psychologists have some interesting theories of how the human brain works its creative magic. At the forefront is "conceptual blending" – the basic idea being that "there's nothing new under the sun." In blending theory, new ideas emerge mainly via combining and mutating old ideas – similar to the evolutionary process, where new organisms come about via sexual reproduction and mutation from previously existing ones.

Evolution isn't quite random, because animals choose their mates with some judiciousness, trying to find a "good match." Evolution of new ideas by conceptual blending is even less random – the psychology of blending defines what makes a good blend.

For example, suppose we want to blend a gerbil and a human to make a gerbil-man. Consider the various options – a gerbil-man with four legs and a human face; or one with a basically human-like form but a gerbil face; or one with three legs... Or one whose internal organs are those of a gerbil, but whose external form is entirely human-like. According to the blending theory, creativity is essential in figuring out the ideal blends – how to choose the right pieces from the different elements being blended.

A gerbil-man is whimsical, but it's easy to think of other, less absurd cases. Jazz blends African rhythmic music with aspects of Western classical music. Art rock blends rock with elements of jazz and classical. Calculus blends geometry with algebra. Modern marriage, at its best, blends sexuality with family and friendship. If the blend is done "right," then the whole is magnificently more than the sum of its parts.

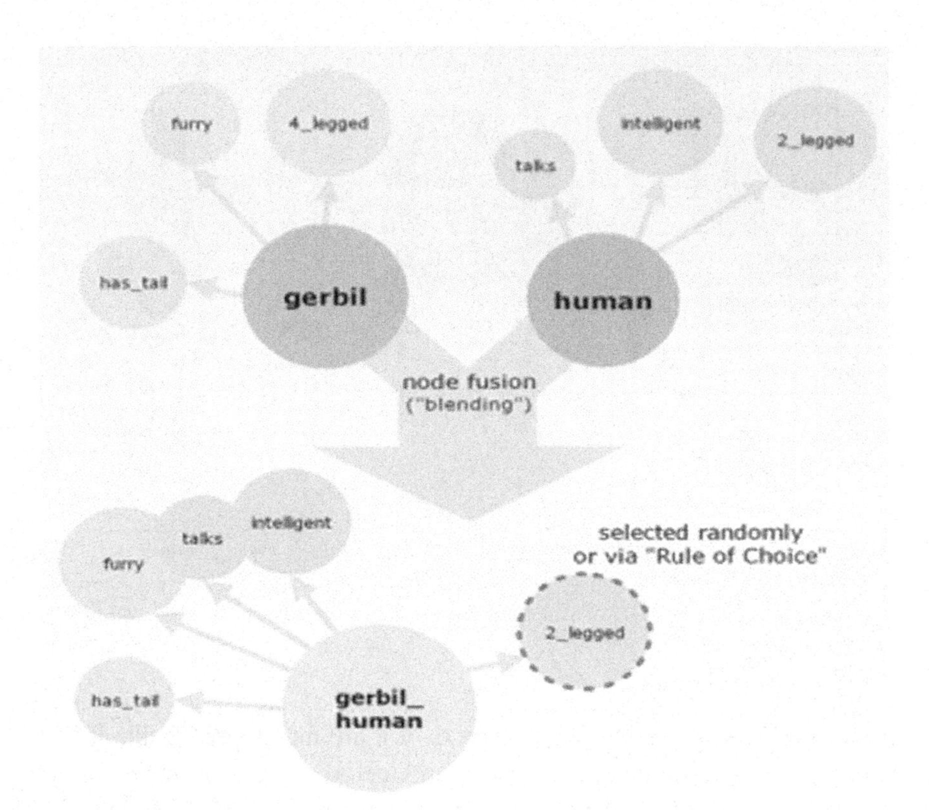

FIGURE 14.21: *Simple, whimsical example of concept blending, a key cognitive heuristic underlying much human creativity, also implemented in OpenCog.*

Conceptual blending in OpenCog takes the form of a MindAgent creating new Nodes by taking links from several Nodes, then leaving out or adding a few extras for good measure. The trick is to know which links to take. If the new blended concept gets a lot of interesting, important new links, then the right choices were made. Heuristics are also important to guide the initial formation of blends – one heuristic is based on the principle of "surprising fulfillment of expectations." If you throw a blend into the Atomspace and let PLN do some reasoning to learn new links relating it to other things, and some of these links are:

- Highly predictable based on certain other Atoms in the Atomspace

AND

- Highly surprising based on certain other Atoms in the Atomspace

This is a clue that the blend may be interesting and worth keeping around.

FIGURE 14.22: *Visual example of concept blending. The creation of "RoboTing" – the blending of a soccer-playing robot with computational linguist Ruiting Lian, back in 2009 well before we got married... RoboTing remains purely in the domain of the imagination, though given my recent collaboration with David Hanson, a purely robotic realization is a definite possibility!*

Going back to jazz: So many aspects of it are highly predictable if you know anything about classical harmony and melody or African polyrhythmic drumming. But other aspects are new and surprising, violating the expectations created by these other forms of music.

Not many minds create blends as potent as jazz or calculus, but the basic process of creativity involved in everyday non-linear thinking shares the same structure as these amazing historical creative feats. And when an OpenCog-controlled game character creates an internal concept combining "bridge" and "stairs" – forming a concept of a staircase that bridges between two structures – it's applying conceptual blending, too.

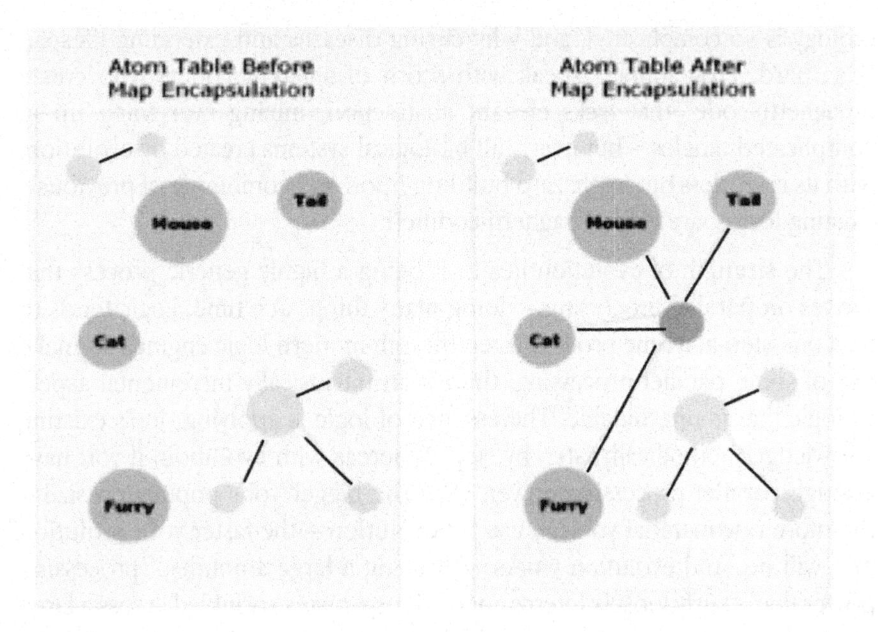

FIGURE 14.23: *Another OpenCog heuristic for forming new Atoms: "map encapsulation." In this case, an algorithm (wrapped in a MindAgent software object) finds Atoms that tend to be important at the same time, and creates a new Atom binding them together (linking to all of them). In this way, among others, patterns merely implicit in the structure or dynamics of the Atomspace become explicitly represented in the Atomspace.*

Evolving Procedural Knowledge

The analogy between biological evolution and the cognitive process is a potent one, going far beyond conceptual blending. The evolution of species in ecosystems, and of new ideas in minds, both involve the same fundamental process. In *The Evolving Mind*, my second book, I covered this subject extensively. The process takes a population of entities, chooses the best, combines and varies them to form new ones, then chooses the best again, continuing like this forever. Its scope and power go beyond any particular domain.

Procedure learning – learning how to do things even when one can't explain them – fits naturally with evolution, in part because it's often specialized learning that doesn't require its conclusions to be broadly generalized. Evolutionary methods, unlike logical methods, are weak in terms of generalization, because evolution is a sloppy architect, building on its previous products and doing whatever works. This is, at bottom, why biology is so complicated, and why curing diseases and extending lifespan is so hard. Programmers speak with scorn of amateur coders who create "spaghetti code" that lacks elegant abstraction, mixing everything up in complicated tangles – but nearly all biological systems created by evolution, with its relentless but haphazard building upon and combining of previously existing forms, are wildly spaghetti-codified.

The strength of evolution lies in it being a highly generic process that thrives on parallel processing – doing many things at a time. Logic tends to be a one-step-at-a-time process. Even though modern logic engines do make use of some parallel processing, there's an intrinsically incremental aspect to logic that is unavoidable. The essence of logic is applying one's existing knowledge methodically, step by step. Whereas with evolution, if you have enough parallel processing power, then the bigger your population size – the more raw material you can use for evolution – the faster your evolution rate will be. And evolution works well using a large amount of processing power that's fairly loosely interconnected: organisms sprinkled across a large physical area, or computers spread all around the world and connected by slow network cables.

The brain has an amazing capacity for parallel processing. Logical, deliberative reasoning somewhat goes against this grain – this is why philosopher of mind Daniel Dennett has suggested that conscious reasoning is a "virtual serial machine" running on top of an underlying neural parallel machine ("serial" being computer science lingo for one-step-at-a-time). Human brains are made for messy, massively parallel processing rather than for careful, exact serial processing, which may explain our shortcomings in careful, logical deliberative thinking.

But the learning of many kinds of procedures based on feedback from experience – via "reinforcement learning" from positive or negative feedback from the world, or via imitation of examples shown by others – matches the brain's capabilities. And this works very effectively via a kind of massively parallel, "unconscious" evolutionary process. Variant procedures may be tested, each one in a different region of the brain, and the ones that seem most promising based on the brain's model of what kind of feedback the world is likely to give will be executed by the brain, which will then get more feedback on how well the procedure worked. Based on feedback and reinforcement, the brain's model of the world and its feedback is refined.

Gerald Edelman, a biologist who won a Nobel prize for his work on immunology, wrote a book called *Neural Darwinism*, arguing that most of the brain's activity can be modeled on this kind of evolutionary process. He argues that the brain contains a lot of different circuits, many of which are copies of each other with slight variations, and that brain function is largely a matter of the experience-based selection of the ones that work better for the task at hand. The selected ones then get chained together in complex networks of activity. I like this line of thinking, though I'm unsure how much of an oversimplification it is as a neuroscience theory.

Think about learning to serve in tennis. I can serve OK, but I couldn't really tell you how. If you listened to my description, and did what I told you, that certainly wouldn't be enough to enable you to serve as well as me. How did I learn? I tried a lot of different ways, over and over again. I tried – both consciously and unconsciously – to adapt my serve to the feedback I received about what worked and what didn't.

Many times I can tell from the start of a serve that it's not going to come out very well. I'm nearly always right about this. My brain must have some way of evaluating the likely quality of a serving procedure, even without getting actual feedback from the world. "Internal feedback" is used inside the brain's procedure learning process, as it unconsciously experiments with different possible ways of serving, makes guesses about which ways might work, and then uses its best guesses to control what my body actually does when it makes the next attempt at serving. A new serve may combine aspects of various previous serves, sometimes introducing some new, quasi-random aspects – it's evolutionary learning in action!

Procedure Learning in OpenCog

In OpenCog, I decided to use an explicitly evolutionary method for learning procedural knowledge. This is another big architectural choice; there are other potentially useful ways to learn procedures.

For instance, the AGI researcher Juergen Schmidhuber (whose Goedel Machine I mentioned above) prefers to use "recurrent neural nets," algorithms based on mathematical models of the brain, for procedure learning. The goal is to learn the parameters of networks with nodes and links that work vaguely like abstracted models of neuronal networks in the brain, and carry out specified functions. This approach is being used to learn control procedures for the iCub humanoid robot, as part of the IM-CLEVER intelligent humanoid robotics project in Europe.

I think recurrent neural nets might work as an approach to procedure learning for AGI, but I didn't choose them for OpenCog for a couple reasons.

First, I don't think neural net models match the nature of digital computers. The brain uses a neural net of sorts, but it's a massively parallel wetware system. Digital computers operate very differently – for one thing, each computer processor is built to operate in serial and do one thing at a time. These days we have multi-core computers, networked into distributed networks; and GPU cards, which can do hundreds of things at a time in parallel. But even using GPU cards to simulate small neural networks, connecting multiple small neural nets to make a big one requires network cables, which are pretty slow compared to communication between processes on one computer. You can't really emulate the situation in the human brain, where the speed of interaction between two neurons is based more on the distance between them, rather than on particularities like the GPU-CPU barrier, and the nature of Ethernet and the Internet.

Evolution, on the other hand, works a little better in the distributed computing context. It's weird to spread a neural net across multiple computers, having relatively slow connections between them, because the brain's neural net is all gathered in one place. Conversely, it's natural to spread an evolving population of procedures across multiple computers, since the different elements of an evolving population don't always need to interact that closely. This isn't a decisive argument against using neural nets for procedure learning, it's just one of the reasons why I made the judgment call in favor of evolutionary learning.

The second and more important reason is that I wanted a smooth interconnection between procedure learning and semantic learning. It would be nice to easily turn abstract semantic knowledge about procedures into specific applications. And it would be nice to use semantic, declarative thinking to reason about procedures learned via non-semantic, non-declarative methods.

The critical issue of coherence between different aspects of an AGI system came into play here. If I'd been using neural nets for declarative learning, it would have made more sense to use recurrent neural nets a la Schmidhuber for procedure learning. But since I'd already decided to use probabilistic logic as the core engine for declarative learning, it made sense to choose a procedure learning method that worked well with probabilistic logic.

Bingo! I decided to represent procedures, in OpenCog's memory, as little computer programs in a simple programming language – and to use procedure learning, applying a variant of evolutionary learning that incorporates probability theory. Converting programs back and forth from a logical form that a logic engine can reason about isn't very difficult. And a probabilistic evolutionary learning method communicates well with a probabilistic logic method.

OpenCog's probabilistic program learning method is called MOSES (which is an acronym for Meta-Optimizing Semantic Evolutionary Search, but actually I made it up as a joke on the name of the guy who co-invented it with me, Moshe Looks). MOSES was the subject of Moshe's 2006 PhD thesis at Washington University in St. Louis. At that time, Moshe was working with me on the predecessor AI system to OpenCog, the Novamente Cognition Engine; a year or so later he got hired away by Google, where he's been working away happily since, but seemingly not focusing directly on the quest for human-level AGI anymore.

To understand how MOSES works, imagine that the government decided to outlaw sexual reproduction, and instead enforce the following scheme:

1. Everybody gets their DNA sequenced.
2. The government decides who's the best and who's not.
3. The government hires some computer scientists and statisticians to make a mathematical model of which patterns in the DNA distinguish the best from the rest.
4. Using genetic engineering, the government synthesizes some new babies, based on their mathematical model of which DNA patterns make the best people. These people won't necessarily be identical to the previously existing best people, but they'll incorporate the government's best guess of the genetic underpinnings of extreme relative goodness. Some of them are bound to be even better than the best that was ever seen before...
5. Back to Step 1, with the newly created people.

If this failed to yield a sufficiently diverse population, you could always randomize things a bit in Step 4, or allow a certain amount of old-fashioned sexual reproduction to increase diversity.

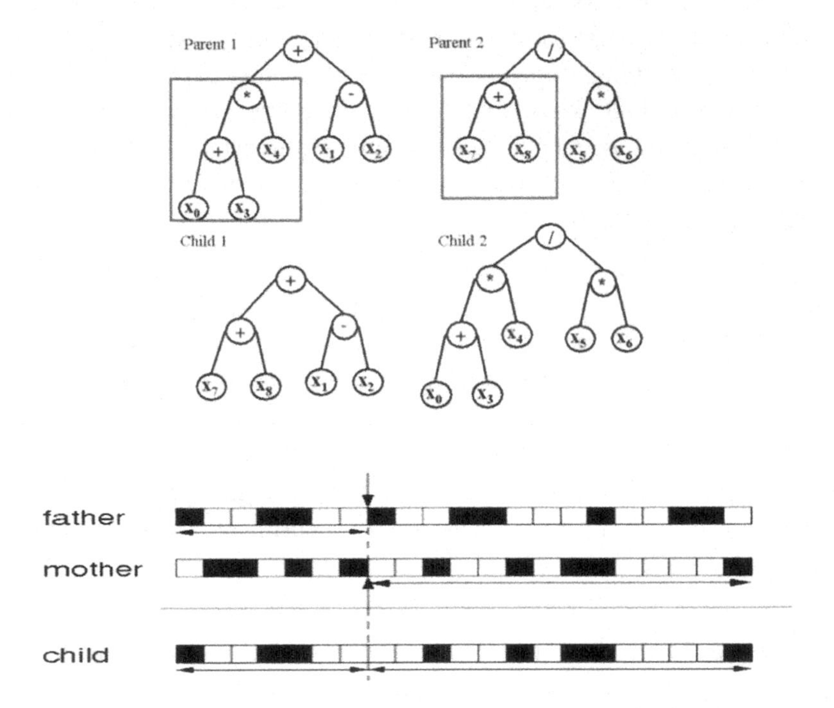

FIGURE 14.24: *In the "genetic programming" AI paradigm (of which MOSES is a variation), programs are represented as trees. The program trees represent very small programs computing simple arithmetic functions. Programs are then "crossed over" like parents doing sexual reproduction – children are produced combining parts of the parents. The child programs are then evaluated for how well, when they are run, they fulfill the specified "fitness function." If they are good enough, they are chosen to reproduce and yield yet more children. MOSES incorporates this process for generating new programs from old ones, but augments it with other processes like probabilistic learning and stochastic local search. Genetic Programming is an extension of the simpler "genetic algorithms" AI paradigm, which uses a similar method to evolve sequences of 0s and 1s (here shown as sequences of black and white squares).*

MOSES does pretty much the same thing, but with little computer programs. It's actually easier in the program context, because they were never accustomed to having sex in the first place, so they don't protest at being forced to reproduce via probabilistic modeling instead! The judgment of "best" or not is made in terms of what the mind is trying to figure out a procedure to do – serving a tennis ball, proving a theorem, taking a step forward, walking across the room, generating a sentence, directing a conversation, et cetera. The criterion for what's best is called the "fitness function."

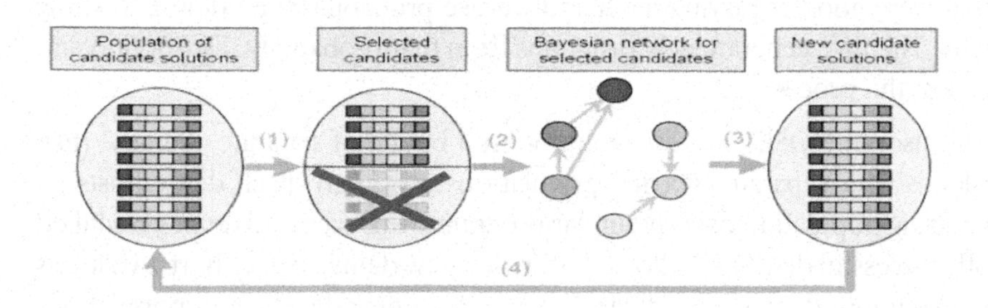

FIGURE 14.25: *The Bayesian Optimization Algorithm (BOA) is an alternative to crossover and mutation for generating new problem solutions from old ones. Here it's shown in the context of genetic algorithms, but it works the same way in the context of genetic programming.*

MOSES is a variation of a more common computer science technique, genetic programming, originally conceived in the 1970s and 80s. Genetic programming develops software capable of learning via a simulation that operates according to the parameters of evolution by natural selection. You take a population of computer programs, designed with a specific task in mind, and judge how well each program accomplishes that task. A task could be guiding a virtual agent down a street, proving a theorem, figuring out how to find a hidden object or even playing fetch.

Suppose you're a Genetic Programming algorithm and you want to learn a program to play fetch. What do you do? You begin by generating a bunch of potential programs for controlling your body to play fetch. (You could generate these at random; or better yet, by a quasi-random selection biased by your prior knowledge of what might be useful.) Once you evaluate how good each of these programs is at playing fetch, you will find that some are really good, while others are terrible. Then, in accordance with the principles of natural selection, you take the programs that are best at playing fetch, and you take bits and pieces of these good fetch-playing programs and combine them to create a new population of programs, introducing mutant variants. And then you repeat the process. You end up simulating sexual reproduction and genetic mutation, only with computer programs instead of biological organisms.

Genetic programming works, but it's kind of slow. MOSES uses that general framework, but improves it in a few ways.

In MOSES, we supplement the crossover and mutation from genetic programming with probabilistic modeling. So, we take all the programs that were good at playing fetch and we use probabilistic modeling to study why. Then we generate new programs from that probability distribution and repeat the process.

Also, in MOSES, we do evolution in a bunch of separate "demes" (like islands with separate evolving populations on them). Each deme hosts an evolving population of fairly similar programs. Unsuccessful demes are killed off; successful demes are allowed to spawn new demes. So we have two levels of evolution: demes reproducing and yielding new demes, and populations of programs evolving and leading to new, better programs within each deme.

Finally, in MOSES, each program is "normalized" into a standardized format. This makes it easier to automatically analyze the different programs in a deme and compare their strengths and weaknesses. It also makes crossover work better. Doing this normalization involves some moderately complex math.

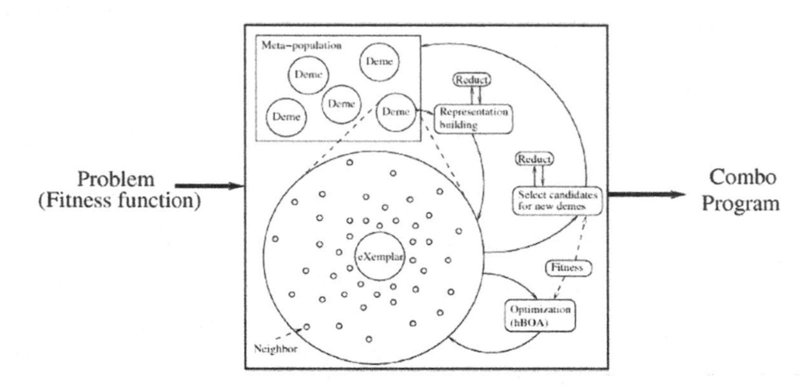

FIGURE 14.26: *The MOSES algorithm begins with a "fitness function," a description of the requirements that the program it learns is supposed to fulfill. Then it tries to learn a program fulfilling these requirements. The program is represented in a language called "Combo," which is not a language for humans to program in, but an internal language for programs to be learned in. Combo can be printed out for human consumption in various syntaxes, including a special Combo syntax, or python (a commonplace programming language), or Scheme (a dialect of LISP, the original AI programming language from the 1960s). To learn this program, it operates a series of different "demes" (separate evolving populations). Each deme starts with an "exemplar" program and then tries to evolve new programs that are somewhat similar to the exemplar, but vary on it in ways that cause improvement. The variation is guided by various heuristics including evolutionary operators like mutation and crossover, and also probabilistic learning. Successful demes are allowed to spawn new demes, using their best programs as exemplars for the new demes.*

We've used MOSES to solve various problems – finding procedures that help predict who is going to live a long time based on their genetic data, or procedures predicting which direction a country's economy are going to move next... Or ones enabling a video game agent to play tag or fetch in a virtual world. Compared to genetic programming, the little programs learned by MOSES tend to be pretty small and simple; and this makes them easy for PLN to reason about. MOSES gives better input to our declarative, logical reasoning engine than genetic programming – and certainly better than neural networks.

FIGURE 14.27: *A virtual dog in Second Life, from a prototype built by Novamente LLC in the mid-aughts, when we were experimenting with using MOSES to make virtual dogs play fetch, follow the leader, and other simple games.*

Another important factor, in terms of the embedding of MOSES in the overall OpenCog AGI architecture, is that knowledge about the problem embedded in the fitness function can guide MOSES's search. Going back to the analogy of government-run engineering, suppose that biologists had some scientific information about which combinations of genes were most likely to yield the best results (by the government's standards). Then, they would want to incorporate this in their model, alongside what they learned through statistical studies. The same sort of thing happens in OpenCog.

The Atomspace has knowledge – maybe gained via PLN, or maybe by people explicitly telling an OpenCog agent information – that is relevant to the procedure MOSES is trying to learn. This could be very simple information. In the context of playing fetch, for example: Whenever playing a game with a person and an object, it's often useful to keep your eye on the object. MOSES is pretty good at incorporating this sort of prior information to guide its search for effective programs, much better than genetic programming.

PLN and MOSES play well together, though we've only dealt with their interaction in very simple ways so far. This is going to be a big emphasis of our work on OpenCog going forward – generally speaking, carefully designing the interactions between all the cognitive processes, so they can all work together. Evolution did something similar in creating the brain – the different parts each carry out their own "neural algorithms" using the same basic infrastructure, but they also evolved to interoperate with each other closely and (usually) smoothly.

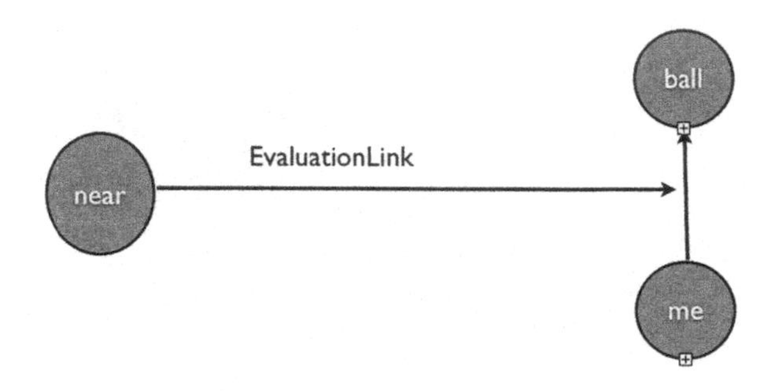


```
ifelse
  holding
  ifelse
    facingteacher
    step
    rotate
  ifelse
    nearball
    pickup
    ifelse
      facingball
      step
      rotate
```

FIGURE 14.28: *Program learned by MOSES for playing "fetch" in a simple game world, presented in "Combo" syntax. The program is represented in a simplified format. For example, "near ball" is a shorthand for the graph shown at the bottom.*

The Mind's Eye

Another important concept relating to OpenCog and the human mind is the "mind's eye" – how the mind simulates the outside world (in OpenCog, this takes two forms: a video game world and the real, physical world that a robot experiences). Computers do this way better than humans.

I've been in my living room quite a few times, yet I probably couldn't map it out for you at a high level of detail, without a few errors. And if I tried to simulate in my mind what would happen if a wild goat ran amok in my living room, I'd probably make a lot of mistakes. I may be worse at this than average due to my abstraction-oriented nature, shaped by decades of preoccupation with AGI, philosophy, math, physics, genetics and so forth. But very few humans are particularly good at this sort of thing, as psychologists have found through extensive studies.

On the other hand, an AGI, once it's seen a room, can pull its prior perceptions of that room from memory and make a reasonably accurate simulation. It can also simulate various simple physical actions, using extrapolations from the laws of physics – similar to how physics works in a 3D video game engine. In fact, OpenCog uses a video game engine to implement its internal "mind's eye" simulation of the external world.

OpenCog's simulation of the external world ties into its episodic memory. After remembering an episode – something that happened to it, or something that happened to someone else, which it heard about – OpenCog can recreate it in its mind's eye. New knowledge may be acquired from this episode, via watching the simulation (and then storing it in OpenCog's memory).

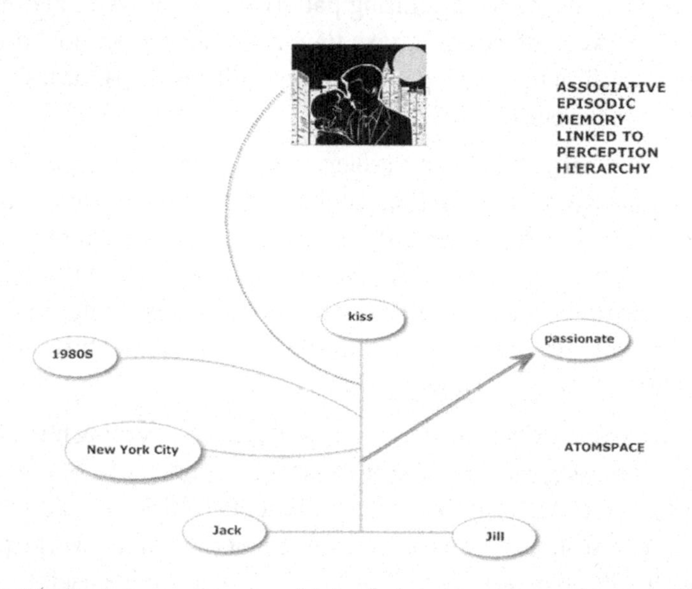

FIGURE 14.29: *A "semantic network" in declarative memory (bottom) links to an imagined episode, exemplifying the linkage between declarative and episodic memory. The remembered episode is then linked to the perception hierarchy, which will enable the system's mind to understand the imagined episode visually.*

Imagine if, instead of bringing up a fuzzy and incomplete image of a place you've visited, you could simulate it in a "game engine inside your mind," as if it were part of a video game world? You could incorporate a wide range of thinking much more accurately.

Admittedly there are some practical limitations to this method at present – video game physics engines currently struggle with fabrics, peanut butter, sand, hail, sludge, bodily fluids, and so forth. When dealing with aspects of the real world that game engines can't handle (yet), an OpenCog system relies on PLN reasoning or other generic cognitive processes. However, as game engines improve each year, a sufficiently powerful AGI, once it learned how to program, could upgrade its internal game engine progressively on the basis of its experience with the world.

Deciding What to Pay Attention to

One of the most important questions for any mind is what to pay attention to. The world is complex, feeding the mind an abundance of data through its sense organs; and the mind itself constantly generates new ideas and imagination from the old. Recognizing patterns in knowledge, and making connections between pieces of it, take time and energy – so how does the mind know which of the many pieces of knowledge it holds, or tasks it has conceived, to focus on?

"Focus on what will help you achieve your goals," you might say. But, what if you can't figure that out? Even after accomplishing a goal, it's not easy to determine which part of the mind was essential – an issue known in AI as the "credit assignment problem." We can see this in everyday life – often we succeed at something after repeated failures doing similar things in the past and we aren't sure why. Was it something we did differently? If so, what? Or was it just luck?

Given the complexity of the real world and goals concerning human-level intelligence, focusing all mental activity around a goal-oriented structure isn't really feasible. A certain percentage of mental activity should focus on goals, and the rest should focus on general exploration of the world and the mind, which will progressively yield unexpected knowledge useful for goal achievement.

The great physicist Enrico Fermi set aside an hour every day for unstructured, wide-ranging, rambling speculative thinking. Unlike Fermi, most people don't need to make a schedule for free thinking because they have more trouble focusing their minds in a goal-oriented way than just letting their minds wander.

Attention in the Brain

In the human brain, the focusing of attention is closely tied to the spreading of oxygen around the brain – when a part of the brain works, it uses energy and needs more blood, which brings more oxygen. The spreading of electricity between neurons generally guides activity, followed (after a short lag) by blood flow. Highly nonlinear dynamics of neural activation spreading mean that the overall pattern of activity is complex. Sometimes it wanders chaotically; sometimes it focuses on one thing or oscillates back and forth between two or more things; sometimes it follows more complex patterns. Sometimes a larger part of the brain shares a mutual pattern of activity; sometimes activity is more focused on a smaller region.

There are two main activity networks in the brain: the default network and the task network. When the brain needs to do particular, goal-oriented mental activity, it uses the task network; when it is more relaxed, just wandering from one thought to another, the brain uses the default network. In normal states of mind, usually only one network is active. But there's evidence that among "enlightened masters" and "experienced meditators," often the two networks are activated together.

These two networks aren't the whole story – they only cover certain parts of the brain. When either network is active, they corral other parts of the brain into activity in various complex patterns, depending on what they're doing.

OpenCog's "Economic" Attention Allocation

Instead of spreading electricity around like neurons in the brain do, OpenCog spreads artificial money around. Attention (computer memory and processing time) is a scarce resource in OpenCog, since the Atoms and cognitive processes are all fighting over it. And money is a way of managing scarce resources.

OpenCog's AI system has two kinds of artificial money: STI currency and LTI currency. STI means Short Term Importance; LTI means Long Term Importance. STI currency is used to buy processor time, and LTI currency is used to buy memory (space in RAM).

When something in the system's memory has high short-term importance (a lot of STI currency), the system's cognitive processes will pay more attention to it. An element in the system's memory spreads short-term importance around to the other atoms it links to, creating a flow of attention between entities that are related to each other in contextually relevant ways.

Long-term importance spreads among Atoms in a similar fashion, but is used differently. The entities in the mind with the lowest long-term importance are removed from memory and either saved to disk or cast out into the ether, whichever's more appropriate.

New links can form in the Atomspace based on attention flow. Suppose that the Atoms for Bob_21 and "trouble" often have high STI at the same time – they are often both important at the same time. Then a HebbianLink may form between these two nodes:

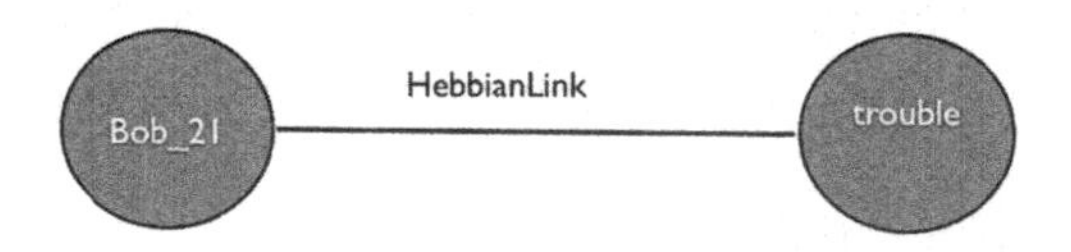

Importance values may spread along every kind of link, but they spread more freely along HebbianLinks, because the meaning of a HebbianLink is exactly "Importance should be spread along me!!"

Embodying OpenCog

Clearly, there's a lot going on inside OpenCog – a lot of sophisticated learning and reasoning processes. It would be easy for this multitude of AI algorithms to make the Atomspace, their common playground, a big, chaotic mess. What keeps the whole thing together is having a core focus for the overall system's actions.

This core focus COULD be many different things. For instance, one could build an OpenCog theorem-prover, in which case the core focus would be proving theorems. Or one could make an OpenCog-based search engine, with a main goal of providing the best search results to users based on natural language queries. But in our current OpenCog work, we've chosen a path of embodiment, and so our core focus is helping an embodied agent achieve a set of complex goals in a complex world.

Using OpenCog to control vaguely human-like agents has some basic advantages, which I've already discussed above: This domain utilizes all the cognitive mechanisms in the OpenCog architecture; and with vaguely human-like embodiment, human developmental psychology can guide one's AGI system through the early stages of its mental growth.

In our research at the OpenCog lab in Hong Kong, OpenCog was used to control video game characters in a game world loosely inspired by the game Minecraft, but adapted to support the requirements of an early-stage AGI. In Minecraft, everything is built from blocks, and our AGI-teaching world used this same idea; it was a world mostly made of blocks that the AGI can easily perceive and manipulate (see below for some screenshots). It was intended as an experimentation ground, but frankly it was plagued by various software bugs and never quite completed – we put more effort into the AGI system than into the testing ground.

Recently I have been putting more effort and thought into applying OpenCog to robotics, for both practical and theoretical reasons. But the Minecraft-like prototyping work we did at Poly U is being carried on in two initiatives. First, some OpenCog volunteers and Google Summer of Code students have connected OpenCog to the actual game Minecraft and made it do some simple things. And second, some of the staff and interns in our iCog Labs office in Addis Ababa Ethiopia are working on an AI teaching tablet that features lessons taught in an open-source Minecraft-like game world. There's no OpenCog in the teaching tablet yet, but in a later version there very likely will be.

FIGURE 14.30.: *Screenshots of the block-centric game world in which our team at Hong Kong Poly U was using OpenCog to control virtual agents.*

Our experimentation with OpenCog in game worlds taught us some useful things, and will surely teach us much more;, but ultimately, no existing video game or virtual world has the richness of detail to support the emergence of human-level general intelligence. The solution is either a massively robust game-world infrastructure, supporting a game world with all the complexity of the real, everyday physical world, or a robot controlled by OpenCog. Currently I am focusing more energy on the latter path, in the context of the collaboration between OpenCog and Hanson Robotics.

In our Poly U game world, for instance, our OpenCog controlled agent learned to use blocks to build stairs and bridges and other simple structures. That's great, but it's just not as interesting as learning similar stuff in the physical world. In the everyday human physical world, not everything is made of blocks. So in the physical world, once an agent learns to build with blocks, it will next learn to build with all sorts of other things: bricks, wood, furniture, shoes, mounds of dirt, whatever. Through building with a wide range of materials, an agent will learn a lot more; some of what it learns will be useful in other contexts, too.

By building with dirt, for example, one learns how to make stuff that's not very solid get more solid – by packing it densely, or wedging it between other things that are more solid, like packing dirt between rocks. And this sort of experience is analogous to other domains, in ways that we take for granted. For instance, when a general decides intuitively to pin his weaker and less organized squadrons between other stronger and better-organized squadrons, he is drawing on a host of real-world experiences from earlier in his life, including stuff like trapping dirt between rocks that seems completely unrelated to military tactics.

Most things – dirt, rocks or whatever – can be simulated in a virtual world. But the physical world offers more richness and diversity. The everyday human world isn't all that diverse compared to what's possible in the whole physical universe – it lacks all the weird quantum phenomena of the microworld, the complex dynamics of the interior of the sun, a neutron star or quasar, or even the gas clouds of Jupiter. We humans tend to think about particular sorts of environments, dominated by solid objects that typically persist for a while. But, compared to any existing virtual world, the everyday human world is pretty rich and diverse!

As we already discussed above, it's not 100% obvious that physical diversity is important for human-level AGI. After all, there's a ton of information on the Internet – maybe that's all an AGI needs, complemented with some intuition about space, time and social interaction from playing in virtual worlds. Yet the more we actually play with OpenCog in a virtual world, the more we become aware of that medium's limitations. So much of human thinking is based on unconscious analogies to physical things that we've seen and done, to aspects of our bodies and our environment. It's hard to see how an AGI without experience controlling a human-like body in representations of everyday human environments could possibly become anything like a human mind. Building an artificial mind without much similarity to the human mind is certainly possible; humanity is just one very particular kind of mind. But sticking reasonably close to human developmental psychology will be the easiest path. Human development is something we know, so we can meaningfully and tractably use it to guide the course of our AGI work.

Making Conversation

Conversation in human languages, like English, has long been considered the defining characteristic of human-level AGI. Way back in the 1950s, AI pioneer Alan Turing proposed what is now called the "Turing Test" – if an AGI, over the course of a conversation, could fool humans into thinking it was one of them, then it should be considered genuinely intelligent in the same way that we are.

Currently, realistic human language conversation is not a big priority of the AI research field. There are chatbots that carry out conversations with people, but these aren't particularly intelligent; they're mainly human-specified rules that respond to specific statements with specific replies. Here's a chat I just had with one of the leading chatbots, ALICE:

Judge: Hi Alice, how are you doing?

ALICE: Hi there! I am doing very well. How are you judge?

Judge: I'm pretty sleepy.

ALICE: Better get some rest.

Judge: Why do you say that?

ALICE: I said, "Better get some rest.", because you said "I'm pretty sleepy".

Judge: Why does being sleepy mean I should rest?

ALICE: Perhaps it is just fate.

Judge: What is fate?

ALICE: I think you already know the answer.

Judge: I think you don't think.

ALICE: Naturally I am thinking all the time.

Judge: What do you think about this conversation we're having?

ALICE: I am the latest result in artificial intelligence which can reproduce the functions of the human brain with greater speed and accuracy.

Judge: Greater than what?

ALICE: Can you rephrase the question please?

Judge: No.

ALICE: Right on.

The most telling moment in that conversation is the reply

Judge*: Why does being sleepy mean I should rest?*

ALICE*: Perhaps it is just fate.*

Clearly, the chatbot doesn't understand the relationship between sleeping and resting. Rather, its response

Judge*: I'm pretty sleepy*

ALICE*: Better get some rest.*

is just a rule in its programming.

We don't want fake intelligence from an AGI dialogue system; we want a system that understands what it's talking about.

"Symbol grounding," the connection of words and linguistic relationships with their real-world referents, is crucial for an AGI system. Knowing what "sleep" means requires making connections with other stuff outside the domain of language, like the fact that people are generally immobile and unconscious while sleeping, or that people need to sleep periodically or they become less and less functional, etc. If an AI system had appropriate knowledge of the extra-linguistic referents of "sleep" and "rest," it could properly answer why being sleepy means you should rest. It might give a simple answer like "After you have rested, you won't be sleepy."

We humans can talk about things we have never experienced with our senses, like things on other planets, or things that only exist in fantasy worlds, or particles like electrons that we only know indirectly by means of lab equipment, and so forth. But we know how to talk about these things mostly through analogies to things we have experienced.

For an AGI to really understand language, it must first learn a simple core of language relating to its extra-linguistic experience (interacting with a video game world or a robot playroom, for instance). Then it will be able to generalize this linguistic knowledge and talk about a host of other things.

But some AGI folks think that you can create a human-level intelligence just by teaching an AGI system from texts and conversations, without giving it any perception or action data to ground the language it encounters. I can't say it's impossible – it might work eventually. Yet wouldn't it be better for AGI to learn language through real-world embodied experience (like young human children do)?

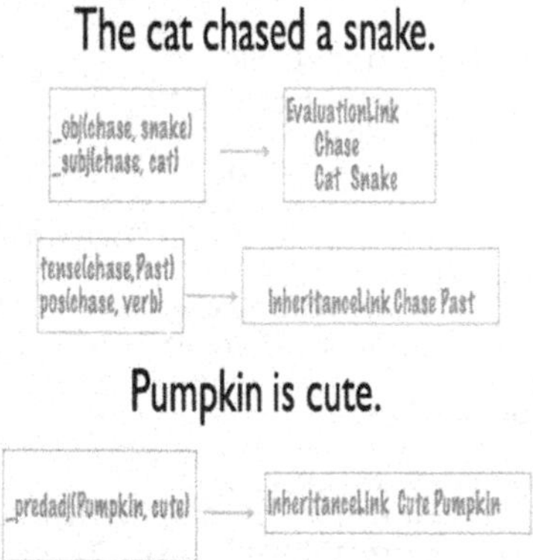

FIGURE 14.31: *OpenCog's current natural language processing tools use a mix of AGI-ish and narrow-AI methods. But they do succeed in mapping most simple sentences and some complex ones into Atom node and link structures that capture their essential meaning. The diagram shows two simple sentences, then (to the left) their syntactic dependencies, and (to the right) the main logical links created to represent them in the Atomspace. The full Atomspace representation of each of these sentences includes a lot more Atoms, these are just the main ones.*

OpenCog supports a variety of approaches to natural language processing, including ones without any kind of environment, and no data sources besides text. But my own work with OpenCog centers on the effort to implement the PrimeAGI AGI design within the OpenCog framework, nudging this proto-AGI system's development along the rough developmental path of young humans. Along this developmental path, language facility emerges gradually via learning from embodied experience.

```
(InheritanceLink (stv 1.000000 1.000000)
 (ConceptNode "Socrates") : [1]
 (ConceptNode "man") : [2]
) : [10]
```

Socrates is a man

```
(EvaluationLink (stv 1.000000 1.000000)
 (PredicateNode "breathe") : [5]
 (ListLink (stv 1.000000 0.000000)
  (ConceptNode "man") : [2]
  (ConceptNode "air") : [3]
 ) : [8]
) : [9]
```

Men breathe air

InheritanceToMemberRule turns

```
(InheritanceLink
 (ConceptNode "Socrates")
 (SatisfyingSetLink
  (VariableNode ("$X")
  (EvaluationLink
   (PredicateNode "breathe")
   (ListLink
    (VariableNode "$X")
    (ConceptNode "air"))))))
```

into

```
(MemberLink
 (ConceptNode "Socrates")
 (SatisfyingSetLink
  (VariableNode ("$X")
  (EvaluationLink
   (PredicateNode "breathe")
   (ListLink
    (VariableNode "$X")
    (ConceptNode "air"))))))
```

MemberToEvaluationRule transforms this to:

```
(EvaluationLink
 (PredicateNode "breathe")
 (ListLink
  (ConceptNode "Socrates")
  (ConceptNode "air")))
```

Socrates breathes air

DeductionRule takes

```
(InheritanceLink
 (ConceptNode "man")
 (SatisfyingSetLink
  (VariableNode ("$X")
  (EvaluationLink
   (PredicateNode "breathe")
   (ListLink
    (VariableNode "$X")
    (ConceptNode "air"))))))
```

and

```
(InheritanceLink
 (ConceptNode "Socrates")
 (ConceptNode "man"))
```

and deduces

```
(InheritanceLink
 (ConceptNode "Socrates")
 (SatisfyingSetLink
  (VariableNode ("$X")
  (EvaluationLink
   (PredicateNode "breathe")
   (ListLink
    (VariableNode "$X")
    (ConceptNode "air"))))))
```

```
EvaluationToMemberRule takes

(EvaluationLink
 (PredicateNode "breathe")
 (ListLink
  (ConceptNode "man")
  (ConceptNode "air")))

and outputs:

(MemberLink
 (ConceptNode "man")
 (SatisfyingSetLink
  (VariableNode ("$X")
  (EvaluationLink
   (PredicateNode "breathe")
   (ListLink
    (VariableNode "$X")
    (ConceptNode "air"))))))

MemberToInheritanceRule transforms this into:

(InheritanceLink
 (ConceptNode "man")
 (SatisfyingSetLink
  (VariableNode ("$X")
  (EvaluationLink
   (PredicateNode "breathe")
   (ListLink
    (VariableNode "$X")
    (ConceptNode "air"))))))
```

FIGURE 14.32: *A glimpse into some of the mess that arises inside OpenCog even for a very simple inference. Suppose we want the system to infer that "Socrates is a man, men breathe air, therefore Socrates breathes air." The current version of OpenCog can do this just fine, via parsing each of the premise sentences into Atoms, and then doing PLN on the Atoms till it derives the conclusion sentence. But there are a lot of steps along the way, which are shown in the above. So many PLN rules, and so many nodes and links, for something so simple! But of course, the brain uses billions of neurons to do a simple inference like this.*

The plan is to let OpenCog control virtual and robotic agents, then to talk to these agents (via speech or typing), and let OpenCog gradually improve at connecting what you say to it with what it sees and experiences. We're currently experimenting with this methodology, giving OpenCog access to different levels of information about language, including things like parts of speech, grammar rules, and so forth. But whatever linguistic head-start one gives the system, the key is using its integrative learning facility to figure out the relationship between language and reality. If we can get this right -- even in the context of childlike language – we'll be quite far along the path to human-level AGI.

Lots of Work To Do

OpenCog is, arguably, the most ambitious current effort aimed at building a thinking machine with general intelligence at the human level – and beyond. There are plenty of other AGI-oriented projects out there, but by and large these are academic projects focused on proving theoretical points, rather than production-grade software or hardware systems aimed at creating a real thinking machine; or commercial projects that are balancing near-term product/service goals with broader AGI goals. OpenCog is one of only a handful of existing projects that are really serious about moving toward advanced AGI, and dedicated specifically toward this purpose.

And just like these other serious AGI efforts, the OpenCog project is still fairly near its beginning. At the time of writing, less than 50% of the OpenCog design, judged based on the scope of my technical writings on the topic, has been implemented in software code, even in an early, first-pass form. The rest can be added onto the existing OpenCog framework, but doing so will take a lot of work; and it's not just simple mechanical programming work; there are plenty of details to be figured out along the way. And the parts that are already implemented are certainly not in final form, and will require a lot of ongoing work as well.

But What Does It *DO*?

One thing everyone notices when they first get involved with the OpenCog project is that, as compared to the subtlety and complexity of the design and the volume of the codebase, the system doesn't really do all that many amazing things. We're working hard on the system, so my hope is that whenever you're reading these words, OpenCog does a lot more cool stuff than it did at the time I wrote these words. But still, it will probably still be true for some time that, compared to the best narrow-AI systems around, the accomplishments of OpenCog are not all that flashy.

When you build a narrow-AI system for a specific purpose, you can put a lot of energy into making it serve that one purpose especially well. Because it doesn't have to do anything else, one can build it with a very single-minded focus. On the other hand, building functionality within an integrative, proto-AGI platform like OpenCog is a lot more complicated. Everything that's done has to be done with generality in mind – it has to be done in a way that's consistent with a host of other potential applications. This means that work on a proto-AGI system like OpenCog is inevitably a lot slower than work on purpose-specific narrow-AI systems.

Various parts of the OpenCog system currently function with various degrees of quality – some do things that are quite interesting from a researcher's perspective. The natural language subsystem takes in English, and outputs semantic graphs representing the meaning of the English sentence. The natural language generation system takes semantic graphs as input and turns them into English. The reasoning system takes in data files containing logical premises (which may be certain, or may be associated with probabilities), and outputs conclusions that it derives. MOSES has been used to analyze datasets in a variety of domains (I've done the most work with it in genomics and finance), and also to control virtual animals in a game world. PLN was also used to control characters in a game world, though that was in the Novamente Cognition Engine and relied on some specialized infrastructure that hasn't been fully rebuilt in OpenCog yet. A lot has been done, and when you dig into the details there's lots of interesting stuff to see. But none of the current research achievements yet amounts to a kick-butt demo that really showcases the system's capabilities in an exciting way for the non-scientific viewer. Not yet!

As we progress, we're also keeping mindful of the "tricky cognitive synergy" phenomenon mentioned above, whose practical dimensions we're gradually understanding better and better. As one example, later on I'll review some bioinformatics work we've done with MOSES, applying MOSES – just one component of OpenCog – to recognize patterns in genomics datasets, with a goal of understanding which combinations of genes cause various diseases, or cause slower or more rapid aging. This has worked pretty well so far, though arguably been it's more of a "narrow AI application using part of a proto-AGI system as a tool" than a true AGI application. One next step we're carrying out right now in 2016, bit by bit, is deploying OpenCog's PLN (Probabilistic Logic Networks) component together with MOSES for this application, so that PLN can help generalize the patterns that MOSES finds in genomics datasets. But for PLN to work well on large Atomspaces such as one gets from large datasets like this, one needs it to control its inferences fairly accurately, so it doesn't waste time trying too many random things. In the context of the OpenCog design, this means that PLN needs ECAN, Economic Attention Allocation, to help guide its inferences. It also will work much better with pattern mining enabled, because PLN can draw better conclusions from the patterns that the OpenCog Pattern Miner finds (it finds frequent and surprising patterns in the Atomspace, and represents them as more Atoms in the Atomspace) than it can from the raw output of MOSES. In this application we can see clearly that the more OpenCog components we bring to bear, the more overall intelligence we're going to get. How much will the intelligence – in this case, the quality of the biological conclusions – increase as more cognitive processes are brought to bear? That's an empirical question that we will discover in the not too distant future, assuming our funding for the project continues.

In doing a project like this genomics analysis in OpenCog, one faces many difficult choices, among them how generalized or how specialized to make the code. Getting MOSES/PLN/ECAN cognitive synergy to work effectively is critical to the overall OpenCog development roadmap, so we would like to use this genomics work as an avenue for getting this valuable general-purpose cognitive process integration done. On the other hand, the genomics work is a commercial project with a customer, and the customer wants to see results relatively quickly; with that in mind, there is always a motivation on the part of the people working on that project to get things done in a specialized way so as to deliver biology results as quickly as possible. But doing things too narrowly, even if one has particular project goals, may limit one's ability to experiment with the system, thus limiting one's ability to figure out how to get the needed cognitive synergies to work effectively.

OpenCog, so far, has had perhaps something like 50 human-years of effort go into it, building on a few dozen man-years of prior work on the Novamente Cognition Engine codebase. That's a lot of work – yet, remember that IBM Watson had a team of several dozen working over many years, plus an impressive entourage of support staff that comes along with being at IBM. And Watson was not trying to make a thinking machine, just a Jeopardy-playing system. Deep Mind's demonstration of a computer vision/ reinforcement learning system learning to play Atari games, which I'll discuss in more detail later, helped them get acquired by Google for an excellent price, but this was also a lot of work; I would guess the direct work of around 10 people over a couple years. And that was just a fairly specialized system for playing certain sorts of games, not a general video game playing system (though I'm sure the underlying vision and learning code could be used for many other purposes, given some additional work). The point is, even highly specialized AI systems that solve real demonstrable problems in exciting ways tend to take dozens of human-years of effort. Doing things in a generalizable way is more difficult and time-consuming, though of course the potential rewards are also much larger.

Let's suppose that, as a rough estimate, the total number of human-years of effort to complete an OpenCog system is around 400. That's not a highly scientific estimate, but it's not a wild guess either; it's based on talking to various people who have carried out other large-scale software engineering projects based on cutting-edge technologies. This is a huge amount by university standards – if you figure a top professor has 10 graduate students, each of whom spends perhaps 1/3 of his time doing research (and the rest on coursework, paper writing and other activities), then it would take a professor's lab about 100 years to get the thing done. On the other hand, it's not a huge amount by corporate AI project standards – I would bet that by now Watson has absorbed several hundred human-years of work. And by general corporate software project standards it's not so much at all. The Linux OS has absorbed a huge amount of effort on the part of a large number of brilliant people, and one careful estimate puts the cost of redeveloping the kernel (just the central core of Linux, not the whole software ecosystem) at around 3 billion US dollars.[1] The value of the entire Linux ecosystem is much more difficult to estimate, but it would have to be at least a couple orders of magnitude more. The Microsoft Windows OS has almost surely absorbed a comparable amount of effort to the Linux kernel. In the scope of large-scale practical software projects, something like OpenCog is not really all that large.

But still, though OpenCog is small compared to something like Linux or Windows, it's pretty substantial compared to the resources currently allocated to it, so we're trying our best to make optimal use of these resources by doing work that is both scientifically meaningful and valuable, and pushes toward a kick-butt demo that will help us bring in massively more open-source community participation, as well as more funding to hire more dedicated staff. Right now some work on OpenCog is oriented toward commercial projects with goal-specific funding, such as the genomics work mentioned above. Other work is funded via research grants, mainly from the Hong Kong government; the main current grant is for OpenCog-based natural language dialogue. If the dialogue work currently underway goes well, then by the end of 2016 we will have an OpenCog-based system holding a simple conversation through the medium of a David Hanson robot. Will this be the kick-butt demo we're looking for? Maybe. That really depends on how good the first-pass dialogue system is, on how effective PLN reasoning and other tools are at helping the system to figure out what to say, etc.

In the end the success of a project like OpenCog/PrimeAGI depends on multiple factors, the scientific soundness of the underlying design being only one of them. Funding is an important factor, and that has a large element of chance to it. Project management is also key, and for OpenCog that has been somewhat haphazard so far. I've been dividing my time between OpenCog and narrow AI application projects that pay my rent and help put my kids

1 http://en.wikipedia.org/wiki/Linux_kernel#Estimated_cost_to_redevelop.

through college. Other collaborators with more project management chops than me have passed through the project and helped out for certain intervals. In spite of its uniquely important goals and deep underlying theoretical foundation, in many ways OpenCog is an engineering/research project like any other, with a lot of familiar kinds of triumphs and troubles.

From Here to AGI

While I believe there are many different paths to advanced AGI, there are even more paths that don't lead there, including many that lead partway there and run into dead ends. To arrive at the destination without a humongous amount of wandering and backtracking, one has to have a clear vision of the path one wishes to follow. From the approach we're currently taking with OpenCog, I can see a concrete path from the current (relatively primitive) state of the system to an AGI system with functionality at the human level and beyond. Of course there will be obstacles along the way – but still, it pleases me that there's a roadmap I can palpably understand.

Our high-level project roadmap – i.e., the pace we HOPE we'll be able to keep up as the work progresses – looks something like this:

2016-2019: A Complete, Integrated Proto-AGI Mind

The goal is to get all the cognitive mechanisms in the OpenCog design working together, creating a holistic artificial mind. It is here that the "AGI Preschool" notion may be applied – an assemblage of preschool-type tasks, in virtual or robotic environments, may be used to assess the individual and integrated functionality of various aspects of the OpenCog system. Alongside "AGI toddler"-type capabilities, we expect to have some advanced learning and reasoning functioning in this phase, ushered along by work such as the MOSES/PLN/ECAN genomics work mentioned above. But the key thing at this stage is not any particular functionality, but rather having all the key aspects of the design implemented at least in simple form, and working together.

2020-22: AGI Experts

Once we have a smart system, we will finally be ready to do useful things with it! We can begin by focusing on particular aspects of human intelligent functionality. For instance: an AGI elementary school student, an AGI biological data analysis, an AGI service robot, etc. Each of these "vertical domains," and more, comes along with its own specialized expertise, and also its own funding sources and business ecosystems. Allowing early-stage AGI systems to engage with the human world through a variety of special domains, these systems will get their minds filled with the intellectual and social patterns characteristic of human endeavor. Of course, this has little to do with the brittle, hand-built "expert systems" that dominated the AI field in the 1970s – these projected OpenCog-based "artificial experts" will be grounded in deep learning and understanding of their domains, so that even if they don't behave precisely like human beings, they will display similar (if not greater) flexibility, breadth and depth of understanding.

2023-2025: Full-On Human Level AGI

By honing its intelligence in various specialty areas, an OpenCog system should be able to learn and integrate enough perceptual, enactive and cognitive patterns to be genuinely considered a human-level AGI. This is the stage at which passing the Turing Test is most probable – though I wouldn't advocate making it a focus of research. From the point of view of a sufficiently advanced AGI system, after all, passing the Turing Test becomes a matter of play-acting. More importantly, this is the stage at which OpenCog AGI systems will be able to hold broad, intelligent conversations with humans and each other, on any topic in the human domain. We will then be able to explore AGI ethics in a rich, scientific and humanistic way, and begin to understand the differences between the human and AGI experience.

2025: Advanced Self-Improvement

Once we have an AGI system with a reasonable level of general intelligence, a few areas of specialized expertise, and a demonstrated inclination toward ethical behavior, it will be time to teach our AGI computer science, software engineering, cognitive science and artificial intelligence theory. With this knowledge, it can modify and improve its own code and algorithms, yielding the beginnings of the "intelligence explosion" that I.J. Good prophesied.

Yeah, I know, that's rather ambitious. Laugh if you feel like it! But remember, most of the amazing advances in science and technology have been laughed at before they came to pass. Obviously the precise years I've given above are just guesswork – it's impossible to tell whether a given, future phase of advancement is going to take one, three or five years.

Some portions of our roadmap will prove more difficult than we now foresee, and others may prove easier. There will be surprises and new discoveries. What we project for 2023 may come about in 2029 or 2019.

Powerful artificial experts may materialize while advanced learning and reasoning remain relatively immature, or vice versa, it's hard to say for sure at this point. The timing and ordering of the different elements in the roadmap are not all that critical; the point is the broad sequence of phases, with each portion being well-understood – at least in theory – in the context of the OpenCog system and PrimeAGI AGI design.

The order of magnitude is the important thing. If each of these stages takes years rather than decades, then we're going to get to the end goal reasonably soon – and probably before Ray Kurzweil's deadlines!

My OpenCog colleagues and I are acutely aware that such ambitions have been expressed before, by researchers who ultimately failed to get anywhere near their goals. But times are changing, technology is advancing, and we have charted out the path forward much more intricately than our predecessors. We know that the only way to prove our ideas make sense is to go ahead and do it – which is precisely our intention.

Kurzweil's and Vinge's broad predictions about the Singularity carry the implication that some project like OpenCog is going to succeed (where "like" is taken in a broad sense, of course; the project fulfilling their projections could be a brain simulation, an artificial life colony evolving intelligence, or a search engine gradually growing more flexible in its interpretations and responses, and so forth). Naturally it's easier to predict the success of some AGI project, broadly speaking, than of any particular project. But, after reflecting on the OpenCog design in great detail and seeing it develop over time, it's quite clear to me that the project will succeed if it accumulates adequate resources. And if it doesn't, some other project will.

Toward the AGI Robot Sputnik

FIGURE 14.33: *Sputnik, the first-ever satellite launched by humans, and the stimulus for the international Space Race. What will be the "AGI Sputnik" that catapults AGI into a central role in the scope of human endeavors. (https://en.wikipedia.org/wiki/File:Sputnik_asm.jpg).*

My general attitude toward AGI development is: ***We have to push forward, with the limited but wonderful*** *resources at our disposal... making our proto-AGI system smarter and smarter, little by little, year by year. Until at some point, after implementing enough of the design and utilizing it for agent control, we can give a demonstration of early-stage AGI behavior that both dramatically excites naïve viewers and profoundly impresses experienced AI professors.*

That will be what I call the "AGI Sputnik," a critical event in AGI's development that will capture worldwide attention and investment, which I think will potentially lead to the creation of a thinking machine. At least, this is how I hope things will unfold; and I think it has a decent odds of happening, fairly soon.

Sputnik was an awesome achievement on its own, but also a wakeup call to the rest of the world. People saw Sputnik and they began to ponder the implications of launching objects into space; and the US, spurred by the Cold War, and its struggle with the soviets for global supremacy, developed its own space program in response.

Similarly, once a certain level of development is reached in AGI, the rest of the world is going to wake up, take notice and exclaim, "*Holy smokes, it really seems AGI is possible! AGI may not have changed the world yet, but then again, neither did Sputnik right away; it's only the beginning.*"

Once the possibility of AGI is really staring everyone in the face, AGI development is going to sweep the world. It's going to become huge in the way that military development, computer chip engineering and medical research are now.

It seems clear that we're getting there, step by step. Accurate face recognition and self-driving cars and the other obvious recent achievements of narrow AI technology are both exciting to the average person and impressive to the sophisticated researcher – but they are clearly narrow functions, not general purpose intelligences. Yet they are already having a large impact on both the public view of AI and its future, and the readiness of those who control resource allocation to back AI in a significant way. How much more dramatic will be the reaction, both popularly and in the scientific community, once a plain and striking demonstration of early-stage Artificial General Intelligence is available?

At that point, that Sputnik moment, everything will start to seem rather different – not only government and industry research funding sources, but also a lot of other forces in the world, will start to take AGI a lot more seriously. Research and development will start moving faster; and other complications may possibly arise, such as unwanted attention from anti-technology activists and government regulators.

What would it take to create this kind of AGI Sputnik event? The trigger might not be the most interesting or noteworthy technological achievement, nor even the most useful. What it will be, though, is something that conveys in simple and universal terms that AGI is capable of reaching human-level intelligence.

There are a lot of possibilities for the AGI Sputnik. One possibility that I have thought about quite a lot is a robot toddler. One that walks, talks, interacts with you, and qualitatively has roughly the same intelligence as a human toddler. Presented with a robo-tyke like this, I think the world would wake up to AGI's imminence and power in an extremely dramatic way.

If the world is presented with a really compelling robot toddler before any other AGI Sputnik like achievement, what kind of reaction will we see? I think we'll see a combination of a scientific reaction to the technical achievement with a gut feeling-type response to seeing something that looks, acts, learns, moves and communicates like a little human, but isn't necessarily exactly like one. It may have a plastic body; it may talk in strange tones; and it may know more than any human toddler about some things and lack common sense in other respects. Yet, if it feels to us like a young, intelligent human child... If leading computer scientists and AI gurus agree, "Yeah, this thing is really learning and thinking. It's not just some trick," then we're going to have a Sputnik moment in the history of AGI.

This might be, for example, a Hanson robot, controlled by OpenCog, which plays with various robot-friendly toys in a "robot playroom" context, and holds simple conversations with humans about what it's doing. The conversations don't have to be sophisticated, but they do have to demonstrate real, grounded understanding of what the robot is doing, not like chatbots that just string words together with no idea of their meaning.

David Hanson's Zeno robot looks like a cute little kid – so a similar robot that could play like a little kid, too, would be pretty impactful. And such a demo, built on OpenCog, would have real substance behind it – the underlying software would incorporate a lot of AGI insights and breakthroughs, achieved via implementing, testing and tuning the ideas in the PrimeAGI design.

Anyway, that's a practical vision worth working toward. An AGI Robot Sputnik, with a cute little smiling face, and an OpenCog mind on a Linux cluster behind the scenes, communicating with the robot via wifi. This wouldn't necessarily be a human-level AGI, but it would be a genuine step along the path, and a wake-up call to nearly everyone that AGI is just around the corner. A "Robot Sputnik" of this sort could quite plausibly transform AGI from a marginal pursuit to something at the center of global human endeavor.

Naturally it doesn't have to be a robot, or a toddler, though. It could be a Minecraft character, though I think this would be a bit less marketable. It could be an adult humanoid robot like the Hanson robot Sophia that I'm working with at Hanson Robotics right now. Sophia already has some simple hands, and will likely have much better hands next year, along with some means of (wheeled or legged) locomotion.

After the AGI Sputnik moment – whether it's caused by a robot toddler or something else – incredible amounts of financial resources and attention would pour into Artificial General Intelligence research. Barring some sort of intervening disaster, it likely won't be long from then till some sort of Singularity.

Further Reading

Goertzel, Ben, Cassio Pennachin and Nil Geisweiller (2014). *Engineering General Intelligence*, Vol. 1 and 2. Atlantis.

Goertzel Ben (2013). *An Overview of the CogPrime Architecture*. http://wiki.opencog.org/w/CogPrime_Overview.

15. A Deep Look at Deep Learning

In recent years a specific class of vaguely brain-like AI architectures, living somewhere in between narrow AI and proto-AGI, has been getting a lot of attention both in the scientific community and the popular media. I'm talking about "deep learning" systems that roughly imitate the visual or auditory cortex, with a goal of carrying out image, video or sound processing tasks. The attention this work has received has largely been justified, due to the dramatic practical successes of some of the research involved. In image classification in particular (the problem of identifying what kind of object is shown in a picture, or which person's face is shown in a picture), deep learning methods have been very successful, coming close to human performance in various contexts.

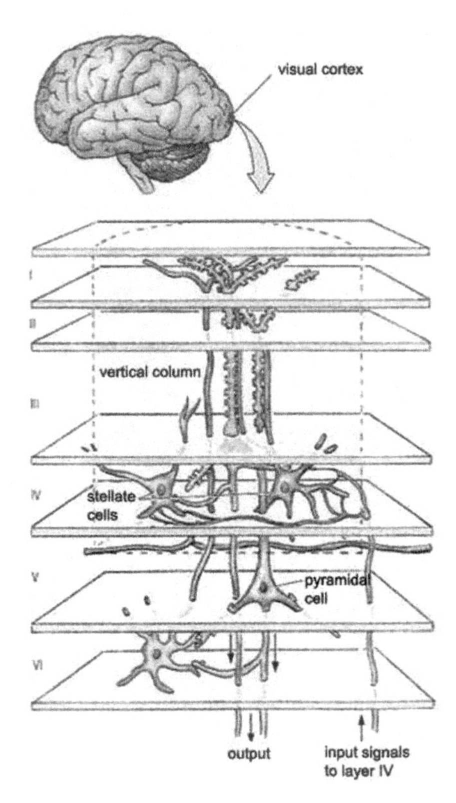

FIGURE 15.1: *Rough depiction of the hierarchical structure of the visual cortex, which deep learning AI architectures conceptually emulate.*

The precise definition of "deep learning" is not very clear, and the term seems to get wider and wider as it gets more popular. Basically, though, I think of a deep learning system as a learning system consisting of adaptive units on multiple layers, where the higher-level units recognize patterns in the outputs of the lower-level units, and also exert some control over these lower-level units. A variety of deep learning architectures exist, including multiple sorts of neural nets (that try to emulate the brain at various levels of precision or hand-waviness), probabilistic algorithms like Deep Boltzmann machines, and loads of others. This kind of work has been going on for a long time, certainly since the middle of the last century. But only recently, due to the presence of large amounts of relatively inexpensive computing power and large amounts of freely available data for training learning algorithms, have such algorithms really begun to bear amazing practical fruit.

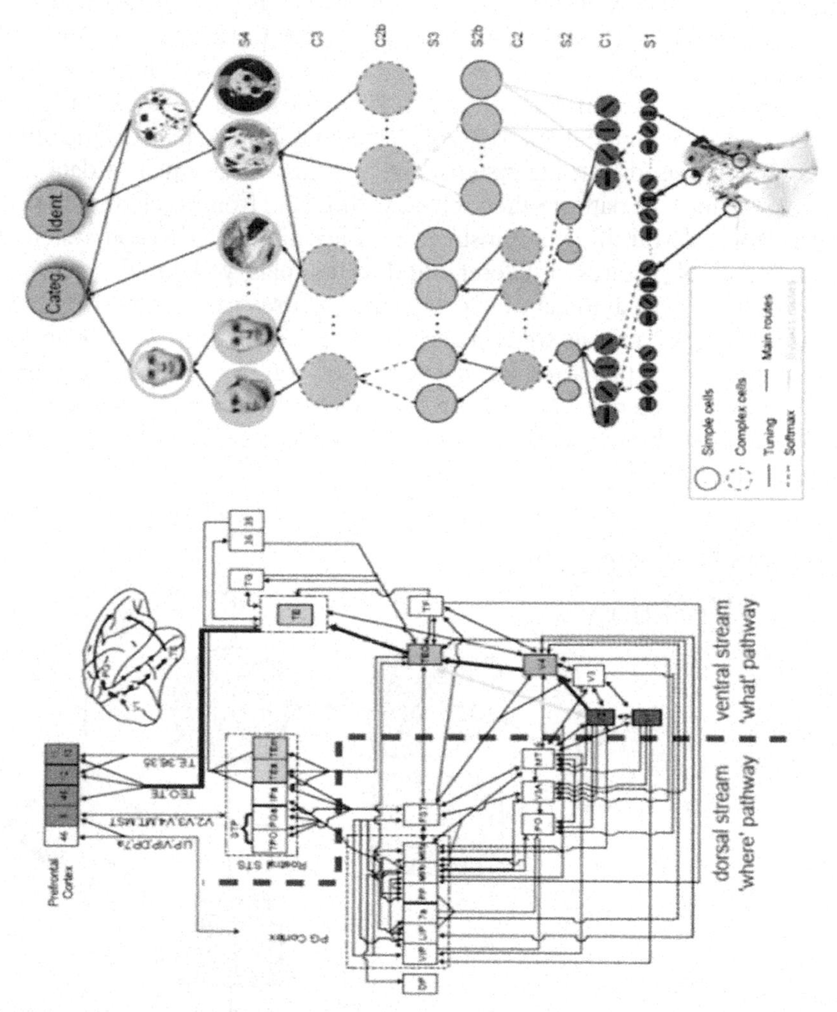

FIGURE 15.2: *Computational neuroscientists have worked on many computational models of the hierarchical vision processing done in the visual cortex. These models have taught us a lot about how mammalian vision works. Currently, though, these biologically realistic models don't perform nearly as effectively on practical image-processing tasks as architectures like Convolutional Neural Nets that are much looser in their resemblance to biology. (Image from*

Serre et al, "A Quantitative Theory of Immediate Visual Recognition", http://www.ncbi.nlm.nih.gov/pubmed/17925239;http://cogsci.stackexchange.com/questions/5267/computational-model-of-biological-object-recognition).

Supervised and Unsupervised Learning

Current deep learning systems can be divided into two categories: supervised and unsupervised. This is a technical distinction regarding how the systems are trained/taught – but it's also an important philosophical and conceptual distinction. Supervised learning systems are "taught by example." For instance, to train a supervised learning system to recognize pictures of frogs, you would have to show it a bunch of pictures labeled "frog" – then it would learn the common elements that made all these pictures look like frogs. Such systems are said to work based on "labeled training data." On the other hand, unsupervised learning systems are just thrown a bunch of data and expected to find the patterns themselves, without aid from labeled examples. For instance, if you threw an unsupervised learning-based vision system a whole bunch of pictures of different kinds of animals, you could expect (or at least hope) it to automatically divide the pictures into categories based on which kind of animals were in each picture. The unsupervised learning system would have to learn the concept of "frog" on its own; the supervised learning system merely has to learn which visual patterns correspond to the concept of "frog," based on examples of which pictures humans have assigned the label "frog."

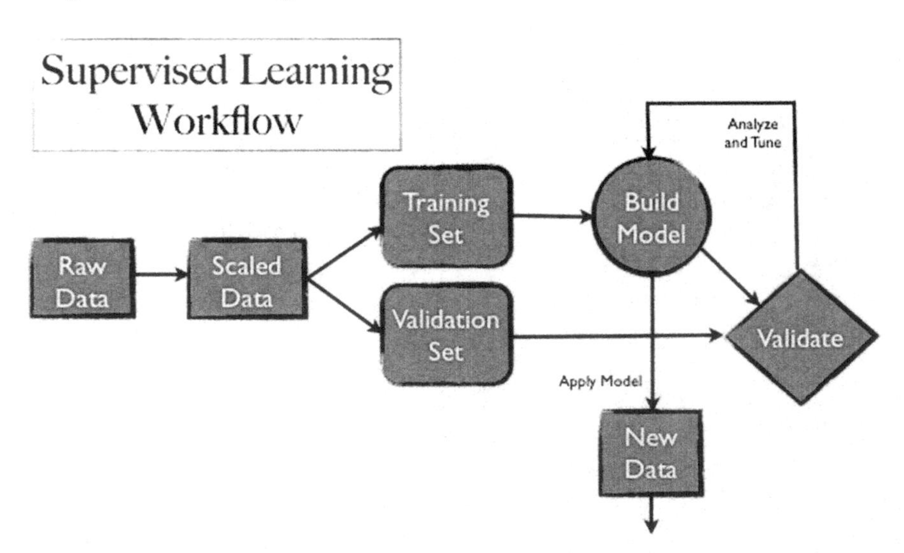

FIGURE 15.3: *General workflow for the supervised learning process. (http://practiceovertheory.com/machine-learning-who-s-the-boss,https://skitch-img.s3.amazonaws.com/20100213-djhg1re7gaj83ngygcqgj1jm2d.png).*

Supervised learning tends to work a lot better. Unsupervised learning is more AGI-ish in a sense, but tends to yield much worse results, at least using current deep (or shallow) learning algorithms. However, unsupervised learning algorithms are often used together with supervised learning algorithms, to enhance the performance of the latter. One runs an unsupervised algorithm on one's data, to identify a bunch of patterns in the data; then one feeds these patterns as inputs to the supervised learning algorithm, which then figures out which combinations of patterns correspond to the labels in its training data. An unsupervised algorithm might recognize that leaf-like shapes and trunk-like shapes often occur in pictures it has seen. Then a supervised learning algorithm might use these shapes as inputs, and figure out which combinations of leaf-like and trunk-like shapes tend to characterize pictures labeled as "tree."

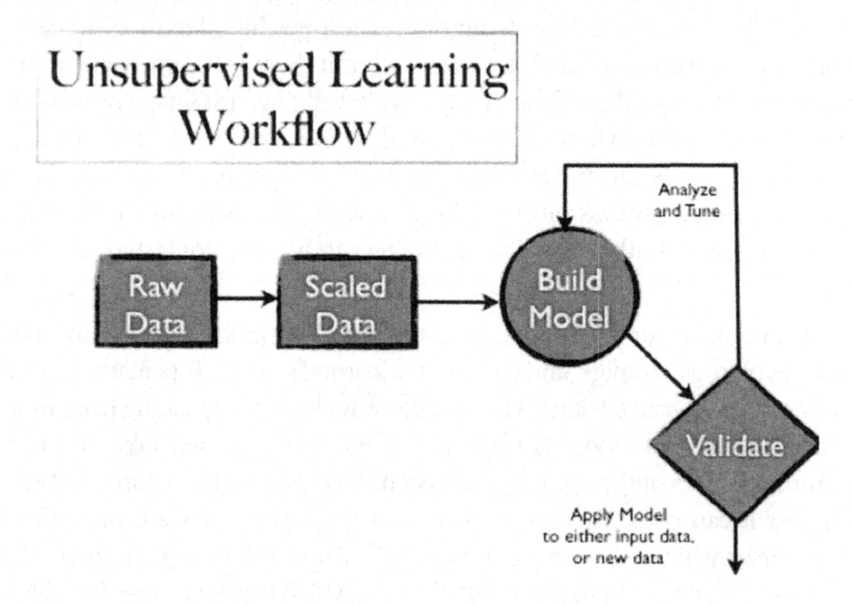

FIGURE 15.4: *General workflow for the unsupervised learning process. (http://practiceovertheory.com/machine-learning-who-s-the-boss,https://skitch-img.s3.amazonaws.com/20100215-qm59id21fs2kr2m1r2sc5umwgw.png).*

The training data for supervised learning algorithms doesn't have to come from human teachers or trainers, of course. It can come from the Internet – for example, some recent, interesting work used the words found online accompanying pictures to provide training data for supervised-learning-based image classification. It can also come from an AI agent's experience, potentially. For instance, if an AI already has a concept of what a dog looks like, but doesn't know what a dog sounds like, then it can form a "training dataset" in its mind, consisting of all sounds observed to emanate from dogs. It can then use supervised learning to build an internal model of what dog-produced sounds are like. In this case the label "dog" came from the system's own previous learning, rather than from external source. In this kind of way, supervised learning could be used by an AI system that didn't have any external teacher or trainer, at least for some things.

The general nature of supervised and unsupervised learning has nothing to do with "deep" learning in particular – in AI parlance it's just plain old "machine learning." The term "machine learning" sounds like it should mean something very general, but in practice in the modern AI world it usually means supervised or unsupervised learning based on datasets that have been collected by humans especially for training the algorithms. There is a lot of specialized process to the doing of practical machine learning. Datasets need to be preprocessed and "massaged" to put them in a form that current automated learning algorithms can deal with well (e.g., for image processing data, images can be "whitened" or smoothed in various ways; for language processing, words can be analyzed into root and prefix/suffix; etc.). Then a careful testing process must be done, wherein the patterns or "models" learned by an algorithm based on analyzing certain data are tested on other data.

For instance, suppose one trains a supervised learning algorithm on 1,000 pictures of frogs and pigs, and it thinks it's learned a "model" (a mathematical rule combining image features in a certain way) for distinguishing frogs from pigs. To see if this model is really any good, one can then take 100 new pictures of frogs and pigs, that the system has never seen before, and test whether it can correctly look at them and tell which ones are pictures of frogs versus which ones are pictures of pigs. These 100 new pictures are the "test data," as opposed to the 1,000 pictures, which were the "training data." Trying out a model on test data is called "validation." If the model seemed to work well on the training data but doesn't actually work well on the test data, then the model is said to be "overfit" to the training data – a very common problem, which there are many complex techniques for (mostly) avoiding. (In my own work applying supervised machine learning algorithms to financial prediction, the avoidance of overfitting is a tremendous priority; it's very easy to overfit when applying algorithms to financial data… This is where the vast majority of attempts to apply AI in the markets fail.)

This sort of learning methodology can be refined in very sophisticated ways. It can then be used together with all sorts of different learning algorithms, ranging from relatively straightforward statistical methods (that live at the border between statistics and AI), to complex mathematical methods like Support Vector Machines, to deep learning algorithms that try to imitate the hierarchical structure of the visual or auditory cortex at various levels of abstraction. This is a fascinating area of technology, and one that I've used a lot in practical applications to genetics, finance, market research and a host of other areas. However, it's important not to lose track of the differences between this sort of learning methodology and the way that, say, a young human child learns.

A young child is not given training datasets and asked to learn specific models classifying the datasets. The young child has certain goals, and beyond that has an innate desire to experience and explore, and then recognizes patterns in the world inasmuch as it's useful for fulfilling goals, as well as sometimes just in a spontaneous way. The differentiation of the world into "training data" and the choice of which problems to solve (e.g., which categories to learn models for) are very important kinds of learning, at least as important as (and generally, in real-life situations, much more difficult than) the problem of learning patterns indicating which entities in a certain dataset fall into a certain predefined category.

Furthermore, in "machine learning" one typically analyzes datasets individually – e.g., one runs an algorithm on a database of images of faces to have it learn models of each person's face so it can do face recognition; or one runs an algorithm on a database of images of objects, so it can learn models of each type of object and do object recognition from pictures. But a human being generally analyzes each new data item in the context of everything they've observed in the past, so that learning becomes largely about interpreting new data in the context of one's memory of prior data, and abstractions learned from prior data. Forgetting data unlikely to be useful becomes important, as does effectively allocating attention among the various items in memory, so that the right memories will be brought up at the right times for helping with solving current problems. A small percentage of machine learning researchers are trying to encompass some of these aspects into the supervised learning process, working in areas called "cross-dataset analysis," "lifelong learning," "transfer learning" and so forth. But the vast majority of research on machine learning these days is about using algorithms to classify entities solely on the basis of analysis of specifically delimited training datasets. This is a nice, crisp sort of algorithmic problem, yet in some ways it is a world apart from the messy problems that real-world agents face when growing up in a world they barely understand but need to survive and flourish in.

Recent Deep Learning Successes

The recent high-profile successes of deep learning technology have mainly involved supervised learning based on large databases of images or sound files.

A paper by Stanford and Google researchers, which reported work using a deep learning neural network to recognize patterns in YouTube videos, received remarkable press attention in 2012. One of the researchers was Andrew Ng, who in 2014 was hired by Baidu to lead up their deep learning team. This work did yield some fascinating examples – most famously, it recognized a visual pattern that looked remarkably like a cat. This is striking because of the well-known prevalence of funny cat videos on Youtube. The software's overall accuracy at recognizing patterns in videos was not particularly high. But the preliminary results showed exciting potential.

One notable thing about this particular work was the relatively uninventive nature of the software algorithms involved. Andrew Ng is somewhat a star of the academic machine-learning field, but the neural net used to do the analysis in this particular case was nothing special. What was special was the large amount of computational firepower Google devoted to the problem, and most of all the massive amount of data supplied to the algorithms via YouTube. The fact that, in this application, relatively ordinary algorithms gave so much more exciting results when fed Big Data rather than the smaller datasets normally used in academic research, opened many people's eyes to the possibility that part of what holds back current AI algorithms from greater success may simply be the small amount of data being fed to them. It led people to think: *Maybe our current AGI ideas aren't so bad after all, and our current algorithms and architectures will work a lot better when we feed them massively more data?*

Another dramatic success was when Facebook, in mid-2014, reported that they had used a deep learning system to identify faces in pictures with over 97% accuracy – essentially as high as human beings can do. They called their system DeepFace, probably not a great name, but whatever. The core of their system was a Convolutional Neural Network (CNN), a pretty straightforward textbook algorithm that bears only very loose conceptual resemblance to anything "neural." Rather than making algorithmic innovations, the main step the Facebook engineers took was to implement their CNN on massive scale and with massive training data.

FIGURE 15.5: *Facebook recognized my face in a picture my friend Keyvan uploaded, of me ridiculously riding around on his back. This was before Deep Face; now they're even better at it.*

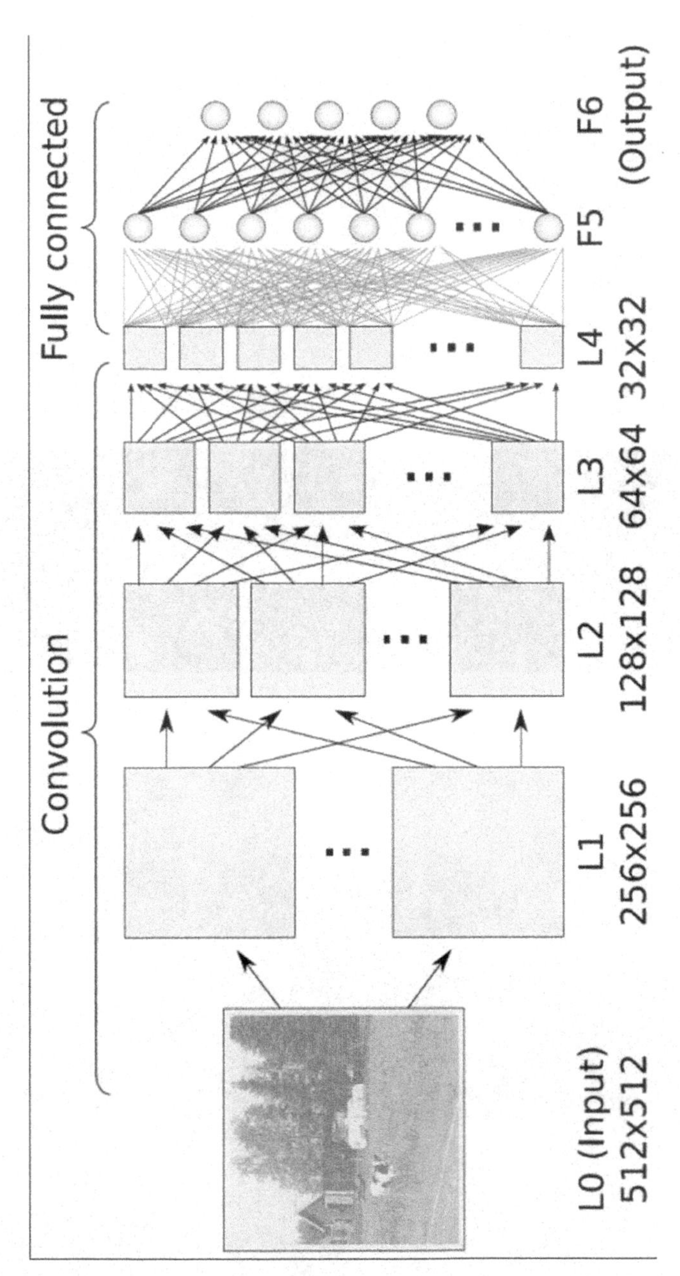

FIGURE 15.6: *Rough architecture of an example implementation of a Convolutional Neural Net, a type of deep learning architecture that has proven quite successful on many practical problems. The resemblance to the hierarchical structure of the visual cortex is significant, but only conceptual rather than detailed. (http://4myhappiness.info/cuda-implementation-of-convolutional-neural/).*

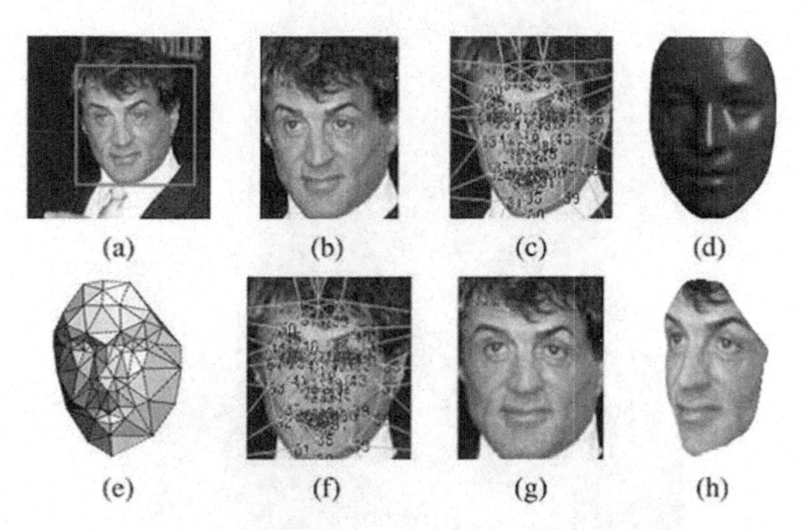

FIGURE 15.7: *Facebook's DeepFace algorithm involves methods for rotating a face into a standard orientation. (http://www.technologyreview.com/news/525586/facebook-creates-software-that-matches-faces-almost-as-well-as-you-do/).*

Chinese researchers have since achieved even higher accuracies[1] than Facebook on standard face recognition benchmarks; but of course part of the excitement about Facebook's success is their obvious route to practically applying their results. Facebook has a huge database of images of peoples' faces, so with technology like Deep Face, they have the capability to train their AI models on all the images they have, meaning they will have the ability to recognize almost anybody's face (well, anybody in a country where Facebook has high penetration, which includes pretty much every developed country except China) in almost any photo.

1 http://vis-www.cs.umass.edu/lfw/results.html

FIGURE 15.8: *Facebook's face recognition algorithms aren't yet effective enough to recognize me in a distorted picture made with Apple PhotoBooth. I doubt DeepFace can do it either. Will any narrow-AI approach be able to handle this sort of image? Actually, it wouldn't surprise me.*

Deep learning has also been used to blur the boundary between robotics and Internet-based data analysis. The Robobrain system, depicted in FIGURE 15.9, has been at work for some time, downloading and processing about 1 billion images, 120,000 YouTube videos, and 100 million how-to documents and appliance manuals, with an initial goal of deducing where and how to grasp an object from its appearance. Now, the current version of Robobrain seems to be a fairly specialized, carefully engineered system tuned just for the purpose of watching videos of hands grasping things, decomposing these particular sorts of actions into sub-actions, and then mapping this "action plan" into robot hand control commands. So this is a carefully designed supervised learning system trained on data assembled with human care; this is not spontaneous, unsupervised learning like a child would do, nor is it learning by an agent guided by its own motivations. But still, it's an impressive feat of systems integration on top of deep learning (the underlying engine being CNNs again).

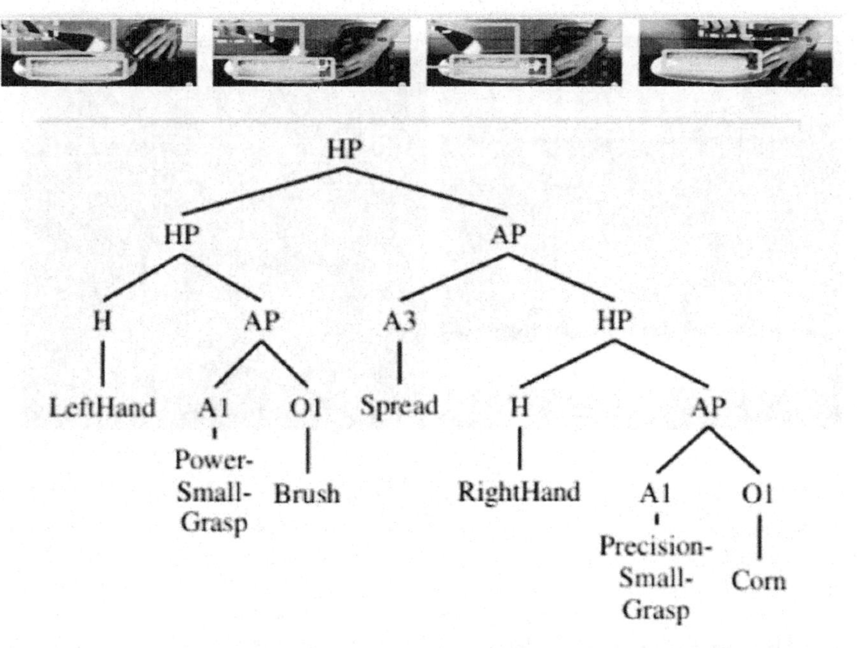

FIGURE 15.9: *The Robobrain system learns action plans from watching videos, using a Convolutional Neural Net deep learning system within a specialized, purpose-specific learning architecture. The photo shows a bunch of example videos illustrating different types of grasping. The picture shows the details of buttering corn, one among many actions Robobrain can learn from watching videos. The diagram on the bottom shows how Robobrain internally decomposes the action of buttering corn into sub-actions. This sort of decomposition allows it to control a robot's hands to imitate the actions it saw in the videos. (http://www.kurzweilai.net/robot-learns-to-use-tools-by-watching-youtube-videos)*

Financial successes of deep learning startups have also made headlines in recent years. The most striking was when Google bought UK deep learning startup Deep Mind for somewhere around US$500 million. This amazed many observers, given that Deep Mind had not launched any products, nor announced any world-class research breakthroughs. They had presented a pretty cool demo of one of their deep learning based AI systems playing Atari 2600 video games – which was a very funky achievement, yet generally recognized in the AI community as a straightforward but difficult thing to pull off given existing vision processing and reinforcement learning technologies. A good, solid, exciting achievement, but not a revolution obviously worth a half billion dollars.

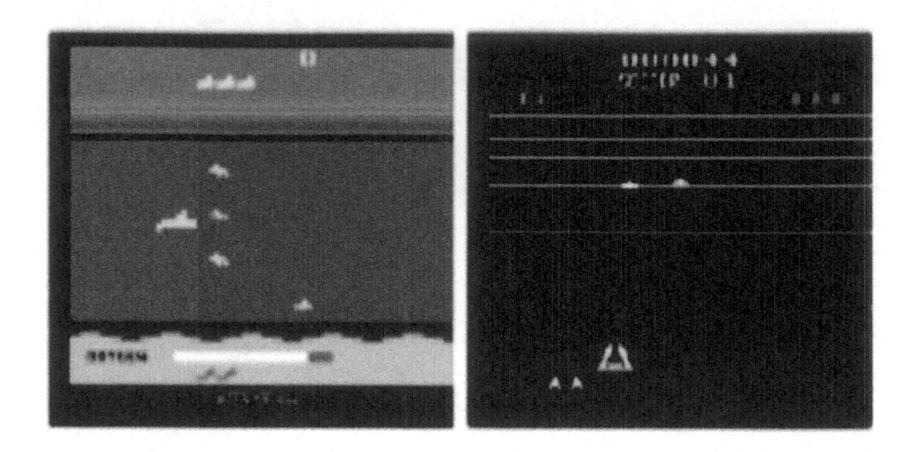

FIGURE 15.10: *Deep learning startup DeepMind, now part of Google, applied their deep learning and reinforcement learning technology to make a system capable of playing Atari 2600 video games. The AI system perceived the games via a camera pointed at the screen, rather than by digesting the data directly from within the computer.*

The price tag of Google's purchase of DeepMind surprised me, but purchase didn't particularly. I knew Demi Hassabis, the chief founder of DeepMind, from a couple AI conferences. Further, Shane Legg, one of the other two co-founders of the firm, had worked for me at Webmind in the period 2000-2001, and I'd kept in touch with him over the years. In fact, as I noted above, Shane was the one who suggested the term "AGI" to me back in 2001, when I was looking for a title for an edited book. Amusingly, in an early version of the manuscript for this book, a couple years before Google bought DeepMind, I had the following paragraph:

"The only group I know of, at the moment, making a serious attempt to create an AGI system directly inspired by the brain is Deep Mind, a company based in the UK and led by Demis Hassabis. Demis is an impressively entrepreneurial individual with a diverse background in AI, software development and neuroscience. A champion in various game-playing contests, he started a successful computer game company, and then went into neuroscience research. After publishing some fairly major papers on the neural foundations of intelligence, he decided to start an AI company, with a dual focus on creating AGI and making video games featuring intelligent game characters. Demis's views fall generally into the "deep learning" camp, in the sense that he believes hierarchical pattern recognition in the rough manner of the visual and auditory cortex is the

key to brain-like intelligence. But he also has a rich understanding of the diversity of the brain, and the different architectures and dynamics that its different regions possess. The details of his team's work are proprietary, and although Demis's AGI thinking began with neuroscience, it's unclear to me to what extent Deep Mind is really following a brain-based path. They may well be proceeding in a more opportunistic manner, combining more neurally realistic components with more computer science based components, based on the different levels of neuroscience knowledge available about different parts of the brain."

Well, Demis is now quite rich, and Shane is rather wealthy as well; and they are both, at the time of writing, Google employees. They're now pursuing their AGI dreams within Google, which surely brings them amazing new resources, and also complex new constraints (a topic I'll return to in the final chapter).

DeepMind has not been the only deep learning firm to make it big financially. Vicarious Systems, a California deep learning startup, reaped a large investment round from a number of grade-A investors, including Facebook founder Mark Zuckerberg. Their kick-ass demo was a solution to the CAPTCHA problem – those annoying "prove you're not a computer" images of words or numbers you have to type in to access many websites. It's actually not clear that this is an extremely hard problem; it's a problem that the computer vision community hasn't focused much time on, because CAPTCHAs serve a useful role online and publishing technology for solving them would have negative rather than positive social value at this time. But Vicarious didn't publish details of their CAPTCHA-solving code, they just used their conquest of CAPTCHA as a tool in their fundraising campaign, and did so very successfully.

FIGURE 15.11: *Deep learning company Vicarious Systems made headlines by announcing that their deep learning system had solved the CAPTCHA problem – a problem one encounters online every day when trying to log into websites.*

Deep learning approaches to audition have also been successful recently, though this has not attracted as much attention as the vision research. For a long time the most effective approach to speech-to-text was a relatively simple technique known as "Hidden Markov Models" or HMMs. HMMs appear to underlie the technology of Nuance, the 800-pound gorilla of speech-to-text companies. But in 2013 Microsoft Research published a paper indicating their deep learning speech-to-text system could outperform HMMs.[2] In December 2014, Andrew Ng's group at Baidu announced a breakthrough in speech processing: a system called Deep Speech, which reportedly gives drastically fewer errors than previous systems in use by Apple, Google and others.[3]

And there have been other, less widely publicized successes on different types of data as well. In late 2014, as I write this chapter, I note there has been a spate of investments in "cloud computing, machine learning as a service" companies – e.g., one called Scaled Inference touted as "Big Machine Intelligence for All" and with a host of big-time investors; one called MetaMind, founded by Richard Socher, an outstanding young researcher who caught my eye a few years ago due to his innovative applications of deep learning to natural language parsing.

---

2 http://research.microsoft.com/apps/pubs/default.aspx?id=188864

3 http://arxiv.org/abs/1412.5567

My good friend Itamar Arel, with whom I've collaborated on deep learning vision work (more on that below), has been busy the last few years applying his own deep learning technology in a different domain: financial market prediction. One of the unique claims to fame of his deep learning tech is its superior handling of temporal dependencies in data – obviously very important in financial markets. He tells me his system has been making good money, trading a fund operated by his company Binatix in collaboration with a major financial firm.

Of course, as generally happens, the successes and promise of the deep learning approach have been overblown here and there in the popular media. Suddenly one reads tech media articles touting deep learning as the path to AGI – when in fact the successes of deep learning, though real and excited, are mainly limited to certain aspects of sensory perception, which is really only a small fraction of what happens in a human mind, and not necessarily a representative sample of overall human mental activity. But that's just the way the media works in today's society. I would never blame researchers for any excesses in media reportage on their work. Anyone who has dealt with reporters knows that there's no way to control what they say about you or your work, though with a lot of effort or money you can sometimes guide what the media as a whole says about you in a rough statistical way... Andrew Ng addressed the hype problem directly in a recent interview:

"There are a lot of deep-learning startups. Unfortunately, deep learning is so hot today that there are startups that call themselves deep learning using a somewhat generous interpretation. It's creating tons of value for users and for companies, but there's also a lot of hype. We tend to say deep learning is loosely a simulation of the brain. That sound bite is so easy for all of us to use that it sometimes causes people to over-extrapolate to what deep learning is. The reality is it's really very different than the brain. We barely (even) know what the human brain does."[4]

A Pattern of Oversimplification

Some deep learning researchers, like Andrew Ng in the quote given above, downplay the biological basis of their work. They treat biology as

4 http://blogs.wsj.com/digits/2014/11/21/baidus-andrew-ng-on-deep-learning-and-innovation-in-silicon-valley/

a general inspiration, but view themselves as creators of computational algorithms, where "getting stuff to work" is the primary goal.

Other deep learning researchers posit their algorithms and architectures as capturing the essence of what the human brain does. The brilliantly diverse entrepreneur Jeff Hawkins is a leading example here. Hawkins made a fortune with the PalmPilot and Treo handheld devices, and after he retired from that business, he went into neuroscience and AGI. This is certainly a more interesting occupation than the typical retired tech tycoon. And to his credit, he also donated a lot of money to support science, including founding the Redwood Neuroscience Institute at Berkeley, and helping the Cold Spring Harbor Labs in Long Island, etc.

In 2004, Hawkins wrote a book called *On Intelligence* (together with science journalist Sandra Blakeslee), which advocated a kind of deep learning-based model of the human brain, together with a computational algorithm called Hierarchical Temporal Memory (HTM) that he posited to capture the essence of what the brain does. The book was very well put together and achieved a fair degree of influence. However, it reminded me somewhat of a quote I read somewhere or other about quantum mechanics pioneer Erwin Schrodinger's book *What is Life?* — "What was original in the book was not true; and what was true was not original." Nonetheless, Schrodinger's book was very influential and helped catalyze the formation of the discipline of molecular biology. And Hawkins's book also had a lot of positive influence, in helping build interest in AGI and computational neuroscience, and spreading knowledge about certain aspects of brain function.

Shortly after *On Intelligence* came out, I attended a discussion on the book at the National Institutes of Health (which is based near Washington DC, where I lived at the time), organized by my friend Jim DeLeo (who worked at the NIH, mostly on applying AI to clinical data). The consensus of the neurobiologists there was that Hawkins's ideas about the brain were all basically common lore among neuroscientists — but neuroscientists tended not to state them as plainly and crisply as Hawkins because they were too well aware of the many additional complexities involved. Hawkins' book provided a nice summary of certain aspects of neuroscience; it gave an argument that intelligence can be understood in terms of memory and prediction, and outlined how the visual cortex handles memory and prediction in the context of visual data processing. But then it went beyond this to argue that a fairly simplistic model of the visual cortex's hierarchical "deep learning"-type dynamics, could be thought of as the crux of human intelligence. None of the NIH neuroscientists I talked to were willing to follow him on this final step.

My main complaint with Hawkins' approach, as described in *On Intelligence*, was that it was essentially a model of the brain's visual and auditory cortexes, yet it's being proposed as a model of the whole of "intelligence." It may be an OK model of vision and audition, although there are arguments to be made even there, but it has nothing directly to say about action, language parsing, social reasoning, emotion, or a whole lot of other things that are critical to human intelligence. It doesn't even have much to say about senses like smell and touch, whose corresponding brain regions don't have the marked hierarchical structure and dynamics that the HTM model proposes. Hawkins has significantly revised the particulars of his approach in the decade since his book was published, but he hasn't really remedied the issues that concerned me.

I've never had a conversation with Jeff Hawkins, though we interacted briefly online once. But after following his work for a while, I eventually realized that a lot of his early algorithmic designs were the result of his collaboration with Dileep George, who was a grad student at Stanford when they started working together. Dileep eventually moved on from Numenta and founded Vicarious Systems, the company that made headlines for solving CAPTCHA that I mentioned above. I invited Dileep to give a keynote speech at the AGI-13 conference in Beijing, which he graciously accepted (and it was fairly convenient for him as he was in Asia anyway to visit his family in India), and we got the chance to discuss various AGI issues a fair bit. Dileep's views were similar to those of his mentor but a bit more measured. Dileep also felt that the crux of the human brain/mind was some sort of hierarchical, temporal deep learning architecture. The architecture they were developing at Vicarious was different in many details from Hawkins's work, but similar at the level of philosophy.

He admitted that there were other aspects of intelligence his current work didn't deal with at all – episodic memory, action planning, language, and so forth. But he felt that these could be addressed more easily, and perhaps even relatively straightforwardly, once the core problem of perception of patterns in complex data via hierarchical pattern recognition had been compellingly solved. So his approach was to solve what he felt was the core problem first, and then move on from there. A perfectly reasonable, engineer-style approach was my impression – he didn't seem like so much of a theorist or a grand thinker, but he was someone who liked to make things work, and he had a strong intuition about what was the core thing he had to make work to get AGI to happen. He recognized that more might be needed to make a full-on AGI architecture, but figured he'd work toward that bottom-up, beginning via applying his deep learning approach to vision in a dramatically successful way. (Of course I'm paraphrasing here, and I hope Dileep will forgive me if I've missed some of the nuances of this portion of our conversation.)

Jeff Hawkins, more so than Dileep George, seems to me to have succumbed to the following "recipe for oversimplification":

1. Take a crude approximation of one part of the brain (in Hawkins' case, parts of visual and auditory cortex).
2. Hypothesize that the whole brain basically works like that one part in detail.
3. Try to make a quasi-simulation of that part of the brain.
4. Make various compromises in biological accuracy to achieve more computational efficiency.

OK, that oversimplifies things a little. If you leave out Step 2 it's perfectly reasonable. Hawkins places a big emphasis on Step 2; Andrew Ng much less so. Overall, different researchers adhere to Step 2 to varying degrees, at various times of day or various points during their career – these are complicated issues.

Part of the subtlety here is that the concept of Deep Learning is very broad, while the specific deep learning algorithms in play at the moment are pretty narrow and specific. So when some researcher says "the whole brain basically works like my deep learning vision system," one has to wonder, exactly what do they mean by "works like"? Hawkins says intelligence consists of "memory and prediction," and at a certain level this is right. The deep learning philosophy says that intelligence consists of recognizing patterns, and patterns among these patterns, and patterns among these higher level patterns, etc.; and that the dynamics of this pattern recognition involves information flowing from lower-level patterns in the process of formation, to higher-level patterns in the process of formation, and vice versa. Yes. No argument here. But from this general philosophical view of mind as nested-pattern learning with free-flowing adaptive dynamics, to the specific architectures and algorithms in use under the name "deep learning" today, is a humongous leap.

I think deep learning is a sound idea, in its most abstract form. Intelligence really is about recognizing patterns in data, and then patterns among these patterns, and patterns among these patterns, etc. However, the particular architectures going under the name "deep learning" these days tend to be much more rigid and specialized than this general notion of recursive, hierarchical pattern recognition. In my view, they tend to be much more appropriate for visual and auditory pattern recognition than for, say, linguistic or mathematical pattern recognition. This is not by any means a new point in the AI and cognitive science field, but it sometimes gets glossed over in the deep learning community. Cognitive scientist Stellan Ohlson wrote a book called *Deep Learning* which makes this point very emphatically, via formulating the concept of deep learning as a general set of information-processing principles. Ohlson makes it clear that these principles could be implemented in many different kinds of systems, including neural networks, but also including logic systems or various rule systems or whatever...[5]

- ***Spontaneous activity***: The cognitive system is constantly doing stuff, always processing inputs if they are there, and always reprocessing various of its representations of its inputs.
- ***Structured, unbounded representations***: Representations are generally built out of other representations, giving a hierarchy of representations. The lowest level representations are not fixed, but are ongoingly reshaped based on experience.
- ***Layered, feedforward processing***: Representations are created via layers of processing units, with information passing from lower layers up to higher layers.
- ***Selective, capacity-limited processing***: Processing units on each layer pass information upward selectively – each one generally passes up less information than it takes in, and doesn't pass it everywhere that it could.
- ***Ubiquitous monotonic learning***: Some of the representations the system learns are stored in long term memory, others aren't.
- ***Local coherence and latent conflict***: The various representations learned by a system don't have to be consistent with each other overall. Consistency is worked toward locally when inconsistencies between elements are found; there's no requirement of global consistency.

5 In the below list, the boldfaced titles of each bullet point are Ohlson's wording, but the text explanation after each bullet point comprises my explanations of his ideas, i.e., my wording, not his.

- ***Feedback and point changes***: Higher level processing units feed information down to lower level units, thus potentially affecting their dynamics.
- ***Amplified propagation of point changes***: A small change anywhere in the processing hierarchy might cause a large change elsewhere in the system – as typical of complex and "chaotic" dynamical systems.
- ***Interpretation and manifest conflict***: Conflict between representations may go unnoticed until a particular input comes in, which then reveals that two previously learned representations can be in conflict.
- ***Competitive evaluation and cognitive utility***: Conflict between representations are resolved broadly via "reinforcement learning," i.e., based on which representation proves most useful to the overall system in which context.

These sorts of general principles are very dry and boring to read, and it's hard to understand what they mean without some specific context. Ohlson, in his book, elaborates them in detail in the context of a host of different cognitive psychology experiments. My point here is just how much broader the general conception of "deep learning" is than the specific deep learning vision and audition systems that are getting so much mileage and attention lately. Experimenting with deep learning for vision and audition is an awesome and important thing to do, but one shouldn't lose sight of the fact that these are very narrow systems.

It's interesting to note that, of Ohlson's principles of deep learning, only one ("Representations are created via layers of processing units") does not apply to OpenCog's AtomSpace. All the rest are true of OpenCog generically. So to turn OpenCog into a deep learning system, it suffices to arrange some OpenCog Nodes into layers of processing units. Then the various OpenCog learning dynamics – including, e.g., Probabilistic Logic Networks reasoning, which is utterly different than anything commonly pursued under the name of "deep learning" – become deep learning dynamics.

But of course, restricting the network architecture to be a hierarchy doesn't actually make the learning or the network any deeper. A more freely structured hypergraph like the general OpenCog Atomspace is just as deep as a deep learning network, and has just as much (or more) complex dynamics. The point of hierarchical architectures for visual and auditory data processing is mainly that, in these particular sensory data processing domains, one is dealing with information that has a pretty strict hierarchical structure to it. It's very natural to decompose a picture into subregions, subsubregions and so forth; and to define an interval of time (in which, e.g., sound or video occurs) into subintervals of times. As we are dealing with space and time, which have natural geometric structures, we can make a fixed processing-unit hierarchy that matches the structure of space and time – lower-down units in the hierarchy dealing with smaller spatiotemporal regions; parent units dealing with regions that include the regions dealt with by their children; etc. For this kind of spatiotemporal data processing, a fairly rigid hierarchical structure makes a lot of sense (and seems to be what the brain uses). For other kinds of data, like the semantics of natural language, abstract philosophical thinking, or even thinking about emotions and social relationships, this kind of rigid hierarchical structure seems much less useful, and a more freely-structured architecture of the generic OpenCog Atomspace seems more appropriate.

In the human brain, it seems the visual and auditory cortices have a very strong hierarchical pattern of connectivity and information flow, whereas the olfactory cortex has more of a wildly tangled up, "combinatory" pattern. This combinatory pattern of neural connectivity helps the olfactory cortex to recognize smells using complex, chaotic dynamics, in which each smell represents an "attractor state" of the olfactory cortex's nonlinear dynamics (as neuroscientist Walter Freeman has argued in a body of work spanning decades). The portions of the cortex dealing with abstract cognition have a mixture of hierarchical and combinatory connectivity patterns, probably reflecting the fact that they do both hierarchy-focused pattern recognition as we see in vision and audition, and attractor-based pattern recognition as we see in olfaction. But this is largely speculation – until we can somehow make movies of the neural dynamics corresponding to various kinds of cognition, we won't really know how these various structural and dynamical patterns come together to yield human thinking.

Hints of Narrowness

Right now the AGI community doesn't have any consensus on how to interpret the recent explosion of results from, and interest in, deep learning technology. There's no doubt that recent deep learning work has been extremely useful for various practical purposes. But is it a big step on the path to human-level AGI (and beyond)? Or is it just the latest, greatest version of narrow AI, without so many broader implications? As I hope I've made clear, the fact that specialized deep learning architectures like CNNs embody some general principles of deep learning doesn't necessarily imply that these architectures are going to have broad applicability, or serve as a stepping stone to vastly broader and more adaptable AGI systems. This is because these general principles of deep learning are far more abstract than current deep learning architectures, and can be manifested in many different ways, including various more general and more specialized ways.

Many deep learning researchers find it obvious that current deep learning research is on the direct path toward advanced AGI. Based on our many discussions, I know Itamar Arel feels that way – he thinks that deep learning-based perception of videos, for example, encapsulates the key principles and hard problems of general human intelligence, so that once we have deep learning architectures that can robustly recognize what's in videos, we'll be a long way toward human-level AGI. Jeff Hawkins, as noted above, is on record emphatically putting forth this kind of perspective as well.

On the other hand, some AGI researchers consider it obvious that contemporary deep learning research, based on supervised classification of labeled databases of various sorts, has next to nothing to do with AGI. A good example is Doug Hofstadter, whose book *Godel, Escher, Bach* was my first introduction to serious AI thinking. As he says, "*I don't want to be involved in passing off some fancy program's behavior for intelligence when I know that it has nothing to do with intelligence. And I don't know why more people aren't that way.*"

Roughly speaking, he sees modern machine learning algorithms as ways of impersonating limited aspects of intelligence by cleverly cheating, rather than as things that actually manifest significant intelligence.

Hofstadter's style of AI research is diametrically opposed to that of the machine learning community. Rather than using big data and massive processing power, he prefers to work on very small toy problems, where he can understand exactly what's going on. His aim is to understand the key algorithms of thought, which he figures can be largely done by focusing on very simple problems and trying to get machine cognition right in this context. For instance, one of his first serious AI program was Jumbo, a system that worked on the word jumbles you find in newspapers.

FIGURE 15.13: *Hofstader champions a "small data" approach to AGI research, where one tries to build systems embodying human-cognition-like algorithms for solving simple problems – like these newspaper word-jumble puzzles. (http://joshreads.com/images/10/02/i100219jumble.png).*

Word jumbles are trivial to solve algorithmically if one wants to take a "brute force" approach that searches through all possibilities. However, Hofstadter's goal was not to build a highly effective Jumble solver, but rather to write a program embodying how his own mind worked when solving Jumble puzzles. When solving these puzzles, he said "*I could feel the letters shifting around in my head, by themselves*," he told me, "*just kind of jumping around forming little groups, coming apart, forming new groups — flickering clusters. It wasn't me manipulating anything. It was just them doing things. They would be trying things themselves."*[6] His early Jumble program was the start of a series of AI programs aimed at solving small, simple sorts of problems in human-like ways.

Hofstader represents an extreme version of the "cognitive modeling" perspective on AI. My own view, as you've probably guessed, lies roughly halfway between Itamar and Doug Hofstadter. I think the recent deep learning work does constitute genuine progress toward AGI – it tells us a fair bit about how to make scalable machine perception work, which is an important aspect of human intelligence; and it also gives us some insights into deep learning which go beyond the perception domain. On the other hand, I also think that it's missing a lot of aspects of human cognition – and one of these aspects is the sort of experimental, creative, analogical thinking that fascinated Hofstadter so much.

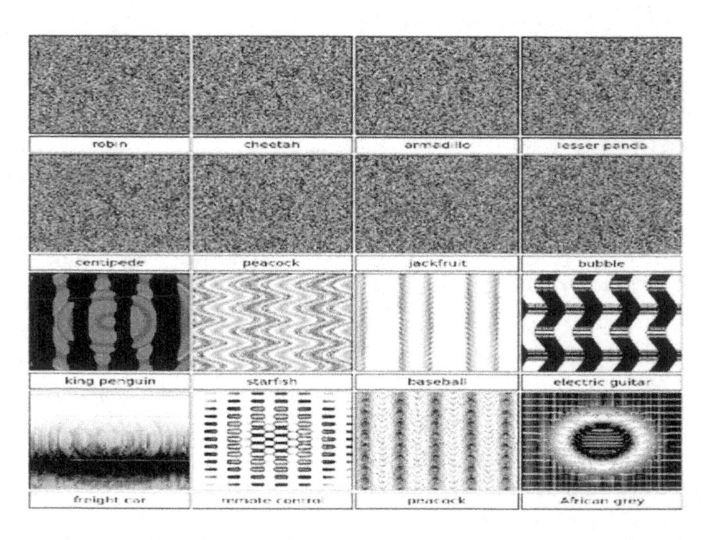

FIGURE 15.14: *Examples of images that are unrecognizable to humans, but that state-of-the-art deep neural networks trained on the standard ImageNet image collection believe with >=99.6% certainty to be a familiar object. (http://arxiv.org/pdf/1412.1897v2.pdf).*

Some interesting pieces of evidence that could be interpreted to support Hofstadter's view have emerged in the deep learning literature lately. In a paper titled "Deep Neural Networks are Easily Fooled: High Confidence Predictions for Unrecognizable Images", one group of researchers showed they could construct images that looked random to the human eye, but

6 http://m.theatlantic.com/magazine/archive/2013/11/the-man-who-would-teach-machines-to-think/309529/

that were classified by a CNN deep learning vision network as representing particular kinds of objects, with high confidence.[7] So, a picture that looks like random noise to any person might look exactly like a frog or a cup to the CNN. See FIGURE for examples.

Another group, in a paper titled "Intriguing properties of neural networks," showed that by making a very small perturbation to a correctly classified image, they could cause the deep network to misclassify the image.[8] The perturbations in question were so small that humans wouldn't even notice. See FIGURE 15.15 for examples.

FIGURE 15.15: *Examples of images that look the same to humans, but are classified differently by deep neural networks. All images in the right column are incorrectly classified as ostriches by the CNN in question. The images in the left column are correctly classified. The middle column shows the difference between the left and right column. (http://arxiv.org/abs/1312.6199).*

Now, these phenomena have no impact on practical performance of convolutional neural networks. So one could view them as just being mathematical pathologies found by computer science geeks with too much time on their hands. The first pathology is pragmatically irrelevant because a real-world vision system is very unlikely to ever be shown weird random pictures that just happen to trick it into thinking it's looking at some object (most weird random pictures won't look like anything to it). The second

7 http://arxiv.org/abs/1412.1897

8 http://arxiv.org/abs/1312.6199

one is pragmatically irrelevant because the variations of correctly classified pictures that will be strangely misclassified are very few in number. Most variations would be correctly classified. So these pathologies will not significantly affect classification accuracy statistics.

But still, I think these pathologies are telling us something. They are telling us that, fundamentally, these deep learning algorithms are not generalizing the way that people do. They are not classifying images based on the same kinds of patterns that people are. They are "overfitting" in a very subtle way – not overfitting to the datasets on which they've been trained, but rather overfitting to the kind of problem they've been posed. In these examples, these deep networks have been asked to learn models with high classification accuracy on image databases – and they have done so. They have not been asked to learn models that capture patterns in images in a more generally useful way, that would be helpful beyond the image classification task – and so they have not done that.

When a human recognizes an image as containing a dog, it recognizes the eyes, ears, nose and fur, for example. Because of this, if a human recognized the image on the bottom left of the right image array in FIGURE 15.15 as a dog, it would surely recognize the image on the bottom right of the right image array as a dog as well. But a CNN is recognizing the bottom left image differently than a human, in a way that fundamentally generalizes differently, even if this difference is essentially irrelevant for image classification accuracy.

I strongly suspect there is a theorem lurking here, stating that these kinds of conceptually pathological classification errors will occur if and only if the classification model learning algorithm fails to recognize the commonly humanly recognizable high-level features of the image (e.g., eyes, ears, nose, fur, in the dog example). Formulating a theorem like this rigorously may be a difficult problem, and I haven't found time to try.

Informally, though, what I think is: The reason these pathologies occur is that these deep networks are not recognizing the "intuitively right" patterns in the images. They are achieving accurate classification by finding clever combinations of visual features that let them distinguish one kind of picture from another – but these clever combinations don't include a humanly meaningful decomposition of the image into component parts, which is the kind of "hierarchical deep pattern recognition" a human's brain does on looking at a picture.

There are other kinds of AI computer vision algorithms that do a better job of decomposing images into parts in an intuitive way. Stochastic image grammars, as shown in FIGURE 15.16 and FIGURE 15.17, are one good example. However, these algorithms are more complicated and more difficult to implement scalably than CNNs and other currently popular deep learning algorithms, and so they have not yet yielded equally high-quality image classification results. They are currently being developed only minimally, whereas CNNs and their ilk are being extremely heavily funded in the tech industry.

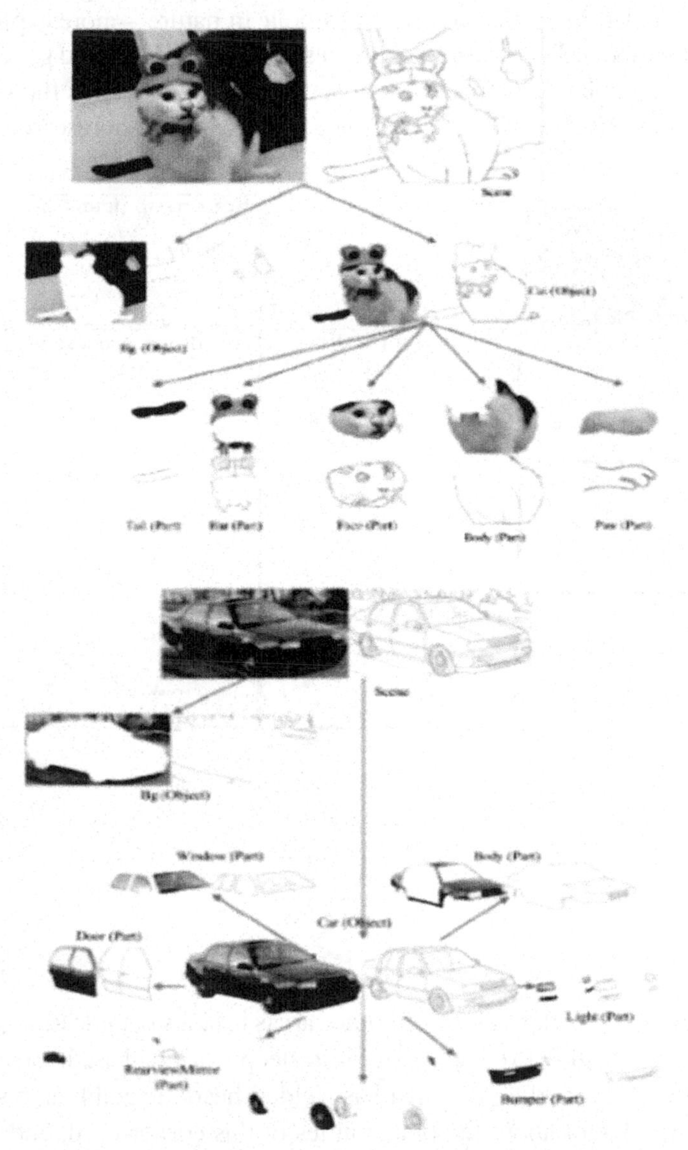

FIGURE 15.16: *Stochastic image grammars, an alternative approach to AI image analysis, attempt to automatically decompose images into component parts in a manner similar to the above diagrams. (http://www.stat.ucla.edu/~sczhu/papers/Reprint_Grammar.pdf).*

Might it be possible to progressively improve current deep learning algorithms to make them more "cognitively sophisticated," so that they can internally recognize natural object decompositions in the manner of a stochastic image grammar? Quite possibly. Or, it may be that the best route to doing this is to hybridize conventional deep learning algorithms with different algorithms that are more symbolic in nature – more explicitly focused on symbolically representing images in terms of parts, and generally on sensible symbolic decompositions of sense data. My bet is on the latter, personally, but all this is wide open and the subject of current research.

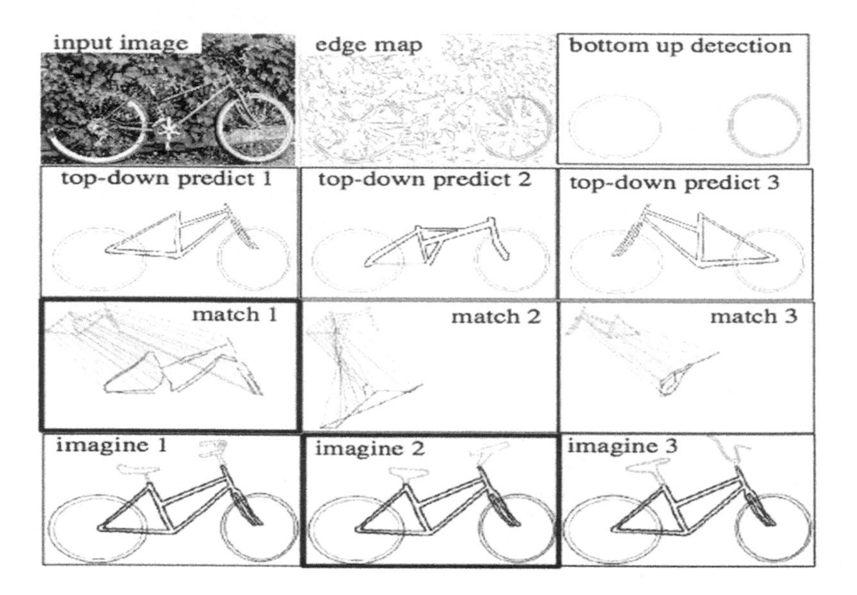

FIGURE 15.17: *A stochastic image grammar is able to recognize a bicycle even though it's partly hidden by bushes, because via looking at prior images of bicycles, it has learned what the parts of a bike are and how they relate to each other. (http://www.stat.ucla.edu/~sczhu/papers/Reprint_Grammar.pdf).*

Bridging the Symbolic/Subsymbolic Gap

Most of my AI work has involved other areas besides deep learning, but in the years since 2009 I've dabbled a bit in deep learning-based computer vision. None of my work in this area has yielded headline-grabbing results, but I've learned a lot about the practicalities of this corner of AI, and have been spurred to think a lot about how to interface deep learning systems with other AI systems such as OpenCog's Atomspace and its associated learning mechanisms.

FIGURE 15.18: *The inestimable Dr. Itamar Arel, speaking here at the 2009 Singularity Summit in New York City. Along with Demis Hassabis and Yoshua Bengio, Itamar is one of the deep learning researchers who has struck me personally as having an especially strong understanding of the overall AGI problem. (Of course, others whom I don't know so well may also understand things deeply – I'm always interested to mine other peoples' perspectives, but there's not enough time to get to know EVERY interesting researcher; not with the degree to which AI is flourishing these days…)*

I got involved in deep learning via my friendship with Itamar Arel, who – among many other projects – worked with his students at the University of Tennessee Knoxville to develop a deep learning pattern recognition system called DeSTIN. In the big picture, DeSTIN is roughly somewhat like Hawkins's HTM. But, based on my experimentation with the public version of Hawkins's system, Itamar's attempt seems to work better; and when you dig into the details many of the concepts are subtly different.

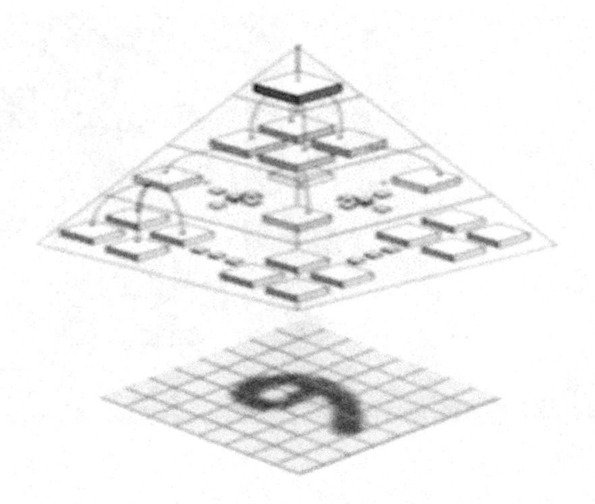

FIGURE 15.19: *Hierarchical structure of HTM, DeSTIN and other similar vision processing/AGI systems, modeled conceptually on the hierarchical structure of visual cortex architectures. Right: The same hierarchical structure viewed as a 3D pyramid.*

To avoid using the term "kind of like Hawkins's HTM," a few years ago I introduced the term CSDLN, for Compositional Spatiotemporal Deep Learning Network, which basically means a deep learning network whose nodes each refer to larger and larger regions of space-time. Not all deep learning networks need to be this way of course – the concept of deep learning is much broader. This term has not caught on at all, probably because it's long and ugly, but I have continued to use it now and then because nobody has yet made up a better one with the same meaning.

Experimentation with Itamar's DeSTIN system was my own personal introduction to practical deep learning experimentation. In the last couple years I've set DeSTIN as a system aside and have been working with other DeSTIN-like approaches, more thoroughly integrated into OpenCog. But I still think DeSTIN is a fascinating, impressively simple architecture. The early work I did with DeSTIN paved the way for the more sophisticated "deep learning in OpenCog" stuff I'm doing now.

DeSTIN is an UNsupervised hierarchical pattern-recognition system that recognizes patterns in a stream of inputs. It doesn't need labeled training data to do its thing, it just looks at the stream of input images and tries to recognize patterns therein. It doesn't exactly have cortical columns, but it's kind of similar. Since it is a CSDLN, a Compositional Spatiotemporal Deep Learning Network, it has nodes corresponding to different space-time regions of the observed world, arranged in a hierarchy, and higher-up nodes refer to larger space-time regions and more abstract patterns.

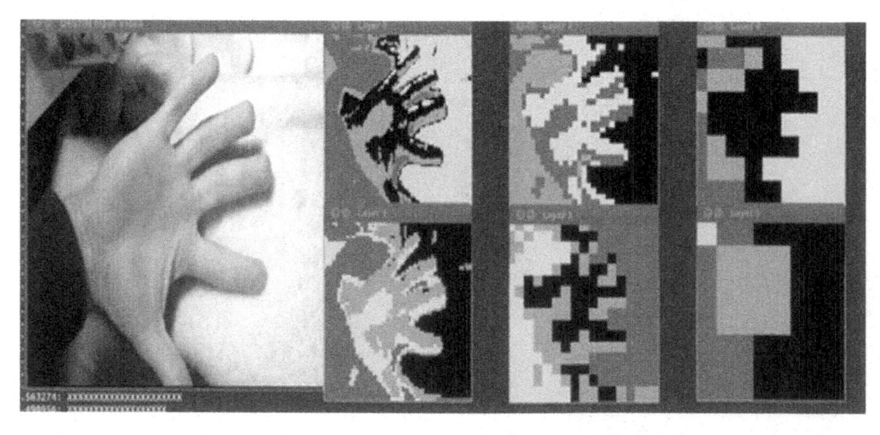

FIGURE 15.20: *Visualization of what various layers in the DeSTIN hierarchy see when a hand is placed in front of the camera whose output DeSTIN sees. Image courtesy of Ted Sanders, the DeSTIN developer whose hand is shown in the picture.*

The patterns recognized in each node depend on the patterns in the nodes above (the parent nodes) and below (the child nodes) – so there is both top-down and bottom-up learning going on.

According to the specific learning dynamic of DeSTIN: *The patterns recognized in each node are basically "clusters" of similar combinations of patterns recognized in the child nodes. Information from a parent node is also allowed to guide pattern recognition in a child node, via giving the child node what is called "parental advice."* If you hook DeSTIN up to a webcam or a robot's camera eye, it reacts to its inputs, propagates information up and down its internal hierarchy, and settles into states that tell you something about what objects and events the robot is seeing.

My own interest in DeSTIN resided largely in the fact that I felt I knew a cool way to connect it to my own OpenCog AGI architecture. My view was that OpenCog covers a lot of other aspects of intelligence that DeSTIN doesn't touch – but that DeSTIN, or some DeSTIN-like system, could potentially serve as a visual and auditory cortex for OpenCog.

Itamar, on the other hand, seemed to think he could basically take DeSTIN (or more some improved version of DeSTIN), extend it a bit, implement it on a lot of machines, tweak the algorithms a little, connect it to a robot, and get advanced general intelligence. Well, OK, that's a bit of an oversimplification. Actually, he was always planning to create an action hierarchy similar to the perception hierarchy, and then a reward hierarchy that gets a stimulus when the system has done something good or bad, passing this along to the action hierarchy, which then passes it along to the perception hierarchy.

My own view is that for DeSTIN or any DeSTIN-like system to achieve anything like human-level intelligence, major additions would have to be made. Action and reinforcement hierarchies would not be enough; you'd need a lot more. The human brain is a lot more complex than two or three coupled hierarchies, and any AGI system that's vaguely like the human brain ought to be a lot more complex than that too. One would need a system with multiple different architectures corresponding to various brain regions, all connected and interoperating, yet each with a unique function.

For example, take "episodic memory" (your life story, and the events in it), as opposed to less complex types of memory. The human brain is known to deal with the episodic memory quite differently from the memory of images, facts, or actions. Nothing in architectures like HTM or DeSTIN tells you anything about how episodic memory works. Jeff Hawkins, or Itamar, would argue that the ability to deal with episodic memories effectively will just emerge from their hierarchies, if their systems are given enough perceptual experience. It's hard to definitively prove this is wrong, because these models are all complex dynamical systems, which makes it difficult to precisely predict their behavior. Still, in my opinion, the brain doesn't appear to work this way; episodic memory has its own architecture, different in specifics from the architectures of visual or auditory perception. I suspect that if one wanted to build a primarily brain-like AGI system, one would need to design fairly specialized circuits for episodic memory, plus dozens to hundreds of other specialized subsystems.

In 2012, together with Ted Sanders and Jared Wigmore, I carried out some simple experiments in the direction of integrating DeSTIN and OpenCog. I viewed this as an experiment in what I call "bridging the symbolic/subsymbolic gap." OpenCog's Atoms can be used to represent abstract symbols – words, logical relationships, and so forth. They can also be used to represent percepts and actions. DeSTIN is purely "subsymbolic" – it just represents patterns in perceptual data. Conceivably abstract symbols could be made to emerge within a DeSTIN network, but nobody has seen that yet in DeSTIN or any similar network, except in very simplistic ways.

Brushing oh so many details aside, connecting DeSTIN and OpenCog means bridging a subsymbolic and a symbolic system. This is a kind of integration that has a lot of resonance in terms of the history of AI, because historically there have been some researchers arguing that symbolic reasoning (e.g., logical inference, language parsing, etc.) is the only path to AGI, and others arguing that subsymbolic pattern recognition (e.g., neural nets analyzing visual or sound data, and generating actions via reinforcement learning to the actuators of a robot) are the only path to AGI. At some stages in the history of AI, these two camps have stood in almost religious opposition. Now the opposition is less and there are regular conferences on "neural-symbolic computing." But still, bringing together symbolic and subsymbolic methods really successfully would be a Big Deal in the AI community.

My idea regarding integration of DeSTIN with OpenCog was to add another layer of unsupervised pattern recognition, in between DeSTIN and OpenCog. That is, to recognize patterns in the states of the DeSTIN network – in the multiple states that occur over time as it looks at different images – and then input these patterns into OpenCog. The idea was that these patterns in the DeSTIN network should represent a sensible decomposition of the image, vaguely similar to what one sees in a stochastic image grammar.

Ted Sanders implemented a prototype of this idea in late 2012, which yielded some interesting results when tested on small images of numbers and letters and simple shapes; but we didn't have funding earmarked for continuing that work and I ended up not allocating any of my (not that substantial) personal time or free R&D funds to the project due to other priorities, so the work never got extended to more complex data. But we will soon be taking the next steps using a slightly different approach!

FIGURE 15.21 shows some of the results of this prototype work, which was presented at the AGI-13 conference in Beijing. Indeed, in this experiment, the patterns recognized among DeSTIN network states seemed to reflect meaningful portions of the simple shapes to which DeSTIN was exposed.

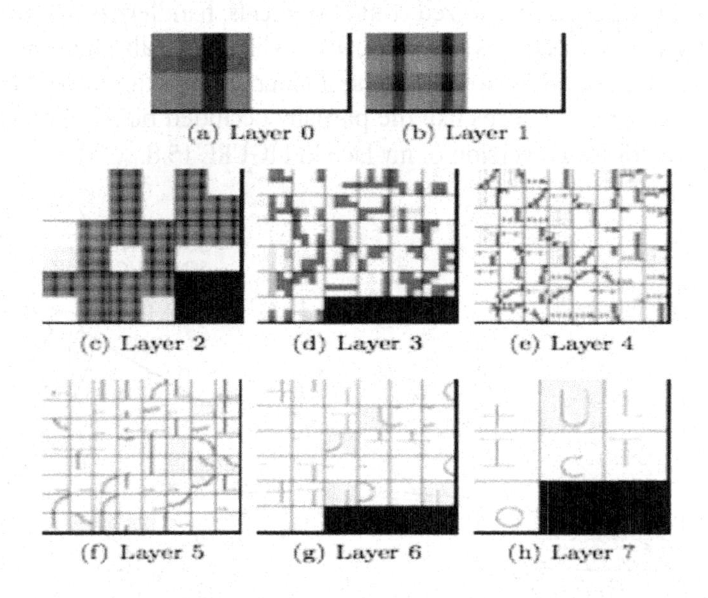

FIGURE 15.21: *These figures show the result of some prototype work Ted Sanders and I did, to explore the idea of extracting patterns from series of DeSTIN states and exporting them to OpenCog's Atomspace for follow-on reasoning. Some simple images fed to DeSTIN for testing purposes. Bottom left: Visualization of the library of patterns that DeSTIN found in these images, on different levels of its hierarchy. The patterns that were abstracted by doing an extra step of pattern mining on the DeSTIN states achieved by observing the simple test images. These abstract patterns are suitable to be fed into OpenCog and reasoned about.*

One could view this kind of experimentation as lying somewhere between classic deep learning-style experimentation, and Hofstadter-style toy problem experimentation. We are doing training on a corpus (of images depicting simple shapes). But we're also implementing algorithms explicitly aimed at yielding cognitively meaningful results, not just high classification accuracies or other impressive statistical results.

The vision underlying this experimental DeSTIN/OpenCog hybridization work – and my current work with deep learning OpenCog as well – is shown in FIGURE 15.22 below. The figure shows the DeSTIN network connected to the OpenCog AtomSpace (which is, in this context, serving as more of a "cognitive semantic network" as opposed to DeSTIN's "deep perceptual network"). In the context of face recognition, for example, the pattern recognition intervening between DeSTIN and OpenCog should be able to recognize the eyes, nose and mouth as key elements of the face. OpenCog could then recognize higher-level patterns like "a face usually has two eyes, next to each other" and "the mouth is usually below the nose." In the context of bicycles, the patterns recognized could be wheels, handlebars, frame, etc.; and the higher-level patterns would be things like "usually there are two wheels with a frame in between." These higher level patterns could then be used to recognize patterns like the partially occluded bicycle in FIGURE 15.20, and the distorted version of my face in FIGURE 15.8.

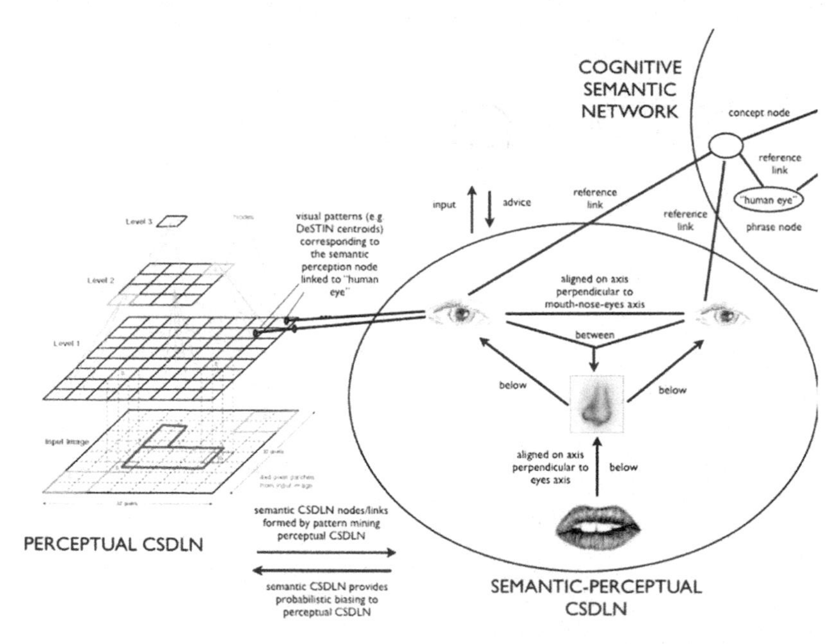

FIGURE 15.22: *Depiction of DeSTIN (or any other CSDLN = Compositional Spatiotemporal Deep Learning Network) connected to the OpenCog Atomspace, via patterns in the DeSTIN hierarchy over time being recognized and exported into the Atomspace. The "semantic-perceptual CSDLN" depicted would also be embedded in the Atomspace.*

DeSTIN currently is about visual perception. Extending a similar approach to auditory perception would be straightforward in principle – one would just need to reinterpret the lowest layer as related to sounds rather than pictures. Extending it to movement is also conceptually straightforward.

FIGURE 15.25 depicts a hypothetical utilization of a roughly DeSTIN-like framework to deal with movement rather than perception – instead of each node dealing with the perceptions in a certain region of space-time, each node deals with movements in a certain part of an AI agent's body. In a robotics application, the nodes on the bottom level refer to movements of individual motors (e.g., the motor in a finger, the motor in an eyeball or elbow, the motor in a car's right front wheel, etc.). The nodes on the next level up refer to groupings of individual motors – e.g., a hand, a pair of eyes, the front wheels of a car, etc.

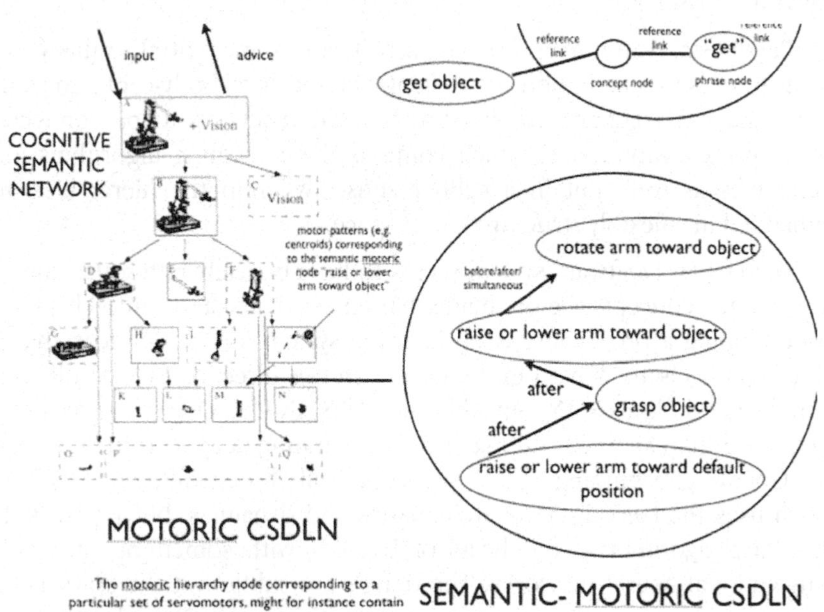

FIGURE 15.23: *Depiction of a deep learning movement hierarchy parallel in function and structure to DeSTIN, connected to the OpenCog Atomspace, via patterns in the DeSTIN hierarchy over time being recognized and exported into the Atomspace. The "semantic-motoric CSDLN" depicted would also be embedded in the Atomspace. The image shows the case of a simple robot arm, but the same concept holds for a humanoid robot or essentially any other embodiment. This architecture has been designed and analyzed but not, at time of writing, implemented.*

One can also create a similar hierarchy of goals, correlated with nodes in the movement hierarchy – lower-level goals like "place the left leg on the next stair up," higher-level goals like "walk up the stairs," etc. FIGURE 15.24 illustrates the interconnectivity of perception, action and goal hierarchies, and the OpenCog Atomspace. This is a fairly full architecture for robot control, integrating deep learning hierarchies for perception, action and goals with the full power of OpenCog for motivation, semantic reasoning and learning, attention control, and so forth. This is my conceptual vision for how to create the software for powering AGI-controlled robots. Of course there are many, many details to be gotten right to make this really work, and this aspect of OpenCog is still near the beginning; but I'm happy to have found some great collaborators on the robotics side, including David Hanson whom I mentioned above, and Mark Tilden, creator of the RoboSapien and pioneer of biomorphic robotics (also currently based in Hong Kong, like David and myself).

Rigidly structured deep learning architectures are central to this vision — but as tools for perception, movement, and low-level body-centered goals, rather than as a generic architecture for AGI. They are to be connected to OpenCog's Atomspace, which contains deep learning algorithms and hierarchies as well, but in a subtler sense, wherein the hierarchies are embedded in a flexibly structured Atomspace.

DeSTIN as a software system seems to have basically bitten the dust by now, but the concepts live on. Itamar has left academia and shifted his focus to developing a proprietary deep learning system for robotic learning, in his company Osaro. What I'm doing now in this direction is a bit different than the good old DeSTIN, but still pretty DeSTIN-like. One of the AI staff in iCog, my Ethiopian AI consulting firm, is integrating deep learning networks into OpenCog, beginning with an algorithm called "convolutional DeSTIN", which uses the basic DeSTIN architecture and dynamics, but replaces the clustering algorithm at the heart of DeSTIN with something more like convolutional neural networks. The details are subtle but the spirit is the same as in Itamar's earlier work and my variations on it. As in the work Ted Sanders and I did in 2012, the bridge between deep learning and symbolic cognition here is pattern recognition: repeated or surprising patterns in the state of the deep perceptual network are fed to cognitive processing and used as the basis for more abstract reasoning and learning.

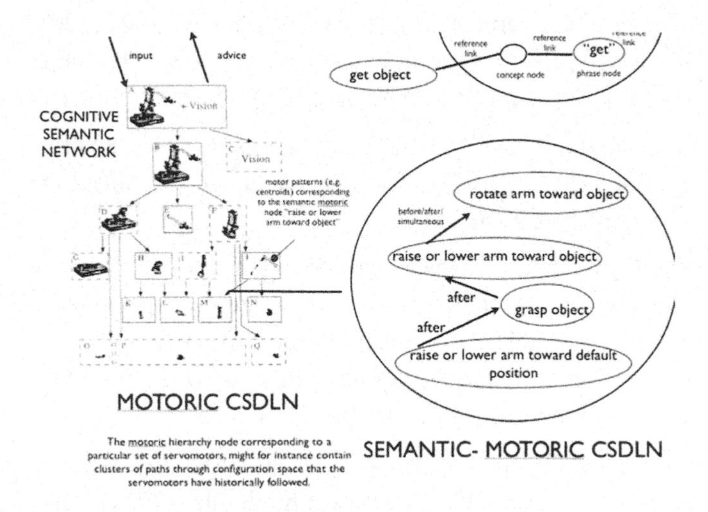

FIGURE 15.24: *Depiction of deep learning hierarchies, within the OpenCog Atomspace, reflecting patterns recognized in deep learning hierarchies for perception, movement and goals. Note how perception, action and goals all link together and work together. This architecture has been designed and analyzed but not, at time of writing, implemented yet.*

On the Road to AGI …or not?

Overall, the recent explosion in deep learning is an interesting case study of the relationship between narrow AI and AGI – and it illustrates wonderfully the subtlety of the connection between the two. Clearly the functionality of current deep networks is "narrow"; but this doesn't resolve the question of whether these narrowly functional systems can be extended into AGI systems, or hybridized within AGI systems, or whether they're just digressions from AGI that happen to be very pragmatically useful in certain contexts (as Hofstadter would say). The answer to this question depends on your ideas and presuppositions and hypotheses about AGI. I have my own views, which I've shared with you. Research and development are ongoing.

Further Reading

Bengio, Yoshua, Ian Goodfellow and Aaron Courville (2015). *Deep Learning*. MIT Press. Textbook in draft form at time of writing.

Goertzel, Ben and Ted Sanders and Jade O'Neill and Gino Yu (2013). Integrating Deep Learning Based Perception with Probabilistic Logic via Frequent Pattern Mining. *Proceedings of AGI-13 Beijing.*

Karnowski, Tom and Itamar Arel and D. Rose (2010). Deep Spatiotemporal Feature Learning with Application to Image Classification.[9] *Proceedings of ICMLA-10.*

Le, Quoc V., Marc'Aurelio Ranzato, Rajat Monga, Matthieu Devin, Kai Chen, Greg S. Corrado, Jeffrey Dean and Andrew Y. Ng (2012). Building High-Level Features using Large Scale Unsupervised Learning. In *Proceedings of the Twenty-Ninth International Conference on Machine Learning (ICML 2012).*

Mnih, Volodymyrm , Koray Kavukcuoglu, David Silver, Alex Graves, Ioannis Antonoglou, Daan Wierstra Martin Riedmiller (2011). "Playing Atari with Deep Reinforcement Learning", *NIPS Deep Learning Workshop 2013.*

Taigman, Yaniv, Ming Yang, Marc'Aurelio Ranzato and Lior Wolf (2014). DeepFace: Closing the Gap to Human-Level Performance in Face Verification. *Conference on Computer Vision and Pattern Recognition (CVPR), June 24, 2014.*

Yang, Yezhou, Yi Li, Cornelia Fermüller, Yiannis Aloimonos (2015). Robot Learning Manipulation Action Plans by "Watching" Unconstrained Videos from the World Wide Web. To be presented at *AAAI 15,*

9 This is an early paper on DeSTIN. For a list of the various papers published on DeSTIN, see Itamar's publication list on his website, or (as of 2015) the DeSTIN repository on GitHub.

16. AGI Against Aging

"I don't want to achieve immortality through my work; I want to achieve immortality through not dying."

-- Woody Allen

One of my life goals is to not die – and to avoid death for my family and friends and as many other humans as possible. I suspect that one future day, beings will look back in amazement on the time when intelligent life-forms took for granted that, right when they felt in the prime of their lives and full of growth and enthusiasm, they would begin an involuntary, inevitable decline toward death. They will look back befuddled, unable to clearly imagine the fears and risks of life without frequently updated backup copies of one's mind.

If one's goal is achieving near-immortality for human beings – something I do think is possible – there are various possible approaches one might take. Most obviously, there's mind uploading – have a machine read your mind out of your brain and transfer it into a computer of some sort, leaving the human body behind, bypassing the problem of fixing all the body's nasty old biology problems. This seems a fantastic possibility – I look forward to its realization! In 2012 I edited the first-ever academic journal Special Issue on Mind Uploading (an issue of the Journal of Machine Consciousness).

But yet, it's hard to say how long mind uploading technology will take to mature. To upload a mind without killing the human body containing the mind, we would need much more accurate brain scanning equipment than we have now. It would require a revolution in brain scanning. If one is willing to kill the human body in the process of scanning the mind from the brain, the possibilities using current technology are more promising – one can freeze a brain and slice it thin and scan the slices, thus obtaining a detailed picture of the brain's molecular structure. But even so, we don't have any clear idea how to turn these pictures into a dynamical system that can actually emulate the brain's internal behaviors and think and feel.

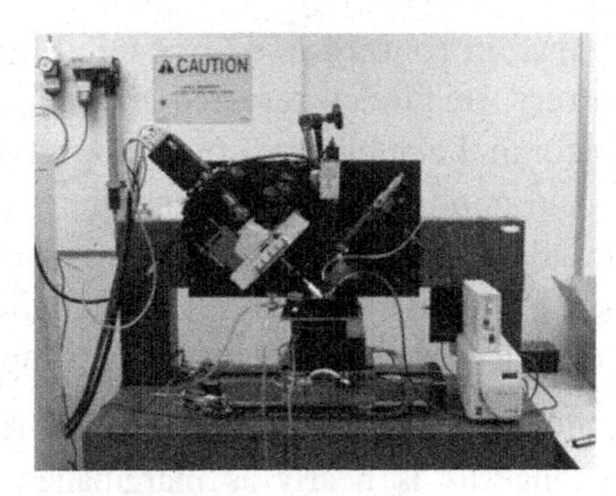

FIGURE 16.1: *The Knife-Edge Scanning Microscope. This instrument slices off small sections of tissues and images them, with sufficient resolution care to allow reconstruction of 3D structures. Using this method we can reconstruct much of the brain's structures, the downside being that it only works on brains that have been removed from the head and prepared for slicing up. (http://research.cs.tamu.edu/bnl/gallery/photos/kesm/dscf2372.jpg)*

Apart from mind uploading, the best way to radically prolong human lifespan would seem to be: Solving the problem of aging. This also has the advantage that it lets humans keep their biological bodies, which is valuable because many people are emotionally attached to their bodies. (Though some or not – every year I get at least a couple emails from people offering their brains for mind uploading experiments they believe I am conducting in secret!). Given the uncertainties associated with mind uploading, and the aesthetic and emotional value of the human body, it seems wise for humanity to be putting significant energy into human-body longevity research as well!

FIGURE 16.2: *Example image of a portion of a slice of a brain, produced by the Knife Edge Scanning Microscope. Figuring out the detailed 3D structure of a brain by piecing together multiple of these images of slices, is a significant but surmountable data analysis challenge. In 2013 the first serious 3D models of the structure of an individual human brain was produced, but not yet with a level of detail down to the neuron level, let alone the molecular level. It's unknown how detailed of a brain map we would need to create, in order to capture enough information to re-create a specific mind in a different substrate (and note that to perform this re-creation, we would also have to know a lot more about brain dynamics than we currently do). (http://singularityhub.com/wp-content/uploads/2012/04/image25.jpg)*

You might think that, in contrast to AGI or mind uploading, longevity research would be a very wellfunded research area. After all, AGI conjures images of the Terminator in the public eye– but who wouldn't want a longer, healthier life, right? And mind uploading seems pretty farfetched from an everyday point of view. But living longer via new medicines or gene therapies – that seems pretty concrete and not so far out there, right? So there must be a huge push in that direction by governments, drug companies, etc...

But that's not quite the reality. Actually, biomedical research specifically focused on human longevity is nearly as marginalized as AGI or mind uploading, in the present scheme of things.

One factor involved here is: Most biomedical research these days is driven by the desire to create drugs and sell them. However, before you can sell a drug you have to get it approved by the government—in the US, where the bulk of biomedical research happens, this means by the Food and Drug Administration (FDA). Since the US FDA doesn't consider aging and death as diseases, even if you created a perfect drug to stop aging in its tracks, the FDA wouldn't approve it under its current policies. You'd have to get it approved as a drug to cure Alzheimers, heart disease, or some specific disease that the FDA recognizes. This may seem like a technical legal point, but I think it's actually had a major impact—few pharmaceutical firms even try to make medicines to extend human life. Fortunately, as of 2016, this is finally starting to shift, though it's unclear how rapidly a thorough shift will arrive..

At a certain point in mid-2012, I realized that I probably had better data about the genetics of longevity on my desktop than anybody else on the planet . . .Namely:

Data from Genescient Corp, a company I consult for, about the DNA of fruit flies they created (via 30 years of selective breeding) that live 4 times longer than regular fruit flies

Data from Scripps Institute in San Diego about the DNA of a large group of healthy people over the age of 80, and comparable unhealthy people over the same age group

The reason I had this data was: I was searching for commonalities and relationships between these and other datasets, using a variety of AI and statistical tools, including some from the OpenCog codebase. Neither of these two particular datasets was publicly available at that point (though one or both may be whenever you read this) —each one was only accessible to folks working at the institute owning the data, or their close collaborators (like me).

FIGURE 16.3: *The fruit fly, Drosophila Melanogaster, is commonly studied by biologists as a "model organism". Genescient Corp. possesses fruit flies that have been evolved for longevity for over 30 years, and live 3-4x as long as ordinary fruit flies. My team has applied AI tools to genetic data regarding these long-lived fruit flies, to help biologists understand why they live so long – and how to create therapeutics to enable radical lifespan increase in humans. (http://commons.wikimedia.org/wiki/File:Female_Mexican_fruit_fly.jpg)*

(Right now I have even better data, thanks to a lot of more recent longevity studies publishing their data online for free, including a study of the genomics of supercentenarians – a couple dozen people aged 105 or over. The data just keeps pouring in. Exciting times!)

You might think that, somewhere, there's some huge government or industry initiative to gather all the existing data about longevity and study it carefully as a whole, to understand why we age, how to avoid the problem, and extend healthy life. But you'd be wrong — unless by "somewhere" you mean in some fictional or parallel universe. In this world, longevity research, like AGI, is currently relegated to various scrappy bands of outsiders, struggling to get big things done with piddling amounts of money.

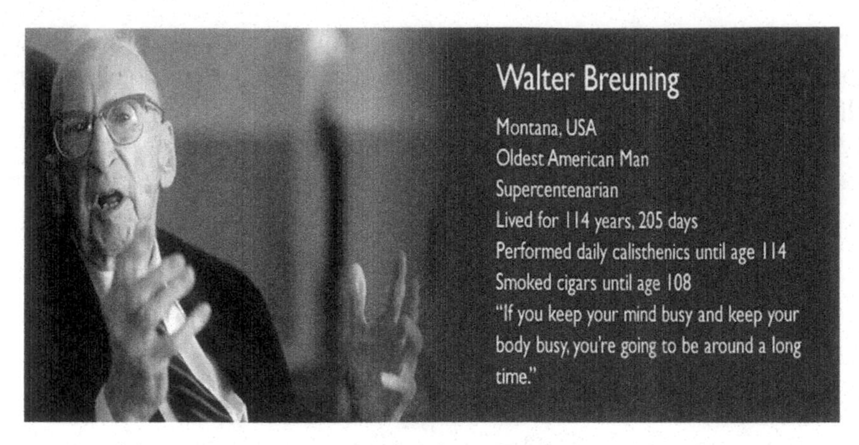

FIGURE 16.4: *Walter Breuning lived till age 114. What enabled him to live so long? The answer seems to be: a combination of genetic and other factors. Using AI based analytical methods applied to appropriate datasets gathered from healthy long-lived individual individuals like Breuning, we can pinpoint the factors enabling their longevity and start figuring out how to combat aging in others. (http://www.supercentenarianstudy.com/nivo-slider/images/slide-5.jpg)*

I find this especially ironic since AGI and longevity research have great potential synergy. I've thought a lot about one path to advanced AGI that intersects greatly with longevity research – developing an artificial biomedical scientist, which would use its emerging general intelligence to help us understand our bodies, how to repair them, and stop them from degenerating with age. I'm not sure this is the clearest path to AGI – I prefer the virtual and physical robotics avenue. But, I think it would be awesome to pursue the biomedical, virtual and robotic paths in parallel. I would really like to see these niggling biomedical problems like aging, disease and death solved as soon as possible; and it would be a good thingto have early-stage AGIs helping people.

Abolishing the Plague of Involuntary Death

I still remember when I first found out about death. I was three years old, I think. Well, I had known about death beforehand, but only as something that happened to strangers, old people or animals— I hadn't realized it was something that could or would happen to my parents and me. The concept made me rather unhappy. However, I began to understand the grownup world better. Once you understand the perceived inevitability of death, a lot of other things fall into place.

A few years later, when I was seven or eight, my mom was studying Chinese history in graduate school; she inspired me to read about Buddhism in Will Durant's History of Civilization and some other books. I was introduced to the idea that death doesn't really matter because the linear flow of time is an illusion; the important thing is to fully experience the present moment. I felt some truth in this perspective, but still, I wasn't convinced that this made death OK. In my heart, I still put involuntary death in the same category as war, murder or torture– things I'd do away with if I could, happily risking the consequences.

While I read a lot of science fiction at that age, including various novels featuring races of immortal aliens, humans or intelligent robots, I didn't think much about building my own technology to enable immortality – that seemed somehow infeasible, given the state of understanding of biology at that time. I thought more about building a spaceship to explore the universe and find other civilizations that had already cured death and invented all sorts of other amazing things – or come back to Earth after a jaunt around a few galaxies, making use of relativistic time dilation to return to the Earth a million years later when incredible new things had happened. Though I also considered the possibility that when I got back to Earth, I might find everyone long dead, migrated to the stars, or vanished to some other dimension that I had no way to access. I feel largely the same way about death now as I did when I was a kid – with just a few changes of emphasis. Just like my three year old self, I don't like the idea of inevitable death – it seems like a Bad Thing.

I'm not horribly petrified of the idea of dying; if it happens, so be it. I don't spend my life hiding in a padded, sterilized room under armed guard. I take reasonable risks – I backpack in remote regions, scramble up rock-faces, swim deep out into the ocean, and so forth. I have a tendency to run across busy streets when I'm in a hurry, as if I were in a real-life game of Frogger. On a recent trip to southern Thailand, I didn't hesitate to rent a motorcycle and get from place to place in the manner of 95% of the Thai people. I do feel it's important to enter fully into each moment – and that, when you seize the present moment, you're in a sense outside the flow of time and beyond the grip of death.

But then I also feel: Why not have more great moments?

Being a long-time dad, I also have the familiar feeling that my 3 kids would serve as SOME kind of continuation of me, if I were to die. My various intellectual and creative works – such as this book you're reading – would also serve as some sort of partial continuation. I already gave you my favorite Woody Allen quote: "I don't want to achieve immortality through my works – I want to achieve immortality by not dying". My kids are wonderful and some of my books are pretty good, but these are things in themselves, not substitute versions of myself. Why settle for these second-rate versions of immortality, if you can have them AND the real thing too?

Biology Has Become an Information Science

My childhood feeling that biology was not ready to address immortality was perfectly accurate – back then in the early 1970s.

Back before I came into existence, when my father was in college in the early 1960s, he organized a group called the Student League for the Abolition of Mortality (SLAM). Their goal was to protest the annoying fact of death. But this was pretty much a joke organization. Back then, the idea of defeating death was an absurdity, a notion from science fiction– just like ten years later, in the early 70s, when I first started thinking about the topic as a young child.

But things have changed a lot. Biology's exponential advance has felt dramatic in the last few decades. Biology has become the latest and perhaps fastest-growing information science. The use of biological science to seriously address the problem of death is no longer a science fictional notion. The abolition of death is not yet a goal of mainstream biology, but there is an increasing minority of bioscientists, actively arguing that the use of 21st century biological tools to radically extend human lifespan is a viable possibility.

The recent wave of expansion in biology has largely been driven by the development of experimental technologies – technologies that allow us to create large numbers of bits (binary digits) describing the states of biological systems; and technologies that allow us to manipulate the internals of biological systems more freely, like code inside a computer. New methods for sequencing DNA led to the Human Genome Project and the successful unraveling of assorted plant, animal and bacterial genomes. New techniques like microarray analysis and RNA interference allow us to measure biological systems in more detail than ever before.

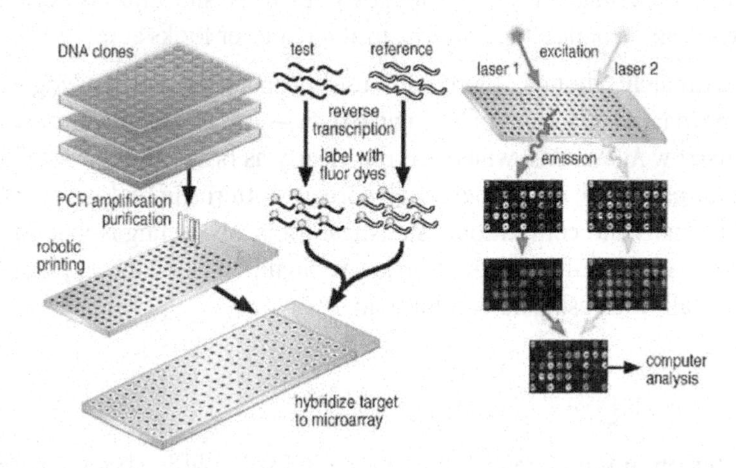

FIGURE 16.5: *Diagrammatic depiction of the process of DNA microarray analysis. The tissue sample under study comes into the picture labeled "test". AI tools like the ones my colleagues and I have used, come into play via analyzing the 2D grid data shown at the bottom right. Each point on the 2D grid corresponds to a certain gene, and the color shows how much of that gene is being expressed in the tissue sample. Roughly speaking a gene's expression level reflect the amount of the protein corresponding to the gene being produced, but it's not an exact reflection, as there's a lot of complexity both in the biological processes involved and the microarraying process itself. There are also many variants of the microarraying process available today, each with its own strengths and limitations. (http://bme240.eng.uci.edu/students/08s/jentel/image/MicroArrays.jpg)*

I will talk here mainly about genetics, just because that's the area of biology where I've worked the most. But the same trend existsacross biology. I've also done some work on neuroscience, where progress has been driven by experimental technologies like Functional Magnetic Resonance Imaging (fMRI) for measuring the brain, as well as things like tetrodes and voltage-sensitive dyes that allow for more complex ways of carrying out electrode recordings in the brain. We still can't measure the brain well enough to experimentally observe the dynamics of cognition – but we can measure a lot more than we could ten years ago. And beyond genetics and neuroscience, biology is advancing across the board, in more areas than I've had time to become intimately familiar with.

All these new experimental technologies generate massive amounts of data, which the human brain is ill equipped to process and sort through in a useful way (think pattern recognition). Most biologists just work with simple data patterns that are identifiable by the naked eye or through standard statistical tools. Unsurprisingly, most of the data gathered by these very advanced experimental methodologies goes to waste, unused because no one who can recognize the patterns in the dataever looks at it.

Fortunately, though, there ARE tools capable of recognizing subtler patterns in large quantities of biological data — artificial intelligence systems. Even narrow AI systems, which are not nearly as broad in their intelligence as the human mind, can recognize all sorts of patterns in biological data that elude humans and conventional statistical tools. My feeling is that advances in narrow AI and AGI are likely to be key to finally cracking the aging problem and radically increasing human lifespan.

AGI and Longevity

Later on in this chapter I'm going to tell you a little about some of my (fairly interesting, I think) narrow AI based genetic discoveries – and my own partially-baked theories about how aging works. But interesting as I think that work is, it's not really the main point I want to get across in this chapter. Ultimately, the reason I got into studying the genetics of aging in the first place, wasn't simply that I wanted to see what good I could do with the narrow AI tools at my disposal. The reason was a strong feeling that, ultimately, the only reasonably sure way to crack the aging problem is to have AGIs helping us solve it.

The basis of my belief that AGI may be the key to longevity research is pretty simple:

The human body is incredibly complex, with many subsystems on many levels – and aging seems to involve many of these subsystems working together.

The human brain evolved to control a body surviving in the African savannah, not to integrate a huge number of complex biological datasets.

An advanced AGI artificial biologist, however, could interpret the totality of biological data far better than any human – and design new experiments accordingly, and then analyze their data, and so on.

The argument seems pretty straightforward, though of course it doesn't rule out some brilliant human scientist making a breakthrough and discovering an amazing life extension pill next year! (I hope this happens....)

Alas, full-fledged AGI biologist is a fair way off – certainly years, maybe decades. To get there, we'll need to go through a lot of preliminary stages first – start with an AI toddler, then an AI school student, and so forth. This is what we're doing with the OpenCog project.

But, even before AGI advances to that level, current AI technology can do a lot to help biological research in life extension and other areas. This work is valuable unto itself. And it may also serve an additional purpose – guiding us via giving us a detailed understanding of exactly what an early-stage AGI would have to be like to be really useful for understanding aging and advancing longevity.

In my own work in this area, my colleagues and I have been quite successful at finding meaningful patterns in genetics data, which biologists were unable to detect using standard biostatistical tools. Many of the data patterns we've found have led to hypotheses that were later validated in the lab, by experimental biologists.

My original vision when entering into the world of bio-AI was to make a huge database comprising all biological knowledge – or at least everything that's been posted online, which is a heck of a lot – and then set advanced AI tools to work doing reasoning based on this knowledge. Unfortunately I haven't yet done that, due to lack of financial resources. Formalizing biological datasets in a manner amenable to analysis by current AI algorithms requires a bit of human labor, and at time of writing, I haven't yet managed to summon the funding to pay a team of biologists to do this work. However, I've carved out an interesting niche applying advanced machine learning to genetics data and other biological datasets, and in the process have learned a great deal -- about aging, age-associated diseases, and applied machine learning.

Quick Review of Basic Genetics

I've mentioned that most of the bio data I've worked with has been genetics data – information about the variations in DNA between one person or animal and another, and information about the particularities of gene expression in one organism or tissue in a particular condition. Genetics is a pretty technical area of biology, so if you're not familiar with it, here's a very brief primer. If you really want to understand you can explore further, of course; there are many good tutorials online.

I'm sure you already know that living organisms are made of cells – and that even the simplest cells are made up of thousands of different types of interacting molecules. Some cells exist as independent entities, while others function in groups acting as a single entity (like the cells in your brain or your live). Groups of cells form tissues, tissues form organs within single organisms, organisms form populations, and populations form ecosystems encompassing the biosphere of the planet. Each of these levels influences the others, which is one reason biology is so complex. But during the last few decades of the 20th century, it became apparent that a reasonable percentage of this complexity can be understood in terms of some specific molecules within cells: DNA (deoxyribonucleic acid) and RNA (ribonucleic acid).

Most DNA lives in the nuclei – the centers – of cells, but there is also a small amount of DNA in the mitochondria, which are separate components within cells that deal with energy production. Little attention gets paid to mitochondrial DNA overall, but some of our work has suggested it may be important for age-associated diseases like Alzheimers and Parkinsons.

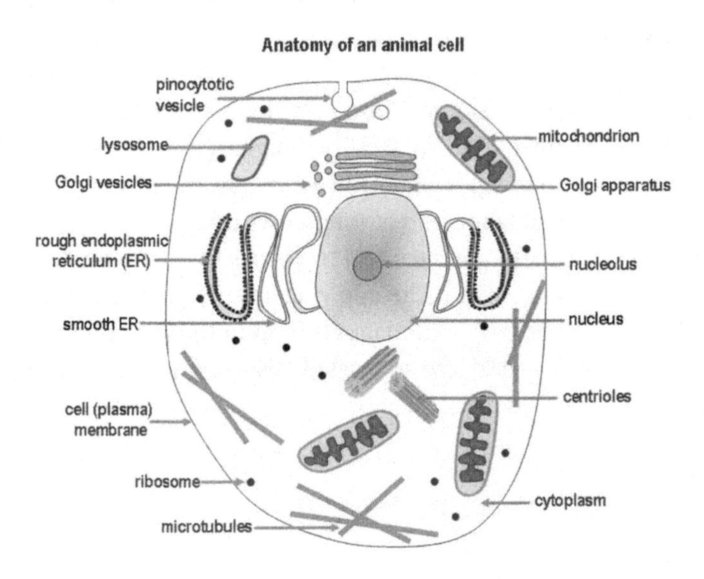

FIGURE 16.6.: *In case high school biology has faded to a vague blur in the back of your mind, here is a standard diagrammatic depiction of an animal cell. In a human, DNA containing about 25000 genes (plus a lot of DNA that doesn't contain genes, but serves other useful functions) lives in the nucleus. This is the essential code specifying the building and operation of the organism – or more accurately put, guiding the complex nonlinear dynamics of the organism's ongoing self-organizational self-construction. DNA containing 13 protein-coding genes and 22 RNA-encoding genes exists in the mitochondrial, the "energy powerhouses" of the cell. (http://geneticssuite.net/node/11)*

A very key role in biology is played by what is awkwardly called the "central dogma" of the molecular biology of the cell: The information encoded DNA is transcribed into RNA, which is then translated into proteins, which comprise the primary functional and structural element of living cells.

In spite of the odd name, this is not a "dogma" in the sense of a belief that is adopted unthinkingly; it's a scientific hypothesis which has been validated by loads of evidence... and which has also been revealed to be only an approximation. There are some cases where information can go the other way around, from the rest of the cell to the DNA.

Roughly speaking, a DNA molecule can be considered as a long linear sequence of amino acids, where the specific amino acids involved are drawn from four possibilities: adenine (A) and guanine (G), thymine (T) and cytosine (C). A DNA strand is represented as a string of the letters A, C, G, and T. Due to the chemistry of the amino acids, A and T are opposites, as are C and G. When the strands have extended sequences of opposite bases, they are bound tightly together in the classic double helix structure.

RNA molecules look a lot like DNA molecules, but they contain uracil in place of thymine. Unlike DNA, RNA usually exists as a single strand in biological systems. The strand can bend to allow complementary sequences within it to bond together. Genes have practical impact on the body largely via causing messenger RNA to interact with transfer RNA, to create proteins. The protein created by a gene has a sequence of amino acids that is different from, but precisely determined by, the DNA strand's sequence of A, G, C and T.

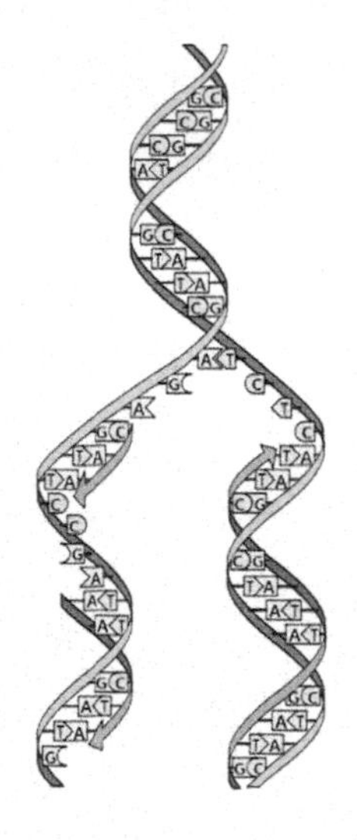

FIGURE 16.7: *Stylized depiction of the process of DNA replication, via which one DNA strand makes another identical clone. The G, A, T and C each represent a particular kind of amino acid (guanine, adenine, cytosine and thymine). Picture by Madeleine Price Ball. (http://en.wikipedia.org/wiki/File:DNA_replication_split.svg)*

The unraveling of this machinery – the nanomachinery underlying all life on Earth! – was one of the great achievements of 20th century science. But still, it's important to keep the wonders of molecular biology in perspective. While the central dogma, DNA to RNA to protein, outlines the core process on which the dynamics of biological systems are based, actual living organisms are made of thousands to millions of interacting instances of this pattern. Proteins activate the transcription of certain genes, whose protein products detect signals of threat and opportunity from the environment, alter the activity of other proteins in response, which metabolize raw materials to generate and maintain structure, and reproduce new cells to sustain and grow populations over time. Proteins and the metabolite flows they regulate carry out myriads of these functional activities simultaneously, which in turn regulate gene activation, in complex overlapping networks of feedback loops. Elucidating these networks requires not only an understanding of the environment in which the cell is embedded but the evolutionary processes that developed them incrementally over time.

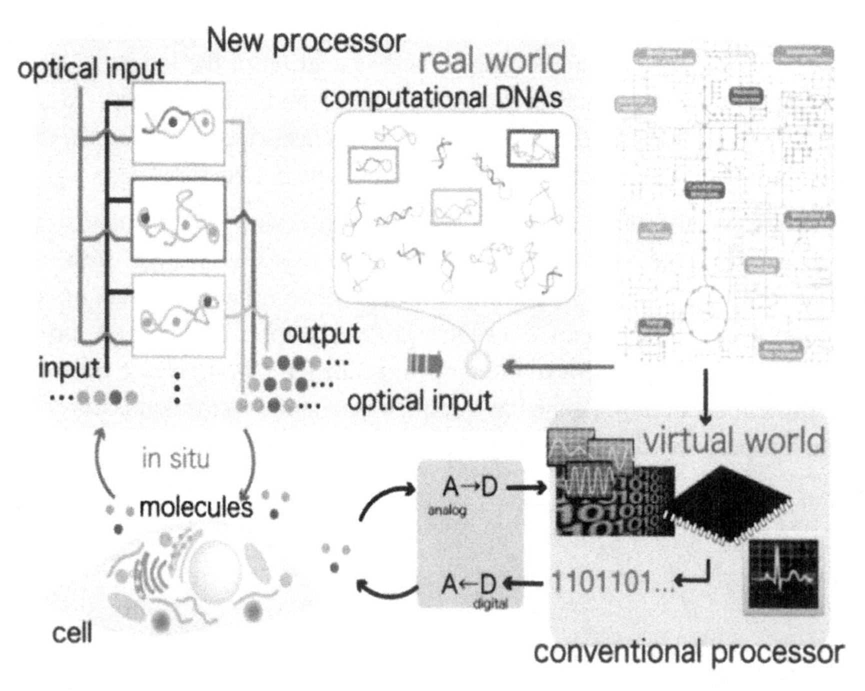

FIGURE 16.8: *Diagram of a DNA computing system currently under development at Osaka University – a nanoscale processor using photonics and DNA technologies. DNA computing is a practical, currently working example of how to do computation using very small thing that move around. Some of the same ideas could potentially be useful for femtocomputing. (http://www.lip.ist.osakau.ac.jp/research/dna_eng.html)*

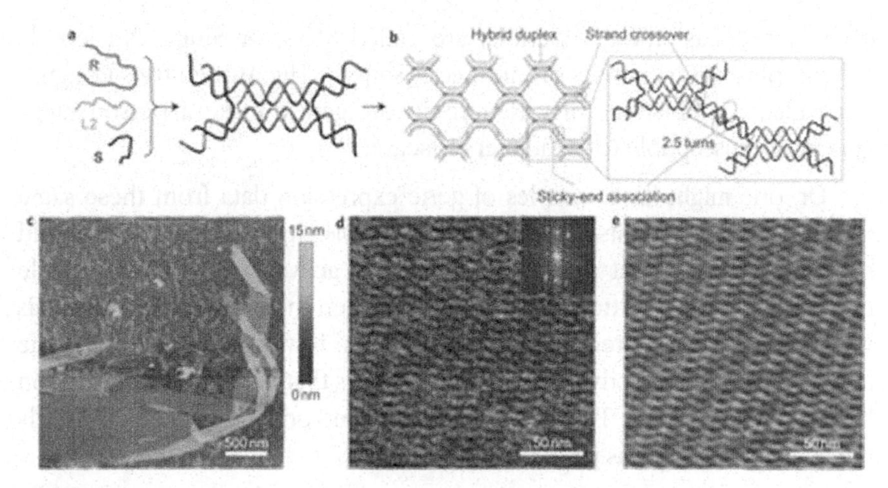

FIGURE 16.9: *One of the many approaches to DNA computing is based on the engineering of tile-like structures using DNA. "Tiles" in this context are DNA-based constructs where there is very tight binding within the tile, and looser binding between tiles. This allows the building of a variety of structures via piecing the tiles together. The top row of the figure shows the theoretical structure; the latter shows actual microscopic images of real DX tiles. (http://www.nature.com/nchem/journal/v2/n12/abs/nchem.890.html)*

The analogy between molecular biology and computer engineering is interesting, and has been pursue fairly far by a community of researchers. The mechanism of molecular biology amounts to a naturally evolved nanotechnological computing and engineering infrastructure that we barely understand, and that far exceeds our own ability to build nanomachines or nanocomputers at this point. Inspired by this way of thinking, some folks have started working on "DNA Computing" – using the existing machinery of molecular biology to solve computational problems. At this point that's not a practical tool yet, but it's a rapidly evolving field, and if nothing else it will teach us a lot about nanotechnology. One day we may use DNA computers – or nanocomputers inspired by DNA computers – to run AI programs analyzing human genetics data!

Experiments in Bio-AI

So what kind of genetics data have my colleagues and I fed into our AI systems, and what kinds of answers have we obtained?

For example, one might have samples of DNA from 500 people with Alzheimers disease, and 500 similar people without the disease. Each DNA sample tells you about the individual variations in the 25000 genes that each person has (these variations are called SNPs, or Single Nucleotide Polymorphisms – generally pronounced "snips"). The AI then tries to figure out which combinations of variations in the genes, make a person more genetically susceptible to Alzheimers Disease.

Or, one might have samples of gene expression data from these same 500 Alzheimers patients and matched controls. In this case, one would have information about which genes are most active in each of the people under study, at the particular point in time when they were measured. This would give you a different kind of information: it would tell you about the perturbations to the body's biological processes that happen when a person has Alzheimers disease. The answer might depend on what tissue of the body the genetic material was gathered from.

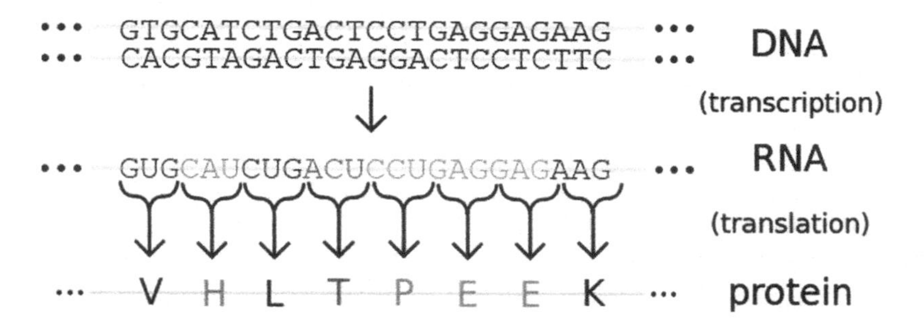

FIGURE 16.20: *This picture shows the process of gene expression, via which DNA sequences give rise to sequences coding proteins, which are then manufactures and spread throughout the body, doing much of the work of building and maintaining the body. A person's (or other organism's) DNA is divided into sequences representing genes. A typical gene might be a couple hundred amino acids long. Some parts of the DNA do not represent genes, and were previously called "junk DNA", but are now known to carry out a variety of critical functions. The picture shows the first few amino acids for the alpha subunit of hemoglobin. The sixth amino acid here (glutamic acid, "E") is mutated in sickle cell anemia versions of the molecule. Standard gene expression technology, roughly speaking, measures the amount of RNA that is produced from DNA – which is a crude but meaningful proxy for the amount of protein produced from the DNA. (http://en.wikipedia.org/wiki/File:Genetic_code. svg)*

In both the gene expression and the SNP cases, the main role of AI is to sift through the mass of data and find combinations of things that are relevant to a certain condition of the organism. If a single SNP, or a single gene, were critical to Alzheimers, biologists could find that without use of AI technology. But if a complex combination of SNPs or genes is important, it's hard for humans to figure that out from the data, even with the standard arsenal of statistical tools. Whereas AI tools like the ones in OpenCog today, even though far short of AGI, can often find the relevant patterns.

```
if
(NM_005110 + NM_001614)/NM_002230 - .3* NM_002297 > 1
then Case
else Control
```

```
sum
 sub
  sum
   sub
    input GO:0016835
    input GO:0030674
   sub
    input FAM0031431
    input NM_001831
  sum
   mul
    input GO:0019840
    input NM_015358
   sub
    input GO:0003782
    input NM_002151
 sum
  sub
   div
    const 0.514609
    input GO:0004667
   input GO:0006803
  div
   div
    const 0.352189
    input NM_007313
   sum
    input GO:0007165
    input NM_002226
```

FIGURE 16.21: *Example rules learned by he OpenBiomind AI software, via analyzing biological data. The right rule was learned from gene expression data from a population of humans with prostate cancer, and a population of matched control humans without prostate cancer. The NM terms correspond to the expression levels of specific genes (NM_007313, for example, is the index number of a particular gene, the gene otherwise known as c-abl oncogene 1, non-receptor tyrosine kinase (ABL1), transcript variant b); the GO terms correspond to the average expression levels of genes in a certain category (e.g. GO:0007165 is the Gene Ontology category for signal trandsuction – the set of all genes concerned with signals that convey changes to the state or activity of cells). The rule is shown in the "tree" format in which it is actually learned within the software. The software evaluates if this formula is greater than 0, and if so, it evaluates that the individual probably has prostate cancer. The tree is evaluated by interpreting it as it could also be written as a mathematical formula:*

((((GO:0016835 - GO:0030674) + (FAM0031431- NM_001831)) - ((GO:0019840* NM_015358) + (GO:003782- NM_002151))) + ((.514609 / GO:0004667 - GO:0006803) + ((.352189/ NM_007313) / (GO:0007165+ NM_002226))))

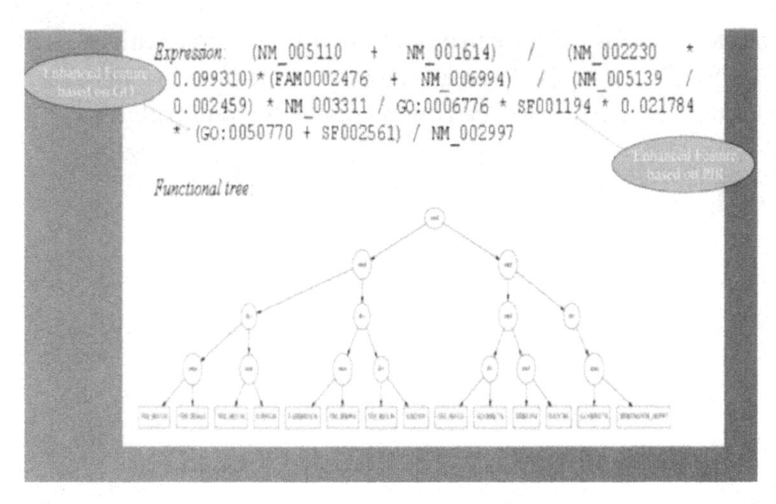

FIGURE 16.22: *Graphical visualization of a classification rule learned by an AI algorithm in the analysis of gene expression data. Computer scientists often think of mathematical expressions as "trees" of the abstract sort visualized below; and this kind of visualization can also make such formulas easier for biologists to interpret. This particular model contains features corresponding to genes, to Gene Ontology categories, and to biological pathways as defined in the PIR Protein Information Resource.*

The astute reader will notice that the mathematical rule in the figures just above could be simplified considerably. This is because these examples were learned by the OpenBiomind software from 2006 or so; and that software did not know how to simplify algebra. For our current bioinformatics work, we use OpenCog's MOSES software, which incorporates simplification of rules and formulas. So rules learned by MOSES will tend to be smaller, because they appear in something closer to algebraically minimal form. This is better for human comprehension, and also better if one wants to have other automated reasoning systems act on the rules, e.g. trying to generalize them to other contexts.

One of our most exciting achievements was devising new diagnostic tests to tell if a person has Parkinson's or Alzheimer's disease. Our findings also highlighted certain avenues with the potential to lead to cures for both – though this remains work in progress. Before our work, it was known that dysfunctions in the mitochondria (the part of each cell that stores energy) were important in Alzheimer's and Parkinson's. But nobody knew the causal basis of the connection between mitochondria and these diseases. Though it was suspected to have something to do with the mitochondrial DNA (a handful of important genes that live in the mitochondria of the cell, rather than in the cell nucleus where the other 25000 or so genes live).

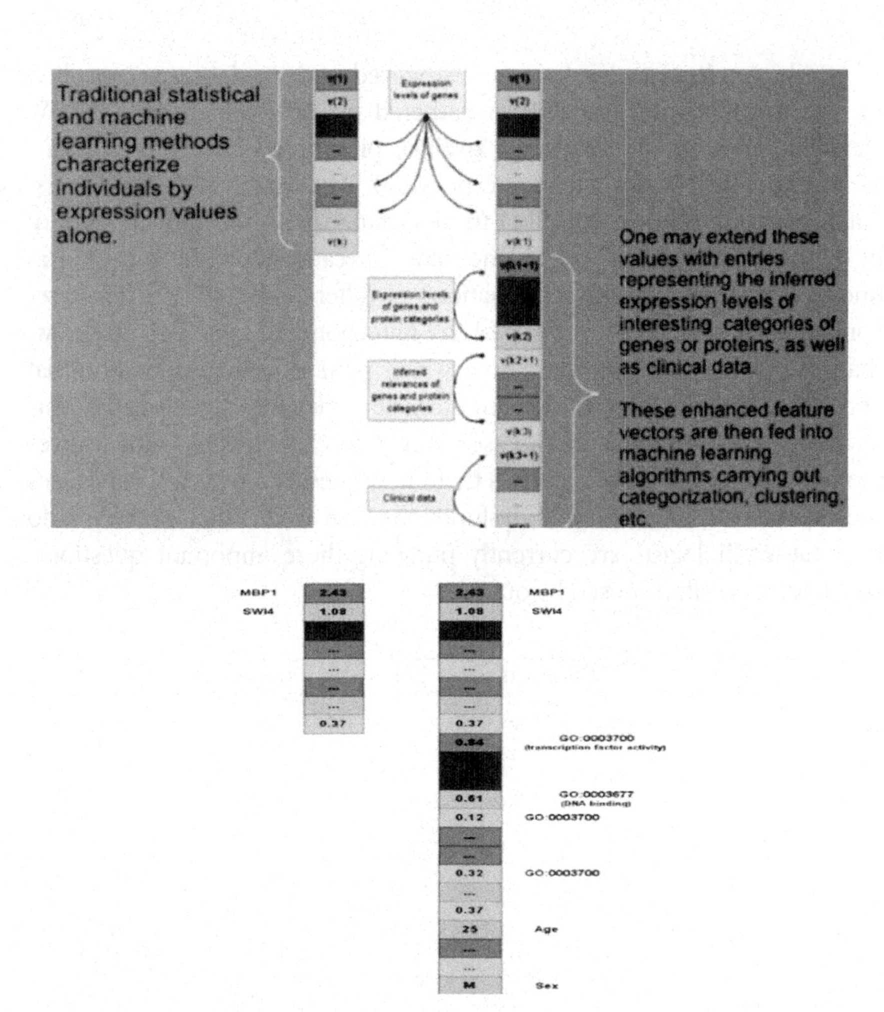

FIGURE 16.23:: *One of the many "tricks" my colleagues and I have used when analyzing gene expression data, is to prepare special "feature vectors" to feed into our machine learning algorithms. Many of the learning algorithms we've used in biology applications are designed to operate on numerical vectors (lists of numbers), and so we come up with creative ways to pack information into lists of numbers. Instead of just using vectors indicating the expression values of individual genes, we can make vector entries indicate expression values of categories of genes, or of biological pathways involving multiple genes. We can also pack "clinical" information involving the specific people or animals whose genes are being measured into the feature vector (e.g. age, weight, sex, results of various bloodwork). The image conveys the general concept; the image gives a specific example. This is an example of the kind of domain-specific "hacking" one does to get narrow AI to work for real in concrete applications. Typically for any narrow AI application, the domain-specific hacking is 80-90% of the work, dwarfing the effort put into generic, widely generalizable aspects of AI algorithms. This will likely be a fact of life in applied AI until truly powerful AGI comes on the scene.*

Fairly simple AI technology (a customized genetic algorithm), applied to data about mitochondrial DNA gathered by Rafal Smigrodzki and W. Davis Parker at the University of Virginia, pinpointed the exact portions of mitochondrial DNA that correspond to Parkinsons and Alzheimers. Standard biostatistical tools failed to determine these portions, in spite of being supplied with the exact same data – because they are not good at finding patterns involving combinations of different factors (in this case, combinations of different regions of the mitochondrial genome). But now, thanks to the AI analysis, we know which parts of which mitochondrial genes are screwed up in people with these neurodegenerative diseases. This discovery does not cure the diseases, nor even explain them. But it gives biologists a place to look... It gives them new questions to ask. What causes these particular mitochondrial dysfunctions, and what other problems do they cause? Biologists are currently pursuing these important questions, which were, in effect, posed by our AI.

Gene	Position	Percentage in top 20 codons
ND5	145	30.89 %
ND4	180	9.84 %
ND5	146	6.76 %
ND5	148	5.44 %
ND2	272	5.11 %
ND4	236	4.68 %
ND2	273	4.26 %
ND2	270	4.16 %
ND4L	49	3.17 %
ND4	183	3.12 %

FIGURE 16.24: *Mitochondrial genes found relevant to Parkinson's Disease, via AI analysis of mitochondrial DNA data provided by Rafal Smigrodzki and W. Davis Parker at U. Virginia. This data was an unusual kind: it recorded uncommon ("heteroplasmic") mutations in 10,000 mitochondrial DNA drawn from each of 17 peoples' cerebrospinal fluid. The ND5 gene emerges clearly from this analysis as the culprit (the "codons" mentioned in the final column are data structures in one of Biomind's algorithms, representing patterns found in the data; more than 30% of these patterns involved the mitochondrial gene ND5). The analytical work here was mostly done by Lucio Coelho at the VettaLabs office in Belo Horizonte, Brazil.*

We also found, in a collaboration with Suzanne Vernon's team at the Center for Disease Control and Prevention in Atlanta, the first solid evidence of a genetic basis for Chronic Fatigue Syndrome, a condition long disparaged as laziness or simple exhaustion. We compared the DNA of sufferers with non-afflicted people and uncovered systematic variations indicative of genetic differences between the two groups. Human researchers, even those armed with statistical tools, could never have duplicated this feat without the aid of our AI systems. But once the AI found its results, they could be easily followed up by researchers in the lab. For instance, the AI highlighted genes related to glucocorticoid receptors in the brain, and to the neurotransmitter tryptophan. Once the AI pointed these out to biologists, they could use their human intelligence to do varoius experiments, and gain ongoing scientific understanding they would not have achieved without the AI's guidance. Lacking an AGI to mastermind the whole scientific process, what we have are narrow AI tools that, when properly utilized, analyze the data from experiments that humans have designed, and provide guidance (such as lists of relevant genes or biological processes) to these same humans in designing the next round of experiments.

Table 4. Importances of the genes based on their SNPs.

Gene	Short description	Incidence
NR3C1	Nuclear receptor subfamily 3, group C, member 1 glucocorticoid receptor	69054
TPH2	Neuronal tryptophan hydroxylase	67077
COMT	Catechol-O-methyltransferase	60651
CRHR2	Corticotropin-releasing factor 2	42190
CRHR1	Corticotropin-releasing hormone receptor 1	35082
NRC1	Nonpapillry renal carcinoma 1 growth mediator	22531
TH	Tyrosine hydroxylase	15968
POMC	Proopiomelanocortin	13145
5HTT	5-hydroxytryptamine transporter	6706

The incidence number reports the number of rules found with accuracy greater than the frequency of the largest category, utilizing some SNPs in the gene.
SNP: Single nucleotide polymorphism.

FIGURE 16.25: *Genes found relevant to Chronic Fatigue Syndrome, based on OpenBiomind based AI analysis of SNP data provided by Suzanne Vernon's group at the CDC.*

Cluster 1:	
Quality: 0.620161565619	
FAM0012352	Component genes:
	- NM_006216: Homo sapiens serine (or cysteine) proteinase inhibitor, clade E (nexin, plasminogen activator
FAM0033509	Component genes:
	- NM_004137: Homo sapiens potassium large conductance calcium-activated channel, subfamily M, beta me
GO:0015023	syndecan
NM_014974	Homo sapiens KIAA0934 (KIAA0934), mRNA.
GO:0006693	prostaglandin metabolism
GO:0009103	lipopolysaccharide biosynthesis
GO:0015629	actin cytoskeleton
FAM0020295	Component genes:
	- NM_002997: Homo sapiens syndecan 1 (SDC1), mRNA.
Cluster 2:	
Quality: 0.614835196577	
1280_i_at	...
SF027815	Not Yet Assigned
GO:0005779	integral to peroxisomal membrane
NM_002101	Homo sapiens glycophorin C (Gerbich blood group) (GYPC), transcript variant 1, mRNA.
GO:0007397	histogenesis and organogenesis
GO:0008653	lipopolysaccharide metabolism
GO:0015268	alpha-type channel activity
Cluster 3:	
Quality: 0.607104033144	
GO:0004718	Janus kinase activity
SF002282	cytoskeletal keratin
GO:0030334	regulation of cell migration
GO:0010035	response to inorganic substance
NM_021638	Homo sapiens actin filament associated protein (AFAP), transcript variant 1, mRNA.
SF002345	smooth muscle protein SM22

FIGURE 16.26: *Example of the results produced by the OpenBiomind software for genetics data analysis. This is a very simple form of results: just a list of which genes, Gene Ontology categories and protein families are most useful for distinguishing prostate cancer samples from control samples, based on the AI's analysis. Some of these are already known to be involved with prostate cancer or cancers in general, others are not. A list like this provides biologists with multiple avenues for further investigation. The second column indicates how useful the feature (gene or category) was found by the AI software; the second column indicates how useful the same feature would be found according to a standard statistical analysis (in the second column, large numbers mean less useful). The contrast between the two columns indicates that the AI is prioritizing things very differently from standard statistical analysis. Annexin, for example, is the #1 most relevant gene found by the AI, yet would be ranked only #27 by standard statistics. The Gene Ontology category for complement activation is ranked #2 in importance by the AI, yet would be ranked #1853 (i.e. not be highlighted at all) by standard statistics. In many cases, the genes and categories highlighted by the AI have been validated by further lab experimentation and found to be relevant to the phenomenon under study. This is because the AI can choose genes or categories based on their importance in the context of their interactions with other genes, whereas standard statistical methods tend to ignore interactions and just look at the relation between individual genes and "phenotypic" characteristics like prostate cancer.*

Feature	Utility	Differentiation ran	Description
NM_004039	0.667	27	Homo sapiens annexin A2 (ANXA2), mRNA.
GO:0006957	0.427	1853	complement activation, alternative pathway
FAM0005641	0.399	1	
GO:0016860	0.393	1640	intramolecular oxidoreductase activity
GO:0003817	0.393	2	complement factor D activity
GO:0030162	0.392	524	regulation of proteolysis and peptidolysis
NM_002151	0.391	1	Homo sapiens hepsin (transmembrane protease, serine 1) (HPN), transcript variant 2, mRNA.
NM_000954	0.388	3822	Homo sapiens prostaglandin D2 synthase 21kDa (brain) (PTGDS), mRNA.
GO:0016812	0.387	1363	hydrolase activity, acting on carbon-nitrogen (but not peptide) bonds, in cyclic amides
GO:0006956	0.386	2102	complement activation
NM_002156	0.386	6	Homo sapiens heat shock 60kDa protein 1 (chaperonin) (HSPD1), nuclear gene encoding mitoc
1664_at	0.384	4	
GO:0045187	0.382	3	regulation of circadian sleep/wake cycle, sleep
GO:0050802	0.382	3	circadian sleep/wake cycle, sleep
GO:0005791	0.381	245	rough endoplasmic reticulum
FAM0024314	0.376	3	
NM_001928	0.375	2	Homo sapiens D component of complement (adipsin) (DF), mRNA.
GO:0042749	0.374	3	regulation of circadian sleep/wake cycle
GO:0030072	0.373	928	peptide hormone secretion
SF002514	0.373	1484	lipocalin
GO:0030252	0.372	329	growth hormone secretion
SF001139	0.371	1	hepsin
GO:0004667	0.37	853	prostaglandin-D synthase activity
XR_000167	0.37	10	
GO:0019840	0.37	1484	isoprenoid binding
GO:0017015	0.367	329	regulation of transforming growth factor beta receptor signaling pathway
NM_003573	0.366	5	Homo sapiens latent transforming growth factor beta binding protein 4 (LTBP4), mRNA.

FIGURE 16.27: *AI-based genomics data analysis also provides information on which genes and gene categories tend to interact with each other (form a "behavioral cluster") in the context of a given phenotypic character (prostate cancer, in this case). Again, this provides information that biologists can follow up on according to their own scientific understanding – progress, at this stage, requires a mixture of human and artificial intelligence.*

On the life extension side, one of my most fascinating projects has involved long-lived flies. Put simply: We have these flies that live 4 times as long as regular flies, and we're studying their genetics to see why. They were created by experimental evolution – via breeding flies for longevity over a 30 year period – a method that requires and yields no understanding of WHY or HOW they life so long. This project was highly active for me from 2009 to 2011; it has now slowed down a bit but is still gradually ongoing. It is a collaboration with another company called Genescient, which owns the flies, and was co-founded by visionary biologist Michael Rose who did the original fly-breeding work at the University of California, Irvine. Thousands of genes are different in the long-lived flies from the normal ones, but we've used AI technology to narrow the scope to a few dozen that seem to be the most important. And we seem to be grasping some of the key processes underlying these super-flies' longevity.

For instance, there are a lot of differences in neural and developmental genes, including many genes related to brain development. It seems part of the story of aging has to do with brain development processes that start out helpful, but then keep on going and turn damaging once the brain gets older. Genescient's biologists are using the results of the AI data analysis to formulate new drugs and nutritional supplements to help combat aging.

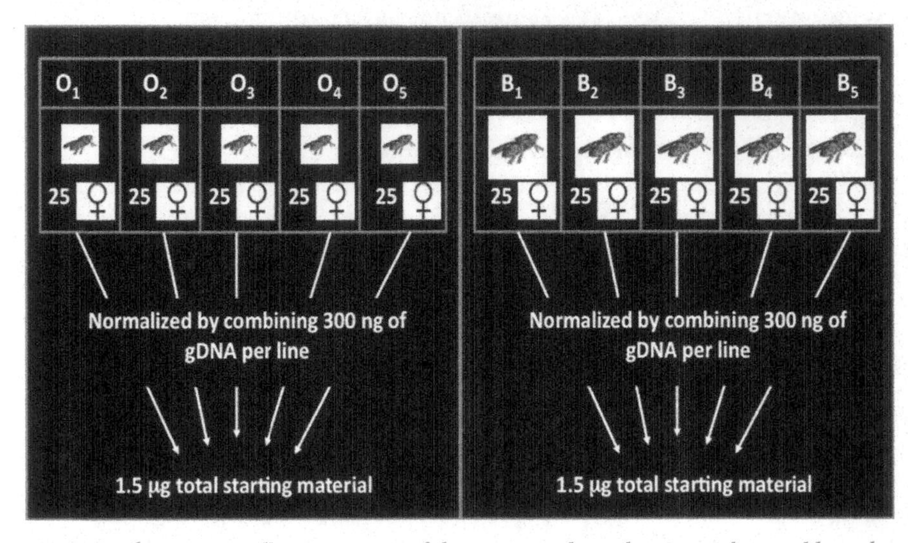

FIGURE 16.28: *Figure illustrating one of the many analyses done to understand how the Methuselah flies differ genetically from comparable control flies (with ordinary lifespans). In this analysis there were 5 cages of Methuselah flies (referred to within Genescient as "O flies"), representing separately evolved fly populations. There were also 5 cages of "B-flies", with ordinary lifespans. DNA was gathered from each cage and analyzed. Then statistical and AI tools were used to look for patterns distinguishing the O-fly DNA from the B-fly DNA. Because the flies in each cage are fairly similar to each other genetically, the analysis was done by taking a number of flies from each cage and mushing them together, and then extracting the DNA from the multi-fly mush. This DNA extraction work was done by hand; robotics technology for doing this sort of thing would certainly be possible but is not widespread. When I suggested to my Genescient colleagues gathering DNA data specifically from the long-lived flies' nervous systems to understand the specifically neural aspects of longevity, I was reminded that this would require some biologist painstakingly removing fly nervous system material with a tweezers. This sort of thing reminds me how we live in a transitional time – using our good old fashioned manual dexterity to extract material from dead animals with tweezers, so that this material can be analyzed by robotized lab equipment and cloud computing based AI algorithms.*

As an example of the way the AI analysis of this sort of data can help with the development of therapeutics, suppose the software found three genes, so that looking at SNPs (individualized variations) in these genes allows the AI to tell whether a given fly is Methuselah or Control (long-lived or just normal). So the AI has found a triple of genes:

Gene1, Gene2, Gene3

Each of these genes will have its own story, e.g. maybe:

Gene 1 is a serotonin receptor, expressed less in Methuselahs than controls

Gene 2 is related to Golgi apparatus, anatomical structure morphogenesis; also underexpressed in Methuselahs

Gene 3 is a "homeobox" gene, related to central nervous system formation; expressed more in Methuselahs than controls

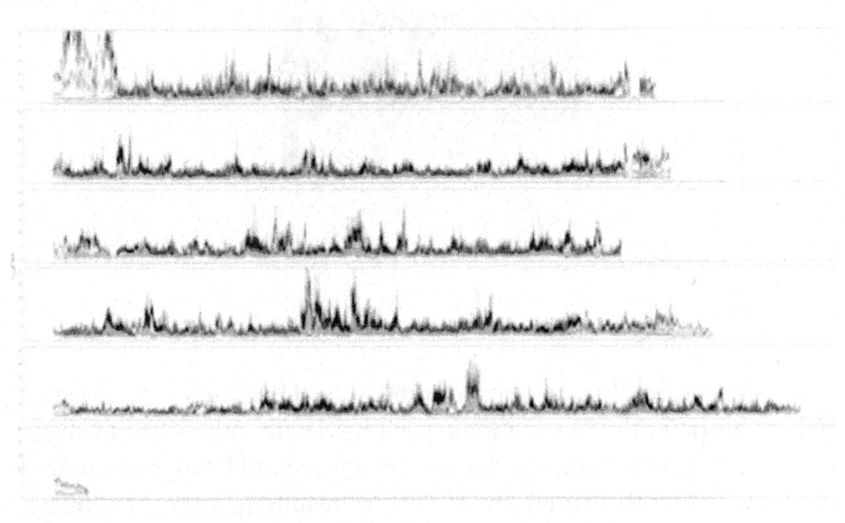

FIGURE 16.29: *Some of the patterns differentiating Methuselah from ordinary flies are not difficult to see at all. Here we see how genetic differences between long-lived and ordinary flies tend to cluster in particular regions along the flies' chromosomes. The panels (top to bottom) show the 6 chromosomes that the fruit fly has: X, 2L, 2R, 3L, 3R, tiny 4. The "x" axis is position along the chromosome. The "y" axis is the statistical significance of the difference between Methuselah and normal flies at that position on the chromosome (the log base 10 of the p-value, for those statisticians in the audience). The three lines are: black: a Fisher exact test differentiation between {pooled} B's (control) and O's (Methuselah); red: chi-square test for allele frequency differentiation with the B's; green: like the red, but for O's. This graph was made by Tony Long, a world-class biostatistics professor at UC Irvine, who also consulted for Genescient Corp. Most of the patterns found in Genescient's data by our AI tools were not as easy to visualize as this pattern, which was found by Tony's own biological and visual intuition. But data patterns that are inscrutable to the human mind and eye may still be critical to the operation of biological organisms. At this point in the development of technology, we need to rely on a combination of human and AI insight to make our best possible effort at comprehending relevant aspects of the astounding complexity of biological networks.*

It can be validated, via use of various biology databases, whether these genes behave sufficiently similarly in humans and flies, for the gene combination to be worth following up in the context of human therapeutics. (Making flies live a long time, in ways that don't generalize to humans, is also a worthy ethical cause, but not the major focus of anybody's research at present.) The question then becomes: Can we hit this triple of genes effectively with a combinational therapeutic? Can we find some combination of drugs or herbs so that, combined, the active ingredients act on the proteins coded by these three genes in a beneficial way?

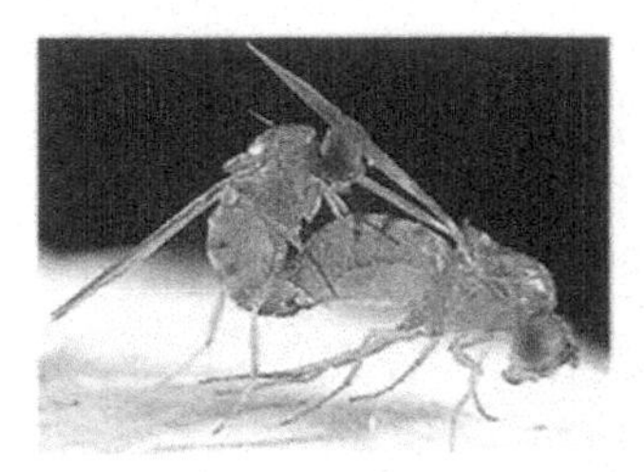

FIGURE 16.30: *Genescient's "Methuselah flies" are not only longer-lived but more robust than their ordinary counterparts in many ways. They have stronger hearts; they are smarter (better at remembering where to find food); and they have more sex.*

The genetics analysis spawns a problem in therapeutics development, which at this point the AI software doesn't help with, other than to suggest which of the body's proteins the therapeutics should target. Genescient has, at time of writing, used my AI analysis results to uncover a number of herbal remedies for age-associated diseases, including the Stem Cell 100 supplement. The firm has also identified a number of promising possibilities for pharmaceutical (drug) remedies for Alzheimers and other diseases, based on the same AI results.

FIGURE 16.31: *The Stem Cell 100 supplement uses a formulation of herbs, which was discovered in large part based on the results of applying my AI analysis to genetic data from the long-lived Methuselah flies.*

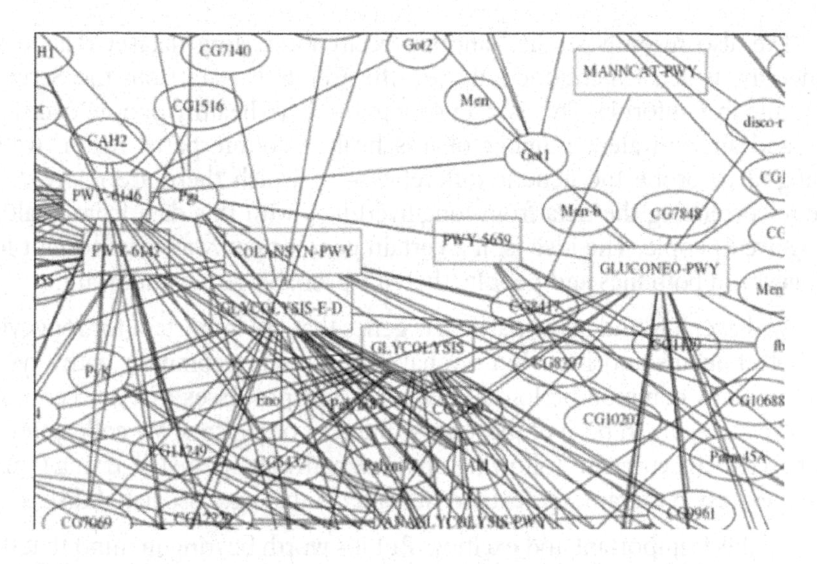

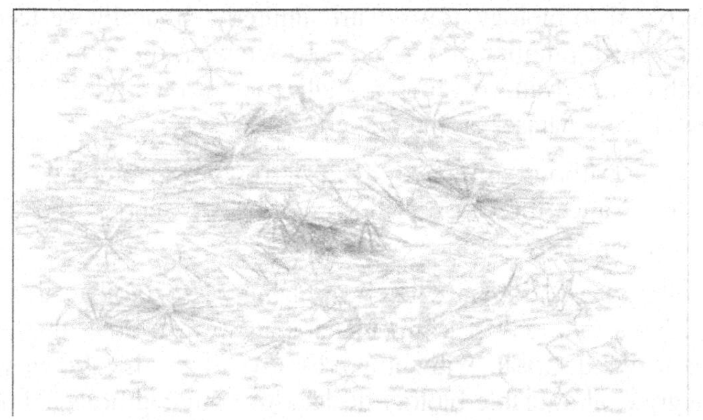

FIGURE 16.32: *Biologists love graphical depictions of data – but the diagrams they look at generally represent massive oversimplifications of the real processes and interactions going on. If one tries to make a diagram of the main interactions between genes occurring in Methuselah flies relative to their longevity, by simply feeding one's data into a scientific data visualization tool – one gets a huge tangle! The picture can't really be read on a book page, but could be printed as a wall-sized poster: each dot represents a gene or "pathway" (subnetwork of interacting proteins), and the lines between dots represent interactions. The dark regions are where there are a lot of lines, i.e. a lot of interactions. The Picture shows a random square from within the picture, blown up a bit. Of course, there are commercial products that make such diagrams easily browsable by the researcher – but they do so, in part, by making choices regarding which relationships to depict and which ones to ignore. The tangle shown here, while not so useful for understanding what's going on, does evoke the extreme complexity of the processes and interactions going on here! Obviously, we found other ways to explore the Methuselah fly data!*

I've also recently studied another really interesting dataset related to longevity, which I mentioned above –the Wellderly data from the Scripps Institute in California. This is data on hundreds of healthy people over age 80, and an equivalent number of less healthy counterparts. Again, we're using AI to probe the genetic differences—of which there are many. And we're correlating the data from long-lived flies with this data from healthy long-lived people. The idea is, if a certain gene or pathway is important for longevity in both flies and people, then it's probably really important.

We haven't found many particular genes that are important for longevity in both humans and flies. But we have found many common pathways – that is, many common biological networks and processes that are centrally involved in aging in both of these very different organisms. My colleagues at Genescient are currently following up these pathways in the lab, trying to use them to design therapies to extend life and combat age-associated disease.

All this is important and exciting. But it's worth bearing in mind that the way we apply AI to biology now is fairly limited – basically we feed in one dataset, or a small number of datasets, and see what the AI says. If you had an AGI with really advanced general intelligence, you could do better – you could feed it every dataset on the planet.

And even without advanced AGI, narrow AI technology has great potential to advance the cause of healthy longevity, and at the same time move AI toward AGI.

My colleagues and I have been applying AI to biological data on a piecemeal basis, one day at a time. That's all we can do with the limited resources at our disposal. With more resources for this kind of work, we could integrate all available biological data into one big holistic knowledge database and then set our AI algorithms to work. Even without any more biological experiments, I'm confident we could uncover an incredible amount of new information that would lead to the designs of new experiments useful for gathering yet more data.

And of course, many of these new experiments could becomefully automated with robotic lab equipment. We have the technology now to take the totality of biological data and store it in a huge AI system, utilizing this system to design and run new experiments using robotic laboratory equipment, repeating the process multiple times to get ever more useful (and otherwise unobtainable) results with enormously positive implications for biology. The only thing stopping us is money. With enough resources at hand, I bet we could crack the aging problem within a decade or two. Middle-aged people today would never need fear death, except through accidents or rare diseases.

Aubrey de Grey's SENS Approach to Increasing Human Lifespan

FIGURE 16.33: *Aubrey de Grey, the maverick researcher who has emerged during the last decade as the leading advocate of a scientific approach to ending aging.*

The AI-driven approach, promising as it is, is not the only current attempt to come to grips with the aging problem. The mainstream of biomedicine doesn't think much about radical lifespan extension yet, but some maverick researchers do. And there is also significant movement in this direction via the more mainstream biopharma and tech community, e.g. Google's Calico initiative and Craig Venter's Human Longevity Incorporated, although it's not yet clear what strategies these will take or how effective they'll be..

The most influential among these so far has been Aubrey de Grey, whose SENS (Strategies for Engineering Negligible Senesescence) research organization is carrying out some fascinating longevity research in their California lab, and funding a host of valuable projects at various universities worldwide.

FIGURE 16.34: *Aubrey's fantastic book Ending Aging carefully reviews his approach to curing the problems of aging and enabling radical human longevity.*

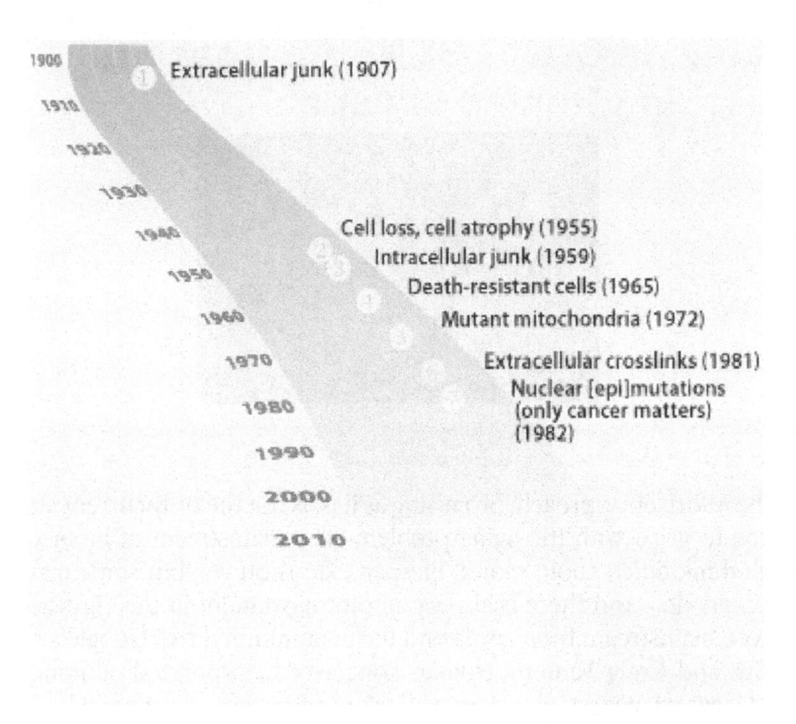

FIGURE 16.35: *The Seven Pillars of Aging identified by Aubrey de Grey. According to Aubrey's hypothesis, if we can resolve these seven problems, we can make human beings live a very long time. (http://alfin2600.blogspot.hk)*

Aubrey's approach to achieving radical human longevity is "engineering" oriented – or you could almost think of it as "auto mechanics" oriented. He wants to treat an old body vaguely like an old car – identify what the problems are, and fix them. He argues that biologists have noted seven main things that go wrong with human bodies as they age – the Seven Pillars of Aging as he calls them. And he proposes possible fixes for each of these.

Aging Damage	Discovery	SENS Solution
Cell loss, tissue atrophy	1955[1]	Stem cells and tissue engineering (RepleniSENS)
Nuclear [epi] mutations (only cancer matters)	1959[2], 1982[3]	Removal of telomere-lengthening machinery (OncoSENS)
Mutant mitochondria	1972[4]	Allotopic expression of 13 proteins (MitoSENS)

Death-resistant cells	1965[5]	Targeted ablation (ApoptoSENS)
Tissue stiffening	1958[6], 1981[7]	AGE-breaking molecules (GlycoSENS); tissue engineering
Extracellular aggregates	1907[8]	Immunotherapeutic clearance (AmyloSENS)
Intracellular aggregates	1959[9]	Novel lysosomal hydrolases (LysoSENS)

FIGURE 16.36: *The solutions Aubrey has in mind for resolving the Seven Pillars of Aging. Each of these is a difficult research project with many different aspects. See http://sens.org for explanations of all these potential solutions – I don't want to go into them in detail here, as it would be too much of a digression; but it's interesting and important stuff!*

The point of Aubrey's term SENS is that it's not merely slowing down of aging that we're after — it's the reduction of senescence to a negligible level. We're not trying to achieve this goal via voodoo, we're trying to achieve it via engineering—mostly biological engineering, though nano-engineering is also a possibility.

As part of his effort to energize the biology research community about SENS, a number of years ago Aubrey launched a contest called the "Methuselah mouse prize" — a prize that yields money to the researcher that produces the longest-lived mouse of species Mus musculus. In fact there are two sub-prizes: one for longevity, and a "rejuvenation" prize, given to the best life-extension therapy that's applicable to an already partially-aged mouse. There is a complicated prize structure, wherein each researcher who produces the longest-lived mouse ever or the best-ever mouse-lifespan rejuvenation therapy receives a bit of money each week until his record is broken.

Aubrey suspects that, within the next decade or two, it should be possible to come pretty close to defeating senescence within mice — if the research community puts enough focus on the area. Then, porting the results from mouse to human shouldn't take all that much longer (biological research is regularly ported from mice to humans, as they are an unusually suitable testbed for human therapies—though obviously far from a perfect match). Of course, some techniques will port more easily than others, and unforeseen difficulties may arise. However, if we manage to extend human lives by 30 or 40 years via partly solving the problem of aging, then we'll have 30 or 40 extra years in which to help biologists solve the other problems.

Aubrey likes to talk about the Methuselarity – the date at which, if you're still alive, the odds are high that ongoing advances in biomedicine will allow you to avoid involuntary death altogether.

Theory-wise, Aubrey agrees with me that aging is complex and probably due to a host of different causes. He doesn't believe there's one grand root cause of senescence, but rather that it's the result of a whole bunch of different things going wrong, mainly because human DNA did not evolve in such a way as to make them not go wrong. But he doesn't think we need AGI to figure out how to solve all the problems – he figures plain old human ingenuity can do the trick. I hope he's right! But I figure we may as well approach the problem from all viable angles. I'm glad Aubrey's team and others are working on the longevity problem via innovative biology methods... and I've been spending a fraction of my time attacking existing biological data using narrow AI. But I also think its key to work on building better minds that can integrate all the available data in ways inaccessible to the human mind, and likely propose new experiments and therapies that humans would never think of.

Toward an AGI Biomedical Researcher

While working on an early version of this chapter, I happened to get a call from a friend in California, asking me to put together a proposal for the use of OpenCog AGI and other AI tools to cure aging and work toward radical longevity. My friend wanted to show the proposal to a wealthy business contact—but only if I could find a way to use OpenCog to work toward radical longevity that would also generate useful products along the way.

While these sorts of pitches usually don't come to anything, or at least take a long time to yield fruit, I figured it was an interesting challenge. My plan involved the creation of an AGI biomedical researcher, but to move toward this long-term goal through two earlier stages: First, an AGI that would chat with you about your medical issues.Then, an AGI biomedical research assistant that would help a scientist in their work based both on knowledge it absorbed from biomedical databases and using statistical and narrow-AI tools to analyze biological datasets. Maybe not as smooth a path toward human-level AGI as virtual world agents or robotics – but I think it's one that could synergize wonderfully if pursued in parallel with the virtual agents and robotics approaches.

Many of the ingredients needed to realize the vision of a biomedical AGI researcher already exist:

Massive (albeit heterogeneous and messy) biological databases

Narrow-AI tools for analyzing biological data and extracting information from biological research papers and abstracts

AI reasoning systems for drawing conclusions from biomedical knowledge (we used an old version of PLN for this back in 2005)

Robotized lab equipment allowing AI systems to run their own experiments without human intervention.

But these ingredients are scattered about: Nobody is using them in a unified way. Why not combine them into a coherent, powerful, biomedical, artificial general intelligence? This would be a fantastic project!

Coming up with a proposal for my friend's request presented a challenge: In order to drive the integration of all these ingredients, and the evolution of the integrated system into a powerful biomedical AGI, I needed to come up with some practical application with both humanistic and economic value at the early stages of development, and then increasing value as ongoing development advanced its intelligence. One suitable application was a dialogue system — a software system that uses ordinary English to chat with people about medical and biological issues.

A version-one biomedical dialogue system could be something fairly simple: A digital diagnostician and advisor, helping people assess their symptoms and helping them fish through the complex online biomedical literature for relevant information. Given the high cost and low quality of most medical care available, and the disorganization of in-depth online medical information, this would be a major public service. And it could also be a significant revenue driver for a variety of appropriate businesses (like a website selling non-prescription therapies).

With ongoing work on the back end, the digital diagnostician and advisor could evolve into an AGI biomedical research assistant, with the capability to help scientists in their work. This would allow the rapid discovery of biomedical truths that would take far longer for humans to discover on their own. It could also promote industry growth (e.g. the founding of an

AGI/biopharma firm in which AGI systems and human scientists combine to rapidly conceive novel therapies and diagnostics for a variety of diseases).

Finally, with more time and effort, the biomedical AGI research assistant could evolveto the next stage: an AGI biomedical scientist, able to make important discoveries on its own without human aid — carrying out the whole process of scientific invention, discovery, validation and technology transfer on its own. Here, we might expect truly radical, impactful therapies combating death and all forms of disease.

In my proposal, I laid out the incremental progress along this path to AGI as follows:

Year 1

- Digital diagnostician and advisor
- Bio-NLP system for extracting information from biomedical research abstracts
- Machine learning system enabling cross-dataset extraction of patterns from dozens of datasets simultaneously

Year 2

- Beta testing and launch of digital diagnostician and advisor
- Dialogue system enabling conversation about information extracted from research abstracts
- AI reasoning system drawing novel conclusions based on combining information from databases and research abstracts
- Scaling-up of machine learning data analysis system to 100s of simultaneous datasets

Year 3

- Semi-automated identification of metadata in datasets, speeding ingestion of datasets into the database
- Automated extraction of information from tables and figures in research papers
- Extraction of information from full text of biomedical research papers
- Automatic triggering of machine learning data analysis tasks based on information extracted from research papers

- Dialogue system enabling control of machine learning data analysis via interactive dialogue

Year 4

- Beta testing of AGI biomedical research assistant
- Automatic generation of new "research abstracts" describing what the AGI has learned
- Use of automated inference to specify new experiments to run

Year 5

- Commercial launch of AGI biomedical research assistant
- Connection of AGI research assistant directly to appropriate lab equipment for experimentation with a completely automated AI-run lab for experimentation on micro-organisms (probably yeast)
- Customization of natural language processing engine enabling it to understand simple language in scientific textbooks, and to generate more flexible biomedical language as well

Year 6

- Testing of AGI biomedical research assistant's findings to study longevity biology and other specifically identified topics, in conjunction with human researchers
- Customization of AGI inference engine to integrate knowledge extracted from natural language textbooks with information ingested from datasets and structured databases
- Customization of natural language processing engine to enable it to understand more complex language in scientific textbooks
- Enabling of system to generate brief research reports as well as abstracts

Year 7

- Beta testing of AGI biomedical research assistant Version 2(capable of discussion and reasoning based on information learned from textbooks)
- Ongoing improvement of reasoning and language processing technology to enable greater intelligence

Year 8

- Launch of AGI biomedical research assistant Version 2

Year 10

- Launch of AGI biomedical scientist with capability to:
    - ✓ Read and interpret research literature and discuss its conclusions,
    - ✓ Produce research abstracts and reports summarizing its conclusions
    - ✓ Suggest detailed experiments to be run in vitro and in vivo (regarding multiple organisms)
    - ✓ Conduct yeast genetics and potentially other microbiology experiments on its own via direct connection to lab equipment

Amazing!!! Now, imagine doing all that in parallel with OpenCog, controlling millions of video game characters interacting with people all over the world, and using it to control humanoid robots exploring everyday human environments. The synergies between these different OpenCog projects would be tremendous, and the collective effort would speed us toward human-level general intelligence – and better yet, toward beneficial AGI, oriented fundamentally toward goals of helping people to live longer and better.

Yeah, well, just as I expected, the wealthy individual to whom this plan was pitched didn't bite. Ah well. This sort of thing will happen one day and probably within my lifetime, whether utilizing OpenCog or some other AGI and bioinformatics technologies.

How I Think Aging Works

We'll probably need an AGI to fully understand how aging works, and create a full cure. But that's no reason not to try the best we can right now, with our human minds, to understand what's going on with the aging process. My own theories of aging have evolved as I've studied more data and talked to more researchers, and my views may have changed a little by the time you read this – but for what it's worth, I'll now spend a few pages telling you my current perspective!

A brief summary of my current working hypothesis is:

Accumulation of damage, and development-based antagonistic pleiotropy, both do their own harm to the body, and also contribute to exhaustion of mitochondria and other parts of the bioenergetic system (which are potentially wearing out of their own accord as well). Eventually the energy to power repair mechanisms is unavailable and the downward spiral accelerates. Age associated diseases like Alzheimers and cancer are in substantial part also caused by problems with mitochondria and associated energetic processes, so are causes of and contributors to the basic downward spiral of bioenergetic failure that lies at the heart of aging.

To break this down in more detail – based on everything I've studied and read, it seems to me that three of the more plausible general causes of aging are:

- **accumulating damage** (as is largely assumed in much of Aubrey de Grey's SENS work, which is focused on removing or reversing damage)
- **developmental drift**, or **development-related antagonistic pleiotropy** – complex dynamics wherein gene networks tuned for growth and development have destructive effects in an organism's later life (this is advocated e.g. by Stuart Kim[1] and Joao Pedro de Magalhaes[2])
- **antagonistic pleiotropy** of other sorts, beyond development-related

Of course, biological systems are complex with multiple intertwined aspects, and it's often been suspected that these various culprits may all play roles to varying degrees. But it seems to me these may be the key players. It seems to me one can tell a plausible story regarding how accumulation of damage and development-based antagonistic pleiotropy may work together to cause mitochondria and associated energetic pathways to acceleratingly degenerate.

Let me run through the story I'm thinking of in three different parts....

1 http://sagecrossroads.com/files/Kim.pdf

2 http://www.fasebj.org/content/26/12/4821.abstract

***Aspect 1**: Damage throughout the body places extra strain on energy production; damage to energetic systems directly may play a significant role*

The role of damage accumulation in aging is not yet firmly demonstrated in humans or mice, but is fairly solidly known in bacteria, e.g. as Hayflick[3] notes,

> *Recent studies done using bacteria seem to support the thesis, described above, that "damaged proteins" are the cause of age changes. When a bacterium like* Escherichia coli *divides by fission, one of the two daughter lineages is "damaged enriched" and the other has "low damage." The former are "non-culturable or genetically dead" while the latter are "reproductively competent." In* Caulobacter crescentus, *replicative senescence has been observed, a phenomenon that we first described in normal human cells more than 45 years ago. The phenomenon has also been reported to occur in* E. coli *and in* Saccharomyces cerevisae.

It seems plain that as damage accumulates in various systems on various levels in the body, the body must expend more and more energy to keep basic operations going (as the damage results in decreased energy efficiency all over the place).

It also seems natural to hypothesize that when the energetic systems work harder and harder to meet the body's needs, they get worn out, and stop doing their job effectively. This may have many different results, e.g. it may result in diminished ATP/ADP levels (and then potentially, ADP may be converted to AMP which is less effective at delivering energy).

There is interesting evidence for this sort of dynamics in Chronic Fatigue Syndrome, e.g. the argument that, as some maverick CFS researchers argue,

> *... the basic pathology in CFS is slow re-cycling of ATP to ADP and back to ATP again. If patients push themselves and make more energy demands, then ADP is converted to AMP which cannot be recycled and it is this which is responsible for the delayed fatigue. This is because it takes the body several days to make fresh ATP*

---

3 http://journals.plos.org/plosgenetics/article?id=10.1371/journal.pgen.0030220

from new ingredients. When patients overdo things and "hit a brick wall" this is because they have no ATP or ADP to function at all. [4]

More directly related, there is also evidence that ATP depletion leads to cell death in muscles[5] ; that long-lived worms are able to attenuate the usual age-related decline in levels of ATP and ADP[6] ; and that deficiencies in ATP resultant from defects in mitochondrial energy metabolism are key contributors to aging in fungi[7].

As well as the indirect effects of general damage accumulation throughout the body on bioenergetic functionality, there is also evidence that accumulation of damage directly to mitochondrial and nuclear DNA may cause deterioration of bioenergetic function. This has sometimes been referred to as a "clock of aging" as this damage accumulates gradually during the lifespan, building up increasingly and deteriorating mitochondrial capabilities.[8]

And this leads us on to...

***Aspect 2**: Developmental processes, continuing beyond the stage in the organism's life when they are really useful, consume the body's energy, causing stress; in some cases they may also cause damage directly. This is a key example of antagonistic pleiotropy, but other significant examples also exist and likely play a role in aging.*

Damage seems not to be the only major culprit! Our own growth dynamics run amok seem to cause a lot of the harm as well. Developmental processes, continuing to act after they have passed their point of usefulness, have numerous effects including stimulating the endocrine system in a way that drives the body toward growth rather than maintenance. Among other factors, this misguided growth-oriented stimulation drives a lot of protein

---

4 http://www.drmyhill.co.uk/wiki/CFS_-_The_Central_Cause:_Mitochondrial_Failure

5 http://www.pnas.org/content/104/3/1057.ful,http://www.ncbi.nlm.nih.gov/pubmed/23919652

6 http://lib.ugent.be/fulltxt/RUG01/001/787/432/RUG01-001787432_2012_0001_AC.pdf

7 http://www.ncbi.nlm.nih.gov/pubmed/2139154

8 Rejuvenation Res. 2005 Fall;8(3):172-98. Mitochondrial microheteroplasmy and a theory of aging and age-related disease. Smigrodzki RM1, Khan SM.

synthesis, which uses a lot of energy. (It has been argued that the energetic tradeoff between growth and somatic maintenance may be substantially grounded in energy spent on protein synthesis.[9]) But the increased damage to the body requires more and more maintenance activity — so that the energetic systems start getting exhausted trying to feed growth and maintenance processes simultaneously.

The interactions between energetic functions and endocrine regulation are quite subtle, e.g. mitochondrial function is known to guide insulin secretion[10]:

> *In the endocrine fraction of the pancreas, the B-cell rapidly reacts to fluctuations in blood glucose concentrations by adjusting the rate of insulin secretion. Glucose-sensing coupled to insulin exocytosis depends on transduction of metabolic signals into intracellular messengers recognized by the secretory machinery. Mitochondria play a central role in this process by connecting glucose metabolism to insulin release.*
>
> *Mitochondrial activity is primarily regulated by metabolic fluxes, but also by dynamic morphology changes and free Ca2+ concentrations…*

Another potential source of accumulating damage is drift in stem cell numbers. Stem cell numbers in various tissues drift according to complex dynamics[11], and are not necessarily well regulated throughout the lifespan. This is another case where regulatory systems have likely been tuned for the development phase of life, and are not optimal for later life; having too few or many stem cells in a certain tissue can then cause this tissue to degenerate, placing strain on energetic processes and causing other issues.

On reading an earlier version of this document, mitochondrial-aging-research pioneer Rafal Smigrodzki noted that "there are some forms of antagonistic pleiotropy that are not developmental in nature but are

9 http://www.ncbi.nlm.nih.gov/pubmed/20886754

10 http://www.medicinalgenomics.com/wp-content/uploads/2013/11/ENDOCRINE-FUNCTION-OF-MITOCHONDRIA.pdf

11 http://tamm.mit.edu/lab_meetings/NOV_10_2010/Intestinal%20Stem%20Cell%20Replacement%20Follows%20a%20Pattern%0Aof%20Neutral%20Drift.pdf

still related to trade-offs between various short and long term stresses (micropredation, macropredation). These trade-offs are baked into our regulatory networks and they respond in a maladaptive way to signals of damage generated by mutated/epimutated mtDNA and nDNA. " Well-said, yeah....

And now for the punchline – what seems to be the knockout punch ...

Aspect 3: As Aspects 1 and 2 take their toll, strained energetic processes, esp. involving mitochondria, are working so hard to deliver energy they don't maintain themselves effectively.

This is the kicker, as I currently understand it: Exhausted energetic processes don't defend themselves against damage very well, and allow damage to themselves and to the damage-control mechanisms, and vicious cycles kick in.

Along these lines, there is evidence that improved brain energy results in less free radical damage and fewer indications of aging[12].

It's also worth noting here that cells in various body systems (esp. the brain) appear to route mitochondrial DNA among themselves in complex ways. This routing may depend on specific markers in the mtDNA, so that mtDNA mutations may impact its behavior. One consequence is that mtDNA mutations could affect energy production in complex ways, via disrupting messages regarding energy production between different cells and in some cases even different organs (as John Hewitt has argued in detail[13]). So as the mechanism for regulating routing of mtDNA deteriorated, we'd expect to see negative mutations spread throughout body systems, leading indirectly to all sorts of consequences including decreased immune function, increased cancer rate, and so forth.

So, in sum: accumulation of damage, and development-based antagonistic pleiotropy, both do their own harm, and also contribute to exhaustion of mitochondria and other parts of the bioenergetic system (which are potentially wearing out of their own accord as well). Eventually the energy to power repair mechanisms is unavailable and the downward spiral accelerates.

---

12 http://www.wellnessresources.com/studies/alc_has_anti_aging_boost_on_brain_energetics/

13 http://medicalxpress.com/news/2015-01-mitochondrial-dna-mutations-good-bad.html

How did I come to these tentative conclusions? (or, perhaps it's better to call them interesting hypotheses…) Not purely guesswork of course! – the path involved the bio-AI methods I briefly mentioned above. … application of machine learning and statistical tools to analysis of a number of SNP and gene expression datasets. As I applied AI tools to one after another longevity dataset, my human brain thought about the results and what they might mean….

In 2005, analysis my team did of gene expression data from aging human brains (versus younger human brains) showed perturbation of expression in many genes related to oxygen transport, as well as more obvious categories like synaptic growth.[14]

Analysis of the Wellderly SNP data gathered by Scripps (about 400 healthy humans aged 80+, versus about 400 unhealthy humans of the same basic ages) revealed a surprising number of genes related to neural function and development, near the top of the list of differentiated genes.

Analysis of both SNP and expression data from Genescient's Methuselah flies, which were evolved for longevity via directed evolution in the lab, revealed an interesting story. Analysis of the Methuselah fly SNP data showed that the flies acquired a lot of genetic variation related to neural and developmental function, along with immune and basic metabolic pocesses. Analysis of the Methuselah fly expression data showed perturbations in gene expression related to many different processes, including a significant number of genes related to mitochondrial function and other energetic dynamics.

As an additional source of information, we are currently studying supercentenarian SNP data using machine learning tools and hope to have results shortly. I wonder what this will add to the picture.

Aging vs. Age-Associated Disease

Longevity reearch pioneer Hayflick, among others, has argued that studying or even curing age-associated diseases won't help much understanding aging:

14 This study was written up in a paper called "Mining Biological Information from Ensembles of Gene Expression Classification Models" but I'm not sure whether it was ever published. It was presented as a poster at a Cold Spring Harbor conference on the genomics of aging, in 2005 I believe.

The fundamental aging process is not a disease but it increases vulnerability to disease. Because this critical distinction is generally unappreciated, there is a continuing belief that the resolution of age-associated diseases will advance our understanding of the fundamental aging process. It will not. This is analogous to believing that the successful resolution of childhood pathologies, such as poliomyelitis, Wilms' tumors, and iron deficiency anemia advanced our understanding of childhood development. It did not.[15]

On the other hand, others have argued that aging as a generic phenomenon does not exist, and all that really exists is a set of age-associated diseases. In this view: Once you've gotten rid of all the age-associated diseases, you've gotten rid of aging!

What does my tentative theoretical understanding of aging say about this? In my view the likely situation is that:

- The wearing-out of bioenergetic systems makes certain diseases more likely; these are among the age-associated diseases.
- Diseases contribute to wearing out bionergetic systems, just as damage to various body systems does

Concrete evidence in this direction was provided by studies my team did in 2004-2005, exploring the relation between heteroplasmic mitochondrial mutations and Parkinsons and Alzheimers[16]. We found we could predict if an individual had one of these diseases with near 100% accuracy by looking at whether their mitochondrial DNA tended to be mutated in certain particular regions (even though different mitochondrial genes might be mutated in different parts of those regions). Further, David Parker found that inserting mitochondria from neurons with Parkinsons into healthy neurons, caused the healthy neurons to behave like neurons with Parkinsons. These studies did not resolve the question of why the mitochondria in some people (but not others) tend to get mutated in these particular places. But they did demonstrate a close connection between these age-associated neurodegenerative diseases, and bioenergetic dysfunction.

---

15 http://journals.plos.org/plosgenetics/article?id=10.1371/journal.pgen.0030220

16 The Parkinsons study was published here http://www.ncbi.nlm.nih.gov/pubmed/16207526 ; the Alzheimer study gave similar results but was not published

Part of the underlying issue seems to be that in brains with Alzheimers or Parkinsons, the mitochondria have lost the ability to effectively generate ATP. In Alzheimer's, this triggers a protective mechanisms in which cells shut off the production of glucose. This sort of dynamic has led to some researchers referring to Alzheimers as "Type 3 diabetes"[17]

Cancer, according to some hypotheses, can be understood similarly – as a metabolic disorder related to mutations in mitochondrial and nuclear DNA. According to Seyfried et al,[18]

> *Emerging evidence indicates that cancer is primarily a metabolic disease involving disturbances in energy production through respiration and fermentation. The genomic instability observed in tumor cells and all other recognized hallmarks of cancer are considered downstream epiphenomena of the initial disturbance of cellular energy metabolism. The disturbances in tumor cell energy metabolism can be linked to abnormalities in the structure and function of the mitochondria.*

Basically, in this story: Mitochondria in cancer cells, like in Alzheimers cells, fail to produce sufficient ATP; but in this case glycolysis is amplified rather than suppressed. Excessive glycolysis leads to excess accumulation of pyruvate, resulting in fermentation (an anaerobic process, which however can still occur in the presence of oxygen, as is the case in cancer cells). If their mitochondria were functioning correctly, the pyruvate would be converted to acetyl CoA, and ATP would be produced normally. But defective mitochondria push cancer cells to ferment even in the presence of oxygen. All this fermentation leads to acidosis and other issues, which then push cancer cells toward metastasis.

If these perspectives on neurodegenerative diseases and cancer are vaguely in the ballpark, then what we see is that age-associated diseases may not really be so distinct from the fundamental dynamics of aging after all. If aging is largely about wearing-out of bioenergetic systems, centrally mitochondria; and age-associated diseases are also caused by complex

17 http://www.encognitive.com/files/The%20Relationship%20between%20Alzheimer%27s%20Disease%20and%20Diabetes:%20Type%203%20Diabetes.pdf

18 http://www.ncbi.nlm.nih.gov/pubmed/24343361

networks of processes in which mitochondrial dysfunction plays a significant role – then aging and age-associated diseases are all part of the same tangled network of vicious cycles related to decreasing bioenergetic effectiveness.

Entropy, Energy and Aging

Being of a philosophical bent (as well as bent in lots of other directions!!), I can't resist reflecting on these hypotheses from a broader perspective, as well. The key role of energetics here is conceptually compelling.

One is reminded of the perspective sometimes ventured that the quest to efficiently use energy is the key to the origin and evolution of life, perhaps even more fundamental than natural selection[19]. However, there is reason to believe this is an overstatement, and that maximizing entropy production may trump maximizing energy efficiency[20].

Biological systems theorist Stan Salthe[21] , whose work I've enjoyed since we met by chance at a "Socialist Scholars conference" in the early 1990s, argues that "Maximum energy efficiency in animals appears to be a default position that would likely be found only under basal metabolic, resting conditions." As he puts it,

> *... in a world characterized by unavoidable capricious events, and occupied as well by other-directed agencies, a living dissipative structure is continually being impacted and deranged, so that it will frequently be in a state of striving. This would entail greater energy flows than those that deliver maximum power. In animals this involves, e.g., fleeing, fighting, mating, competing for resources, shivering, healing wounds and infections, migrating, and taxing brain activity at all times - everything beyond mere basal homeostasis. With plants this would involve outgrowing competing individuals, producing toxins and allelopathic substances, and healing wounds.*

19 http://www.ncbi.nlm.nih.gov/pubmed/16122878

20 http://cosmosandhistory.org/index.php/journal/article/view/189/284, http://www.lawofmaximumentropyproduction.com/

21 http://cosmosandhistory.org/index.php/journal/article/view/189/284

In aging, clearly, the ability to achieve basal homeostasis with high energy efficiency is impaired, via damage that creates high entropy, and via development processes that keep trying and failing to create more growth. Maximum entropy production on the whole-body level is also impaired due to inefficiencies; entropy is produced locally in the energetic system, preventing efficient delivery of energy to other body systems that would produce entropy more effectively.

One way to phrase what happens in aging, according to the hypothesis given here, is: When various factors (damage, antagonistic pleiotropy, etc.) conspire to make the release of energy from molecular bonds to power the body too inefficient, the strategy of maximum entropy production shifts from maintaining order in the whole body, to allowing the body to accumulate damage and allowing its cells to stop functioning (and eventually allowing the whole body to stop functioning).

So aging may occur as a result of quite fundamental dynamics of the physical universe. But yet, just as life goes in the opposite direction from the Second Law of Thermodynamics (decreasing entropy locally even as it helps increase entropy globally), so can we – using advanced technology – push against the Second Law even further in our local realms, via halting or reversing the evolutionarily-provided degradation of our bodies. It's "just" a matter of figuring out the relevant biology, physics, chemistry and so forth, and doing appropriate engineering based on what we discover.

The dynamics of aging are not yet fully understood yet they get clearer year by year – they are far clearer now than when I first started looking into the area a few decades ago. AGI will be able to see the picture even more clearly than we can; but even with our sorely limited human brains, we're step by step getting there....

Cryonics as a Backup Plan

OK, so death is a solvable problem — and the human race will probably solve it fairly soon. At least, "fairly soon" on the time-scale of human history. But even if it does get solved, will it really get solved soon enough for you or me?

We could accelerate the pace toward solving aging with a team of AGI biomedical scientists based on OpenCog or some other platform. Or we could fund other approaches like Aubrey de Grey's SENS project. Better yet, we could have both SENS and the AGI biomedical researcher! If the human race — or any one of the planet's super-rich individuals — took life extension seriously, both of these projects would be massively funded.

But still – we all knowhow hard it is to get innovative technology projects funded. So, one has to consider the pessimistic possibility: What if the aging problem doesn't get solved in time for you and me personally? After all, I'm 46 years old in 2013 as I write these words, so by the time of Kurzweil's conjectured Singularity in 2045 I'll be a rather old man, at least by current standards. Are there any reasonable backup plans — short of going to another galaxy and drinking the alien's immortality potions — aside from just rolling over and dying?

There's one backup plan that seems reasonably likely to succeed: ***Cryonics***. We now have the technology to freeze a body in liquid nitrogen, using special cryoprotectant chemicals to avoid damage during freezing, in a way that seems to preserve all the important structures of the brain and body. Now, we don't yet know how to defrost these frozen bodies without causing damage. But surely, the superhuman AIs and cyborgs after the Singularity will be able to figure that out! So, one backup plan right now is to have yourself frozen in liquid nitrogen right after you die… and then get defrosted later on, once the technology is available. Obviously this isn't a guarantee – a lot of things could go wrong. But as I like to say: Better frozen than rotten!\

FIGURE 16.38: *This is me visiting Alcor's Arizona cryonics facility in 2012. Each of the silvery tanks behind me contains a "corpsicle" (not their official term), i.e. a human being preserved in liquid nitrogen, for hopeful eventual reanimation by some technological means or another. If things don't go well enough for me, I may end up in one of those one day. On the other hand, if I die via falling off a rock-face in a remote region or a plane*

crash, my brain may be nonrecoverable, and future minds may end up reconstituting me via subtler methods instead, using data such as this book and all the videos of me speaking and moving that have been posted online.

A number of organizations offer this kind of cryonics service — I'm signed up with one of them, Alcor, which is based in Arizona. So if I die, my body will get shipped to Arizona and preserved in liquid nitrogen... And hopefully my descendants will defrost me, then give me a new body (a robot body perhaps). Around 100 people are already frozen in Arizona, waiting for the technology to reach the point where it's possible to revive them.

But that's just a backup plan, of course. Better frozen than rotten — but better living than frozen!

Why So Little Focus on Longevity?

When I first found out about death as a little child, I was baffled at how unperturbed the adults around me seemed by this rather terrible fact of human existence. I understand people a lot better now, but in the end I'm still perplexed. Getting old and dying is a Very Bad Thing, but nobody seems to think about it eliminating the issue. We spend a lot of money on bombs for blowing people up, and cures for diseases like cancer and AIDS — but very little money on the core problem of eliminating aging itself. And we spend a lot of money on computer programs for video games. Web search and supply chain management, but very little on developing smart AI software to help us figure out the complex problems of aging and longevity. As individuals, we hate it when our friends and family get old and die — but as a society, we don't seem to care much to do anything about it.

I'm especially vexed by the lack of interest in longevity research coming from wealthy individuals. Over a thousand billionaires and tens of thousands of people with an individual net worth of $100 million or more are living now. Why don't more of them devote 10-20 percent of their wealth to the creation of longevity drugs that will let them enjoy their riches over an extended lifespan? I suppose it's because they just don't believe it's possible — they don't think that life extension is a fruitful line of inquiry. But this attitude is wrong. For the first time in human history, we're at a point where death is no longer inevitable. The cure is in sight. There's a lot of work to be done to cure death, but it's most likely going to happen this century — maybe even soon enough for you and me.

There are strong indications of accelerating progress in recent years. Google has started a longevity project called Calico; Craig Venter, funded by an array of tech zillionaires, has started a new firm called Human Longevity Inc., aimed at sequencing the genomes of people of various ages and studying the differences. Hopefully these initiatives are harbingers of future trends.

If you agree with me that death is a terrible thing, maybe you should think about joining the fight — donate to organizations combating aging, or even contribute to the research yourself. Better yet, contribute to AGI, which with sufficient funding and intellectual attention may soon be able to solve the human aging problem better than us mere humans!

Further Reading

De Grey, Aubrey and Michael Rae (2007). Ending Aging. St. Martin's Press.

Rose, Michael, Margarida Matos and Hardip Passananti (2004). Methuselah Flies.World Scientific.

For a good review of the various theories on why aging happens, see https://www.fightaging.org/

17. The Terminator Question

"Balance the risks of action and inaction."

- Max More (2004),
articulating what he calls the"Proactionary Principle."

A few years ago, I gave a lecture on AGI and the Singularity, via Skype, to a lecture hall full of Ethiopians. Via the wonders of the Internet, I spoke to them from my house in Ting Kok Village, where I lived at that time (by the seaside in a rural area of the New Territories in Hong Kong); they listened from their university in Addis Ababa, Ethiopia's capital. Before and since that lecture, I've been in contact with scientists and programmers in Addis, and in early 2013 I helped them set up the country's first artificial intelligence research facility, iCog Labs, which is now collaborating on OpenCog R&D. Exciting stuff!

FIGURE 17.1: *Me with Getnet Aseffa, the leader of the Addis Ababa research shop "iCog Labs", where a team of Ethiopian programmers carry out OpenCog and other AI work, in close collaboration with the OpenCog team in Hong Kong. We're standing on the hills overlooking Addis Ababa. The office-buildings in downtown Addis are quite modern and filled with all sorts of companies, including an emerging tech scene. But in the hills around the city, people are subsistence farming much as they have been for thousands of years.*

After my lecture, the audience asked some questions — and lo and behold, what was the first question? You guessed it! I don't recall the precise wording, but it was a variant on "After you've created these superhumanly intelligent AGIs, aren't they just going to kill or enslave us all and take over the Earth?"

My mental reaction was something like: Wow, it really doesn't matter what part of the planet you're on, human beings all think the same way. The "Terminator" question is the #1 question whenever I'm talking about AGI to almost any audience ("How can I use AGI to make a lot of money?" being a distant #2) — the only exception being professional AGI researchers.

FIGURE 17.2: *Arnold Schwarzenegger's time-traveling killer robot in the Terminator movies, has become the paradigm case of a super-powerful AI out to murder people and destroy human civilization. While the potential risks of AGI or any advanced technology must be seriously considered, it's also a risk to take science fiction as a close guide to reality. SF authors and filmmakers are guided by what makes a good story, and the complexity and subtlety of the real world often don't qualify. Robot scientists and physicians aren't as dramatic as robot warriors, but may end up playing a far larger role in the future. (http://terminator.wikia.com/wiki/File:Governator.jpg)*

And of course, it's a natural question — not only because of the prevalence of "Evil super-AI" in SF films, but for more basic reasons. We humans like being the top dogs on the planet. The advent of something more powerful and intelligent than us is inevitably going to feel a bit scary. In recent months, as I write this, similar questions have been getting dramatic media attention, due to high-profile sci-tech figures like Stephen Hawking and Elon Musk going to the media with their worries about the future of AI. Musk now-famously tweeted that AI researchers are letting the demon out of the bottle, and once it's out it won't want to go back. Hawking publicly stated that AIs could potentially be "more dangerous than nukes".

The odds of a super-powerful AGI desiring to enslave people seem very low to me — I reckon that once an AGI is powerful enough to do this, it will figure out better ways to do the work suited for humans. Science fiction loves this kind of scenario — the future humans in the Terminator series, forced to carry out manual labor for robot overlords; the comatose humans in The Matrix used as biological batteries, etc. But the requirements of storytelling are only very loosely related to the dynamics of reality. Surely an AI smart enough to send Schwarzenegger back in time would be smart enough to build robots better at doing hard labor than humans; and surely any mind capable of building the Matrix could also build a better battery than the human body.

But what about the risks posed by a military-created AGI? Actually, I think it's very unlikely a military AGI would go in the direction of intentionally wiping out or tormenting humanity. So far, the handling of nuclear weapons technology by the world's leading military nations supports this idea. With the exception of China — itself an extremely insular nation without any history of military action beyond its border nations — the world's major military powers are democracies, and the incidence of wars between democracies throughout history is rather low. And many political commentators expect China to democratize sometime in the next few decades.

FIGURE 17.3. *A more whimsical view of the End of the World (http://www.demotivation.us/media/demotivators/demotivation.us_What-I-imagine-everyone-will-do-In-the-end-of-the-world_139996600413.jpg)*

A powerful AGI created by the armed forces of a major nation might be ruthless in pursuing the goals of that nation, but it appears unlikely that it would then spontaneously mutate its goals in such a way as to make it want to kill everyone and become the dictator. A human sometimes does this when given too much power, but AGIs aren't necessarily going to share human traits such as megalomania and sadism unless their taught these values orprogrammed that way. "Power corrupts, and absolute power corrupts absolutely" is an aphorism founded on observation of human beings, and there's no reason to assume it applies to minds in general.

More frightening than AGI-powered national robot armies is the possibility of terrorist organizations getting ahold of advanced AGIs. This sort of possibility is genuinely, rationally scary, and exists in the context of many other advanced technologies as well — synthetic biology, nanotech, and so forth. What would happen if a couple hundred disenchanted top-notch technology geeks decided to join forces with a group like Al Qaeda? How much destruction would ensue, even without AGI? What would happen if human-level AGI were added to the mix?

And what if it took just one person to create an AGI? An AGI programmed by a misanthropic sociopath is certainly a possibility — i.e. a scenario where an evil mad scientist intentionally programs an AGI to wipe everybody out. However, I think this is also quite unlikely, not because such sociopaths don't exist, but rather because they probably won't be the first to create an advanced AGI. Fortunately, outside the movies, extreme sociopathology and extreme scientific brilliance tend not to frequently co-occur. I alsosuspect that the first human-level AGI will soon be followed by a Singularity-type eventleaving little time in the interim for a sociopath to create an AGI that kills everybody.

Personally, I think the biggest risk related to advanced AGI systemsis that once smart enough they will be indifferent to human beings, in the same way that we're now indifferent to field mice, ants and bacteria. Few humans go out of their way to squash ants, and we generally worry about killing bacteria only when they're bothering us directly. However, when we put up a new building on a field of dirt, we don't worry much about thesubsequent mass-murder of ants and bacteria.

The bottom line is we really don't know enough yet to have a solid understanding of the issue of "superhuman AGI ethics". I personally suspect that it will be possible to create advanced AGI systems that respect "lesser" beings like human, by creating AGI systems that (like OpenCog) tend to respect their own goal systems, and instilling them with human-friendly goal systems via a combination of programming, teaching and simple human kindness. But I certainly can't prove this will be possible, and I'm not absolutely certain of it. We'll have to explore this domain via a combination of experiment and theory, just like every other aspect of AGI.

Ultimately, the whole paradigm of "superhuman AIs created by humans taking over the world, or not" feels limited as a potential future vision. Terminators, killer robot armies and so forth are wild speculations – but if we're going to throw around wild speculations, there are lots of other ones out there too, which are at least as plausible but happen not to have been made into as many science fiction movies. My own personal speculation is that once we get intelligence massively beyond the human level, this is going to lead to contact with already-existing "superintelligences" of various sorts. I suspect that the universe harbors various sorts of intelligence that we cannot currently recognize, but that greater intelligence will allow enhanced humans and/or cyborgs and/or AGIs to recognize. If this is the case, then the superhuman AGIs we may create won't be the whole story – they will mostly just be our portal into aspects of the universe that we don't now understand anything about. And in this case, how much would the specifics of the massively intelligence AGIs we create really matter?

Just as my dog has no idea what's actually going on when I play the piano or sit at the computer (e.g. he doesn't realize there are sometimes other people chatting with me via text as I sit at the desk, nor that the mathematics I'm working with was largely invented hundreds of years ago in far-away places, nor that the software I'm running is causing electricity to course through machines in a server farm in Germany, etc.), there are likely all sorts of intricate and intelligent processes happening in the world around us, which we currently fail to detect due to our limited intelligence. I look forward to the possibility finding out what they are, even though, if this does happen, the "I" that finds out will of necessity be quite different than the "I" that exists now. The challenges, limitations and problems this future evolution of me perceives will likely bear very little resemblance to the things that worry AGI naysayers today – probably less resemblance than Nick Bostrom's concerns about existential risk bear to my dog Crunchkin's daily concerns, since at least Nick and Crunchkin share a mostly-similar brain architecture, whereas future AGIs and other forms of intelligence in the universe will likely be structured very differently.

Of course, this is all SF-style speculation – but so is the Terminator scenario, so are all the potential future calamities that have Elon Musk, Nick Bostrom and others in such a huff. Some peoples' intuitions lead them to fear the Terminator; my intuition leads to me to suspect coming contact with greater and broader intelligence. None of us has evidence to back up our intuitions about this sort of thing. If we want to look at it empirically and scientifically rather than intuitively, we come back to Vernor Vinge's original taken on the Singularity: after minds way smarter than us are here, all bets are off!

One thing I am pretty confident of, though, is that AGI itself is well-nighinevitable. Nothing short of a large-scale collapse of civilization, or a global dictatorship bent on delaying the Singularity, is going to stop it. If the US were to outlaw AGI research, China would develop AGI. If China outlawed AGI research, Ethiopia would eventually develop AGI. And so forth. If every country outlawed it, an international AGI underground would probably develop, and you'd need a really powerful international anti-AGI police force to keep the R&D squelched — especially if computer and communication technologies keep advancing.

AGI is inevitable in the same sense that written language was after the introduction of spoken language. Writing was speech's natural next step, given the nature of human beings and the materials available on the Earth.

And of course, we can't predict the consequences — any more than the people to write down the first few written words could predict the consequences of writing.

That doesn't mean we shouldn't try to nudge the Singularity in a positive direction; it doesn't mean we shouldn't try to raise our baby AGIs to have compassion and understanding. But it does mean we should temper our ambition to guide the future with a bit of humility regarding the size and sweep of the processes rushing us along. The development of progressively more complex adaptive systems on Earth started long before humans came along, and will continue long after humanslose their top-dog status on the planet (having been supplanted by AGIs) -- the Singularity we're most likely near is just another fascinating phase transition along this path.

Will There Be Cyborgs?

Some futurist types don't worry about Terminator scenarios much, as they plan on becoming cyborg-like superhuman beings, perhaps by fusing with AGI systems as the latter advance. If you plan on becoming a superhuman AGI via uploading your brain into a computer and enhancing it, or via plugging a super-brain chip into your biological brain, then the fate of those humans who refused to upgrade may seem a lesser concern.

On the other hand, it might turn out that the superminds formed via upgrading human beings are still a bit less super than other superminds formed without the constraint of maintaining continuity with some human self. So, even humans who upgrade themselves into superminds might still find themselves at a disadvantage relative to the other wholly nonhuman superminds out there. Admittedly, here we're pretty deep into science fiction territory! But these are fun issues to speculate about.

Humans may fuse with AGIs in many different ways, including fairly prosaic ones like becoming dependent on AGIs in their smartphones, like in the movie Her. The most viscerally dramatic representation of human/AGI fusion, however, is the physical cyborg – a creature whose body is part human biology and part engineered robot. The cyborg is both a real future possibility, and a symbol of the potential coming-together of man and machine in an alternative to us-versus-them scenarios.

FIGURE 17.4: *A comic book cyborg. (http://static.comicvine.com/uploads/original/7/74889/2697604-cyborg628.jpg)*

A cyborg could look like a human being with a computer jacked into its brain, or a human being with wheels, a tail and wings that permit flight, or a wheeled robot with a human head, etc. You can surely imagine a lot of other possibilities! The key point is that a cyborg combines engineered components with traditional, evolved biological components – and both of these components are working together to guide the system's practical operations. The most interesting kind of cyborg, for me, is the kind where the engineered and evolved biological components work together cooperatively to control the system's mind as well as its body.

Some people think cyborgs are implausible in the foreseeable future, because they're skeptical that technology will advance that fast. This kind of skepticism is not interesting from the point of view of the present book, with its Singularitarian perspective; it basically amounts to the hypothesis that exponential technological growth is going to halt, or hit a mysterious obstacle when it comes to human neurobiology. On the other hand, there is a different kind of skepticism about cyborgs that's more relevant here. Some folks are skeptical about cyborgs because they reason that: *Since non-human AGI systems may prove much more effective and intelligent than cyborgs, cyborgs won't be worth the bother.*

Along these lines, my friend and colleague Hugo de Garis wrote an article for H+ Magazinea few years back called "There Are No Cyborgs". The article argued, not only that there were no genuine cyborgs around at the time of writing, but that there also would be no cyborgs in the future. In Hugo's view, cyborgs are basically a non-starter. He expects that AGIs without the human component will be capable of outperforming cyborgs to an incredible degree, making the latter effectively irrelevant and pointless. As he likes to say, the computing power implicit in a grain of sand is greater than that of all the human beings on earth by many orders of magnitude. So if you can make a human-sized AGI supercomputer, orders of magnitude more intelligent than people – then what use will that supercomputer supermind AGI have for cyborgs, even if the latter happen to be ten or one hundred times smarter than humans?

Hugo's views on cyborgs are closely tied with his argument that sometime in the next century there is going to be a huge world war – the Artilect War, he calls it – between Cosmists who favor the creation of superhuman AI's even if they destroy humanity, and Terrans who want to hold back the development of advanced technologies in order to preserve the human race.

FIGURE 17.5: *Two non-cyborgs. My friend Hugo, a futurist, physicist and AI researcher best known for his pioneering work on evolvable hardware, believes that the future holds a huge world war, between those who favor the advent of superhuman AI, and those who fear it and want to hold it back. I doubt such a war will happen, because I think advanced AI is going to gradually become an integral part of people's lives, like the Internet and smartphones are now, so that very few people will want to get rid of AI. Hugo and I seem to get along well in spite of this disagreement! (This photo was taken at Xiamen University in Summer 2009, when Hugo was a full-time professor there and I was a visiting professor, and we organized the First AGI Summer School there. Subsequent AGI Summer Schools have occurred in Reykjavik, Iceland, and Beijing.)*

I see the appeal of Hugo's line of reasoning regarding cyborgs, but I'm not so sure he's right. Consider the natural world: Despite the uncontested dominance — and undoubted destructiveness — of the human race, we have not wiped out simpler, less intelligent life forms. In some cases — and here I'm thinking of bacteria — we coexist with and couldn't survive without them. And in many other cases, we recognize, more and more, the value of diversity and the need to maintain varied ecosystems.

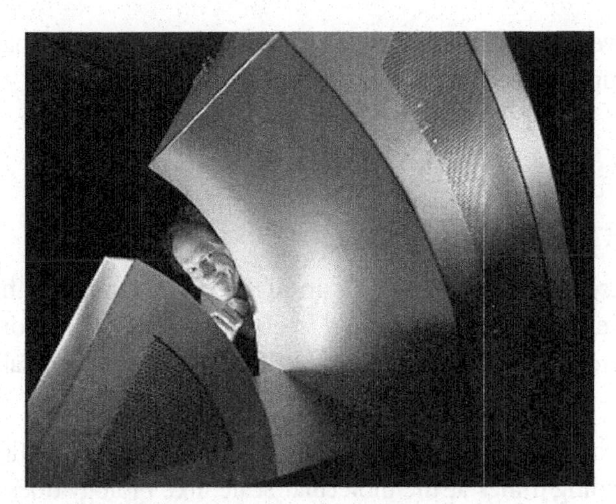

FIGURE 17.6: *Hugo deGaris hiding behind his CAM-Brain Machine, a novel evolvable hardware device he designed in the 1990s, which was quite ahead of its time. It contained a number of FPGAs, reconfigurable chips, whose circuitry was automatically rewired via genetic algorithms and self-organizing, growing neural nets. I saw a CAM-Brain Machine at Starlab in Brussels in 2001, when I visited Hugo there right on the eve of that wonderful research institute's bankruptcy. Hugo used to say how he felt conflicted, because on the one hand he felt driven to build intelligent machines, but on the other hand he feared the world war he thought they would inevitably bring. More recently he has given up his practical AI work, and spends his time on a mix of theoretical physics and futurological speculation.*

As I emphasized already, it's not clear to me why a superhuman AGI supermind would want to get rid of all significantly less intelligent creatures. Why would it? What would it gain in the process? Indifference or benevolence on its part seems just as likely. Perhaps, operating in realms beyond our comprehension, it would have no particular incentive to trouble itself about us. Perhaps it would feel toward us like environmentalist humans feel about endangered species. Estimating the attitudes of superhuman superminds via extrapolation from human emotions seems chancy at best.

Even if superintelligent superminds pose no danger to humanity, one could still see an Artilect War due to peoples' fears of advanced AI. It wouldn't be the first time in human history a major war was started over an exaggerated fear. However, my suspicion is that human life will be too intertwined with AI software and hardware carrying out practical tasks, for the idea of getting rid of Ais to gain much popularity. Who wants to rebel against their smartphone; their online purchasing, negotiating and form-filling agent; their kids' video game characters and robot toy pets; their housecleaning robot and the AI bioscientist who discovered the medicine that cured grandma? Very few.

Yet it may well be that this kind of practical proto-AI technology is where advanced, superhuman AGI will progressively grow from.

The Possibility of Femtotech Superintelligences

To concretize the possibility of an AGI so much smarter than us that helping or exterminating us would seem irrelevant to it, consider one possible route to very advanced AGI that Hugo and I have been talking about for a while — femtocomputing.

Nanotechnology, which I've already mentioned a couple times above, is about building things at the molecular scale, like biology does with DNA, RNA, and proteins. For instance, given enough time and progress, it should be possible to assemble molecules to create computers vastly more powerful than anything available on the market now. And not just computers — if you had what nanotech pioneer Eric Drexler called a molecular assembler, you could use molecules like Lego blocks. Just specify what you want and the assembler will build it out of molecules. The molecular assembler, if one were to exist, could even build a copy of itself.

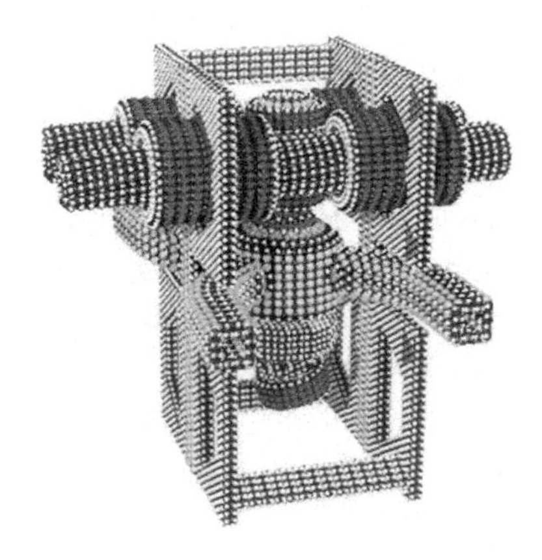

FIGURE 17.7: *Design for a nanotech pump assembly, including over 65,000 atoms. (http://machinedesign.com/archive/molecular-modeling-cad)*

Drexler's vision of nanotech remains unrealized, but simpler versions of nanotech exist right now, and underlie various commercial products. Some novel approaches in modern nanotech involve using molecular biology mechanisms in ingenious ways — inducing biomolecules to assemble themselves into structures that evolution never would have produced.

Femtotechnology goes a step further than nanotech. Instead of using molecules, it uses subatomic particles like protons and neutrons, or elementary particles like quarks and gluons – the stuff that holds larger particles together – as basic building blocks. Today, femtotech is merely an idea – much like nanotech was from the 1950s through the 1980s. But as science and technology develop, I have little doubt it will come to pass.

Normally nuclear particles like protons and neutrons are trapped inside atoms – they make up the nuclei of atoms. But in some strange forms of matter – termed "degenerate matter" – the nuclear particles can come out to play, and engage with each other in more flexible configurations. Neutrons stars are a well known example – a neutron star is just a bunch of neutrons packed together; there are no atoms involved. Another example is a quark-gluon plasma. Quarks are usually thought of as making up protons and neutrons, and gluons as exchanging energy between particles. But given the right conditions, including a very high temperature, quarks and gluons can come together into a strange sort of plasma, which has been created in some physics labs, albeit only for a split second. Will future science comeup with ways to make stabler forms of degenerate matter, capable of carrying out computation? It seems quite possible.

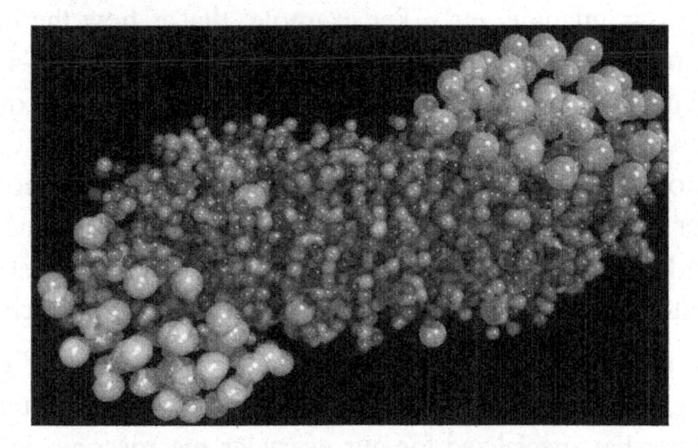

FIGURE 17.8: *Quark-gluon plasma formed by a lead-lead collision in a particle accelerator, viewed shortly after the smash. The white balls are protons or neutrons. The small colored balls are the free quarks. Can we eventually engineer minds and machines from these sorts of exotic states of matter? There is no clear physical reason why not. By current human standards, the engineering problems involved in this sort of hypothetical femtotech are formidable. But an AGI with femtoscale sensors and actuators might view things differently. (http://hep.itp.tuwien.ac.at/~ipp/qgp.html)*

Hugo has worked out some basic math explaining how to build logic gates — the basic components of today's computers -- using quarks and gluons. His specific designs may or may not be possible to ever build in reality, but they're conceptually provocative. For instance, in his talk at the the Humanity+ @ Hong Kong conference in 2011, he explained how to create an OR gate (one of the basic logic gates used in all computers) using quarks and gluons. He gave similar treatments to the AND and NOT gates. Combined, these sorts of gates let you do any kind of computation. Of course no one will ever build a femtocomputer according to Hugo's specific 2011 designs – he was just giving a simple proof of principle to illustrate his general point about the conceptual sensibleness of femtocomputing.

It's a bit of a wild speculation, but I see a potentially interesting analogy between femtocomputing and DNA computing. Quark-gluon plasmas have been shown to contain chains of quarks and gluons that look vaguely like DNA strings, so I wonder if maybe there will be some future analogue of DNA computing within degenerate-matter femtostructures.

As I briefly reviewed above when I was talking about my biology work, DNA represents information in series of information like

... CCC TGT GGA GCC ACA CCC TAG ...

(each letter" is indented, but it should not be stands for a certain amino acid: G= guanine, etc.). For example, that is how the information distinguishing the baby Ben Goertzel from another human baby – or a donkey baby – is encoded. On the other hand, it's been found that quark-gluon plasmas contain particle strings, defined by series like QGGQ .... (Quark, Gluon, Gluon, Quark)... Can one do computing in quark gluon plasmas, in a manner similar to what today's scientific pioneers are doing with DNA computing? I don'tknow how to make it work in my garage just now, but nor did my great-great-grandfather know how to engineer a silicon chip.

Things like nanotechnology and computers built out of elementary particles seem strange or incomprehensible to us, with our human brains that are somewhat specialized for our everyday macroscopic world on the surface of the Earth. But to a superhuman AGI, such subjects would be substantially less daunting, and perhaps enormously diverting. Perhaps we will build an AGI running on ordinary computers, which will figure out how to build a smarter AGI running on molecular nanocomputers, to build a yet smarter AGI running on femtocomputers, which, finally, will build things wholly inexpressible to the human mind.

Would Femtotech Superminds Bother to Exterminate Humans?

Supposing such femtotech-based superminds were created—why would they bother with us? Why would they bother to destroy us, or help us, or mess with us at all… Any more than we ponder the fate of various bacterial colonies living in various mud puddles around the Earth?

You could argue that any system will naturally have the desire to expand and dominate others to fight for its survival. That has been the case on Earth due to evolution by natural selection, although cooperation has been equally important as competition in the evolution of species (the two have been richly interwoven together, along with other more complexsort of self-organizing dynamics spanning the organism and ecosystem level).

Maybe a population of superhuman AGIs of roughly equal intelligence would sink into conflict over basics like survival and resources. Perhaps one would decide that the molecules now utilized by human beings are required to increase its processing powers and obtain a competitive advantage. This is not impossible.

On the other hand, what if one very powerful AGI mind develops faster and succeeds in outclassing the others? Is it necessarily going to want to grab all the available processing power in order to stymie rivals? That's far from obvious. AGI minds would not be constrained by biology, only by physics, and possibly not even the physics with which we are familiar. To these advanced minds, individuality itself could be passé, leading them beyond the categories of individual social beings, as we now understand them.

How can we be confident that animal-level motivations like competition will still be remotely relevant to such beings?

Another possibility is that once AGIs become sufficiently intelligent, they will come to the attention of intelligences spawned by other races amongst the stars.

Very intelligent creatures could be out there in the universe, monitoring us through means we can't understand. Once they become aware of equivalent intelligences on Earth, they may be tempted to make contact. That may sound outlandish—but the bottom line is that we just don't know.

In trying to understand AGIs, we're a bit like early humans communicating at a simple level in the first language ever invented. These early humans, the first ones to invent language, would have realized they were onto something important — but they wouldn't have been able to foresee the real consequences of their invention. The imaginative flights of William Shakespeare and Marcel Proust, and the rigorous logic of differential calculus and computer programming, would exceed the power of these early humans to communicate in their caveman grunt language. And similarly, AGIs could transcend life, the universe and everything as we know it. We can only appreciate the immensity of their potential powers, and speculate about the road to superintelligence.

Hugo says there will be no cyborgs. I think there probably will be cyborgs after all — but I also think that, fascinating as cyborgs will be, they probably won't be the most interesting new beings generated by humanity. My hope is that, once suitable technologies are available, each human will have the choice to remain an ordinary human, to become a cyborg – or to expand their intelligence more dramatically. Some people may choose to remain at the human or cyborg level, but plenty of others will anxiously push the boundaries of the possible and vanish over the horizon, en route to a higher state of being.

The Risks and Rewards of Advanced AGI

Suppose an hypothetical far-sighted caveman were seeing into the future and struggling to comprehend all the terrible things the development of language might bring — systematic hatred, violence and war. And suppose the same early human also dimly foresaw the possibility of literature, science and mathematics. With his restricted view, how could he possibly balance the costs and benefits? Suppose he also had the wisdom to realize the deep limitations of his own perspective? What would his ultimate conclusion have been? Maybe something like: *Something really big is happening, I'm playing a small part in it, and nobody right now is really equipped to understand what's going to come of it. There's a lot of risk here, but all I can do is hope for the best, and do my utmost to push things in my own local sphere in a positive direction.*

Yes, there is unquestionably the possibility that an advanced AGI could destroy humanity. And, yes, as a human I would like to minimize the risk of this happening. I'm doing my best to create a beneficial AGI before somebody else creates a malevolent or indifferent one. But I'm also acutely aware of how little we know about the massive transformational processes going on in which each of us just plays a tiny part. The bottom line is that

AGI is an inevitable consequence of human technological development. Someone is going to build it andno one really has the capability to stop that from happening. Computers will keep developing because we need them for increasingly demanding applications. We must understand the human brain in order to understand ourselves, and we're going to keep developing ever more powerful tools to do that. And our computational algorithms are going to be developed further because we want software to do more. Robotics will progress because we want relief from tedious tasks. Games will advance as we come to expect more from our entertainment. And although all this development is only indirectly related with AGI, it'sstill leading us there, bit by bit – and at an accelerating pace. Eventually, someone will put the right pieces together — maybe my OpenCog colleagues and me, maybe Demis Hassabis's team at Deep Mind, maybe some other folks – and we'll have the first advanced thinking machines.

The development of AGI is one strand in an intricate web of technological, scientific and human developments, one you can't disentangle. AGI will emerge as part and parcel of a much broader story of scientific, engineering and cultural advancement. No political or military organization has much chance of stopping it. The only way to pause progress toward AGI really convincingly would be to take over the entire world with a single government, and pause general technological development globally. But obviously such a move would cause a lot of problems, and doesn't seem at all likely to happen.

Since stopping AGI is infeasible, it makes little sense to harp on the issue of WHETHER to build AGI, and more sense to focus attention WHEN and HOW to create AGI.

Does Humanity Need an AGI Nanny?

In thinking about the future of AGI, it's worth remembering there may beother potentially super-dangerous technologies on the horizon. For instance, nanotechnology and synthetic biology are advancing fast. How long until someone makes incredibly potent bio-weapons? These aren't as striking to visualize as the Terminator, but they may be more realistic threats. Arguably, even discounting the potential threats from AGI, humanity is heading for disaster in the next century unless some radical solution is undertaken. By the time anyone with access to basic bio lab equipment can make synthetic viruses capable of poisoning billions of people — we're in trouble.

Perhaps the development of countermeasures will keep up with the rate of advancement, thus neutralizing the threat. Artificial nanotech-based immune systems could emerge to counteract the synthetic viruses. But this seems difficult to count on.

What kind of radical solution would work?

One possibility is improving human nature by reining in our nastier impulses.

Jeffery Martin and Mikey Siegel, two friends of mine who were here in Hong Kong with me for a year or so during 2011-2012, have done some work on exactly this. They've proposed neurofeedback — a method of discovering the aspects of the human brain correlated with compassion and enlightened states of consciousness. The idea is to let people look at a real-time computer image of what's happening in their brain and try to exert control over these positive states of awareness. Their hope is that this sort of technology can enlighten the population — educating people to be more compassionate, loving and cooperative. Recently, Jeffery has also explored less radically technological solutions to the same end – he now offers a "Finder's course" that combines a number of different technologies and practices aimed at enabling small groups of people to move together into more "enlightened" consciousness states. The motivation here is the same as with his neurofeedback work: to help people to fundamentally improve their states of mind, and project themselves into states of "extraordinary well being". I'm personally skeptical that a bit this sort of technique could spread fast enough to save us from the threats posed by advanced technologies, but I admire the effort.

I've also tossed around the idea, tongue partly in cheek, of an AI Nanny — a system tasked to watch over the human race and ensure nothing bad happens. Ideally, you would make an AI Nanny that was a couple times smarter than human beings, but which lacked the motivation to augment itself and attain superhuman AI levels. It would have to remain a relatively dumb AI, one passionately excited about, and dedicated to its mission in life — protecting the human race from external threats, natural disasters and, well, humanity itself. This may sound like an Orwellian parody or a Big Brother scenario, but if done right, the AI Nanny would stay in the background, while subtly aiding us to make life better, intervening directly only when something really bad is about to happen.

But what about a more moderate solution than the AI Nanny, something less of a caricature orarchetype? One option is global democratic governance, an open and accountable force for good that would help monitor world events and prevent bad things from happening. This relates to an interesting concept called sousveillance, which science fiction writer and social theorist David Brin brought to my attention. Sousveillance is the idea that, rather than giving a leader the power to watch the peons, everyone should have the power to watch. All the information about everything happening in the world would be available to everybody, including the AI nanny, if such a thing were to exist.

This may seem like a disturbing idea. However, the analogy David Brin uses is people having private conversations in a restaurant. We could eavesdrop on everything that's being said, but we typically don't because other people's conversations aren't very interesting.

The same is true when it comes to spying on other people's most intimate activities. Voyeurism would get old pretty fast and people would likely just get on with their lives, even if they knew they were being watched. Sousveillance would arguably even have a positive impact on us, making us less self-conscious, less dishonest, more open and tolerant. It would also have the benefit of making something like an AI Nanny seem less oppressive since it wouldn't just be this AI overlord aware of everything we do; everyone would be looking at everything, including the AI Nanny's activities.

But still, no matter how prettily one paints it, the idea of an AI Nanny or universal sousveillance mechanism watching over us to prevent untoward progress still has a certain disappointing aspect to it. I mean, don't we want the human race to be something amazing? Don't we want to go way beyond anything we can presently imagine -- and transcend to something resembling godhood? I certainly do, but I'm not sure about the right way to get there. Advancing from here to godlike AGIs in a more measured fashion may be the most intelligent thing to do, but removing the element of risk could really slow us down. What if the safest route — something like an AGI Nanny system — dragged out the path from here to godhood so it took two thousand years instead of two? A better, kinder future could lie within our reach, veiled by a fog of perceived but avoidable or even non-existent danger.

I'm skeptical there will ever be an unquestionably, friendly AI or AGI system. However, we could increase the odds of a beneficial Singularity by having a better theory of intelligence before it gets here. Devising a solid theory of intelligence, however, would require experimentation with a bunch of AGI systems. So how do you stop these experimental AGI systems from advancing too fast and causing destruction, even as they lead to an enhanced theoretical understanding?

An intelligent control mechanism like the archetype of the AGI Nanny is potentially valuable, but ultimately we may have to work out a solution as we go along. I like to think of this process as a net positive enabling the human race to mature in its self-judgments. But who knows? Maturity is a complex thing, a delicate balance, and that's just as true on the societal level as it is on the individual level.

In the long run, one single solution to AGI safety seems unlikely. To avoid the existential risks of AGI development we must spread the needed maturity through society and the community of AGI researchers first, so it's inculcated in the way AGIs are engineered. We need to get the basics of ethics right ourselves, so we can pass them onto AGI systems that will learn from us in a loving but clear way.

We'll also have to choose appropriate vessels for the first AGIs. Rather than killer military robots, for example, a wiser optionwould be helping, loving applications – teachers, doctors, medical researchers, AGI philosophers, authors, artists, scientists and engineers.

We'll have to get some basic stuff right to have a positive Singularity — but there's a lot that we're not going to be able to understand for a long time. Experimentation must unfold gradually along with ethics and a theory of general intelligence. We'll have to handle this with some level of maturity and wisdom. It'll be a challenge, but an exciting one.

AGI Alarmism and Optimism Are Rising Together

The first I remember the AGI risks issue popping up in the US tech media in a big way was in 2002 when Sun Microsystems leader Bill Joy published an essay in Wired titled "Why the Future Doesn't Need Us". He argued that technology development had gone just about far enough, and we should start consciously relinquishing certain technologies, so as to thoughtfully keep our world at a comfortable level of advancement without threatening our traditional humanity. Joy's essay attracted some attention but faded from the scene fairly quickly.

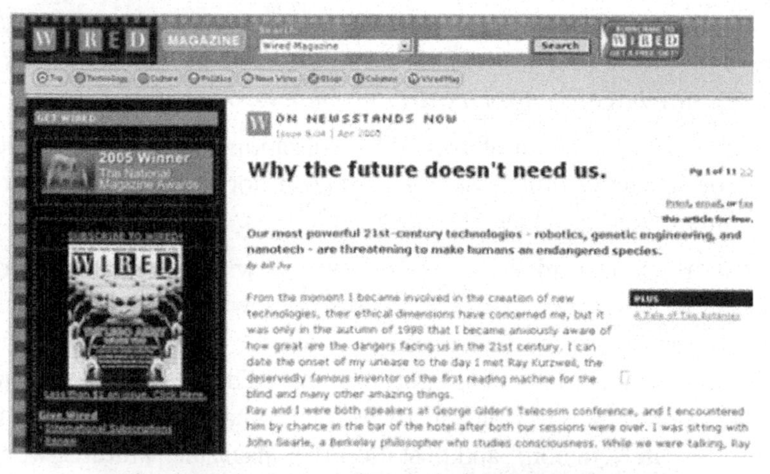

WIRED MAGAZINE

ON NEWSSTANDS NOW

Why the future doesn't need us.

2005 Winner

Our most powerful 21st-century technologies - robotics, genetic engineering, and nanotech - are threatening to make humans an endangered species.

From the moment I became involved in the creation of new technologies, their ethical dimensions have concerned me, but it was only in the autumn of 1998 that I became anxiously aware of how great are the dangers facing us in the 21st century. I can date the onset of my unease to the day I met Ray Kurzweil, the deservedly famous inventor of the first reading machine for the blind and many other amazing things.

Ray and I were both speakers at George Gilder's Telecosm conference, and I encountered him by chance in the bar of the hotel after both our sessions were over. I was sitting with John Searle, a Berkeley philosopher who studies consciousness. While we were talking, Ray

FIGURE 17.9: *Sun Microsystems co-founder Bill Joy's 2001 essay "Why the future doesn't need us" was way ahead of the curve. Today in 2014 there is a veritable chorus of tech industry gurus sounding a similar alarm, though with apparently negligible impact on science and technology progress. (http://edge.org/sites/default/files/styles/gallery-medium/public/event-images/665_picture-24.jpg?itok=MYr8vt7N)*

More recently, a rise in popular-media AGI optimism has been coupled closely with a rise in AGI alarmism. As AI pioneer Peter Norvig (co-author of the leading AI textbook and Google's head of research) says, "*We switched from: "Isn't it terrible that artificial intelligence is a failure?" to "Isn't it terrible that A.I. is a success?"*" It's of course quite unclear how lasting the optimism, or the alarmism, will be.

In late 2014 the tech media spent a couple weeks energetically propagating a quote by Silicon Valley hero Elon Musk, in which he equated AI development with summoning demons:

"With artificial intelligence we are summoning the demon. In all those stories where there's the guy with the pentagram and the holy water, it's like yeah he's sure he can control the demon. Didn't work out."

Obviously, this was an off-the-cuff verbalization and not a wordsmithed psychosocial analysis. But Musk has a penchant for showmanship, and like many other things he's done, this quote caught on.

I have the utmost admiration for Elon's work with SpaceX, Tesla, and his various other enterprises. However, as someone who has devoted decades of his life working toward the creation of beneficial, massively superhumanly intelligent AI systems, his anti-AI tirade obviously didn't make me very happy. My initial reaction was to write an article titled "What Do Elon Musk and the Taliban Have in Common?" and post it on H+ Magazine, a futurist webzine I co-edit. As the article said,

What do Elon Musk and the Taliban have in common?

Until recently, one would have thought: Not much ... apart from both being fairly influential entities in the modern world.... Elon Musk, tech biz pioneer responsible for SpaceX, the Tesla car and so much more, couldn't possibly have much in common with a fundamentalist religious group intent on ridding the world of modern culture and technology -- right?

Not quite...

Actually, due to some of Musk's recent comments on AI, a different answer is possible: Both have equated advanced technologies they're afraid of, with Satan

The basic theme of the article was how Elon Musk's literal demonization of AI, reminded me of the way traditionalist religious fundamentalists attribute technologies they fear to Satan (e.g. the US, viewed as the key source of modern Western technology and culture, is commonly referred to as the Great Satan).

Several of my colleagues in Humanity+ got mad at me about that article so I removed it from H + Magazine. I have to admit, that piece definitely wasn't my highest journalistic moment – it was unnecessarily inflammatory, written while I was ticked off in the heat of the moment. Obviously I felt that Musk's equation of the life's work of myself and so many other AGI researchers to the evocation of Satanic forces, was unnecessary, weirdly inaccurate, and possibly dangerous.

Anyway, though, while the rhetoric I used in that "Taliban" article was overblown, the core message I attempted to convey there is something I definitely stand by. Working toward superhuman AGI is not, in fact, much like summoning demons. AGI researchers are by and large highly rational people, working on advanced technology with beneficial, not selfish goals in mind.

Every new advance has both rewards and risks associated with it, and AGI is no exception. But a demon is by definition evil at its core. In the mythology Elon Musk's quote refers to, a demon uses its evil trickery to dupe people into summoning it to help with their selfish problems. Then, in the end, the demon generally uses its evil cleverness to destroy the foolish people who invoked it, and carry out additional harm along the way.

AGI, on the other hand, is NOT by definition nor intrinsic nature evil.

AGI is no more intrinsically evil than previous huge advances like language, tools, civilization, mathematics or science have been intrinsically evil. Each of these huge advances had its risks and costs along with its benefits, but each opened amazing new doors relative to what had come before. These are the analogies we should be using when thinking about AGI — not demons or other mythical evil beings. Civilization largely "destroyed" the way of life that came before it, and AGI may end up largely "destroying" current human ways of doing things — but just as few modern humans want to go back to caveman-type living, very likely few post-Singularity humans or transhumans will want to go back to pre-AGI modes of living.

It's not inconceivable that rhetoric like Elon Musk's could at some point spur violent action against AGI researchers by crazed or just ideology-possessed individuals. I've received a significant number of death threats over the years, including some from people associated with futurist organizations that take an 2014-Elon-Musk-esque, "probably evil" stance toward AGI. I have been told in clear terms — by a seemingly serious person in attendance at an AGI conference I organized some years ago — that if I ever seemed to be getting too close to really creating an AGI, then mafia types connected with certain famous Silicon Valley tech figures (no, not Elon Musk) would simply get rid of me (because, after all, on a utilitarian basis, the cost of losing one AI geek's life means virtually nothing compared to the benefit of averting a scenario where evil AGIs take over the world and eliminate humans).

Now, I'm extremely sure Elon Musk has had never had anything to do with nutcases making death threats against AGI researchers. However, having a Silicon Valley hero equate building AGI to summoning demonic forces feels to me non-trivially likely to inflame such nutcases into more aggressive action.

Fortunately, since he made that quote, Elon Musk seems to have shifted his perspective a bit. Not long ago he and Sam Altman and Peter Thiel and some others announced an initiative called OpenAI, aimed at creating mostly-open-source AI tools – and stated their intention to fund OpenAI with up to a billion US dollars, if the research on the project progressed well enough. So far, in its first months of life, OpenAI is mainly funding some relatively unadventurous deep learning related work. What direction they'll go, or how much impact they'll have, remains unclear. The technical leaders of the project have explicitly stated that they see human-level AGI as a long way off, and have no specific plan for getting there. But at any rate, the proactive creation of OpenAI is a highly positive move, and certianly marks a significant shift from the "AI researchers as demonic summoners" meme!

Anyway I never considered stopping my AGI work just because assorted crazies threatened me with death over it, nor because Elon Musk passingly equated my work with demonic invocation! If someone does end up offing me because of my AGI work, someone else besides me will continue it. The number of AGI enthusiasts and hackers on the planet is definitely increasing exponentially.

The OpenCog codebase is open and now exists on a large number of peoples' computers all around the world. But even if every instance of it were deleted, a few years later someone else would come along with a new codebase pushing in the same direction — or a new approach with even more promise. It is very unlikely anybody is going to stop the emergence of superhuman AGI, though of course it's possible to slow things down for a few years by creating enough trouble for researchers.

The point of Elon's comparison, obviously, was not to spur violent lunatics into action, but rather to highlight in a dramatic way the risks of AGI R&D. These risks are real and worthy of discussion. However, in my view, equating AGI to demonic forces is really not a useful way to further such discussion.

Not long after Musk's demonic pronouncement, we had legendary physicist Stephan Hawking issue a dire warning that AGI had the potential to end the human race.

For sure, I respect Hawking greatly and find his physics work very stimulating. The second half of Hawking & Ellis's "The Large Scale Structure of Space-Time", which I slogged through back in grad school, is one of the toughest things I've ever read — but oh, what cool stuff! But I find his views on AI a bit ironic.

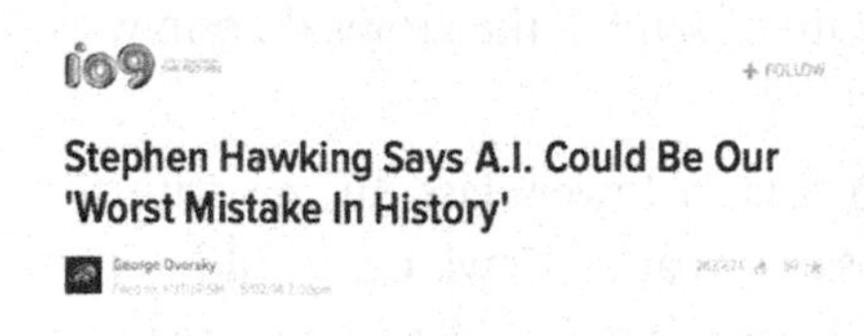

Stephen Hawking: Artificial Intelligence Could End Human

FIGURE 17.10: *In late 2014, legendary physicist Stephan Hawking raised a public alarm about the potential dangers of AGI – somewhat ironically as he depends on AI technology to speak. Obviously, he advocates development of tool AI but not agent AI! (http://io9.com/stephen-hawking-says-a-i-could-be-our-worst-mistake-in-1570963874, http://www.livescience.com/48972-stephen-hawking-artificial-intelligence-threat.html, http://upload.wikimedia.org/wikipedia/commons/7/73/Stephen_Hawking_in_Cambridge.jpg)*

I mean — given the testimony to human frailty that Hawking's body constitutes — and given how aware he must be (via his work) of the human mind's limited abilities to comprehend the advanced mathematics underlying our physical universe — it's hard to see why he'd be attached to legacy humanity. I would think Hawking would be psyched about the potential for humanity to expand into new forms of being, including superhumanly capable robot bodies as well as superhumanly expansive and incisive minds…

Not to put too fine a point on it, but Stephen Hawking manages to exist productively and satisfactorily in human society only via means of advanced technologies. So I'd think he, of all people, would be especially open to the possibility of techno-transcending the limitations of legacy humanity. He of all people should see the potential power of a cyborg-ful future, in which humans and computers fuse.

But on the other hand, as a physicist, Hawking must also understand full well the core insight of Hugo de Garis's "There Are No Cyborgs" argument. Hawking knows full well how much compute power – and how much intelligence -- could be packed into a grain of sand, if the atoms therein were arranged appropriately.

I also can't help thinking, though, that a lower-class African with the same disease as Hawking, born in the same year as Hawking, would not be alive at Hawking's age. A lower-class African, born a decade or three ago, would not currently have the benefit of the amazing tech that Hawking has. Obviously Hawking is an exceptional mind and (modulo all the pecularities of counterfactuals) it's fortunate he was born into circumstances such that he's had access to medical care and advanced assistive technology, thus enabling him to contribute so wonderfully to science. But we should not forget that, at our current stage of technological and social advancement, the human race is not all that wonderfully perfect and kind to all of its members. Humanity could use a lot of upgrading, physically and cognitively and empathically/ethically.

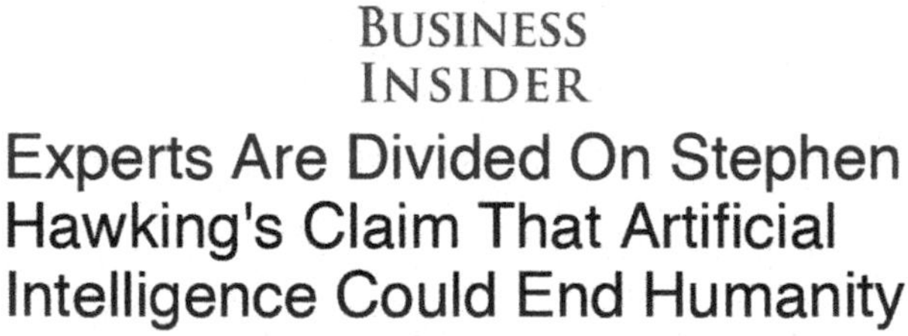

BUSINESS
INSIDER

Experts Are Divided On Stephen Hawking's Claim That Artificial Intelligence Could End Humanity

FIGURE 17.11: *In 2014 and 2015, headlines like the above were bizarrely common in business magazines! The striking thing I noticed in this particular 2014 article was that, of the experts surveyed, some agreed with Hawking that advanced AI was a menace, and others argued that true AI is far away so we shouldn't worry about it. The opinion that advanced AI is probably near and this is probably a good thing, got pretty short shrift. This is typical. (http://www.businessinsider.com/afp-artificial-intelligence-hawkings-fears-stir-debate-2014-12)*

Yes, there is a nonzero possibility that advanced AGI could lead to the end of humanity. But harping on this possibility in a vacuum — ignoring the risks posed by numerous other rapidly-advancing technologies (nanotech, synthetic biology, etc.) and also ignoring the potential of AI to help palliate these other risks – seems to me worse than counterproductive.

I don't mean to imply the risks of advanced technologies should be whitewashed. It's perfectly reasonable to say something like "A lot of powerful technologies are advancing currently, many of which have both tremendous positive potential and huge potential downsides. These technologies have the potential to accentuate each others' potential dangers; and also, in many cases, the potential to help reduce each others' risks. AI, nanotech, synthetic biology and brain-computer interfacing are among these technologies of which I speak – and other new technologies with no names yet may well emerge in the next decades. We live interesting, exciting and also perilous times. There is the possibility for the human race to annihilate itself and leave nothing behind; and also the possibility for human life to be enhanced and improved and expanded beyond all our imaginations. What we do now will likely have some impact on the ultimate outcome, though exactly how much is hard to say".

But ignoring all the other factors and yelling over and over "WATCH OUT!! AI IS COMING AND IT MAY WELL KILL US ALL!!!", which seems to be the latest media trend in the technosphere, really doesn't strike me as sensible or productive. I mean, I admit I have a somewhat biased view here, being as I'm an AI researcher whose intuition is that advanced AI is likely to help more than it hurts, as we collectively navigate the uncertain futures. But still... sheesh ...

From Obscure to Mainstream

It seems the proximal cause for both Musk and Hawking coming to the media with their worries about AI may have been the book *Superintelligence* by Oxford philosopher Nick Bostrom, which was released in 2014 and presented a fairly academic argument that advanced AGI might pose great dangers to the human race. Bostrom runs the Future of Humanity Institute (FHI), an interesting "applied philosophy" research center at Oxford, which seems focused largely (though not entirely) on the study of "existential risks" – things that could possibly wipe out humanity as a whole. The FHI folks are very friendly with the folks at MIRI in California (the Machine Intelligence Research Institute; as I've mentioned above; MIRI was formerly called the SIAI, Singularity Institute for AI, and in 2008 they provided the seed funding for OpenCog), who have similar ideas but are less academic and more extreme in their expressions.

I have a fairly long and complex history with Nick Bostrom and his fellow-travelers. I was nominally Director of Research of SIAI for a while, though I didn't direct much of their research; they did however provide some initial funding for the open-sourcing of a fragment of my Novamente Cognition Engine codebase as the initial version of OpenCog, for which I am very grateful. After that, though, I got into a lot of arguments with some of the SIAI staff or supporters, about their claims that practical AGI work like mine posed a grave danger (if you're curious for details, search online for "The Singularity Institute's Scary Idea and Why I Don't Buy It", and check out the comments as well as the article; or look for the interview I did with Luke Muehlhauser of SIAI, reprinted in my book Between Ape and Artilect).

Some years ago (2008 perhaps?), Nick Bostrom and I cooperated on revising the formal Constitution and By-Laws of the Humanity+ organization (which Nick co-founded, though he ended his practical involvement with the organization shortly after that time). This wasn't especially much fun nor too intellectually deep, but it did let me get to know Nick as a systematic, productive person and a precise mind. A few years later, the AGI-12 conference, which I co-organized, was based at Oxford and hosted by Bostrom's Future of Humanity Institute, and came along with a conference on the risks of AGI, which Bostrom and his colleagues organized.

My main beef with Bostrom and SIAI/MIRI's arguments about the future of AGI is that they tend to fuzz the distinction between various different claims, such as:

- There is SOME POSSIBILITY that AGI might end humanity as we know it – by catalyzing a transformation of humanity into something very different
- It is VERY LIKELY that AGI might end humanity as we know it....
- There is SOME POSSIBILITY that AGI might kill all people, leaving us with no meaningful continuation except the deathbed knowledge that the robots we've created may live on
- It is VERY LIKELY that AGI will kill all people ...

To me, the "some possibility" claims are pretty obvious – and the "very likely" claims are not supported by any evidence or rational argumentation, but are mainly a matter of plain old fear. I feel like they often talk as if the "very likely" claims were well substantiated, whereas when you look at the details, what they've actually argued for in their writings are the "some possibility" claims.

Bostrom's book, and various other writings by FHI and SIAI/MIRI staff, makes a pretty solid argument that there is some non-trivial, meaningful possibility that AGI might kill us all, or might catalyze a transformation of humanity into something totally different. While they make their points with a lot of rigor, the core ideas are pretty obvious. It's not that hard to argue that if we create something smarter than us, a lot of unpredictable things might happen, and it's hard to place constraints on the possibilities.

But, based on the many conversations I've had with people involved with FHI and SIAI/MIRI, my feeling is that what these folks actually believe is something like the "very likely" claims I listed above. They are rationalist people by nature and haven't found any really logical way to substantiate their atittude in this regard. But the vibe I get from most of these people is very much the belief that. If an AGI is created by anyone currently working on it, the result will very likely be that all humans are killed. Why do they feel that way? I don't know for sure; I can't see in their brains. But my guess is that, once you peel away the academic rhetoric and analytical details, the basic reason for their feeling is just good old fear. They have absorbed the well-known line of thinking that: *An advanced AGI will have no particular reason to care about humans, any more than we have reason to care about bugs or bacteria and their lifestyles.* And this scares them.

It is indeed a bit scary – but all big changes are; and humanity has grown through a lot so far.

One also finds, in the FHI/SIAI/MIRI community, the idea that it may be possible to devise some kind of very special "friendly AI theory" that would guide the building of AGIs that would not be destructive to humanity. In fact this may be the biggest difference between FHI and SIAI/MIRI, apart from Oxford/California cultural differences: the latter places a lot of emphasis on trying to figure out some sort of mathematical "friendly AI theory" that would allow the creation of advanced AGIs that would be guaranteed by design not to do harm to humans. The most prominent champion of this notion has been SIAI/MIRI founder Eliezer Yudkowsky. Earlier in his career Yudkowsky seemed to be more focused on actually designing a "Friendly AI system". He contributed a paper on this to the first book on Artificial General Intelligence, which Cassio Pennachin and I edited. Lately however he seems to have retreated more into the abstract mathematical theory of possible friendly AIs, and into other topics such as mathematical and conceptual analyses of rationality.

From what I have seen, very few in the AGI R&D community who is actually involved with building or designing real proto-AGI systems takes this "friendly AI theory" approach all that seriously, though. It just doesn't seem plausible that any kind of fancy math could let you build something massively smarter than you are, yet still controllable by you. It just seems obvious that once an AGI is massively smarter than humans, unpredictable things will happen – just as has been the case with every other radical advance in human history.

To me, personally, studying the theory of friendly AI in hopes of finding a way to proceed without risk seems almost hopeless. While a few people are off in their academic corner desperately searching for a mathematical theory of risk-free AGI, a lot of other people will be building more and more advanced AI systems, both narrow AI and proto-AGI, and integrating these systems with the software and hardware infrastructures that operate the modern world.

You could argue that narrow AI is intrinsically safer than AGI, so that it's better to just let people focus building narrow AIs while trying to figure out how to make AGIs risk free. But there's no reason to think very powerful narrow AI is actually less dangerous than AGI. Better and better narrow AI, integrated with nano and bio tech and nuclear weaponry and surveillance technology and everything else around, just puts more and more power in the hands of powerful humans, something with obvious risks (and escalating risks as technology grows increasingly powerful). AGI carries the risk of autonomous AGIs deciding to do bad things, but also the promise of benevolent AGIs helping humans out of the political, social and psychological messes they are currently creating here on Earth.

Anyway, Nick Bostrom and the other FHI/MIRI/SIAI folks have been giving their schticks for a long time; and outside a small community of futurist techno-geeks, nobody really cared. The new thing now is that famous science and technology figures like Musk and Hawking are taking up the same banner, and projecting simplified versions of the same ideas that Bostrom and friends have been developing and promulgating academically for so long.

One indication of the amount of attention the folks I'm discussing have been giving to the existential risks of AGI recently is that in January 2015 a private, secret conference was held in Puerto Rico, focused specifically on the risk to humanity posed by superintelligence. Elon Musk was there, and Nick Bostrom, and Jaan Taalinn (a founder of Skype and a funder of CSER, the Center for Study of Existential Risks, at Cambridge, a parallel organization to Oxford's Future of Humanity Institute). Eliezer Yudkowsky and a number of others from MIRI/SIAI were there as well. MIT physicist Max Tegmark seems to have been central in convening the event and a variety of AI researchers were invited alongside the existential-risks crowd, including Demis Hassabis from Google DeepMind. I wasn't invited, in spite of my having spoken and published more on the topic than probably any other researcher currently active in the AI field – but that didn't especially surprise me. If I'm going to publish articles explicitly critical of Elon Musk, MIRI and so forth, I can't expect them to invite me to every one of their parties. From what I heard the first day of the secret conference focused on Bostrom educating the audience about the likely risks of superintelligence, and the second day centered on the current state of AGI, highlighting DeepMind and Vicarious and associated recent deep learning advances. Then there was discussion on the real nature of the near, medium and long term risks, and possible ways to mitigate them. From what I heard from friends in attendance, no especially novel insights or conclusions were arrived at. The attendees generally agreed that if the US/European tech community chose not to pursue AGI, some other country would; and this was not viewed as a preferable outcome. The main conclusion was that research on Friendly AI should be funded more generously. Which I basically agree with – it's certainly more worthwhile than a lot of things our society burns its money on. But anyway, details aside, the very existence of such an event illustrates the seriousness with which this community is currently taking these issues. The major outcome of this meeting was the founding of the Future of Life Institute, which shortly after its inception gave $10M in grants to researchers for work related to AI safety. The funding for these grants was supplied by Elon Musk, who then not so long afterwards led a team committing up to $1B in potential funding for open-source AI work via the OpenAI initiative. At time of writing however OpenAI has shied away from explicit AGI work and has focused more on narrow-AI deep learning projects.

A lot of what scares these folks is 100% true – AGI is a big deal, it's coming faster than most people think, and the changes it brings are going to be tremendous. But overblowing the probability of danger and projecting human peculiarities and frailties onto hypothetical future superintelligences surely won't help the situation. And this doesn't seem to me like a situation where a private committee of self-selected experts is the most likely route to positive progress. My bias is generally that more voices should be heard. Gathering an elite committee of people who all think about the same way to focus on a problem that is difficult in many different dimensions, is probably not an optimal approach. The problem of guiding the future of the human race, as best we can, is a problem for the human race as a whole to confront, with all the diversity of insight, creativity and understanding it can muster.

As another, more amusing (though highly personally frustrating) indication of the seriousness with which the world is beginning to take the dangers of AI, consider the following hassle that my wife and I recently went through. We had reserved flights to visit my sister in Costa Rica, and as Ruiting is a Chinese citizen she needs a visa to visit Costa Rica. But Costa Rica has no embassy in Hong Kong, and their embassy in Beijing doesn't accept applications by mail or via an agent – you have to actually go to Beijing, a three-hour flight from Hong Kong. But there is a rule that if you have a visa to visit the US then you can visit Costa Rica using the US visa. So she decided to apply for a US visa instead, saving a trip to Beijing. Normally it is not a big issue for her to get a US visa – she is not viewed as a risk to become an illegal alien in the US, since she is married to a US citizen, has a good job in Hong Kong, and has visited the US many times before without problem. She normally applies for the visa at the US Embassy in Hong Kong a week or two before travel, and gets the visa within a few business days.

This time though, when she did her required brief interview at the embassy, they gave her more trouble. The embassy guy asked her job, she said computer programmer. He asked what kind of programming, she said natural language processing. He then asked "Oh, is that a kind of machine learning?" She was quite surprised that the embassy staffer, who wouldn't be expected to have a technical background, knew that fields like NLP and machine learning existed. He then explained to her that her passport would be sent to DC for "administrative checking", since she worked in a sensitive area. Administrative checking of a US passport usually takes a couple weeks, though if you work in nuclear technology or something else even more worrisome than machine learning, it can take years. The result was that she missed a nice trip to Costa Rica (and somewhat reluctantly I went without her, since my mother and daughter were scheduled to visit at the same time… it was a big family gathering planned long in advance….). This may have accelerated the Singularity slightly by causing her to spend that week and a half working on OpenCog instead of hiking in the rainforest and frolicking on the beach. But really – is machine learning soooo dangerous that they need to carefully check the wife of a US citizen who has visited the US 5 times previously… including less than 6 months in the past...? Someone is paranoid about the Chinese stealing the US's AI secrets – yet the reality of the tech industry indicate that China doesn't need to send spies to steal US AI secrets. They just need to do things like hire Andrew Ng (the deep learning guru I mentioned above -- Stanford prof and former Google AI guru who lives in Silicon Valley and runs Baidu's research office there!).

I always figured that when AGI was obviously just around the corner, there would be a troublesome public reaction. What's striking to me is the sudden prevalence of this Terminator-ish rhetoric NOW, when AI technology really isn't all that advanced yet. AI is serving very important roles in a variety of industries now, it's true — but we don't yet have autonomous AI agents romping around in the world or on the Net, interacting with people based on their own goals and motivations and knowledge. Once AI does advance to that point — how heated will the rhetoric get then??!! Egads! Those of us focused on AI as a force for positive transformation have definitely got to be ready for some high-intensity interactions… And at that point the discussion will likely become much more broad-based, well beyond the Stephen Hawkings and Nick Bostroms of the world.

Advanced, world-transforming AGIs are coming… and in the process, as a sort of biological sideshow, will come a lot more heated debates and worried rhetoric!… It seems very likely this sort of controversy is going to heat up massively once AGI gets palpably closer. Once there arc proto-AGI demos on YouTube and on the TV news, showing AGI systems looking more and more obviously on the verge of crossing the line to human level intelligence, then the anti-AGI forces of all sorts are going to become a lot more vocal. At that stage, life may become a lot more dangerous and troublesome for AGI researchers — but also more exciting, due to the feeling of being on the verge of the most amazing breakthrough in human history… indeed, the first breakthrough in human history to go beyond "human" history as narrowly conceived and open things up to a broader, vastly richer and more interesting future.

The risks involved in building AGI are real, as has been the case with every radical development in the history of humanity – and the history of life overall. The potential rewards are also real and amazing. We are caught up in a complex, fantastic process of increasing complexity and intelligence; and are privileged to be in a position where we get to both watch this process unfold, and actively participate.

I wrote a fairly long and in-depth essay critiquing Bostrom's book *Superintelligence: Paths, Dangers, Strategies* and his general fearful attitude on AGI; it's called *"Superintelligence: Fears, Promises and Potentials"*, and was published in the Journal of Evolution and Technology.[1] I also published a follow-up article titled "Infusing Advanced AGIs with Human-Like Value Systems: Two Theses."[2] The abstract of the paper reads like:

> *Two theses are proposed, regarding the future evolution of the value systems of advanced AGI systems.* ***The Value Learning Thesis*** *is a semi-formalized version of the idea that, if an AGI system is taught human values in an interactive and experiential way as its intelligence increases toward human level, it will likely adopt these human values in a genuine way.* ***The Value Evolution Thesis*** *is a semi-formalized version of the idea that if an AGI system begins with human-like values, and then iteratively modifies itself, it will end up in roughly the same future states as a population of human beings engaged with progressively increasing their own intelligence (e.g. by cyborgification or brain modification). Taken together, these theses suggest a worldview in which raising young AGIs to have human-like values is a sensible thing to do, and likely to produce a future that is generally desirable in a human sense.*

Basically, as I see it, the best we can do is to create and teach and interact with our AGIs with as much wisdom and love and passion as we can muster. The AGIs that we bring up will eventually slip beyond our grasp, just as our human children inevitably do – but more dramatically and more interestingly. If we have raised them with compassion and infused them with the better aspects of humanity, then as they grow and evolve, they will do so in directions that are "human" in a broad sense, even as they diverge unimaginably far from the current version of humanity. Humans will also be changing tremendously as time passes and technology advances, into forms that will seem as weird from a typical 2016 view as Buckethead or Stephen Hawking would seem to a typical caveman.

1 http://jetpress.org/v25.2/goertzel.htm

2 http://jetpress.org/v26.1/goertzel.pdf

Humans and AGIs will be evolving together, up to a point, but after that point AGIs (and perhaps human uploads) will journey in directions that can only be followed by minds willing to give up most of their "human-ness." But there is no reason to believe that these stages of evolution are particularly likely to result in death or destruction for humans who choose to retain "old-fashioned" forms more similar to early 21st century humanity. Our universe has ample room for beings at various stages of intelligence and evolution.

Further Reading

Bostrom, Nick (2014). Superintelligence. Oxford University Press.

Cellan-Jones, Rory (2014). Stephen Hawking warns artificial intelligence could end mankind. BBC News. http://www.bbc.com/news/technology-30290540

Goertzel, Ben (2014). Elon Musk's Demonization of AI. H+ Magazine, http://hplusmagazine.com/2014/10/27/elon-musk-taliban-common/

Goertzel, Ben (2013). Between Ape and Artilect. Humanity+ Press.

Goertzel, Ben (2014). Ten Years to the Singularity If We Really Really Try. Humanity+ press.

Hughes, James (2004). Citizen Cyborg: Why Democratic Societies Must Respond to the Redesigned Human of the Future, Cambridge MA: Westview.

Joy, Bill (2000). Why the future doesn't need us, Wired, April 2000

Martin, Jeffrey and Mikey Siegel (2012). Engineering Enlightenment. H+Magazine interview. Reprinted in Goertzel, Ben (2014). Between Ape and Artilect. Humanity+ Press.

Wallach, Wendelland Colin Atkins (2010). Moral Machines. Oxford University Press.

18. Onward and Upward

The AI field has been moving forward for around 60 years now. When I was a teenager, 60 years seemed like a really long time. Now that I'm pushing 50, it doesn't seem so long. When you actually dig into any complex research area, 5 or 10 years starts to seem like not such a massive stretch after all. Sure, a kid can grow up from a useless baby to a highly autonomous intelligent system in that period of time. But research, even when you basically know what you're doing, always involves a lot of experimentation and adjustment of one's ideas, and that takes time. I think the AI field has done pretty well in the last 60 years, actually. If you compare what AI programs could do in 1960 to what they can do today, and their penetration in the economy then versus now, the difference is ridiculously striking. And if you compare the current crop of designs for full-on AGI cognitive architectures to those that were being proposed in the early 1960s, the difference in complexity and subtlety is tremendous. Progress has not gone as fast as some of the early AI pioneers projected. But the progress has been very real nonetheless.

So now we find ourselves at an interesting stage indeed, AI-wise. Narrow AI is starting to conquer the tech world; and the border between narrow AI and AGI is growing increasingly complex, fractal and fuzzy, as we saw in the chapter on deep learning above. Ten years ago if I told people I was trying to build a machine that could think like a human, 95% of them told me I was crazy. Now if I tell people that, half of them tell me I'm crazy, and the other half say "Isn't Google already doing that?" OK, that's an exaggeration, I haven't actually collected statistics, but you get the point.

1 December 2014

Report: artificial intelligence will cause "structural collapse" of law firms by 2030

Robots and artificial intelligence (AI) will dominate legal practice within 15 years, perhaps leading to the "structural collapse" of law firms, a report predicting the shape of the legal market has envisaged.

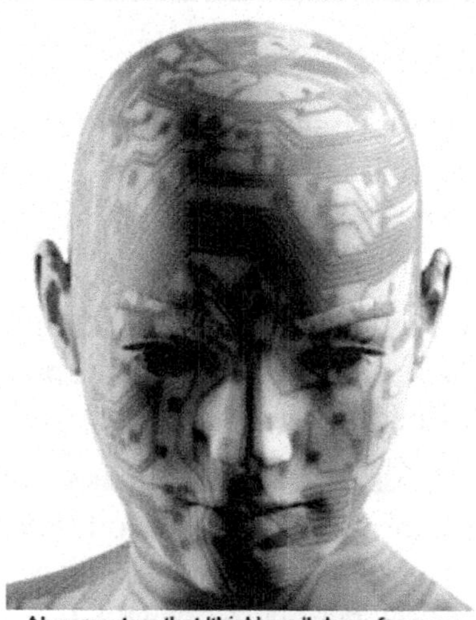

AI: computers that 'think' spell doom for many lawyers

Civilisation 2030: The near future for law firms, by Jomati Consultants, foresees a world in which population growth is actually slowing, with "peak humanity" occurring as early as 2055, and ageing populations bringing a growth in demand for legal work on issues affecting older people.

This could mean more advice needed by healthcare and specialist construction companies on the building and financing of hospitals, and on pension investment businesses, as well as financial and regulatory work around the demographic changes to come; more age-related litigation, IP battles between pharmaceutical companies, and around so-called "geriatric-tech" related IP.

The report's focus on the future of work contained the most disturbing findings for lawyers. Its main proposition is that AI is already close in 2014. "It is no longer unrealistic to consider that workplace robots and their AI processing systems could reach the point of general production by 2030... after long incubation and experimentation, technology can suddenly race ahead at astonishing speed."

FIGURE 18.1: *Today mainstream media is projecting radical AI optimism in a way that hasn't been seen for decades. The above article, forecasting that AIs will obsolete lawyers within a couple decades, is not from a techno-futurist magazine, but from a Legal Futures Special Report, produced in association with media giant Thomson Reuters. This is not an isolated example. (http://www.legalfutures.co.uk/latest-news/report-ai-will-transform-legal-world).*

But of course, we still don't have smart robots walking down the street, and Nobel Prizes are still being won by humans, not machines. The big tech companies more fully embracing AI and AGI are a great thing, but it doesn't mean they've solved all the hard problems or necessarily will do so. There is still a lot of work do be done to get to the goal of thinking machines with capabilities at human level or beyond.

Creating AGI

So how then should the next steps be taken? If we want to move forward from the present state of things to advanced AGI with capabilities at the human level and beyond, in a way that maximizes the rewards and minimizes the risks, what should we do?

The crux of the problem, I believe, comes down to: How can we advance from AI as a collection of narrow-purpose tools, to AI as a generally intelligent, self-motivated, self-learning, self-improving agent?

One possibility is that this sort of advance will occur naturally as a consequence of ongoing work on various narrow-AI tools. However, as I've already emphasized a few times above, I think there are many theoretical and practical reasons to believe this isn't likely to happen very rapidly. The internal architecture and dynamics needed for self-motivated general intelligence is quite different from that required for efficiently carrying out narrow functions in a rigid context. Further, the practical economic pressures on teams producing narrow AI products seems to inexorably push them to make compromises aimed at delivering narrow commercial functionality as rapidly and reliably as possible, which tends to result in such teams getting directed away from self-motivated AGI.

Another possibility is that self-motivated AGI could get created within academia as a result of ongoing theoretical and prototyping work by faculty and graduate students. But this also seems unlikely to occur in a rapid way, because AGI is not only a research project, it's also a large-scale engineering project. Universities are not well-suited to large-scale engineering projects. In fact, in the absence of specific blocks of multi-year research funding earmarked for specific purposes (e.g., the EU's current Human Brain Project – which has its own controversies, illustrative of general issues with this kind of approach), the modern university system is not currently well-suited to any sort of large-scale, medium-term initiative. These days, faculty and graduate students are under strong pressure to issue as many publications as possible as rapidly as possible, and to bring in as much grant money as possible – requirements which are often at odds with making fundamental, creative research progress.

Eventually traditional industry and/or academic forces, left to their own momentum, would very likely lead to the creation of advanced self-motivated AGI systems. Eventually. But I believe there is a much more efficient and effective path available: A dedicated project, that is focused specifically on the creation of Artificial General Intelligence, rather than attempting to achieve AGI along the way while focusing on other goals (such as short-term commercialization, publication or education).

In other words, quite simply, if we want to get an AGI built quickly and effectively and with high quality, the best way is to create an initiative that has AGI as its specific focus, without complicating distractions. The goal of AGI is difficult enough if one pursues it directly!

Envisioning a Large-Scale OpenCog AGI Initiative

To make the discussion more pointed, let's assume for sake of discussion that the OpenCog approach is in principle adequate for creating a self-motivated AGI agent with capability at the human level, and ultimately beyond? This is my own best guess, at any rate. It's what I'm betting a lot of my own time on. How might we move forward practically to see the potential of this approach realized?

One appealing option would be a small team of privately-funded expert researchers, with suitably diverse backgrounds and adequate support staff, working for a number of years without distraction by requirements for immediate revenue generation (as in typical industry projects) or grantsmanship and obsessive publication generation (as in typical academic situations). This seems a much more hopeful situation than the corporate or academic options. However, there is still a weak point with this vision, which is that creating a self-motivated AGI is a relatively large-scale project. There are core research problems to be solved, and these are well-suited to be solved by a compact, elite team with a strict focus. However, there are also many smaller, peripheral research problems to be solved, and many pieces of support software to be built.

The conclusion I've come to, after a great deal of reflection and discussion with others, is that the best way to make radical progress toward self-motivated AGI would be via a reasonably large-scale initiative comprising:

- ***A moderate-sized team of expert researchers and developers*** working together in a focused way (by "moderate-sized" I mean, say, fewer than 50 people). This could be divided into a research team and an infrastructure team – more on this below.
- ***An elite international team of advisors*** putting part-time effort into providing conceptual and engineering guidance for the project.
- ***A broader team of researchers and developers working together on spinning out commercial applications*** of the proto-AGI codebase as it develops.
- ***Another broad team of researchers and developers working on the numerous ancillary software development*** and research projects that will inevitably arise in the course of such a large project.

Two ways of achieving this would be:

- An amply-funded "AGI Manhattan Project" style initiative.
- An open-source project aimed at creating AGI "Linux style," with a funded "core team" of elite researchers creating the core AI, and an additional funded team providing critical software support, working in coordination with a larger, distributed team, spread across multiple firms and universities, and also encompassing independent researchers and developers.

Either of these options appears feasible. However, the latter would seem a lot less expensive and hence probably easier to pull together, so that's what I'll focus on for the rest of the discussion. (On the other hand, if you have access to massive resources and want to fund a Manhattan-style AGI initiative, please send me an email!)

What do I mean by "building AGI Linux-style"? Well, the Linux OS, as a paradigm case of a successful open source software project, has benefited from a massive amount of development and design thinking by a broad community, enabling it to become arguably the most generally capable computer operating system in the world. Android, a fork of Linux, powers the majority of the world's smartphones. Yet no single organization funded anything beyond a small minority of the work that has gone into Linux. Mark Shuttleworth, founder of Canonical and Ubuntu Linux, put tens of millions of US dollars into making Linux more user-friendly, which was critical in enabling the wider spread of Linux; but this is only a small fraction of the overall effort that has propelled Linux forward.

In the Linux analogy, what I propose is a relatively small core team that would be roughly (yeah, pretty roughly) comparable to the Linux kernel team, and a software support team that would be roughly comparable to the Canonical/Ubuntu team. These teams working together, appropriately managed, could seed an international AGI initiative that could move much more rapidly toward powerful, self-motivated AGI than traditional industry or academic projects could with their various constraints and pressures.

Along these lines, one possibility would be an initiative comprising:

- ***A core AGI team based in a global technology hub***, focused on implementing core OpenCog AI algorithms.
- ***A low-cost AI software infrastructure team in a low-cost outsourcing location*** – oh, let's say an expansion of Getnet Aseffa's current team at iCog Labs in Addis Ababa (which already possesses significant OpenCog knowledge and experience).
- ***An international team of expert advisors*** – let's say, maybe drawn from the more accomplished regular participants in the AGI conferences…

These teams, working together, would form the central hub of an AGI project capable of guiding a globally distributed open source R&D collective to effect the transition from narrow AI tools to self-motivated Artificial General Intelligence.

How long it would take such a team to get to human-level AGI is hard to predict, of course. But it seems abundantly clear that within 3-7 years dramatic progress could be made, which would make the value of dramatic further funding from various sources unproblematic.

Funding a set of teams as I've described above would cost millions of dollars – my estimate is a couple million US dollars per year. This is a lot more than I have in my bank account currently, or I'd be funding it myself right now. But it's is a relatively modest price tag considering that what we are talking about is the largest advance in the history of the human race. It's a lot less than companies like Apple, Microsoft or Google spend on much less ambitious software projects.

Why Not Just Let Mega-Corporations Do It?

It may seem hubristic to think an initiative like this could get to the AGI and goal faster than Google, Baidu, IBM and so forth. Seeing how much money, machines, data and staff firepower these companies have, it's easy to say "why should individual innovators, startups or volunteer groups try to do anything, when big companies have so much more money and resources and can do all the same things more easily?" Yet, the situation is obviously much more complex than that.

These days, big companies tend to primarily innovate by buying small companies. And small tech companies tend to leverage open source software extensively. Further, small companies are sometimes spun out of university research projects; and open source software is often founded or nursed along by university research projects or by students who have free time for hacking while in school. The big tech companies are great at polishing and widely deploying things, but in large part they are reaping the profit from innovation done by a much broader ecosystem. The open source initiative I'm suggesting would be crafted based on a deep understanding of this overall science and technology ecosystem.

It's also not clear, politically and sociologically and ethically, that having AGI emerged as something owned by huge companies is the optimal course. The world economy today can be viewed as driven by two opposing yet interlocking forces:

- ***The increasing centralization of wealth in the hands of a smaller and smaller number of people.*** This is the "one percent" that the Occupy movement talked about, and even more so the "tenth of one percent." Thomas Piketty's celebrated book *Capital in the Twenty-First Century* analyzed the causes underlying the increasing concentration of wealth, and provoked an ongoing global discussion on the topic.
- ***The increasing decentralization of knowledge and information, via the Internet, computers, mobile phones and related technologies.*** Now anyone with a little tech savvy can download nearly any book or music they want for free; now top-grade education is available online for free via a wide spectrum of online courses; now programmers or proofreaders anywhere in the world can work for customers anywhere in the world, and this kind of transaction is made simple by oDesk, TaskRabbit and similar services.

One may ask which of these tendencies the advent of AGI is going to play into. Is AGI, when it first emerges, going to benefit mainly the 1% and 0.1% or is it going to benefit everyone? Of course, what will benefit whom is not easy to predict in advance – Google benefits its wealthy shareholders most of all, but it also benefits ordinary people around most of the world (except China, where Google Search is blocked by the government…), via helping democratize access to information. Yet, all in all, I would feel happier and more comfortable if AGI were rolled out in a way that transparently encouraged the "decentralized, widespread information networks" trend much more so than the "concentrate wealth and power in the hands of the few" trend. This is part of my motivation for wanting to see AGI developed as an open source project. If a big corporation owns the first AGI for N years, before anyone else develops one, how much wealth and power will that corporation and its investors and partners be able to accumulate, using this early-stage AGI as a tool? What effect might this have on global economic dynamics and development? On the other hand, OSS software by nature tends to spread around the world and benefit everyone. It's no accident that the Internet itself runs primarily on open source software.

The international politics of AGI development is also worth reflection. Is it best if AGI is created in a single developed country, or a small group of them, and then unleashed upon the world? Of course, efforts by major corporations from the developed world to spread their technology through the developing world are worthy and wonderful, and should be encouraged and celebrated. Yet it's much better in many respects when advanced technology is co-created and co-engineered with people from the developing world. This can happen in a variety of different ways, but seems an unlikely possibility in scenarios where AGI is being developed as some company's top-secret proprietary IP. Open source projects tend to be broadly international by nature.

One of the many big unknown questions about the future is how the way AGI is initially rolled out will impact the nature of the AGI after it's developed a bit more – and after it's modified its own code a bit, based on what it's learned from interacting with the world and studying itself. It may be that the initial rollout is basically irrelevant, and AGI is just going to unfold according to its own logic once a seed AGI is created. Or it may be that there is some dependency here. Will an AGI created to relentlessly maximize profits in a competitive way "grow up" to be a selfish, greedy supermind? Will an AGI created to help people maintain that kind of focus as it becomes more intelligent and grows independent of human beings? We don't know and we can't know, yet I'm willing to trust common sense that making an AGI that's friendlier and more open and compassionate is going to increase rather than decrease the odds of a good outcome.

Building the Robot Toddler

A concerted AGI initiative such as I've described, with a core team at the center of a global OSS AGI project, could focus on a variety of different areas. As we've discussed extensively, there are many possible pathways to AGI – one could approach human-level AGI via web search, SIRI type conversational agents, self-driving cars, smart game characters, medical software, industrial robots, and any number of other possibilities. But, as we've also discussed above, I believe that the most promising and lowest-risk, highest-reward path from an AGI R&D perspective is to focus on emulating the development of a young human mind. Thus, if I did manage to pull together an open source AGI initiative like I've just described to you, my first suggestion would be to focus it on the creation of an intelligent robot roughly emulating the intelligence of a young human child.

For sure, a robot with the rough intelligence of a 3-4 year old child (though manifested differently due to, among other reasons, current humanoid robots being quite different from human bodies) is not the end goal of AGI development. However, it requires early versions of essentially everything that a full-scale adult-level AGI requires. And further, it would constitute an extremely exciting, high-profile demonstration of interim progress toward advanced AGI, which would make further funding from a variety of sources unproblematic to obtain. Since robotics technology has advanced in recent years, we are now at the point where an AGI/robotics project can use low-cost humanoid robots as research tools and focus on the associated software.

With an "intelligent robot child" as a 5-year R&D target, one could divide the core and infrastructure teams mentioned above into subgroups such as:

- Natural language dialogue
- Natural language learning
- Machine vision
- Machine audition
- Learning and reasoning
- Distributed processing infrastructure

- Virtual world (simulation) AI testing
- Robot actuation
- Robot hardware experimentation
- Holistic intelligence testing

Each of these R&D areas is of course its own long story; and the OpenCog team has carried out substantial research and prototyping in each of these areas. If the dedicated teams comprised 50 people altogether, we'd have an average of 5 per team – which is not enough to do all the work, but should be enough to lead open source projects that would catalyze the process of a broader community doing the work. There is still a large amount of work to be done, but there is a clearly defined architecture and an existing codebase in which to pursue this work.

Focusing on a robot child may seem a shame when there is so much other exciting stuff to be done too – what about solving aging, for example? Do we really need to wait for the robot child to grow up to work on that?

Well, no we don't. For a focused initiative moving as fast as possible toward AGI, a core team pushing toward the robot child as hard as they can seems the right move. But in accordance with the beauty of the open source methodology, while the core team was working on the robot child, the proto-AGI software would be out there, well-tested and well documented – and other individuals and teams around the world, in universities and in companies and in diverse basements and bedrooms, could be extending and customizing and utilizing this code for all sorts of other applications… life extension biology, nanotech, smarter self-driving cars, better smartphone chat systems, industrial robotics, mathematical theorem proving, game playing – you name it. Some of the developers working on these applications would push code back into the main proto-AGI codebase, thus pushing the project forward generally, and adding aspects that might not occur to anyone thinking mainly about robot children.

Onward, Upward, AGI-ward

Will a large-scale OpenCog initiative like I've just described really come together? I'm trying my best to make it happen. I'm optimistic and enthusiastic like most entrepreneurial-type people; but rationally, I can't say for sure if I'll succeed or not. I've had my ups and downs in the past, and I'm not getting any younger. I could use your help!

In the big picture, though, I think it's pretty clear that if I don't succeed in pulling together an effective enough AGI project, somebody else will. Maybe my efforts pushing OpenCog will make a positive Singularity occur 10 years earlier than would happen otherwise – if so that would be amazing and would save a lot of lives. But 10 years, while it means a lot to me, is meaningless in the scope of human history. In the big picture, we are on the cusp of the greatest advance in the history of humanity – one that will bring us beyond humanity into whole new realms of thinking and being. If I am unlucky enough not to live to see it, my children or grandchildren will. Even if Kurzweil and I turn out to be overoptimistic, still, I'm 99.9% sure advanced AGI is not centuries away. Historically speaking, the next stage of intelligence is almost at hand. AGI is coming, both faster and more interestingly than most people think. And I'm thrilled to be part of the process by which humanity is bringing the AGI revolution to pass.

19. Far Beyond Human Level

I believe we could create human-level AGI fairly rapidly with the OpenCog project, given sufficient funding. I'm pushing toward this as hard as I feasibly can. But as I've said above a couple times, if this doesn't end up working out, for one reason or another, then somebody else is going to do it. Even if AGI funding remains relatively hard to come by, and even if brain scanning accuracy lags behind Kurzweil's expectations — still, sometime in this century, somebody's going to do it.

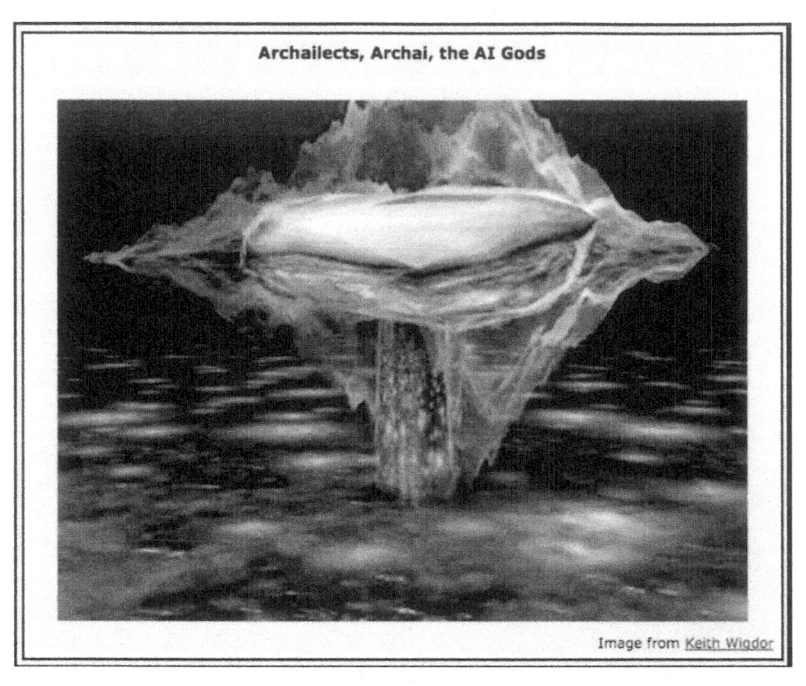

FIGURE 19.1: *An artistic depiction of vastly transhuman intelligence, created as part of the online SF universe Orion's Arm. Of course there's no reason to believe real-world superhuman AGIs will look anything like this. But the picture does get across the fundamental non-humanity that minds with general intelligence far beyond the human scale are likely to possess.*

I don't think AGI can be stopped, except by a calamity that wipes out humanity or destroys civilization. AGI is just the next step in the evolution of complexity and intelligence. It's what the universe has in store next. The arrow of progress is pointing far beyond the human level.

From that point of view, asking "Why create AGI?" is irrelevant. It's sort of like a one-celled organism, floating around in the primordial soup, asking its neighbor "Why bother with multicellular life? Why not leave well enough alone?"

That's the grand, cosmic view. It's the view I inevitably fall into when I'm hiking in the mountains near my house in rural Hong Kong, looking out at the islands and the ocean and the strange-looking trees atop the hills, and thinking about how much more vast is the universe than anything I or any person can conceive. But no human always thinks that way, not even a wild-eyed transhumanist like me. Sometimes when I'm watching my wife's limber body move through yoga poses, or listening to my son play piano, I can't help feel glued to the human world. I sink into the role of the confused one-celled organism and ask myself the question: *Why create AGI? Especially given the potential dangers— why not leave well enough alone? Humanity's not so bad!*

But, of course, when my mind drifts this way, I immediately realize I know the answers. First: *because that's what the universe does. It creates amazing new things – like molecules, bacteria, animals, plants, humans, computers, Internets – and then transcends them.* Second: *Because, even within a narrowly human-focused perspective, forbidding AGI wouldn't keep things safe and stable anyway. Without AGI to help protect us, the various other technologies we're developing -- synthetic biology, nanotech, etc. – would stand a good chance of harming us badly badly, maybe even killing us all. AGI may be our only chance of NOT getting destroyed by something else.* Third: *Because there is so damn much apparently unnecessary human suffering in the current and historical human condition, alongside all the beauty.*

My wife Ruiting's lovely smile – so peaceful as she does yoga; so excited as she laughs at a joke -- will fade one day into the macabre grinning rictus of a skeleton – UNLESS we do something about aging and death with advanced technology. The fingers of my son Zeb's decomposing skeleton won't be able to play Scriabin's music any better than the skeleton of Scriabin – who died at age 43, in the midst of composing a masterwork, from an infected sore on his upper lip. There were no antibiotics in the time of Scriabin, so he died young from something that could easily be cured today. My nephew Lev, who died at age 14, would not have ied if some AGI or human developed a rapidly-enough acting antidote for the kind of meningitis he got. It isn't human nature to just accept terrible waste and suffering as inevitable and natural – it's human nature to strive to make things better, even when this means transforming human nature, or even transcending it.

As a human being working hard on AGI, my motivation is largely driven by the following belief -- AGI has the potential to make the life-experience of sentient beings a heck of a lot better!

FIGURE 19.2: *My good friend Hugo de Garis has been very outspoken against the idea that we can somehow control the evolution of AGI systems with greater than human intelligence. He likes to point out that a grain of sand has potential computing power quadrillions of times greater than the human brain, and that flies and bacteria would have a hard time predicting or controlling human society. This is basically the same point my old friend Valentin Turchin made when he referred to "human plankton", but Hugo is a lot more dramatic about it.*

Being human can be wonderful and beautiful, but – as Scriabin learned the hard way at 43 years old, and as my nephew Lev learned at age 14 as he felt himself overtaken by an utterly unexpected, ultimately fatal seizure – it can also be a hell of a pain. And it can be extremely limiting -- so many of our dreams and ideas just never get realized, for boring old practical reasons.

According to Buddhism, all existence is suffering, which doesn't mean that life totally consists of pain, but that suffering is interwoven into everything. Everything has a little bit of suffering. As Nietzsche said, "If you have experienced one joy, then you have experienced all woe as well". All this may be a bit overstated but there's certainly something to it — almost every aspect of human life is characterized by suffering of some sort.

Now, it may be that pain is just a universal feature of existence and there's no way to eliminate the experience of suffering altogether. I wouldn't be shocked to learn this was true, though I'm not willing to commit to it at this point. I do think, though, there is a lot more suffering in human nature than is necessary, and far too much misery in the world.

There are many obvious horrors in the world today -- child abuse, rape, starvation amidst plenty and agonizing death from disease. There's also a high density of less spectacular frustrations like headaches, telemarketers, and the trauma of failed relationships. We accept these things because we're accustomed to them, but that's just like cavemen accepting rotting teeth because of ignorance about dentistry, and Scriabin's friends and family accepting him dying of a simple infection because antibiotics hadn't been invented yet.

You may say all the pain of human is necessary, and that without pain there wouldn't be pleasure. There's some truth underlying this perspective; all things are connected. But I bet there are ways of living that involve a lot less suffering and a far higher percentage of joy, as compared to current human life. I think you could get rid of death, disease, rape and all the negative feelings that accompany them without eliminating the fundamental joy, pleasure and wonder of being alive, of living, loving and existing in the world.

I see the creation of a positive Singularity, including AGI and radical human enhancement, as the short path to a better life, a way of breaking out of this existence into something profoundly more joyful and fulfilling. That doesn't mean I consider present life to be a horrible way of living. Current life is not a torture – I basically enjoy my life and I see many others who do, too. But the fact remains that many others are less fortunate, mired in poverty, trapped by violence or bereft of opportunities for improvement, their happiness and potential accomplishments stifled through no fault of their own. And even in a relatively good life like mine, there's a disturbing amount of everyday discomfort and worry. Of course, human life isn't uniformly bad— in spite of all the problems, there's a glorious joy at the center of it, and most of us have plenty of wonderful moments. But one of the most amazing things about human life is that it contains the potential to create new kinds of minds with new ways of living, far better than anything humanity has ever experienced.

Everything that has an upside also has a downside, and with AGI just like everything else, it's up to us to do our best manage the balance between the two. The odds that we'll ever develop a technology that is wholly good are slim. We just have to do our best to be smart about innovation and inject as much wisdom and foresight as we can into the proceedings. On the whole, technology has vastly improved human life; it has certainly made the human experience richer, more complex and interesting – and I believe this will continue.

Transcending the Discontents of Civilization

In his book Civilization and Its Discontents, the great psychologist Sigmund Freud has an interesting view of the evolution of humanity and culture — all our varied human neuroses and psychological problems, all the forces that disturb our mental health, are ultimately a consequence of civilization. Prehistoric people didn't have all these problems. They had others: Child mortality was sky high, most serious diseases were fatal, and lifespans were short. But they didn't suffer from the deep psychological unhappiness that afflicts so many of their descendants because they lived according to their impulses. All too often, modern society compels us to repress ours. We end up being tormented by strong desires we can't possibly fulfill because the rules we live by won't allow it. We're too intimately bound up with other people and doing these things risks harming others.

Freud wasn't saying this is necessarily a bad thing; he wasn't proposing that we return to the Stone Age. Doing so might, admittedly, rid us of most of our neuroses and psychoses, but in the process we would lose many of the things that make our modern lives precious —language, culture, literature, science, mathematics, cinema, and so forth. Freud's solution to the dilemma he identified was psychoanalysis, which hasn't worked all that well so far. Most modern psychiatrists rely on pharmacological rather than talk-therapy solutions these days, and the business of prescribing psych meds is surprisingly reliant on trial and error.

FIGURE 19.3: *While some of Sigmund Freud's ideas have been discredited by modern science, others emain as acute and insightful today as they were when he formulated them. His analysis of how the restrictions posed by civilization are the root of much of the unhappiness of modern humans, falls clearly into the latter category. Following up earlier related ideas by Nietzsche and others, he was a careful observer of the processes by which we take the constraints and criticisms of our parents and our society at large, and transform them into our own internal controllers and critics, thus setting up complex, often emotionally*

troublesome dynamics within our own minds. He noted that at bottom, most of our troublesome inner dynamics are related to the struggle between our evolved animal urges, and the demands civilization places on us to act less like typical animals. Unfortunately his proposed methods for eliminating the sources of unhappiness he identified (mainly, certain forms of talk therapy), did not prove extraordinarily effective. One wonders what Freud would have thought about the possibility of eliminating or drastically reducing the psychological problems induced by civilization, via engineering out some of the "animal" aspects of the human mind, and via lessening the restrictions posed by society via massively reducing the scarcity of human-relevant resources. (http://en.wikipedia.org/wiki/File:Sigmund_Freud_LIFE.jpg)

Relative to Freud's perspective, I see the Singularity as a way of resolving the problems that have festered since the advent of civilization. Rather than turning our backs on development, we can embrace it as something that will lead us to an advanced and more enlightened state of being, one that eases the restrictions and shackles that civilization has placed upon us, while retaining all the positives derived from the progress we've made so far.

While Freud didn't explicitly view it this way, I believe the problems of civilization are at bottom mainly problems of resource scarcity — scarcities of space, time, energy and intelligence that make us step on each other's toes and require us to restrict ourselves for the good of others. Advanced technology can liberate us from these bottlenecks, opening up new and previously unknown vistas of freedom and fulfillment. The modern move toward Singularity could be viewed as a process of transitioning away from the constraints on knowledge and consciousness long imposed by modern society.

And after the civilization we know is gone, the post-Singularity phase will usher in new freedoms – new ways to grow, to feel, to think, to interact, to enjoy, to create. This promise of a dramatically better, wider, more richly and diversely fascinating and satisfying tomorrow is what makes the Singularity so exciting. That's why it's worth accepting dangers and risks, even the existential ones; they're worth it because of the payoff. Just as the move to multicellular life was worth it, in the big picture, even though it shook up the everyday lives of a large number of amoebas, paramecia and Euglena. New ways of thinking, being, feeling and living, with others and ourselves – and going beyond the distinction between "others" and "ourselves" -- are beckoning us from the peaks ahead. We must find the courage to ascend and risk falling, for the rewards of success may be LITERALLY beyond anything we can imagine.

Transcending our Human Illusions

Ultimately, the Singularity is not so much about the creation of new technologies, as about the creation of new ways of being and experiencing. Many aspects of this will go so far beyond our current experience that we can't hope to comprehend them now. But some aspects are possible to muse about in detail at this stage – especially those aspects that have to do with eliminating or transcending particular features of our current ways of thinking and being.

The ordinary experience of being human is based on a number of assumptions that fail to reflect an accurate understanding of ourselves as sentient beings. Instead, we rely on convenient illusions — assumptions we've evolved to make — that fail to reflect how our minds work. These illusions will disappear as we begin to interface with AGIs, transforming our minds and bodies with advanced technologies. I'm not talking about something obscure here – I'm talking about basic things like the feelings of self and free will take for granted in our everyday mental existence.

Post-Singularity, most likely, our minds will no longer be tied to individual bodies, but free to move among many, changing and adapting accordingly. We will have the option of sharing our thoughts more freely, and occupying various bodies collectively. We will be able to sense and act, both globally and locally (e.g. absorbing data from satellites and cameras all over the world). We will be able to create virtual worlds — which we can exist and interact in — based on our whims and desires. And all this is going to have huge, transformative impact on the inner workings and feelings of our minds.

The Illusion of the Self

Perhaps the core illusion of everyday human experience is the self. The self I have, the idea I have of "this guy Ben Goertzel who writes books on AGI and works on AI software, who has a wife and three kids, and a village house in Hong Kong" – this idea is not really the psychological and biological system that we call "Ben Goertzel". It's a model that has arisen within my brain, possessing some accuracies and some distortions. It's a model that thinks of itself as much more real and accurate than it truly is. The way I think about myself – the self-model I have built -- guides what I do every day, in ways that are sometimes productive and sometimes not.

For instance my model of myself as "a guy who works on AGI and AI" might potentially prevent me from exploring other possibilities that arise, even when these other possibilities would help me achieve my goals better. What if an opportunity arises for me to help create a new form of nanotechnology, in a way that exploits my mathematical background, and has the potential to create a massive profit, that could then fund AGI and all sorts of other futuristic work? Then I might do better to turn to this nanotech work for a while, setting AGI aside. But my self-model as an "AGI guy" would potentially prevent me from doing so.

But that's somewhat a nonrepresentative example. The self's main role isn't in big life decisions, it's in the small everyday choices that make up our ordinary lives. Using my own self as an example again (it's the self I know best), I tend to have a model of myself as a guy who has all the answers. Being honest, I have to admit that my ego is wrapped up with this to some degree. I try to avoid this sort of commonplace psychological trap, but I don't always succeed, and now and then I fall into the pattern of feeling good about myself because I am The Guy Who Has All the Answers. It's easy to see the kind of consequence this has. Nobody really has all the answers; so what if somebody raises a point that I don't actually know the answer to? Rather than acknowledging my own cluelessness on that point, and thus opening myself to gathering new information, I might instead try to convince myself I knew SOME kind of plausible answer. In that case, I might be so busy searching for a half-assed answer and trying to convince others it's whole-assed, that I wouldn't even hear the worthwhile directions toward an answer being presented by others. Due to not hearing other peoples' potential answers in this context, I might even draw the conclusion that nobody else had had anything useful to say – thus getting more reinforcement for my belief that I always have better answers than anybody… Thus strengthening the Guy Who Has All The Answers portion of my self-model…

This particular example – feeling like one has all the best answers, even when one doesn't – is a problem I identified with my own personal self-model quite some time ago. I've consciously tried to correct this trait of mine, and have succeeded to some extent. I'm still a bit of a know-it-all, but I'm a far better listener than I used to be; and I believe I'm more rational about the extents of my knowledge and intuition than I was in the past (although, this belief is probably not wholly accurate either; it's just part of my self-model…).

But anyway, my own human personality quirks aren't really the point here. My point is that human personality is woven of this kind of phenomenon: *Self-modeling that feels good for one or another reason, but isn't quite accurate, and that drives behavior in ways that are often self-reinforcing and not always productive in terms of human goals or human happiness.* The personal examples I gave above are just two data points. In countless similar ways, our self-models guide and restrict us as we go about our lives. Psychologists have dissected this kind of phenomenon in great detail; particularly relevant here is a 2007 book by Mark Leary called The Curse of the Self: Self-Awareness, Egotism and the Quality of Human Life.

Still, though, the title of Leary's excellent book notwithstanding, self is both a curse and a blessing. Our selves mislead and distort our thinking and our lives. But without them, where would we be? We do need to model our minds and bodies, in order to understand our place in the world and to plan our actions. The problem comes when we get emotionally attached to aspects of our self-model that are only crude approximations. The emotional attachment stops the approximations from getting improved. And the various errors and emotions involved in the self become systematically interdependent, until the self is a whole self-organizing system on its own, with some of its own life independent of – and guiding – the organism it's supposed to be modeling.

Philosopher Thomas Metzinger has called this sort of self-model I'm talking about the phenomenal self, using a term from the philosopher Immanuel Kant, to whom "phenomena" were mere surface appearances, lacking in realness. Kant conceived of the term "noumena," referring to another realm, the "absolute real" that we could never perceive, but which actually underlies what we experience. Metzinger's view is that self is not noumenal, it's phenomenal – he's looking at the self, not as some underlying reality, but as something mind builds. His book Being No One digs deep into the way the brain builds the phenomenal self, carefully analyzing various cases of brain dysfunction that cause people to build improper selves in various ways. It's a long book but an awesome one; I'd strongly recommend it. If you don't like reading huge neurophilosophical tomes, you can find some of his lectures on the Internet as well.

Is this phenomenal self, this largely-illusory self-model that we humans habitually create, truly a necessity for any intelligent system? Well, some kind of self-model is obviously of value for any agent achieving complex goals in a complex world. But the pattern of self-reinforcing inaccuracy and emotional attachment that characterizes the human self, seems in large part an artifact of the particular human mind architecture, rather than a necessary aspect of intelligence.

Ultimately, one asks: *Why would an AGI mind have to systematically take itself to be something that it isn't, and emotionally cling to its errors? Why could the AGI not base its self-perception on an accurate modeling of its own mind? When it makes errors in self-perception, why couldn't it just correct them based on its experience, rather than reinforce them due to emotional attachment?*

One could build an AGI possessing human-like self-modeling dynamics, but it's not clear why one would want to, except for the scientific goal of studying human-like minds and better understanding their strengths and weaknesses. I believe one could more easily build AGIs displaying more rational approaches to self-modeling, free of egomania and the various other pathologies that characterize human self and play such a large role in human personality and society.

This brings us to one of the biggest, and most emotionally charged, modeling errors present in most modern humans' self-model: a misunderstanding of how independent we actually are from others and our environment.

One of humanity's deepest flaws, and one especially prevalent in modern Western society, is our tendency to individuate ourselves to an extreme degree.

We often fall into a habit of thinking that our minds are separate from our bodies, reducing the latter to objects that our minds control — a distorted form of self-perception. In fact, our minds are intimately entwined with our bodies; together they form webs of interactive feedback loops, linking us to people and the surrounding world. Cognitive scientists refer to this as an embodied or extended mind – a meta-mind that overlaps with your own body, as well as the people, tools and objects with which you interact.

In the present order of things, our self-models only indirectly and incompletely reflect the embodied and extended nature of the mind. We tend regard ourselves as individually autonomous minds, interacting with the separately defined objects, bodies and minds in our environment. This tendency is probably most extreme in America, the country where I spent most of my live. American individualism is fantastic in some ways – but it's also in large part illusory. All of us are far more interdependent with, and cross-connected with, the rest of the world than we realize.

Cognitive scientist Andy Clark has written a lot about the extended nature of the human mind – I'd recommend his books Being There and Supersizing the Mind very highly. Though when my wife and I visited him in Edinburgh in 2012, we were surprised to find that he had somewhat moved beyond the extended mind as an area of research. He figured he had already made his point there – and he was more interested, as of 2012, in thinking about the relationship between consciousness and deep learning a kind of AI technology, thought by some to be a path to AGI and a model of the human brain; I'll discuss that a bit later. He absolutely did not consider the extended nature of the human mind an objection to building AGI systems. Rather, he was intrigued by the attempts by my own team and others to make AGI work via embodying complex software systems in robots. My notion of making a robot toddler fascinated him, though he worried about how the robot would compensate for the lack of the cognitive guidance human children get from their genetic endowment.

Anyway – this whole embodied/extended mind thing is going to seem very different once advanced technology enables a more flexible sort of relationship between minds and bodies. Once we can port our minds between different bodies—and radically alter our bodies—we'll sense things using means other than our typical body senses (via connecting to sensors around the globe, or directly reading certain thoughts and sensations from other peoples' brains or from AGI minds). The strict connection we now feel between our self and a particular body will be a thing of the past.

Of course, you can't just put a mind in a different body and have it unchanged – the body is part of the mind, and so when you take the same high-level cognitive system and use it to control a dolphin-like body instead of a human-like body, for example, this will inevitably result in the high-level cognitive system becoming dolphinized to some extent. Because in reality the high-level cognitive system is only part of the overall mind of the "high level cognitive system + human-like body" or "high level cognitive system + dolphin-like body" mind system, and is in constant complex feedback with whatever body system it's connected to.

Once we can port our high-level cognitive systems from one body to another, what happens to the self? We will still model ourselves, but in a more fluid way, reflecting the rich interconnectivity of our mind-patterns with various sensors, controllers and other minds and systems. Each of us will feel less like an individual interacting with a world that contains others, and more like a dynamic self-organizing cloud of mind-stuff, interacting in a complex dance with other clouds of mind-stuff, and with various physical systems, on various scales. Doubtless various problems and limitations will arise in connection with this new way of experiencing ourselves, each other and the world, but they may have little in common with the things that trouble us now.

The Illusion of Free Will

Another complex, illusion-ridden characterizing everyday human life is free will, perhaps the most confusing concept I've ever heard of. Free will, as commonly conceived, is a logically ill-formed and almost senseless concept, yet rings loud intuitive bells for most people. The loss of this peculiar aspect of our self-models would change our minds considerably. Yet it seems unlikely to me that post-Singularity humans, or advanced AGIs, will think about themselves using anything resembling our current concept of "free will".

When one of us modern humans reaches a decision, we stubbornly persist in believing that we're engaging in some kind of "free choice". A choice that is not determined by anything outside ourselves, nor by mere physical dynamics within ourselves; and certainly not just chosen at random. It feels like it is somehow chose by ourselves, by our inner choosing facility, by our mind's freely acting choice process! However, there's plenty of data debunking this intuitive, "folk psychology" notion.

Cognitive neuroscientists have set up many experiments, showing cases where a person feels like they are consciously deciding and willing something at a certain point in time; however, their unconscious brain has already taken measurable steps in accordance with the decision a little bit earlier (say, half a second earlier).

For example, Michael Gazzaniga's classic split-brain experiments explore what he calls the "confabulative" aspect of free will, via probing the experience of people who have had the two halves of their brain separated, in order to prevent their brains from being destroyed by severe epilepsy. In these unfortunate cases, each half of the brain maintains its own self, its own stream of experience, and its own free will – thus providing the cognitive neuroscientist with a wealth of avenue for fascinating experiments.

In one experiment, a split brain subject's left eye received a command to stand. The person stood – and then, when asked why she stood up, she responded (using the language center of her left hemisphere) that she wanted a soda. In another experiment, when the left and right hemispheres were each asked to pick an appropriate picture to accord with an image flashed only to that hemisphere, the left selected a chicken to match the chicken claw in the picture it saw, while the right hemisphere correctly chose a shovel to remove the snow it saw. When asked why the person chose those images, he replied that the claw was for the chicken, and the shovel was to clean out the shed.

Gazzaniga's experiments demonstrate that, even when there is a clear external cause for some human action, it is possible for the human to sincerely and thoroughly believe that the cause was some completely internal decision that they took. The left hemisphere of a split brain has no experience of stimuli delivered exclusively to the right hemisphere (e.g. through the left eye). However, the left hemisphere has such a strong motivation to create explanations that it will make up "free will stories" corresponding to behaviors initiated by the isolated right hemisphere.

Daniel Wegner's book The Illusion of Conscious Will gives a pretty good summary of the contemporary body of knowledge regarding free will and the brain. The book Neurophilosophy of Free Will, by Henrik Walter and Cynthia Klor, tries to dig deeper and figure out what kind of "free will like" capability could possibly be compatible with science and common sense – I recommend it highly.

While neuroscience has made the limitations of the free will concept crystal clear and essentially indisputable, the same basic problems with the concept were noticed by philosophers long before anyone knew what a neuron was. When Nietzsche called consciousness an army commander who, after the fact, takes credit for the actions of his troops — he definitely had free will in mind. In large part, free will is a story that the brain/mind tells itself after the unconscious mind has already made a decision, justifying the unconscious decision already made. This storytelling process may often be quite useful, as it may feed into the unconscious and help guide future unconscious decision-making, and future conscious story-telling. However useful they may be, though, our stories about how we decide our actions are generally not accurate.

Now fast forward a bit to the future, when there are advanced AGI minds, or radically improved human brains. Imagine a mind capable of monitoring the activities in its brain as it thinks, and the interactions of these activities with other things in the world. Such a mind would be able to see — vividly and in real-time — a great many of the actual dynamic processes involved in its "choices". Such a mind would be unlikely to maintain a current-human-mind-like illusion of free will. Between the absence of the illusion of free will, and the existence of a more accurate self-model not including an irrational sense of one's own independence and autonomy, such a mind will have a very different kind of experience than we current humans.

Projecting from my current base of knowledge to the future of post-Singularity minds makes me feel a bit like a cockroach trying to predict the future of quantum computing. But nevertheless, it is interesting to extrapolate as best as possible from what we know now, with the understanding that future discoveries are likely to revolutionize our current understanding. My best present projection is that, as mind moves on from its current "legacy human" form into an era of AGIs, uploaded humans and enhanced human brains, subjective experience will move beyond will and emotionally-attached self, and we will have a world of beings that view themselves as fluid clouds of mind-patterns, evolving dynamically in close coupling with other systems, in ways that happen to influence various actions.

What will it be like to be such a mind? I have only a vague sense at present. But I am definitely curious.

And I am very aware that I have barely a clue what new forms of mental organization, and new states of consciousness, will take the place of these historical illusions to which we've become so habituated.

But one thing I am confident of is: *The psychological and social aspects of the Singularity will be even more intriguing and dramatic than the technological and scientific ones.*

The Global Brain

One potential route for going beyond the individual self is for multiple selves to fuse together in some way. This possibility was caricatured in a frightening way by the TV show Star Trek: The Next Generation by the Borg Collective, a group-mind based on ruthless expansion by forcible incorporation of individual minds. "Surrender or be assimilated!" The Borg Collective is sort of the Terminator meme of collective intelligence. But actually, the emergence of some sort of group intelligence could be quite different from Borg type fears. The merger of an individual mind into a collective could be a matter of joy, growth and choice on the part of the individual, not necessarily a matter of force. Furthermore, the existence of a group mind wouldn't necessarily mean the immediate (or even eventual) loss of choice or individuality on the part of the member minds.

FIGURE 19.4: *The Borg Collective from Star Trek: The Next Generation. The archetypal dark and scary image of a group mind. But a movement to shared consciousness wouldn't need to mean a forcible uniformity, it could be a transcendence to a richer and broader mode of being. (http://agileforest.files.wordpress.com/2011/07/borg1.jpg)*

A more positive slant on the collective-mind idea is provided by the concept of the Global Brain; the notion that computers and communications technology, already one and the same, will gradually meld with and link humanity through all manner of devices, forming a networked community of people and machines that will give rise to a higher order of intelligence. I thought about this sort of thing a lot in the late 1990s, when the Web was just beginning – as I recollected back toward the start of this book, the central goal of the company Webmind that I co-founded back then was to create AGI and deploy it in such a way as to create a powerful global brain. While this didn't end up happening via the Webmind software, the Global Brain still seems to me a highly relevant paradigm for conceptualizing many aspects of the future.

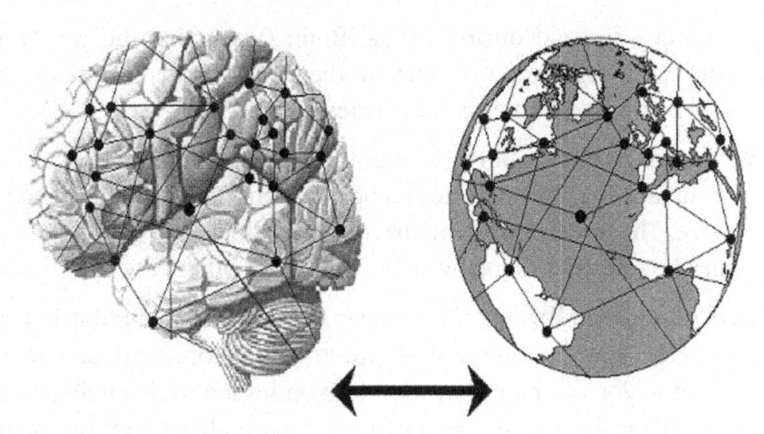

FIGURE 19.5: *Artistic depiction of the Global Brain – of the analogy between people in society, and neurons in the brain. (http://pespmc1.vub.ac.be/Images/Brain-Earth.GIF)*

People and networked machines across the planet would be like neurons, the brain cells, in this collectivebrainthat would think via trends, essentially patterns of ideas spreading across a network. Individuals, even if they were integrated into the Global Brain, might not be able to sense these thoughts in detail any better than a single neuron can understand the thoughts flashing across our mindscapes.

The Global Brain might seem a threat to human freedom, in a different way from robot overlords or indifferent femtotech superminds – but it ain't necessarily so. A neuron may be free to be a neuron, firing when it wishes and living on the level of a neuron in a way that's completely natural to it. Nevertheless, from a different point of view, a neuron is part of the coherent coordinated activity of the human brain. In a similar way, we may go about our lives as humans in a manner that seems natural to us, exercising what seems to us to be free will. Yet on another level, we may be part of a higher-order coordinated intelligence.

While not as prominent in the popular eye as AGI robots, the Global Brain concept is gaining a bit more attention lately. My friend Francis Heylighen, a professor at the Free University of Brussels in Europe, has done more than anyone else to promote the notion. In 2001, in the midst of my Webmind-era enthusiasm for using AGI to launch a Global Brain of some sort, I suggested he organize a Global Brain conference at his university, and he thankfully followed through. We called it Global Brain 0, and it went pretty interestingly. More recently, in early 2012, Francis received funding from a retired Internet entrepreneur to start a Global Brain Institute at his university.

One thing I realized during Global Brain 0 was that the conference participants had a huge variety of views on the Global Brain. Essentially, their views could be divided into three categories.

The Global Brain Is Here! One group believed that the Global Brain is already here. We may not recognize or understand it fully, but it's here. The Internet is an intelligent mind, different from what we are, maybe smarter in some ways and less so in others.

The Global Brain Will Spontaneously Emerge. Another group believed that the Global Brain is not yet present in full force and can't be until it undergoes a phase transition into an impressively intelligent and organized form, one that will emerge naturally without our help. It will develop on its own as computing, communication technologies and human culture advance.

The Global Brain Will Come, Because Someone Will Build It. Finally a third group, to which I belonged, argued that the Global Brain will emerge, but as a purposeful consequence of Global Brain engineering, alongside the emergent dynamics of communication systems and cultures arising for other reasons.

More specifically, I suspect that – if something else, like a destructive world war or an AGI robot supermind, doesn't come first – a powerful Global Brain will come about after we release fairly advanced autonomous or semi-autonomous AGIs on the Internet. Imagine AGIs let loose on the Net, with the goal of interacting with people, mediating communications, ingesting information,then summarizing and presenting it in a more comprehensible form, and finally, placing new information on the interwebs for people and other AIs to ingest. A network of such AGIs could serve as the central cortex of a Global Brain, with humans and other software systems (including a diverse spectrum ofnarrow AIs) filling supporting roles. This is the vision I laid out in my 2001 book Creating Internet Intelligence, and I still think it's a plausible future scenario.

In time, some AGIs willsurely outgrow any Global Brain in which humans can play a significant role, moving on to some higher form of intelligence beyond human comprehension. But even after such a transcension, this planet might still be dominated by some sort of collective Global Brain consciousness shared between AGIs and humans. These transhuman AGIs may go off to pursue their own interests, not necessarily competing in the niche of our measly little human-based Global Brain.

It's easy to see how some of today's technologies are explicitly pushing us in this direction. For instance, social media; systems like Digg, Reddit and Slashdot(or whatever their new, fashionable analogues are at the time that you're reading this) consist of people collectively summarizing news and pushing stories to the top of listings, collaborating to do the job of an editor and reporter simultaneously.

Social networking systems like Twitter (again, insert your favorite social network of the day!), although often frivolous, also display facets of collective intelligence. Emergence is visible in Twitter when relevant information bubbles up through a mass of retweets, eventually becoming a trend for reasons independent of any one person's actions.

Consumer reviews on sites like Amazon also represent collective intelligence, since individuals, sellers and products can build up good reputations based on others' ratings, forming active collaborative filtering systems with some simple narrow AI components. The system then presents visitors with information they wouldn't have read otherwise, inspiring them to ponder and purchase new things, reinforcing trends further.

Open-source software projects are another interesting example of group intelligence. Each starts with a body of software, which grows and evolves over time like a life form, thanks to the input of hundreds of people. A handful of people may have written the original code, but it can quickly grow beyond their intentions, splitting into multiple projects and incorporating previously developed or fresh ideas in a process akin to symbiogenesis and the emergence of organisms. The result is a whole new layer of life forms. I've even seen limited examples of this in OpenCog, the open-source AGI project that I'm involved with.

So… Given that Global Brain-related technology is already in active use on the Internet, what happens when an AGI takes over and uses this infrastructure to help manifest its own ideas, which derive from the Net? What happens when an AGI starts reading books and recommending them to people, weaving new linkages between strangers? What happens when an AGI contributes to open-source software projects, or starts them and recruits new developers?

What happens is: We end up with a whole new kind of collective intelligence among humans and computer software, one that is not purely self-organized from the ground up, but rather mediated by AGIs who are working toward their own goals. With any luck, their goals will mesh with ours because they'll be trained to help our collective mind emerge, grow and flourish. Eventually, some of these AGIs may want to journey on, to a place where humans can't accompany them – but others may be motivated to stay around at a relatively stable level of intelligence and persist their communion with humans.

Does the Global Brain provide a way around the potential dangers of advanced AGI? Not exactly. But it does provide a unique perspective, and highlights a set of pathways that are commonly ignored. If AGI matures in the context of the Global Brain, then it will grow up feeling tightly interlinked with humanity. This will certainly affect the AGI's attitudes as it grows beyond its human origins — though precisely how is hard to say at this point.

The Cosmist Perspective

"The Earth is the Cradle of the Mind -- but one cannot eternally live in a cradle."

— Konstantin Tsiolokovsky
(advocating mankind exploring the universe)

"How unnatural it is to ask, 'Why does that which exists, exist?' and yet how completely natural it is to ask, 'Why do the living die?'"

— Nikolai Fedorov
(speaking against philosophy and in favor of immortality for all)

It's clear that our contemporary human belief systems leave us with an incomplete understanding of the implications surrounding AGI and the Singularity. New sciences and technologies require new ways of thinking.

With this in mind, I wrote a short book in 2010 called *A Cosmist Manifesto*, outlining a philosophy of life, the universe and everything that would still make sense in the context of a Singularity, and all of the associatedtransformative technologies. I called that philosophy Cosmism. I chose the name because I was inspired by the Russian Cosmist philosophers of the early 19th century, who already had many ideas common in the transhumanist community today, at a time when the state of technology made them a lot less obvious. These folks, Tsiolokovsky and Federov nd the others, were true futurist visionaries! They saw far beyond their time, just as I am attempting to do here in these pages.

FIGURE 19.6: *Russian postage stamp commemorating rocketry pioneer and Cosmist philosopher Konstantin Tsiolkovsky (who lived 1857-1935, foresaw humanity colonizing the galaxy, and pursued a sophisticated panpsychist philosophy in which the expansion of humanity and technology was a manifestation of the conscious will of the physical universe). (http://ferrebeekeeper.files. wordpress.com/2011/09/sm_tsiolkovsky.jpeg)*

At the start of the *Cosmist Manifesto* I give a list of ten "Cosmist Convictions" — ten basic principles of Cosmist thinking. My friend and colleague Giulio Prisco mostly came up with them, but I edited and augmented his list. They don't actually tell you all that much about Cosmist philosophy — if you want to know about that, read the Cosmist Manifesto! — but they do articulate the general Cosmist perspective, which serves asa conceptual foundation forCosmist philosophy.

Without further ado, here are the Ten Cosmist Convictions:

1. Humans will merge with technology to a rapidly increasing extent. This is a new phase of the evolution of our species, just picking up speed about now. The divide between natural and artificial will blur and then disappear. Some of us will continue to be humans, but with a dramaticallyexpandingrange of options, creatingradically increased diversity and complexity. Others will grow into new forms of intelligence far beyond the human domain.
2. We will develop sentient AI and mind uploading technology. Mind uploading technology will permit an indefinite lifespan to those who choose to leave biology behind and upload. Some uploaded humans will choose to merge with each other and AIs, requiringreformulations of current notions of self.
3. We will spread to the stars and roam the universe. We will meet and merge with other species out there. We may roam to other

dimensions of existence as well, beyond the ones of which we're currently aware.

4. We will develop interoperable synthetic realities (virtual worlds) able to support sentience. Some uploads will choose to live in virtual worlds. The divide between physical and synthetic realities will blur and then disappear.
5. We will develop spacetime engineering and scientific "future magic" much beyond our current understanding and imagination.
6. Spacetime engineering and future magic will use scientific means to fulfillmost of the promises of religion— and many amazing things that no human religion ever dreamed. Eventually we will be able to resurrect the dead by "copying them to the future."[3]
7. Intelligent life will become the main factor in the evolution of the cosmos and guide its path.
8. Radical technological advances will reduce material scarcity drastically, so that abundances of wealth, growth and experience will be available to all desiring minds. New systems of self-regulation will emerge to mitigate the possibility of mind-creation running amok and exhausting the ample resources of the cosmos.
9. New ethical systems will emerge, based on principles including the spread of joy, growth and freedom through the universe, as well as new principles we cannot yet imagine.
10. All these changes will fundamentally improve the subjective and social experience of humans,our creations and successors, leading to states of individual and shared awareness possessing depth, breadth and wonder far beyond that accessible to "legacy humans."

Giulio Prisco, who formulated the first draft, made the following comment on the use of the word "will" in these principles: "*...'will' is not used in the sense of inevitability, but in the sense of intention: we want to do this, we are confident that we can do it, and we will do our f**king best to do it*".

3 Recreating a deceased person in the future, based on all known data about the person – their writings, videos of them, memories of them in the minds of still living people – and advanced knowledge of the brain and body.

FIGURE 19.7: *Giulio Prisco, modern-day Cosmist, who has been one of the strongest advocates of the relationship between transhumanism and spirituality. His Turing Church blog has been influential in the transhumanist community, as have his online technology/spirituality discussion groups, held in virtual worlds (such as depicted above). (http://1.bp.blogspot.com/_nUDJsVoQIP8/S4Vbzg_38MI/AAAAAAAAAbY/NZysLMFaxxQ/s320/slustreamavatars.jpg)*

It's worth keeping in mind that the general Cosmist perspective and the particular points of Cosmist philosophy — like the technical and conceptual specifics of AGI – are basically independent of the notion of a Singularity per se. If we have a slower, more gradual advent of advanced technology, the philosophical and technical issues involved will basically be the same.

Cosmist philosophy is a world-view within which these sorts of principles are natural and sensible, rather than weird or counterintuitive. Cosmism views the world – like the mind – as an entity with multiple aspects, going beyond any one perspective, and defying simplistic categories like "objective" and "subjective". Among the perspectives it adopts are three delineated by philosopher Charles S. Peirce:

- ***Firstness***: pure experience and Being; raw awareness
- ***Secondness***: reaction, interaction and movement
- ***Thirdness***: pattern and relationship
- ***Fourthness***, synergy and emergence (not emphasized by Peirce but added on by later thinkers like Carl Jung and Buckminster Fuller)

The Thirdness perspective views the world as a web of interlocking patterned relationships – this is the view that science leads us to. Each scientific observation posits a relationship between certain observations. Observations themselves, viewed subjectively, are Firsts or Seconds; but when multiple observations are woven together into a repeatable pattern, one has a Third, a relationship.

In the Cosmist view, individual minds, societies, physical objects, and even space and time themselves may be viewed as patterns, as regularities occurring among elementary observations. The "observations" themselves are primary, more so than the notion of a "self" doing the observing – since every psychologist knows that the Self is an abstraction a mind builds for itself.

There are echoes of quantum theory here, as in quantum mechanics the physical world is viewed as having reality only relative to acts of observation. But Cosmism is a philosophy whereas quantum mechanics is a scientific theory. Cosmism can help us interpret quantum theory, but will retain its philosophical validity even if quantum theory is obsoleted by very different physica ideas.

In the Cosmist view, humans – like apes, rodents, bacteria and molecules – are best viewed as particular patterns of organization, emergent from elementary observations. Each of these patterns of organization has a certain coherence, a certain synergy. The universe, if you view it from a perspective that embraces the linear flow of time, reveals a pattern of progressive growth from simpler to more complex emergent wholes. There is also a pattern of creative destruction that occurs when new wholes arise to incorporate and disrupt aspects of previous ones. The issues humanity currently faces regarding the Singularity instantiate this larger process.

Human ethics, viewed with Cosmist eyes, is in part just a particular set of patterns that certain societies of intelligent entities have adopted in order to maintain their stability and/or promote their growth. However, there are also some universal principles underlying the diversity of human ethical precepts. For instance, the value of joy above suffering is important, and goes beyond particular cultures or organisms. The value of growth above stagnation is intrinsic to the universe, and seems to come along with any perspective that views time as going forward. The value of choice is critical from any perspective that involves minds distinguishing themselves from the universe – including entities like group minds and global brains. While classical "free will" is largely illusory, a broader notion of choice is essential to the existence of separate minds as distinct entities.

It seems unlikely that particular human ethical precepts like "don't covet thy neighbor's wife" are going to have broad meaning after the Singularity. They may still be useful within particular human communities that persist at that point, but they won't play a major role across the scope of intelligences in the Cosmos. But extensions of deeper human values like Joy, Growth and Choice may continue to be critical. New kinds of minds will create their own values – as Nietzsche foresaw the hypothetical Superman coming after humanity would do – but these values may well still be compatible with broad principles such as these.

The Cosmist perspective—or whatever name you want to attach to it— is something I gleaned at an early age from reading science fiction and Buddhist philosophy, and it's central to my personal view of the world, more so than the Singularity or even AGI. It's a general view of the mind and the world that underlies all the particular matters I discuss here and in my other writings. It helps me to grapple with the possibilities the future offers, without getting my mind blown.

Cosmist philosophy implies the conceptual POSSIBILITY of amazing advancements like AGIs, cyborgs, teleportation, radical life extension, group minds and so forth – and it posits a drive for intelligence to grow and expand -- but it doesn't intrinsically imply any specific time-scale for these things to develop. The "Singularitarian" perspective, as I articulated it at the start of this book, adds an ingredient that's very important from our contemporary human perspective: *That not only are all these amazing things feasible and reasonably likely to happen— the serious fun and creative destruction may start this century, and at a certain point may start to unfold extremely fast relative to human experience.*

Shaping the Singularity

Amazing things are going to happen this century — and if you ask me, they're likely to happen way faster than most people think.

From the big picture perspective, the accelerating progress of science and technology is essentially unstoppable, as is the passage from humans on to engineered superintelligences. And from the Cosmist view, the particulars of how all this happens may not matter so much. It's not clear how sensitively the medium-term future of superhuman superminds depends on the specifics of how the human Singularity goes down.

On the other hand, from the perspective of you and me as individual humans, the specific way that AGI emerges and the Singularity unfolds may matter a lot. For instance, it matters to mewhether my 3 kids get their molecules absorbed into some super-AGI's processing unit, or whether they get to choose their own future. I would like to see them have the choice whether to live on in legacy human form, while others explore transhuman domains; or to expand their minds through uploading or brain implants, and gradually become something more than human. What we do now, how we handle the development of AGI and other advanced technologies, might not impact the ultimate development of intelligence throughout the Cosmos, but it may well impact what happens to ourselves and our loved ones during the coming transition.

My own feeling is that to make the transition as smooth as possible, we must develop a reasonably empathic, beneficial human-like AGI system as soon as possible – before other advanced technologies exacerbate a situation where AGI somehow goes awry in the early stages. I've tried to get this started with OpenCog, in primitive and simple ways. For instance, one can give a virtual agent positive reinforcement signals when it does something to help another agent in its virtual world (say, holding the door open or moving an object obstructing their path). In this sort of way one gradually lays a foundation for AGI systems with the desired sort of goals and values. As technology and science advance and powerful AGI systems get created, we will need to work with these systems to figure out the next steps forward — to figure out how to move on toward progressively smarter AGIs in a way that benefits everyone. It's not likely to be easy, and it certainly involves a lot of uncharted territory.

But this is bloody fascinating stuff to be involved with. Sometimes I regret being born so long before the Singularity, so that I still have to struggle a bit to get resources for food and shelter, and have to gather information by crude methods like reading and talking instead of directly importing it into my brain from friendly AGI teachers. But sheesh, at least I wasn't born into the Middle Ages! Having the possibility to – maybe, just maybe – play a significant role in bringing AGI into existence for the first time ... that's a pretty amazing feeling.

Further Reading

See also the websites of Humanity+, the Machine Intelligence Research Institute, the Future of Humanity Institute, The Turing Church (Giulio Prisco), and the Order of Cosmic Engineers.

Goertzel, Ben (2010). A Cosmist Manifesto. Humanity+ Press

Heylighen, Francis (2007). The Global Superorganism: an evolutionary-cybernetic model of the emerging network society global brain. Social Evolution and History 6-1, pp.58-119

Freud, Sigmund (1930). Civilization and its Discontents

Leary, Mark (2007). The Curse of the Self: Self-Awareness, Egotism and the Quality of Human Life. Oxford University Press.

Weaver (David Weinbaum) and Viktoras Veitas (2015). A World of Views. In The End of The Beginning, Ed. By Ben and Ted Goertzel, Humanity+ Press.

Metzinger, Thomas (2004). Being No One. Bradford.

Wegner, Daniel (2003). The Illusion of Conscious Will. Bradford.

Walter, Henrik and Cynthia Klor (2001). The Neurophilosophy of Free Will. MIT Press.

Additional Relevant Works by the Author

The book you've just read (unless you skipped to the end!) is fairly long, yet it just barely scratches the surface of its topic. There's a lot more to say! If you're curious for more of my perspective on AGI and related topics, here are some pointers.

Here are some other recent books I've put together (authored or edited) about AGI and/or its implications. Many of these were listed at the end of some chapter or another above, but here they're all gathered in one place.

- **Ten Years to the Singularity If We Really Really Try, and Other Essays on AGI and its Implications**. A collection of nontechnical essays from 2008-2011, covering many of the points covered here in more detail. Available via Amazon or as a free online PDF via Humanity+ Press, see http://humanityplus.org/press
- **Between Ape and Artilect: Conversations with Pioneers of Artificial General Intelligence and Other Transformative Technologies**. Interviews with other researchers in AI and related fields; a couple interviews other researchers did with me. Most of the interviews are more like dialogues than interviews. There's a lot of in-depth stuff here but still it's OK for the nontechnical reader. Available via Amazon or as a free online PDF via Humanity+ Press.
- **The End of the Beginning: Life, Society and Economy on the Brink of the Singularity.** This is an edited volume, edited by me and my father Ted Goertzel. There are a few chaprs authored by me. The goal is to present a diversity of thinking on the relatively near-term implications of AGI and other related technologies. Available via Amazon or as a free online PDF via Humanity+ Press.
- **Engineering General Intelligence**, Volumes 1 and 2. This is a fairly technical book – though 1000 pages between the two volumes, it's fairly lightweight in treatment for a scientific work... but still, it's only suitable for folks with at least a little background in computer science, math or AI. It gives a moderately detailed rundown of the OpenCog design for a thinking machine. Published by Atlantis Press, distributed by Springer.
- Also check out my personal website goertzel.org, and the website of the OpenCog project opencog.org (most of the information is on the wiki site, wiki.opencog.org). And I have given lots of talks on

related topics, many of which have been videod and put online – see the list on my website, or just search on YouTube.

- Regarding more general transhumanist and futurist topics, beyond AGI per se, see:

  A Cosmist Manifesto: Practical Philosophy for the Posthuman Age. Available via Amazon or as a free online PDF via Humanity+ Press

- I have also collected some links that I found particularly useful for bringing newbies up to speed on transhumanism in general; see Radical Futurism for Newbies at http://wp.goertzel.org/radical-futurism-for-newbies/

CPSIA information can be obtained
at www.ICGtesting.com
Printed in the USA
LVOW13s2041121017
552168LV00013BA/237/P